Aerothermodynamics of Aircraft Engine Components

Edited by
Gordon C. Oates
University of Washington
Seattle, Washington

AIAA EDUCATION SERIES
J. S. Przemieniecki
Series Editor-in-Chief
Air Force Institute of Technology
Wright-Patterson Air Force Base, Ohio

Published by
American Institute of Aeronautics and Astronautics, Inc.
1633 Broadway, New York, N.Y. 10019

American Institute of Aeronautics and Astronautics, Inc.
New York, New York

Library of Congress Cataloging in Publication Data

Main entry under title:

Aerothermodynamics of aircraft engine components.

(AIAA education series)
Includes index.
1. Aerothermodynamics. 2. Aircraft gas turbines.
I. Oates, Gordon C.
TL574.A45A37 1985 629.134'353 85-13355
ISBN 0-915928-97-3

Foreword

Aerothermodynamics of Aircraft Engine Components, edited by Gordon C. Oates, is the third volume in the newly established Education Series of the American Institute of Aeronautics and Astronautics (AIAA). It complements an earlier volume on the *Aerothermodynamics of Gas Turbine and Rocket Propulsion* by Gordon C. Oates, and it will be followed by a volume on the *Aircraft Propulsion System Technology and Design*. These three texts will represent a comprehensive description of aircraft gas turbine theory and technology. They will provide an understanding of the principles of the design of modern aircraft engines and stimulate interest in one of the most important disciplines in aerospace industry.

The Education Series represents the AIAA's response to a need for textbooks and monographs in highly specialized disciplines of aeronautics and astronautics. The Institute's Publications Committee identified this need and endorsed this new series as a service to the aerospace engineering profession. The present volume covers a wide spectrum of topics including combustion, afterburners, axial compressors, turbine aerodynamics, turbine cooling, turbomachinery boundary layers, and engine noise. The comprehensive treatment of each topic makes this volume suitable for the graduate student and, as well, the practicing engineer or scientist concerned with the development and design of aircraft engines.

The publication of this volume would not have been possible without the support of the U.S. Air Force Wright Aeronautical Laboratories, Wright-Patterson Air Force Base, Ohio, and other organizations, notably the California Institute of Technology, Cambridge University, Engineering Research Institute of Iowa State University, Exxon Research Engineering Company, General Electric, Scientific Research Associates, and United Technologies Research Center. Their cooperation and support has made it possible to produce a comprehensive textbook covering the most important aspects of aerothermodynamic principles in the design of aircraft engine components.

J. S. PRZEMIENIECKI
Editor-in-Chief
AIAA Education Series

Table of Contents

Preface

This book was conceived as a fundamental text and reference for advanced engineering students and practicing engineers. It will, we hope, particularly interest and inform advanced students planning to expand their understanding beyond what they would normally attain in senior or first-year graduate classes. In addition, we hope and expect that engineers planning on embarking upon research will find these writings a solid foundation from which to initiate their own programs.

This book complements a preceding volume in the AIAA Education Series, *Aerothermodynamics of Gas Turbine and Rocket Propulsion*, by expanding upon the fundamentals and introducing advanced material leading to identification of the research issues of the day. Each chapter is written by an expert in the given speciality. Because of the advanced nature and complexity of the material, only minor efforts have been made to standardize on notation.

Taken in total, the book presents, we believe, a comprehensive overview of the fundamentals of the major aircraft engine components.

GORDON C. OATES
University of Washington
Seattle, Washington

CHAPTER 1. FUNDAMENTALS OF COMBUSTION

William S. Blazowski

Exxon Research and Engineering Company, Florham Park, New Jersey
(Formerly with Air Force Aero Propulsion Laboratory, Wright-Patterson AFB, Ohio)

1. FUNDAMENTALS OF COMBUSTION

1.1 Introduction

The background information necessary to understand aircraft turbine engine combustion systems is distinctly different from that necessary for diffusers, rotating machinery, or nozzles. Thus, a separate discussion of fundamentals is warranted. The purpose of this chapter is to review the *fundamental concepts* important to aeropropulsion combustion.

While much can be written about aeropropulsion combustion, the scope of this chapter is limited to highlighting the key information. The difficult task of selecting the information to include was made with the objective of providing the reader with the material necessary for understanding the combustion system's operating principles, performance parameters, and limitations. The reader contemplating aeropropulsion combustion as an area of specialization should develop a more thorough background and is referred to a number of readily available texts listed in the bibliography to this chapter.

Studies of combustion involve interdisciplinary investigations requiring consideration of three normally separate topics: chemistry, thermodynamics, and gasdynamics. Interrelationships between these areas, shown schematically in Fig. 1.1, require combustion engineers and scientists to develop a fundamental understanding of each topic. A number of subtopics in each of these areas have been listed to describe further the broad scope of subject matter involved. Each of these subjects will be addressed in this chapter.

The information to be presented in this chapter is organized into five sections. As might be expected, the first three consider chemistry, thermodynamics, and gasdynamics. The fourth is a discussion of the combustion parameters important to the combustor designer. Finally, the combustion properties of jet fuels are briefly described.

1.2 Chemistry

Three combustion chemistry topics will be discussed in this section. The first, chemical reaction rate, addresses the fundamental concepts vital to all chemical kinetics. Important dependencies of the reaction rate on thermodynamic conditions, especially temperature, will be addressed. The second topic, chemical equilibrium, is of importance in relation to the understanding of and ability to analyze high-temperature combustion systems. Finally, the fundamentals of practical hydrocarbon fuel combustion chemistry will

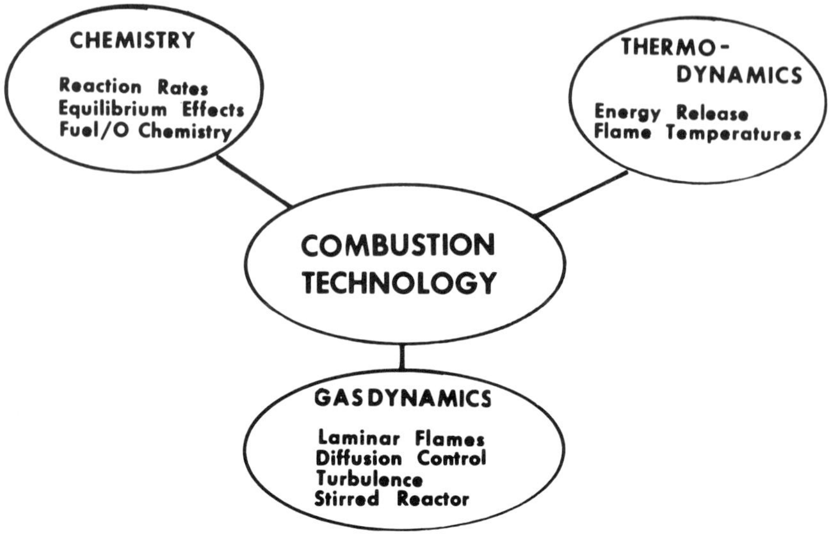

Fig. 1.1 Interdisciplinary nature of combustion technology.

be reviewed. Understanding of the *sequence* of chemical processes leading to H_2O and CO_2 production allows the explanation of many practical combustion characteristics.

Reaction Rate

One of the most basic concepts of chemistry involves the law of mass action, which relates the rate of a reaction (or the time rate of change of the reactant species concentration) to the concentrations of reactive species. This can be illustrated with the use of the following generalized chemical reaction:

$$aA + bB \rightarrow cC + dD \tag{1.1}$$

In this example, a moles of molecule A combine with b moles of molecule B to form c and d moles of products C and D. The reactant stoichiometric coefficients of the atomic balance equation (a and b) are also called the reaction "molecularity." The law of mass action states that the rate of reaction is expected to be proportional to the product of the concentrations of the reactant species raised to their respective stoichiometric coefficients. For this example, the rate of forward reaction r_f would be

$$r_f = k_f[A]^a[B]b \tag{1.2}$$

where the brackets correspond to the molar concentration (moles/volume) of the molecular species indicated and k_f is the rate coefficient for the forward reaction.

Note that the rate of forward reaction r_f could be representative of either rate of disappearance of reactants A and B or the rate of formation of products C and D. These four rates are interrelated by the stoichiometric coefficients a–d. For example, if r_f were representative of the rate of disappearance of A, the following relationships would be valid:

$$r_f = -\frac{\mathrm{d}[\mathrm{A}]}{\mathrm{d}t} = -\frac{b}{a}\frac{\mathrm{d}[\mathrm{B}]}{\mathrm{d}t} = \frac{c}{a}\frac{\mathrm{d}[\mathrm{C}]}{\mathrm{d}t} = \frac{d}{a}\frac{\mathrm{d}[\mathrm{D}]}{\mathrm{d}t} \qquad (1.3)$$

where the variable t represents time.

There is theoretical justification for the observed reactant concentration dependencies of the law of mass action. An analysis based on the assumption that product formation can occur only after the collision of the reactant molecules predicts the same concentration dependencies as described in Eq. (1.2). The rate coefficient k_f in Eq. (1.2) appropriately converts the results of the collision theory to yield the units of the reaction rate. In addition, k_f accounts for reaction rate dependencies due to variations in the molecular energy levels and in the geometrical orientation of the colliding molecules.

For many reactions of importance to combustion systems, k_f is a strong function of temperature. The temperature dependence of the molecular collision rate is minor ($T^{\frac{1}{2}}$) and has only a small influence on k_f. The predominant temperature dependence is a result of the necessity for molecular collisions to occur with sufficient "force" to overcome any energy barrier needed by the reactant molecules to undergo conversion to products. Physically, the formation of an activated complex is assumed to be necessary for the successful conversion of the colliding reactants to products. The height of the energy barrier (the energy necessary to form the activated complex) is often termed the activation energy E_a. Not all of the colliding reactant molecules will have sufficient energy and only a fraction of the collisions will be successful. Since the molecular energy distribution can be described by Boltzman statistics, the fraction of successful collisions is $\exp(-E_{af}/RT)$, where R is the universal gas constant and the subscript af refers to the forward reaction.

The geometrical misalignment of the reactant molecules during collision can also prevent the conversion to products; only a fraction of the collisions occurring with sufficient energy will occur with proper orientation. Consideration of the "steric factor" is a final necessary aspect in the analysis of the reaction rate coefficient. This factor can be thought of as a means of compensating for collisional inefficiencies due to the peculiarities of the geometrical alignment necessary for successful reaction.

An important expression for the reaction rate is obtained by combining the reaction rate coefficient dependencies discussed above with Eq. (1.2). The forward rate of the general reaction described in Eq. (1.1) is

$$r_f = -\frac{\mathrm{d}[\mathrm{A}]}{\mathrm{d}t} = [\mathrm{A}]^a[\mathrm{B}]^b C_f(T)^{\frac{1}{2}}\exp(-E_{af}/RT) \qquad (1.4)$$

where C_f includes the steric factor and the necessary constants to convert the collision rate to the reaction rate. The strong exponential nature of the reaction rate dependence on temperature was first recognized by Arrhenius. Equation (1.4) with the pre-exponential factor taken as temperature independent (i.e., not including the $T^{\frac{1}{2}}$ dependence) is called the Arrhenius equation. Equation (1.4) itself is said to be the modified Arrhenius relationship.

It should be noted that the rate dependency given by Eq. (1.4) is correct only in cases where the written stoichiometric equation represents the entire sequence of events leading to product formation. As will be discussed in Sec. 1.4, combustion of a practical hydrocarbon fuel involves many complex chemical reaction steps before formation of the final products, CO_2 and H_2O. In cases where the stoichiometric equation does not describe the entire reaction sequence, the dependencies of the reaction rate on the reactant concentration may not correspond to the molecularity and even fractional "reaction orders" may be observed. Nevertheless, the form of Eq. (1.4) is valid for each individual reaction step of the complex sequence.

Chemical Equilibrium

As a reaction such as that described in Eq. (1.1) proceeds, changes of concentration with time occur as illustrated in Fig. 1.2. When the concentrations of products C and D become significant, backward or reverse reaction (i.e., conversion of the products back to reactants) can become important. The rate of backward reaction r_b may be analyzed in the same manner as that of forward reaction [Eq. (1.2)] to yield the following relation:

$$r_b = k_b[\mathrm{C}]^c[\mathrm{D}]^d \tag{1.5}$$

In recognition of the existence of both the forward and reverse reactions, the more appropriate convention for expressing the general chemical system

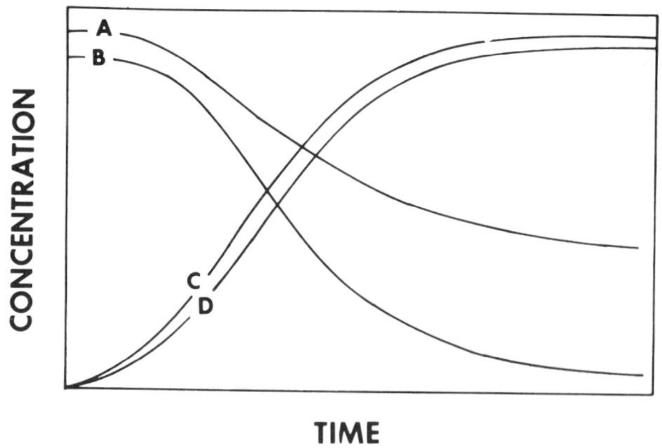

Fig. 1.2 Concentration variations during the course of a reaction.

described in Eq. (1.1) is

$$aA + bB \underset{k_b}{\overset{k_f}{\rightleftarrows}} cC + dD \tag{1.6}$$

Because reverse reactions always exist to some extent, concentrations of A and B will eventually decrease to some finite, nonzero values such that rates of the forward and reverse reactions are equal. These equilibrium concentrations are the asymptotes of Fig. 1.2. Note that the more general case where the reactant concentrations are not in exact stoichiometric proportions has been illustrated in Fig. 1.2, which corresponds to the situation of a large excess of reactant A.

The equilibrium concentrations can be determined from Eqs. (1.2) and (1.5). At equilibrium, the rate of the disappearance of the reactant [Eq. (1.2)] will be entirely balanced by the reactant formation rate [Eq. (1.5)]. Consequently, the equilibrium condition is $r_f = r_b$, or

$$k_f[A]^a[B]^b = k_b[C]^c[D]^d \tag{1.7}$$

Rearranging yields the following useful expression:

$$\frac{k_f}{k_b} = \frac{[C]^c[D]^d}{[A]^a[B]^b} \tag{1.8}$$

Because k_f and k_b are functions of temperature only, Eq. (1.8) provides a convenient means of relating the equilibrium concentration to the mixture temperature. The ratio k_f/k_b is known as the equilibrium constant based on the concentration K_c. An additional equilibrium constant based on the mole fractions K_x can also be developed.

An even more familiar means of characterizing equilibrium involves the partial pressure equilibrium constant K_p. Partial pressure is a concept in which the total mixture pressure is envisioned as a sum of the pressure contributions from each of the mixture constituents. The partial pressure of each constituent is the fraction of the total pressure corresponding to the mole fraction of that compound. Defined in terms of partial pressure, the equilibrium constant is

$$K_p = \frac{(P_C)^c(P_D)^d}{(P_A)^a(P_B)^b} \tag{1.9}$$

where P_A, P_B, P_C, and P_D are the partial pressures of each constituent. By convention, these pressures are always expressed in atmospheres when used in equilibrium chemistry calculations.

Both K_c and K_p are functions of temperature only. The temperature dependencies can be deduced from Eqs. (1.4) and (1.7) as

$$[A]^a[B]^b C_f T^{\frac{1}{2}} \exp(-E_{af}/RT) = [C]^c[D]^d C_b T^{\frac{1}{2}} \exp(-E_{ab}/RT) \tag{1.10}$$

which can be reduced to

$$K_c \text{ or } K_p \sim \exp\left(\frac{E_{ab} - E_{af}}{RT}\right) \tag{1.11}$$

Consequently, the equilibrium constant may have a strong, exponential temperature dependency.

In hydrocarbon/air combustion applications, two equilibrium relations are of paramount importance: the dissociations of CO_2 and H_2O,

$$CO + \tfrac{1}{2}O_2 \rightleftarrows CO_2 \tag{1.12}$$

$$H_2 + \tfrac{1}{2}O_2 \rightleftarrows H_2O \tag{1.13}$$

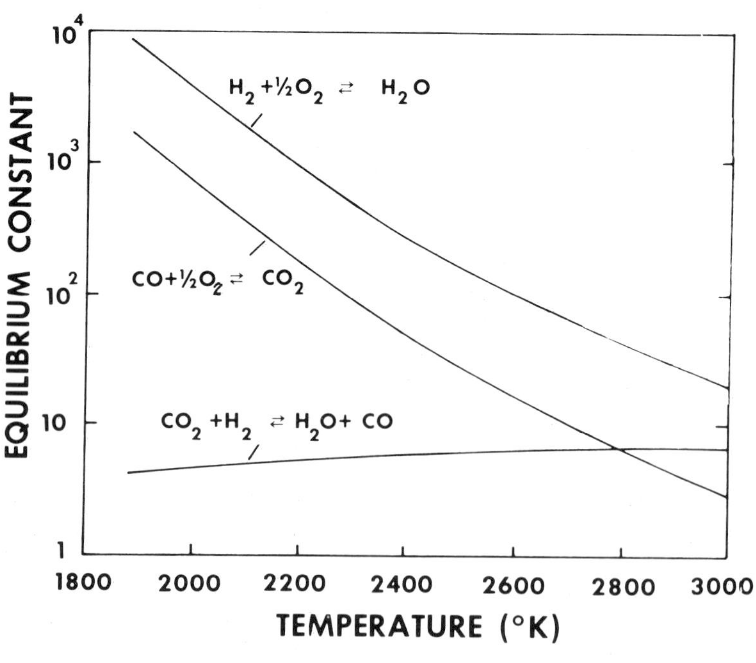

Fig. 1.3 Equilibrium constants for important dissociation reactions.

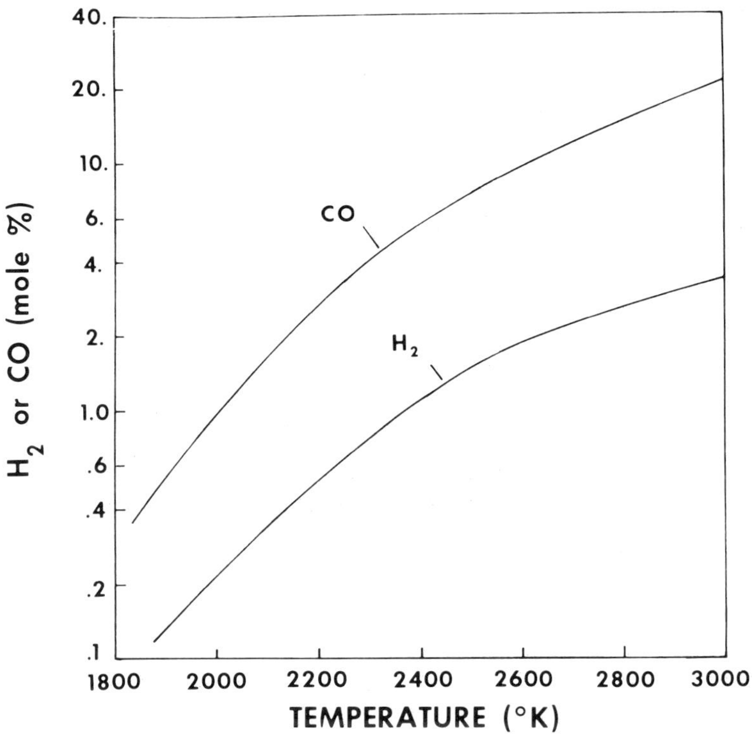

Fig. 1.4 Equilibrium CO and H_2 concentration dependence on temperature.

Mathematical treatment of these equilibrium relationships is often simplified by the use of the water gas reaction,

$$CO + H_2O \rightleftarrows H_2 + CO_2 \tag{1.14}$$

It should be noted that this is not a third independent relationship, but a linear combination of Eqs. (1.12) and (1.13). Partial pressure equilibrium constants for each of these three reactions are illustrated in Fig. 1.3 (data from Ref. 1 have been utilized). Note the relative temperature insensitivity of the water gas equilibrium constant.

Since the maximum chemical energy is released from a hydrocarbon fuel upon conversion to CO_2 and H_2O, dissociation of either of these products results in a decrease of the energy released. As will be shown in Sec. 1.3, the equilibrium flame temperature is strongly influenced by dissociation. Because of the temperature sensitivity of the equilibrium constants, dissociation is more pronounced at higher flame temperatures. Figure 1.4 illustrates the effect of the final mixture temperature on dissociation, using the example of stoichiometric combustion of a C_nH_{2n} type of fuel with air at 1 atm. The influence of the final mixture temperature on the CO and H_2 concentrations in the combustion product is pronounced.

Hydrocarbon Chemistry

The sequence of events occurring during the combustion of a practical hydrocarbon fuel is extremely complex and is not understood in detail. Major aspects of hydrocarbon combustion chemistry involve hydrocarbon pyrolysis and partial oxidation to H_2 and CO, chain branching reactions resulting in H_2 consumption, and CO oxidation by the radicals generated during chain branching. Each of these reaction steps is schematically illustrated in Fig. 1.5. Note that the chronology of these processes is schematically indicated by the flow of mass through the reaction steps. Each process is individually described below.

Pyrolysis is the term given to the process by which fuel molecules are broken into smaller fragments because of excessive temperature and partial oxidation. This molecular destruction is accomplished during the first phase of the combustion process. The predominant resulting products are hydrogen and carbon monoxide. Little detailed information is available concerning the chemistry of these processes for practical fuels such as large hydrocarbons with molecular weights of 50–200. It is well recognized that the hydrocarbon structure and its influence on the pyrolysis chemistry affects the combustion process. For example, low fuel hydrogen concentration leads to excessive carbon particle formation in the early stages of combustion.

Edelman et al.[2] developed a single-step quasiglobal model to characterize the pyrolysis and partial oxidation of any practical hydrocarbon fuel. Their approach is to characterize the kinetics of the numerous complex chemical reactions resulting in the production of H_2 and CO by a single reaction step. An Arrhenius-type expression has been fitted to experimental data involving the variations in temperature and pressure as well as fuel and oxygen concentration. The result is

$$-\frac{d[C_nH_m]}{dt}=\frac{5.52\times 10^8 T}{p^{0.825}}[C_nH_m]^{\frac{1}{2}}[O_2]\exp(-24{,}400/RT) \quad (1.15)$$

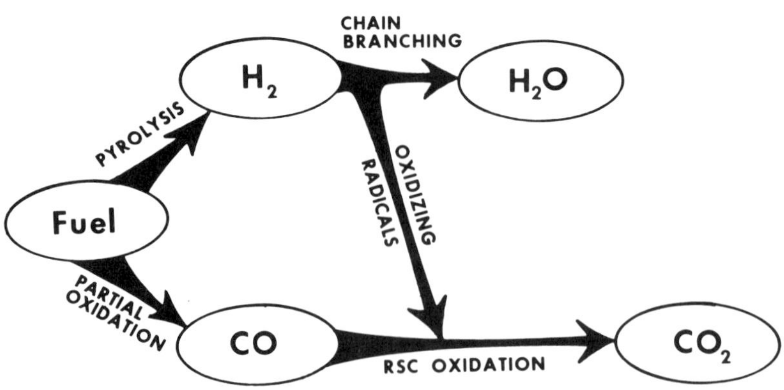

Fig. 1.5 Hydrocarbon combustion chemistry schematic.

where concentrations are expressed in moles/cm^3, T in K, and P in atm. The activation energy of 24,400 is in the units cal/g · mole · K. Although this expression has proved to be useful in some combustion models, additional effort is required to determine the chemical kinetic differences between hydrocarbon fuel types and to study the pyrolysis mechanisms in mixtures. Further, the interface between fuel pyrolysis and carbon particulate formation requires additional study.

The products of pyrolysis are reduced-state compounds (RSC). Oxidation of these species is better understood than their formation. The important oxidation reactions for the reduced-state compounds are of the general form

$$\mathrm{RSC + OR = OSC + RR} \tag{1.16}$$

where RSC is the reduced-state compound, OR the oxidizing radical, OSC the oxidized-state compound, and RR the reducing radical. The rate of oxidation of the RSC may be assumed to be given by the appropriate Arrhenius-controlled mechanism.

While reactions of the nature described by Eq. (1.16) play a role in consuming the H_2 formed during the pyrolysis process, many gross characteristics of hydrocarbon combustion are a result of other chemical reactions involving "chain branching." This type of reaction sequence involves the production of additional radical species during the process. In the case of the H_2 oxidation process, the important chain branching reactions are

$$\mathrm{H_2 + O \rightarrow H + OH} \tag{1.17}$$

$$\mathrm{H + O_2 \rightarrow OH + O} \tag{1.18}$$

Note that in either reaction a single radical (O or H) results in the production of two radicals (H + OH or OH + O). This type of reaction has the potential of producing large quantities of radical species. In portions of the combustion zone having high H_2 concentrations, radical species can reach levels far in excess of equilibrium. During this process, OH radicals also participate in RSC reactions [Eq. (1.16)] to produce H_2O from H_2.

Carbon monoxide consumption is controlled by the following RSC reaction:

$$\mathrm{CO + OH \rightarrow CO_2 + H} \tag{1.19}$$

Since the activation energy of the reaction indicated by Eq. (1.19) is generally low (only a few kcal/g · mole), the carbon monoxide oxidation rate is predominantly influenced by the OH concentration. As previously noted, this quantity is controlled by the chain branching mechanism. Nevertheless, a common method of approximating the radical concentration in a RSC reaction involves assuming local or partial equilibrium. This type of approach has been used in CO oxidation studies by Howard et al.[3] Because the functional relationship between the equilibrium OH concentration and the temperature is exponential [Eq. (1.11)], an Arrhenius-like dependence can be written for a quasiglobal $O_2 + CO$ reaction in the presence of H_2O. Howard et al. determined that

$$-\frac{d[CO]}{dt} = k_0[CO][O_2]^{\frac{1}{2}}[H_2O]^{\frac{1}{2}}e^{-30,000/RT} \tag{1.20}$$

where $k_0 = \text{const} = 1.3 \times 10^{14}\ \text{cm}^3/\text{mole} \cdot \text{s}$. This assumption is not necessarily in conflict with the knowledge that higher-than-equilibrium free radical concentrations may be produced by the reactions of Eqs. (1.7) and (1.18). CO oxidation is much slower than H_2 consumption and, in nonrecirculating systems, occurs predominantly after the chain branching H_2 reactions are largely complete. However, gas turbine combustion systems do employ recirculation and this assumption for that application may provide misleading results.

The production of H_2O and especially CO_2 through the RSC reactions described above results in the release of a great deal of energy. Consequently, the rate of consumption of CO and the predominant energy release rate are strongly connected. Experience has shown that the combustion characteristics influenced by the principal heat release processes (e.g., flame propagation) are correlated by considerations of Eqs. (1.19) and (1.20). On the other hand, those characteristics dependent on fuel breakdown/pyrolysis (e.g., ignition delay) are better correlated by consideration of Eq. (1.15).

The above discussion provides only a simplified description of the complex chemistry of hydrocarbon combustion. Additional detailed treatment has recently been undertaken in efforts to predict pollutant emissions from combustion systems. Table 1.1 is an example of a more complex scheme utilized in Ref. 4.

Table 1.1 Hydrocarbon Oxidation Kinetics Scheme (from Ref. 4)

(1)	$C_8H_{16} + O_2 \rightarrow 2C_4H_8O$	(11)	$C_2H_2 + OH \rightarrow CH_3 + CO$
(2)	$C_4H_8O + O_2 \rightarrow HO_2 + CO + CH_3 + C_2H_4$	(12)	$2H + M \rightarrow H_2 + M$
(3)	$C_8H_{16} + OH \rightarrow H_2CO + CH_3 + 3C_2H_4$	(13)	$2O + M \rightarrow O_2 + M$
(4)	$CH_3 + O \rightarrow H_2CO + H$	(14)	$OH + H + M \rightarrow H_2O + M$
		(15)	$H + O_2 \rightarrow OH + O$
(5)	$CH_3 + O_2 \rightarrow H_2CO + OH$	(16)	$O + H_2 \rightarrow OH + H$
(6)	$H_2CO + OH \rightarrow H_2O + CO + H$	(17)	$H + H_2O \rightarrow H_2 + OH$
(7)	$C_2H_4 + O_2 \rightarrow 2H_2CO$	(18)	$O + H_2O \rightarrow 2OH$
(8)	$C_2H_4 + OH \rightarrow CH_3 + H_2CO$	(19)	$CO + OH \rightarrow CO_2 + H$
(9)	$CH_3 + H_2 \rightarrow CH_4 + H$	(20)	$HO_2 + M \rightarrow H + O_2 + M$
(10)	$C_2H_4 \rightarrow C_2H_2 + H_2$	(21)	$HO_2 + H \rightarrow 2OH$

1.3 Thermodynamics

This section describes the thermodynamic relationships of importance in evaluating the effect of chemical energy release in combustion systems. The first subsection highlights application of the first law of thermodynamics, offers straightforward evaluations of flame temperature dependencies, and describes methodology used in calculating flame temperature. The second addresses important flame temperature dependencies in the practical situation of jet fuel/air combustion.

Energy Release and Flame Temperature

The first law of thermodynamics (energy conservation) is an important factor in any analysis of combustion systems. Adiabatic, flowing, constant-pressure combustion systems, a reasonable approximation for both the main burners and afterburners of gas turbines, are analyzed using conservation of *total* enthalpy. In this case, total enthalpy includes sensible (or thermal), chemical, and kinetic contributions,

$$h_t = h_s + \left(\frac{f/a}{f/a + 1} \right) \Psi + \frac{u^2}{2J} \tag{1.21}$$

where:

h_t = total enthalpy, kcal/kg
h_s = sensible enthalpy, kcal/kg
f/a = mass ratio of fuel to air
Ψ = chemical energy, kcal/kg fuel
u = flow velocity, m/s
J = mechanical equivalent of heat = 4186 J/kcal

Most frequently, standard heats of formation are used to determine the chemical energy released during a combustion process. The standard heat of formation h_f represents the energy addition necessary for constant-pressure formation of a compound from its elements in their natural state at 25°C. The energy required to accomplish any reaction can be calculated by algebraically summing the heat of formation contributions of the products minus the reactants,

$$(\Delta h_r)_{25°C} = \Sigma x_i (h_f)_i - \Sigma x_j (h_f)_j \tag{1.22}$$

where

Δh_r = heat of reaction at 25°C
x_i = stoichiometric coefficients of product compounds
x_j = stoichiometric coefficients of reactant compounds

If Eq. (1.22) is applied to a complete oxidation process of a hydrocarbon where all of the fuel hydrogen is converted to H_2O and all of the fuel carbon is converted to CO_2, the heat of combustion Δh_c will be calculated. Note that this result is normally a large negative value (i.e., the reaction is strongly exothermic).

The amount of heat required to accomplish a reaction Δh_r is a function of reaction temperature. Heat required at temperature T_1, rather than 25°C would be

$$(\Delta h_r)_{T_1} - (\Delta h_r)_{25^\circ\text{C}} = (h_{sp} - h_{sr})_{T_1} - (h_{sp} - h_{sr})_{25^\circ\text{C}} \tag{1.23}$$

where h_{sp} and h_{sr} are the product and reactant sensible enthalpies, respectively. The heats of combustion are generally greater (i.e., less energy is released) as the temperature is increased.

With the important exception of the afterburner nozzle, the kinetic contribution to the total enthalpy in gas turbine combustion systems is relatively small. In such a case, the relationship between the energy released due to combustion and the final flame temperature is

$$-(\Delta h_c)_{T_1} = (h_{sp})_{T_2} - (h_{sp})_{T_1} = \int_{T_1}^{T_2} C_p \, \mathrm{d}T \tag{1.24}$$

where C_p is the temperature-dependent specific heat of the combustion products and the heat of combustion at temperature T_1 is calculated using Eq. (1.23). In this flame temperature calculation, the heat generated in forming the combustion products at temperature T_1 can be envisioned as an energy source for the constant-pressure heating of the combustion products from T_1 to T_2.

The term Ψ in Eq. (1.21) is a temperature invariant representation of a fuel's chemical energy. It may be calculated using the following relationship:

$$\Psi = (-\Delta h_c)_{25^\circ\text{C}} + (h_{sp} - h_{sr})_{25^\circ\text{C}} \tag{1.25}$$

The concept of chemical energy in conjunction with Eq. (1.21) provides a second method for determining final flame temperature.[1] For the case where kinetic contributions are negligible, the conservation of total enthalpy results in the following expression:

$$(h_{sr})_{T_1} + \left(\frac{f/a}{1 + f/a}\right)\Psi = (h_{sp})_{T_2} \tag{1.26}$$

It can be shown that the solutions for Eqs. (1.25) and (1.26) and Eqs. (1.23) and (1.24) are identical.

The case where a hydrocarbon fuel has completely reacted to CO_2 and H_2O results in the maximum achievable flame temperature, as the maximum energy is released upon formation of these products. Note that this can be achieved only for a lean mixture (i.e., more oxygen than is required for a stoichiometric reaction). Conversion to H_2O, CO_2, and CO is often assumed for rich mixtures, as the conversion of H_2 to H_2O is much more rapid than CO oxidation (see Sec. 1.2). The temperature that would result if the reaction were complete is defined as the "theoretical flame temperature." Because of the incomplete combustion, energy losses, and effects of CO_2 and H_2O dissociation, the theoretical flame temperature is never achieved in real combustion systems. Nevertheless, consideration of this simplified flame temperature concept reveals important trends dictated by the first law of thermodynamics.

Equations (1.24) and (1.26) relate the temperature rise to the heat release due to combustion. For a given amount of energy release, it is apparent that the final flame temperature will increase with the initial temperature. Also, since for lean mixtures the heat released will be proportional to the amount of fuel burned per mass of mixture, it is implied that T_2 will increase directly with the fuel-air ratio. However, when the mixture ratio exceeds stoichiometric, CO (and possibly H_2 and unburned fuel) will be present in the exhaust products and a decreasing flame temperature trend will result. Consequently, this analysis indicates a maximum flame temperature for stoichiometric conditions. These trends are illustrated in Fig. 1.6. Rather than considering the fuel-air ratio, the equivalence ratio ϕ has been used in this illustration. The equivalence ratio is the fuel-air ratio of consideration divided by the stoichiometric fuel-air ratio,

$$\phi = (f/a)/(f/a)_{\text{stoichiometric}} \tag{1.27}$$

Values of ϕ less than unity correspond to lean operation, while those greater than unity correspond to rich combustion.

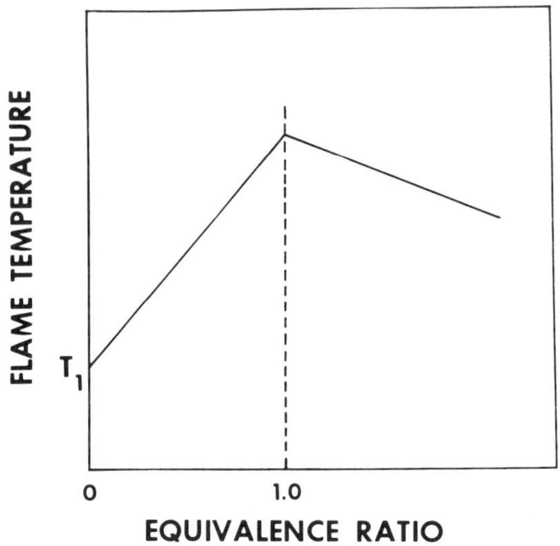

Fig. 1.6 Theoretical flame temperature dependence on equivalence ratio.

Accurate flame temperature prediction requires consideration of the dissociation effects and variable specific heats. The iterative solution of at least four simultaneous equations is involved: (1) stoichiometric chemical equation (mass and atomic conservation), (2) energy conservation, (3) CO_2 dissociation, and (4) H_2O dissociation.

Additional equilibrium relationships may be added to improve accuracy and to predict the concentrations of NO, NO_2, O, H, OH, N, etc. Note further that the water/gas equilibrium equation is usually substituted for either (3) or (4) to simplify the mathematical procedures (see Sec. 1.2).

A number of methods for solving these equations are practical. One technique involves assuming a flame temperature and calculating the species concentrations using the equilibrium relationships. The values are then used to check for balance in the energy equation. Additional guesses and iterations are made until a temperature is determined such that the conservation of energy is satisfied within acceptable limits.

Because of the involved nature of these calculations, detailed tabulated results and computer programs have been established to assist combustion scientists and engineers. Some of the early tabulated calculations are the subject of Ref. 5, while the most popular of currently available computer programs for this purpose is described in Ref. 6.

Important Flame Temperature Dependencies

This subsection presents calculated flame temperature results of practical importance to turbine engine combustion. Important variables to be examined are the fuel-air ratio, initial pressure and temperature, and mixture inert concentration.

The simplified relationship between the calculated constant-pressure adiabatic flame temperature and mixture ratio shown in Fig. 1.6 is significantly altered when the detailed effects of dissociation and specific heat variations are included. This is illustrated in Fig. 1.7, which shows results of the combustion of Jet A fuel with air at 800 K and 25 atm (representative of modern combustor inlet conditions at 100% power operation). The difference between the theoretical and actual flame temperatures as the mixture ratio approaches stoichiometric is due to the presence of significant CO and H_2 concentrations at the higher temperatures (see Fig. 1.4). In addition, note that dissociation causes the peak flame temperature to occur at slightly rich conditions.

An understanding of the influences of the initial pressure and temperature on the flame temperature is important to the combustion engineer, as testing is frequently accomplished at scaled operating conditions. Figure 1.8 illustrates the relationship between stoichiometric flame temperature and inlet temperature at a pressure of 25 atm using Jet A fuel. Note that only one-half of an increase in inlet temperature is translated to a flame temperature at these conditions. Again, the nonlinearity is primarily due to the strong temperature dependence of the equilibrium constants for CO_2 and H_2O dissociation. The effect of pressure is illustrated in Fig. 1.9. An increase in pressure at a constant initial temperature results in an increase in

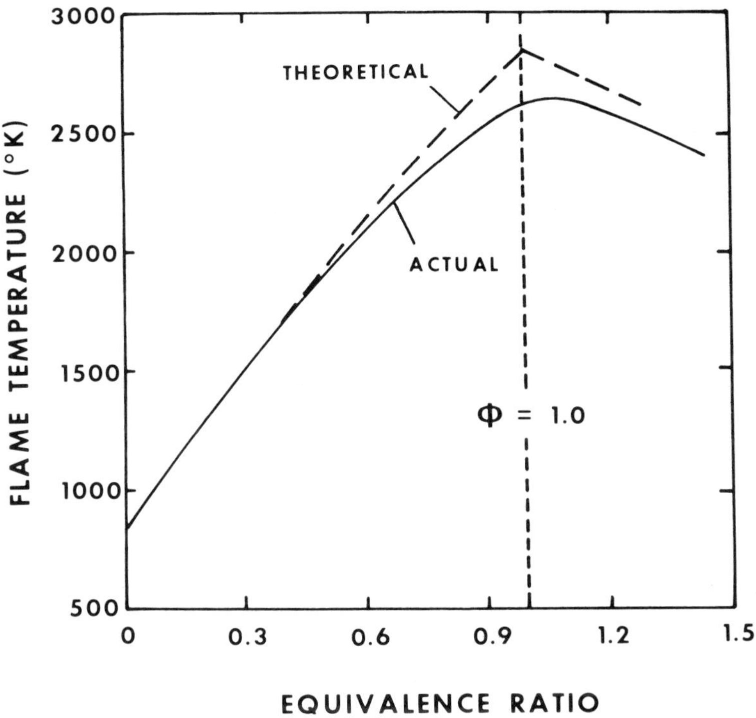

Fig. 1.7 Effect of inlet temperature on adiabatic flame temperature.

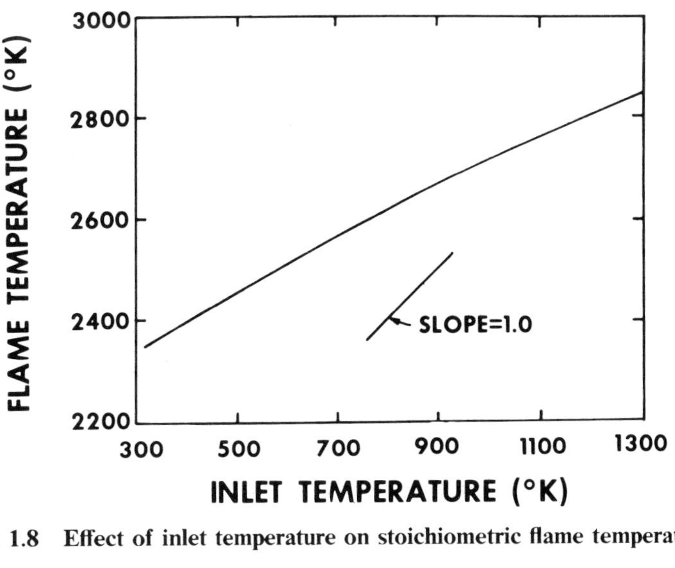

Fig. 1.8 Effect of inlet temperature on stoichiometric flame temperature.

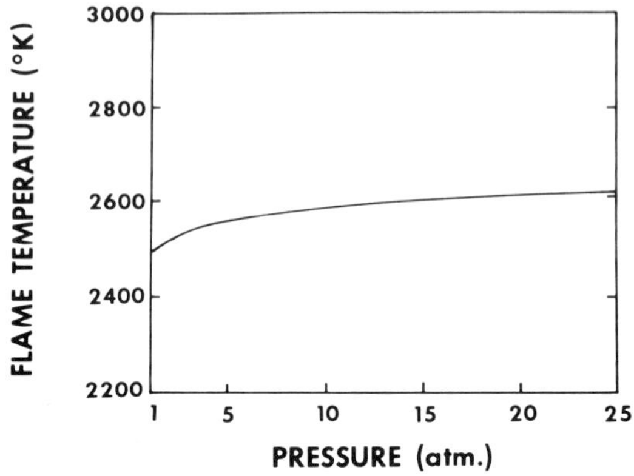

Fig. 1.9 Effect of pressure on stoichiometric flame temperature.

the flame temperature. This dependence can be explained by examining the form of the H_2O and CO_2 dissociation equations. In both cases, dissociation requires an increase in the total number of moles of product. The physics of the equilibrium process, as embodied in Eqs. (1.8) and (1.9), causes an increase in pressure to result in a shift to *less* total moles of product—in this case less dissociation. The increased amounts of H_2O and CO_2 in the combustion products result in greater flame temperatures.

Consideration of the main combustor *exit* temperatures can be somewhat simplified from the complexities of the foregoing discussion. Most of the variations discussed with respect to Figs. 1.7–1.9 occur at the highest values of the flame temperature. Since main burner exit temperatures are generally below 1750 K, the inlet temperature and pressure effects of dissociation are far less pronounced. Figure 1.10 illustrates the relationship between the fuel-air ratio and the total temperature, which may be used for any pressure at flame temperatures less than 1750 K.

Another important factor influencing the flame temperature is the oxygen concentration. Two instances of combustion with an O_2 concentration less than that of air are very important to the combustion engineer. When in actual operation, the main burner of a gas turbine engine encounters air with approximately 21% O_2 (preheating being accomplished by the compressor); however, testing is sometimes conducted using vitiated air (that is, the inlet temperature requirements are satisfied by precombustion rather than compression or indirect heating using a heat exchanger). The other very important example of vitiated combustion is, of course, the afterburner. In this case, some of the energy previously added by the compressor and main burner combustion has been extracted by the turbine. Consequently, the temperatures in the stoichiometric zones of the main combustor significantly exceed the maximum achievable temperatures within the afterburner.

A means of evaluating the effect of vitiation on flame temperatures is to consider the oxygen availability. Vitiated combustion is characterized by

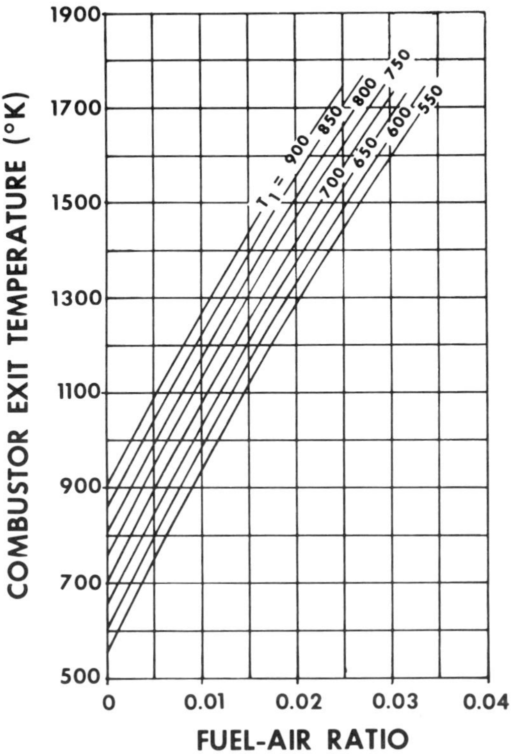

Fig. 1.10 Combustor exit total temperature for Jet A fuel and arbitrary pressure (working chart).

abnormally high H_2O and CO_2 concentrations and lower O_2 concentrations. Therefore, the effect of vitiation is to reduce the amount of fuel per mass of mixture that can be stoichiometrically burned; the reduced final flame temperature would be predicted by Eqs. (1.24) and (1.26). Figure 1.11 illustrates the effect of reduced oxygen concentration as a function of the degree of vitiation. These calculations were performed with combustor inlet conditions of 900 K and 1 atm pressure, values typical of turbine exit conditions. The results include the effects of dissociation.

1.4 Gasdynamics and Diffusion Processes

The influences of gasdynamics, diffusion of species, and heat addition on combustion are extremely complex. Rather than treat these involved processes in detail, this section intends to explain the impact of these phenomena in turbopropulsion combustion and to outline the general approaches utilized to model these processes in turbine engine combustion systems. The topics of consideration include premixed laminar flame propagation, diffusion-controlled combustion, the effects of turbulence on combustion processes, and the perfectly stirred reactor.

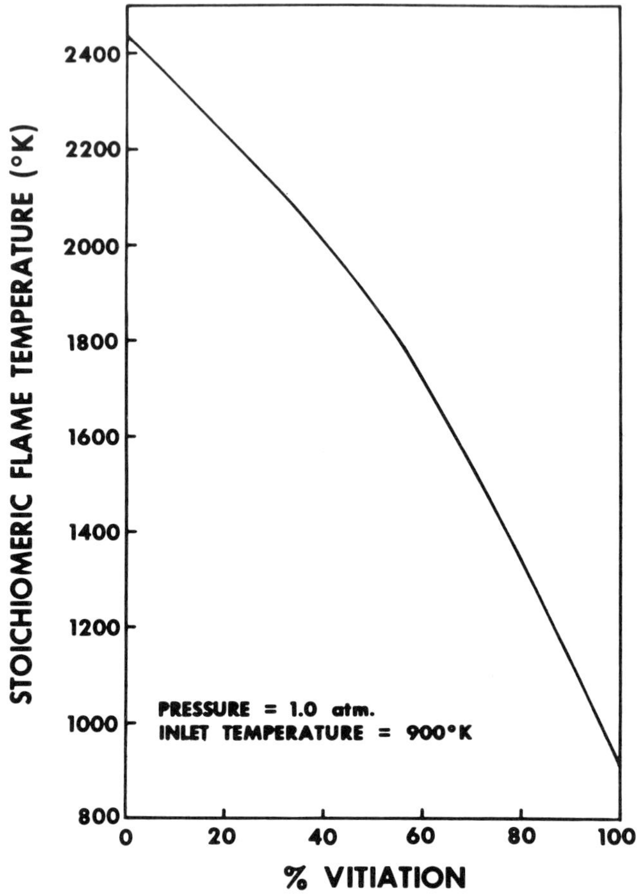

Fig. 1.11 Effect of vitiation on stoichiometric flame temperature.

Premixed Laminar Flames

The simplest situation involving the simultaneous treatment of combustion and gasdynamics is the premixed laminar flame. In this case, a reaction front proceeds through a uniform mixture of gaseous fuel and air with a constant propagation speed. An analysis of the mass, momentum, and energy conservation equations governing this case (called the Rankine-Hugoniot analysis) predicts two types of solutions. First, the reaction front can proceed into the unburned gases supersonically. The velocity of the burned gases with respect to the reaction front can be either subsonic (the detonation case) or supersonic (the supersonic combustion case). Second, the reaction front can proceed into the unburned gases subsonically. In this case, the burned gases must also be subsonic, as the supersonic case would violate the second law of thermodynamics. Subsonic flame propagation is also called deflagration.

While the Rankine-Hugoniot equations predict the existence of deflagration and relationships between the properties of burned and unburned gases, the analysis does not allow calculation of the propagation velocity S_L. Prediction of S_L requires consideration of reaction rate as well as heat conduction and species diffusion across the reaction front. The basic thermal theory of Mallard and LeChatelier (first proposed in 1883) results in the following important temperature dependence:

$$S_L \sim \exp(-E_a/2RT_f) \tag{1.28}$$

This temperature relationship is similar to that for the reaction rate [Eq. (1.4)], except for the factor of two in the exponential denominator. Equation

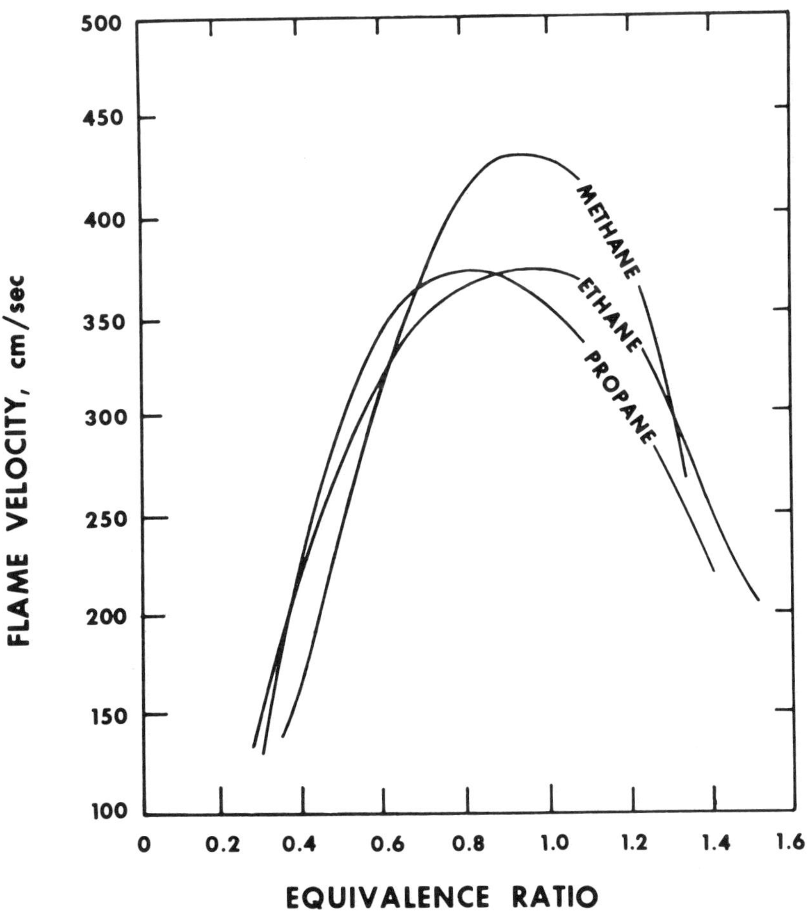

Fig. 1.12 Flame velocities of paraffin-oxygen mixtures at 1 atm pressure and room temperature (from Ref. 8).

(1.28) is consistent with the empirical flame propagation dependencies on the fuel-air ratio as shown in Fig. 1.12; the highest propagation velocity is at approximately $\phi = 1$, where T_f is at its maximum value.

Variations in flame propagation rate with the hydrocarbon types of practical interest to turbopropulsion combustion are not substantial.[8] Practical jet fuels would be expected to behave in a manner similar to that of the fuels described in Fig. 1.12. The pressure dependence of the flame propagation is not straightforward. In mixtures with burning velocities below 50 cm/s, S_L decreases with increasing pressure. Between 50 and 100 cm/s, S_L remains approximately constant. Above 100 cm/s, S_L increases with increasing pressure.[8]

Across the reaction front, a substantial decrease in the gas densities occurs. Consequently, mass conservation requires a substantial increase in velocity. This acceleration results at the expense of some pressure drop across the front. The pressure drop is

$$\frac{\Delta P}{P} = \gamma\left(1 - \frac{T_1}{T_F}\right)\left(\frac{S_L}{a}\right)^2 \tag{1.29}$$

where γ is the ratio of initial gas specific heats, T_1 the initial temperature, and a the speed of sound in the initial mixture. This quantity represents the minimum pressure drop (usually less than 1%) that a combustion system can experience in order to accomplish a given energy release.

The values of laminar flame propagation are usually no greater than 5 m/s. Since the reference velocities in modern combustion turbine engine systems are generally greater than 25 m/s, laminar flame propagation is not considered a predominant mode of combustion. Other modes of combustion are necessary, including diffusion control, turbulent mixing, and the establishment of zones approaching perfectly stirred reactors.

Diffusion-Controlled Combustion

In many practical devices, the fuel and air are not entirely premixed prior to combustion. In these cases, reactions take place in flame zones where the influx of oxygen and fuel occur at a rate corresponding to the stoichiometric ratio. In laminar flows, the mechanism for transport of the fuel and oxygen into these zones is molecular diffusion and these types of systems are called diffusion flames.

Common examples of laminar diffusion flames are a candle, kerosene lamp, or match; in each case, the fuel and O_2 do not premix prior to the flame. An appreciation for the nature of diffusion flames may be gained by examining the early work of Burke and Schumann[9] published in 1928. Experimentally, they utilized a system similar to that shown in Fig. 1.13a. Fuel enters the combustion zone through a central tube. A laminar diffusion flame, the shape of which is governed by the ratio of fuel-to-air flow, is established in the combustion zone.

Burke and Schumann's analysis constituted the first successful treatment of laminar diffusion flames and it continues to represent the approach taken

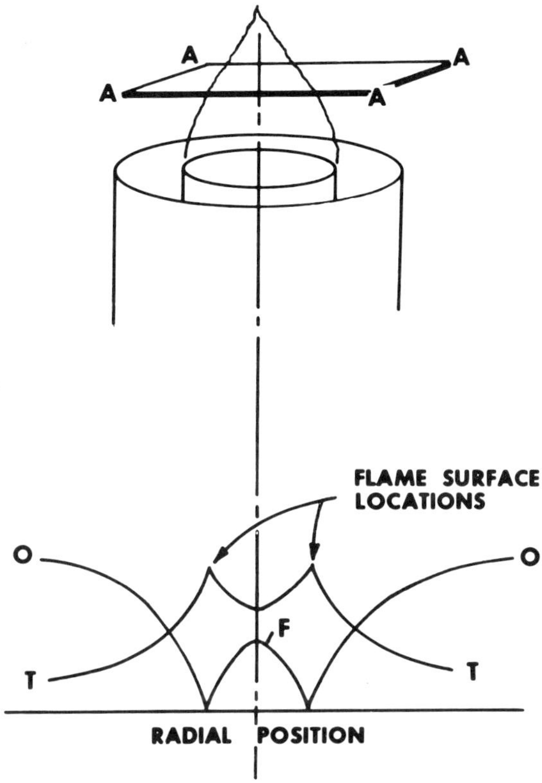

Fig. 1.13 Diffusion flame characteristics: a) experiment; b) important profiles at plane A—oxygen concentration (O), fuel concentration (F), temperature (T).

in modern work. They modeled the flame as an infinitely thin sheet into which fuel and oxygen flow in stoichiometric proportions. All heat release from the chemical reaction occurs at this surface. This approach is the equivalent to the assumption that the chemical reactions are infinitely fast compared to the diffusion processes that, therefore, control the burning rate. In order to mathematically prohibit an infinite gradient of fuel or oxygen at the flame surface (which would indicate infinite flux into the surface), it is necessary to require that both the fuel and oxygen concentrations be zero at the surface. This concept of concentration and temperature profiles is illustrated in Fig. 1.13b.

Important differences between premixed and diffusion flames center around the existence of the flame sheet. In the case of the premixed flame with no flame sheet, the temperatures achieved correspond to that for constant-pressure combustion at the premixture fuel-air ratio and the burning rate is controlled by the chemical kinetic rate and flame temperature [Eq. (1.28)]. However, in the case of laminar diffusion flames, combustion always occurs at stoichiometric conditions at the flame sheet and the burning rate is controlled by molecular diffusion. Moreover, the temperature at the flame does not necessarily correspond to the premixed constant-pressure stoichiometric flame temperature. The thermodynamics at the flame sheet are strongly influenced by the rates of heat and mass transfer near the flame (i.e., mass and temperature gradients) and cannot be accurately calculated through the use of a simple energy balance.

The simplifying assumption of an infinite chemical reaction rate can be eliminated through the utilization of a much more sophisticated analysis. The primary effects of including finite reaction rates are prediction of a flame of finite thickness, lesser gradients of concentration and temperature at the flame, and lower peak flame temperatures. While more accurate prediction of near-flame characteristics is achieved by the more sophisticated analysis, many important properties, notably the burning rate, are predicted nearly as well with the less elaborate approach.

The instance in which the classical thin-flame theory of diffusion flames is of importance to turbine combustion systems involves fuel droplet burning. While many combustion engineers question whether droplet combustion occurs under the high-temperature and turbulent conditions characteristic of high-power operation, droplet burning would be likely during starting and idle operation. In this case, vaporization is caused by heat transfer from a flame surrounding the liquid droplet, assumed to be at its boiling temperature. The gaseous fuel proceeds to the flame and stoichiometrically reacts with oxygen diffusing radially inward from the surroundings. Analysis of this situation, similar to the Burke and Schumann analysis discussed above, results in the following burning rate prediction:

$$d^2 = d_0^2 - kt \tag{1.30}$$

where d is the fuel droplet diameter at time t, d_0 is d at $t = 0$, and k is a constant. This relationship, known as Godsave's law,[10] predicts the fuel effects through variations in k. The influence of convective velocity, air temperature, and oxygen concentration are included in determination of the value of k. Note that this relationship is also applicable to droplet evaporation without combustion, the difference being a smaller value of k.

The foregoing discussion addresses the simplified concept of fuel droplet combustion, wherein the process of each individual droplet is assumed to be independent and the fuel is comprised of a single hydrocarbon component. In reality, the situation is far more complex; droplet interactions and complications of multicomponent fuels cause significant departures from the simplified case discussed above.[11,12] The combustion engineer should utilize empirical information when available.

Diffusion-controlled combustion will also occur in the cases where fuel droplets have vaporized but not mixed with surrounding air, thus forming fuel-rich pockets, or where the gases from a rich primary zone mix with secondary air. Many of the performance characteristics of present-day combustors confirm the existence and importance of such processes. However, the diffusion-controlled reactions in practical gas turbine combustion systems cannot be treated simply by classical laminar diffusion flame theory; turbulence effects are extremely important and the effects of turbulent diffusivity must be included. These will be discussed in the following subsection.

Turbulence Effects

Previous discussions of both premixed and diffusion-controlled combustion have concentrated on laminar systems. In practice, however, the presence of turbulence has an extremely important influence on both premixed and diffusion-controlled combustion.

The rate of flame propagation in a premixed system is greatly enhanced by turbulence. Most available information on this subject has been developed to improve the understanding of turbulent flame propagation in afterburners. However, turbulent flame propagation information is currently of additional importance, as low-emission combustors employ a highly turbulent fuel-air mixing and vaporization zone prior to combustion (see Chap. 2 of *Aircraft Propulsion System Technology and Design*). An important consideration in such systems is the possibility of "flashback" or turbulent flame propagation upstream towards the fuel injection point.

Lefebvre and Reid[13] have reviewed the important turbulent flame propagation literature. The relationships shown in Table 1.2 have been cited as representative of the understanding of turbulence effects. The important parameters influencing turbulent flame propagation S_T are

u' = the fluctuating component of gas velocity
$\bar{y}$ = parameter describing freestream *and flame generated* turbulence intensity
l = turbulence length scale

Generally speaking, the turbulent flame velocity can be the order of the turbulent velocity u', far exceeding S_L.

Turbulence also causes increased burning rates in diffusion flames.[18] Analyses for turbulent diffusion flames are similar to the laminar case, but use an artificially high diffusivity constant. Physically, the increased fuel-air mixing is explained as due to forced mixing of small fuel-lean or fuel-rich elements of gas by turbulent forces. These small elements of gas are called eddies. The analytical adjustment, called eddy diffusivity, accounts for the enhanced mixing at the reaction front. This approach is taken as a convenience and is based on empirical correlations rather than fundamental principles.

The importance of turbulence on many practical aspects of turbine engine combustion has been highlighted by Mellor.[19] He proposes a simplified model for main burner combustion in which important processes are assumed to occur in a highly turbulent diffusion flame stabilized by a recirculation zone behind a bluff body (see Fig. 1.14). His analysis, based on characteristic times for turbulent mixing and chemical reaction and focusing

Table 1.2 Turbulent Flame Theories (from Ref. 13)

Investigator	Relationships	Conclusions
Damkohler[14]	$S_T = S_L + u'$ At high velocities, this approaches $S_T = u'$	S_T is independent of turbulence scale. At high velocities, S_T is determined solely by turbulent velocity.
Shchelkin[15]	$S_T \simeq S_L[1 + (u'/S_L)^2]^{0.5}$ At high velocities, this approaches $S_T = u'$	In agreement with Damkohler.[14]
Karlovitz et al.[16]	For weak turbulence, $S_T = S_L + u'$ For strong turbulence, $S_T = S_L + (2S_L u')^{0.5}$ where $u' = \frac{S_L}{\sqrt{3}}\left(\frac{\rho_u}{\rho_b} - 1\right)$	S_T is independent of turbulence scale. Laminar flame speed is most important parameter.
Scurlock and Grover[17]	$S_T = S_L[1 + C_3(\bar{y}/l)^2]^{0.5}$ where $\bar{y}$ is dependent on approach stream and flame-generated turbulence, l the turbulence scale, C_3 a constant.	S_T is dependent on laminar flame speed and turbulence scale.

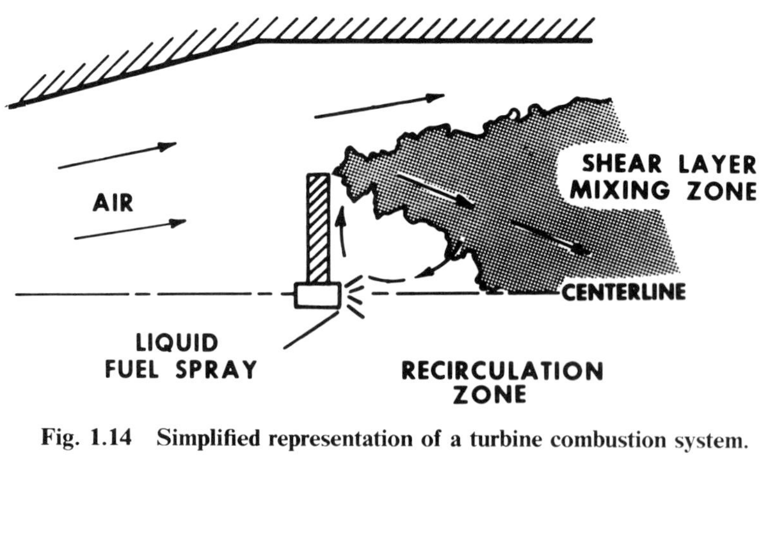

Fig. 1.14 Simplified representation of a turbine combustion system.

Table 1.3 Importance of Turbulence to Chemical Reactions—Values of $(E_a/R\bar{T})^{-2}$

	T (K)		
E_a(kcal/g · mole)	1500	2000	2500
10	0.09	0.16	0.25
20	0.0225	0.04	0.0625
40	0.0052	0.01	0.0156
60	0.0025	0.0044	0.00694
80	0.0014	0.0025	0.0039

on the shear layer mixing zone, has been shown to be successful in correlating a number of combustion characteristics ranging from exhaust pollutants to stability.

A final aspect of turbulence to be discussed here involves its effect on chemical reactions. During fuel-air mixing as well as in the reaction zone, the individual turbulent eddies can have widely differing values of fuel-air ratio and temperature. Because reaction rates are very sensitive to these variables [see Eq. (1.4)], the turbulence characteristics can strongly influence the rates and end products of the combustion process. Gouldin[20] has performed an analysis indicating that turbulence is of importance to chemical reactions when turbulent temperature fluctuations T' are such that

$$\overline{(T')^2}/(\bar{T})^2 > (E_a/R\bar{T})^{-2} \tag{1.31}$$

where the bars indicate average values. Table 1.3 shows the $(E_a/R\bar{T})^{-2}$ for different values of E_a and $\bar{T}$. Turbulence can be expected to play a significant role in all cases except those involving low activation energies (<20 kcal/g · mole) and high temperatures (>2500 K). Because of the obvious difficulties in accomplishing temperature or concentration measurements on the time and length scales of interest to this subject, only limited empirical information is available to provide further explanation of this complex phenomena.

Perfectly Stirred Reactor

The perfectly stirred reactor (PSR) is defined as a combustion region in which reactant and product concentrations as well as temperature are completely homogeneous.[21,22] The fuel-air mixture entering the reactor is assumed to be instantaneously mixed with the combustion products. In principle, this immediately increases the temperature of the entering reactants far beyond the initial state and provides a substantial and continuous supply of the chain carriers that are of paramount importance to hydrocarbon combustion (see Sec. 1.2).

Reaction rates per unit volume are maximized in the PSR. The stabilization characteristics of practical systems—primary zones of main combustors and regions behind flameholders of afterburners—are often modeled using

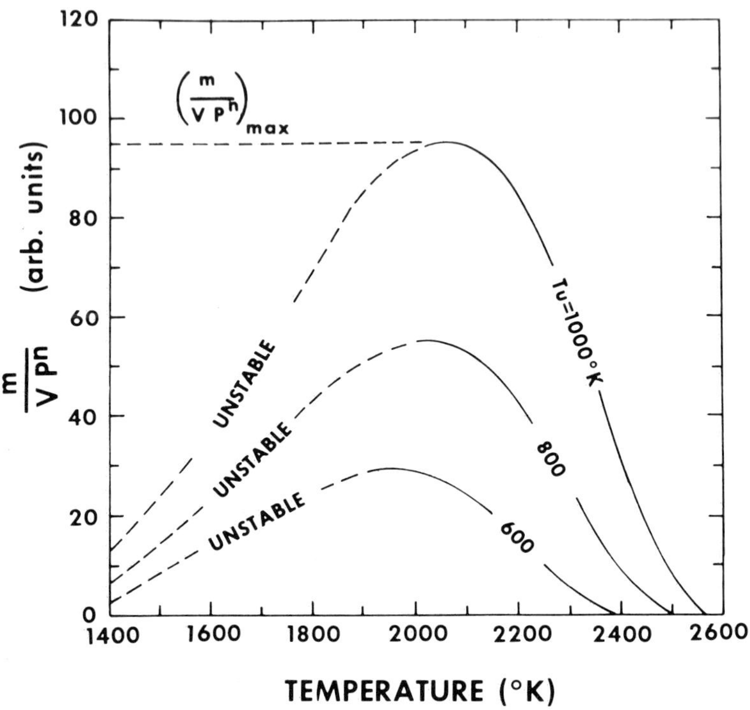

Fig. 1.15 Perfectly stirred reactor operating conditions.

PSR analyses. A simplified version of the analysis presented in Ref. 22 results in the following dependence of the reaction rate on key parameters:

$$\frac{m}{V} \sim \left(\frac{P}{RT_R}\right)^n \frac{(T_F - T_R)^n}{(T_F - T_u)^{n-1}(T_R - T_u)} \exp\left(\frac{-E_a}{RT_R}\right) \tag{1.32}$$

where V is the reactor volume, m the mass flow rate into the reactor, n the total reaction order, T_F the adiabatic flame temperature for complete reaction, T_R the PSR temperature, and T_u the initial temperature of the entering (unburned) mixture.

Figure 1.15 illustrates the relationship between the mass burning rate and the reactor temperature. These results correspond to a case where $n = 2$, $E_a = 40$ kcal/g · mole, and stoichiometric combustion of a fuel yielding T_F values of 2550, 2500, and 2400 K for T_u values of 1000, 800, and 600 K, respectively. Equation (1.32) yields three solutions for any value of m/VP^n —only the two highest T_R solutions are indicated in Fig. 1.15, as the lowest T_R solution, while stable, is not of practical interest here. The mid-T_R solution is also of academic importance as it is unstable. Considering only the highest T_R solution, the analysis indicates that the adiabatic flame temperature for a complete reaction is achieved only at flow rates approaching zero. Further, a maximum value of m/VP^n is indicated and blowout is

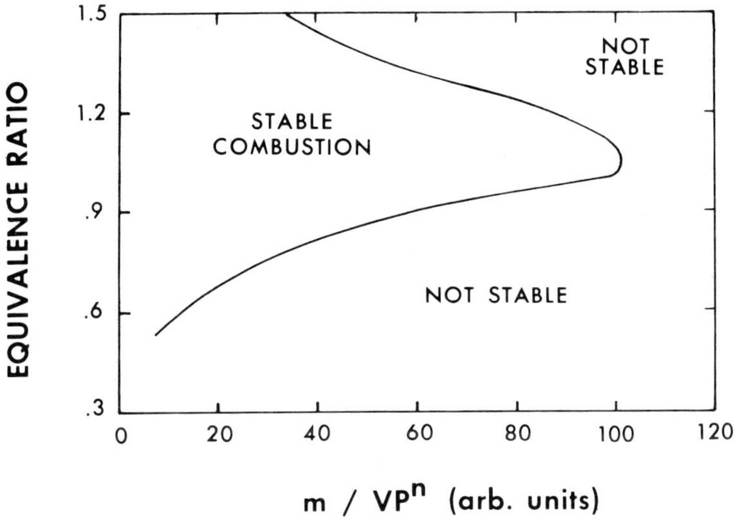

Fig. 1.16 Stirred reactor stability dependence on equivalence ratio.

expected if a further increase is attempted. Typically, this PSR blowout point is imminent when the temperature rise above the inlet conditions is 75% of that corresponding to complete combustion.

The exponential nature of reaction rates [Eq. (1.4)] is directly reflected in Eq. (1.32). For this reason, perfectly stirred reactors have been extensively utilized for high-temperature chemical kinetic studies. Equation (1.32) also indicates the beneficial effect of higher values of T_R on stabilization. This can be achieved by higher initial mixture temperature (as shown in Fig. 1.15) and/or an equivalence ratio closer to unity. Figure 1.16 illustrates the dependence of the well-stirred reactor stability region on the equivalence ratio. Consequently, combustor designers strive to create primary zones that promote stabilization with an approximately stoichiometric fuel-air mixture ratio.

Equation (1.32) can be rearranged and simplified, while maintaining the most important temperature characteristics, to yield the following relationship:

$$m/VP^n \sim \exp\left(\frac{-E_a}{RT}\right) \tag{1.33}$$

This relationship provides some guidance in developing a parameter with which the volumetric heat release of practical combustion systems may be judged. Following the units of Eq. (1.33), a specific heat release rate (SHRR) parameter has been established for aircraft gas turbine combustors with the units of energy/time-pressure-volume. This topic is discussed further in Chap. 2 of *Aircraft Propulsion System Technology and Design.*

1.5 Combustion Parameters

Three important combustion parameters will be discussed in this section: combustion efficiency, flame stabilization, and ignition phenomena.

Combustion Efficiency

Perhaps the most fundamental of all combustion performance characteristics is the combustion efficiency η_c. This parameter is defined as the fraction of the maximum possible energy that has been released during a combustion process. For the case of constant pressure combustion, η_c can be expressed as

$$\eta_c = \frac{(h_{sp})_{T_2} - (h_{sp})_{T_1}}{(h_{sp})_{T_2\,\mathrm{ideal}} - (h_{sp})_{T_1}} \tag{1.34}$$

An excellent approximation of η_c can be made by assuming that the product specific heat is independent of temperature,

$$\eta_c = \frac{(T_2 - T_1)_{\mathrm{actual}}}{(T_2 - T_1)_{\mathrm{ideal}}} \tag{1.35}$$

In cases where significant acceleration occurs during the combustion process, total enthalpy or temperature must be used in Eqs. (1.34) and (1.35). Further, the ideal value of T_2 in Eq. (1.35) or $(h_{sp})_{T_2}$ in Eq. (1.34) is that corresponding to the calculated equilibrium flame temperature. Consequently, the consideration of the dissociation effects is vital when temperatures are in excess of 1650 K.

In cases where the temperature is below 1650 K, combustion efficiency can be related to operating and fuel parameters as follows:

$$\eta_c = \frac{C_p(T_2 - T_1)_{\mathrm{actual}}}{\left(\dfrac{f/a}{1 + f/a}\right)(\Delta h_c)_f} \tag{1.36}$$

where C_p is an average specific heat and $(\Delta h_c)_f$ is the fuel heat of combustion.

In practice, η_c can be determined by measuring the actual T_2. However, this method presents some difficulty in cases where combustion efficiencies are above 90%. The objective of such testing usually involves reduction of the remaining combustion inefficiency and relatively small temperature measurement errors can cause large uncertainties in the determined inefficiency. Consequently, exhaust gas analysis has received wide acceptance as a means of more accurately determining the combustion inefficiency. Exhaust concentrations of species containing chemical energy (predominantly carbon monoxide, unburned hydrocarbons, and hydrogen) must be determined. Inefficiency may then be calculated using the following equation:

$$1-\eta_c=\frac{X_{\mathrm{CO}}(\Delta h_c)_{\mathrm{CO}}+X_{\mathrm{HC}}(\Delta h_c)_{\mathrm{HC}}+X_{\mathrm{H_2}}(\Delta h_c)_{\mathrm{H_2}}}{\left(\dfrac{f/a}{1+f/a}\right)(\Delta h_c)_f} \tag{1.37}$$

where X_i is the mass fraction of species i and $(\Delta h_c)_i$ the heat of combustion of species i. The numerator of this expression represents the unused chemical energy per mass of exhaust, while the denominator represents the chemical energy per mass of the initial fuel-air mixture.

Another common means of expressing Eq. (1.37) involves the use of the emission index EI which represents the mass of CO, hydrocarbons, or H_2 in the exhaust per 1000 mass units of fuel. In this case the equation reduces to

$$1-\eta_c=10^{-3}(0.232EI_{\mathrm{CO}}+EI_{\mathrm{HC}}+2.76EI_{\mathrm{H_2}}) \tag{1.38}$$

Note that in cases where the combustion temperature exceeds 1650 K, Eqs. (1.37) and (1.38) must be modified to account for the amounts of CO, hydrocarbons, or H_2 present due to equilibrium dissociation. The calculated combustion inefficiency should correspond only to that CO, hydrocarbon, or H_2 which is present at concentration levels in excess of equilibrium.

Flame Stabilization

Not all fuel-air mixtures are capable of supporting sustained combustion. The equivalence ratio, temperature, and pressure conditions within which combustion can be sustained in a quiescent, gas-phase system are defined as the flammability limits. Figure 1.17 illustrates the existence of both fuel-lean and fuel-rich limits for a kerosene-air system. While these limits are dependent on pressure, experimental configuration, and the existence of a quiescent system, the fundamental concept of flammability limits is invaluable because it defines the widest possible regimes of combustion. Flammability limit data for some typical hydrocarbons are indicated in Table 1.4. It is noted that variations in the limiting equivalence ratio are not substantial for the lean limit, that of primary practical importance to the combustor designer.

Flammability limits are sensitive to mixture temperature. Correlations have indicated the following approximate relationships between the mixture equivalence ratios for the lean and rich limits (ϕ_L and ϕ_R) and initial mixture temperature[23]:

$$\phi_L(T_2)-\phi_L(T_1)\simeq-9.2\times10^{-4}(T_2-T_1) \tag{1.39}$$

$$\phi_R(T_2)-\phi_R(T_1)\simeq25\times10^{-4}(T_2-T_1) \tag{1.40}$$

Substantial pressure differences may also be observed if the test container is sufficiently small for wall quench reactions to be of importance.

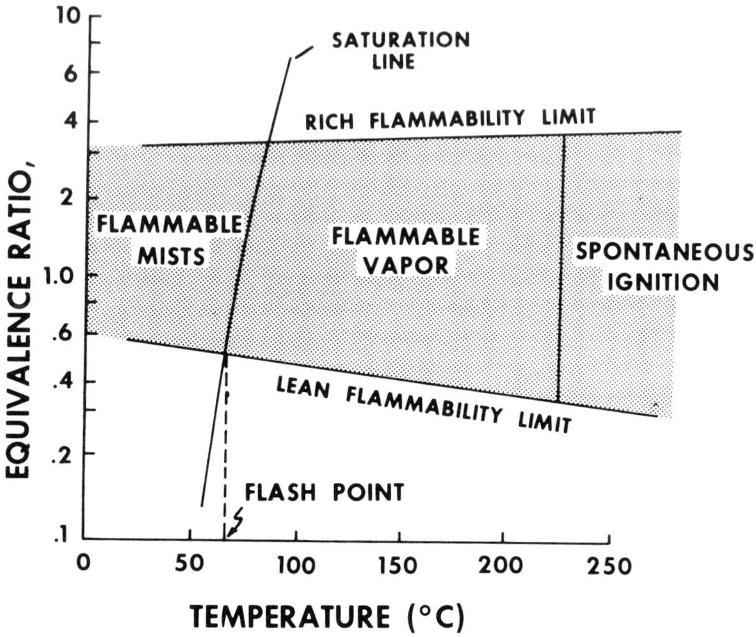

Fig. 1.17 Flammability characteristics for a kerosene-type fuel in air at atmospheric pressure.

Often a flowing system will not be capable of sustained stable combustion under the temperature, pressure, and equivalence ratio conditions within the flammability limits. In practical systems, the limits within which the combustion will be stable are dependent on design details. An example of the onset of instability in a premixed fuel-air system is found in the stabilization process for a turbulent flame at the tip of a Bunsen burner. The blowoff characteristics of such a system have been successfully related to the velocity gradient near the Bunsen burner rim.

To reduce the possibility of instabilities in flowing premixed systems, flameholders and primary zones are often used. In either case, the stability of the system is achieved by creating a region of violent recirculation and thorough mixing. Such a zone approximates the perfectly stirred reactor (PSR), which provides maximum stability for a flowing system. Fuel and airflow distributions are intended to provide a near-stoichiometric equivalence ratio in this region to maximize stability. In many cases, especially the flameholder, the portion of the flow intercepted by the recirculation zone is small and the predominant heat-releasing combustion mechanism involves reactions within the shear layer that bounds the recirculation zone (see Fig. 1.18). The PSR portion of the flow can be thought of as providing a continuous pilot to sustain the shear layer combustion region.

Blowout of these systems can be envisioned as occurring in two phases.[24,25] First, the less stable shear layer is unable to sustain itself as the flow is

Table 1.4 Flammability Limits in Air at Atmospheric Pressure and Room Temperature (from Ref. 23)

	Equivalence Ratio	
Fuel	Lean Limit	Rich Limit
n-Paraffins		
Methane	0.53	1.56
Propane	0.53	2.51
Butane	0.60	2.88
Pentane	0.58	3.23
Hexane	0.55	3.68
Heptane	0.63	3.78
Octane	0.60	—
Isoparaffins		
2,2-dimethylpropane	0.54	2.01
2-methylpentane	0.55	3.42
2,2,4-trimethylpentane	0.66	—
2,2,3,3-tetramethylpentane	0.53	3.45
Olefins		
Ethene	0.45	6.76
Propene	0.52	2.47
1-butene	0.46	2.94
Aromatics		
Benzene	0.50	2.75
Toluene	0.61	—
Ethylbenzene	0.50	—

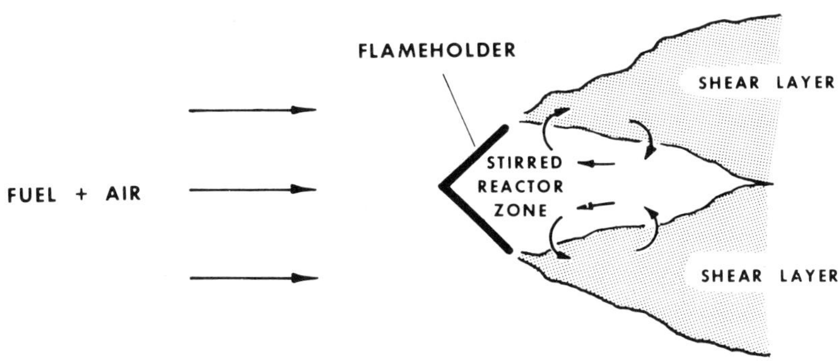

Fig. 1.18 Physical processes in flameholding.

increased and will extinguish. As previously implied, this results in the elimination of the predominant heat-transfer source and causes the combustion efficiency to drop to nearly zero. The recirculation zone, being more stable, will continue to operate until $(m/VP^n)_{max}$ is reached (see Figs. 1.15 and 1.16), at which time this region will also blow out. Since it represents a minor heat release factor, this final blowout point is of secondary interest.

Ignition

By definition, ignition is possible only for those fuel-air mixture conditions within the flammability limits. The entire region within the flammability limits must be further divided into two subregions separated by the spontaneous ignition temperature (SIT). This parameter is usually determined using a standardized test procedure in which a liquid fuel is dropped into an open-air container heated to a known temperature. The spontaneous ignition temperature is defined as the lowest temperature at which visible or audible evidence of combustion is observed. Typical values of the SIT are listed in Table 1.5. Note the trend toward reduced SIT as the length of an n-paraffin chain is extended. Further, the impact of the side methyl groups in the case of isooctane is to increase the SIT to a level consistent with an n-paraffin of much lower molecular weight. Because these data are specific to a particular experiment, direct usage of these data as hazard criteria is not advisable. Most importantly from the combustor designer's standpoint, SIT variations due to pressure are significant. SIT decreases rapidly until approximately 2 atm, with apparently small changes above this pressure.[20]

Above the spontaneous ignition temperature, the key combustion characteristic is the ignition delay time. This parameter is defined as the time lag for a given fuel-air mixture to achieve significant reaction. Zero time may be defined as the incidence of initial mixing (as in a flow reactor) or of near-instantaneous heating (as in a shock tube). While there are many ways to define the onset of significant reaction (dT/dt, $\Delta T/\Delta T_{max}$, etc.), the important point is that the ignition delay is exponentially related to initial

Table 1.5 Spontaneous Ignition Temperatures (from Ref. 20)

Fuel	SIT (K)	Fuel	SIT (K)
Propane	767	Decane	481
Butane	678	Hexadecane	478
Pentane	558	Isooctane	691
Hexane	534	Kerosene (JP-8 or Jet A)	501
Heptane	496	JP-3	511
Octane	491	JP-4	515
Nonane	479	JP-5	506

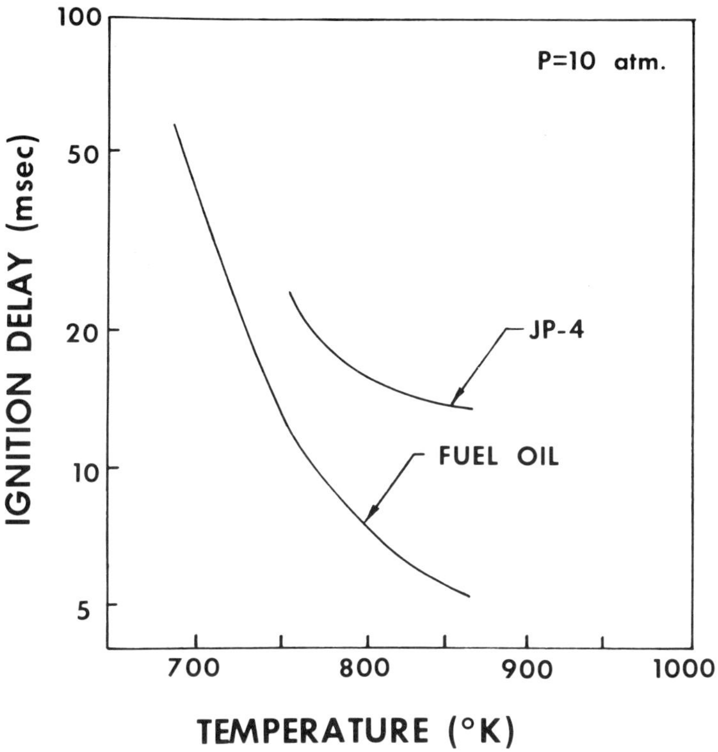

Fig. 1.19 Ignition delay times for practical fuels (from Ref. 26).

temperature T_1,

$$t_{\text{ign}} \sim \exp(E_a/RT_1) \tag{1.41}$$

Because the ignition mechanism is not dependent on the final flame temperature, t_{ign} is not strongly dependent on the mixture ratio within the flammability limits. However, a strong dependence on pressure is usually observed. Ignition delay times for typical fuels are illustrated in Fig. 1.19.

Below the spontaneous ignition temperature, an additional heat source must be utilized to allow temperatures to locally exceed the SIT. The most common method of achieving this is the spark discharge. The necessary amount of energy release to achieve ignition is called the minimum ignition energy (MIE). The quantity varies very significantly with equivalence ratio as shown in Fig. 1.20 for the case of vapor fuel-air mixtures. It is important to note that the minimum condition is not always at a stoichiometric mixture ratio. For heavy fuels, the minimum occurs closer to $\phi = 2$. Other important variables include the initial mixture temperature and pressure. Finally, in the more practical case of liquid fuel spray ignition, the extent of

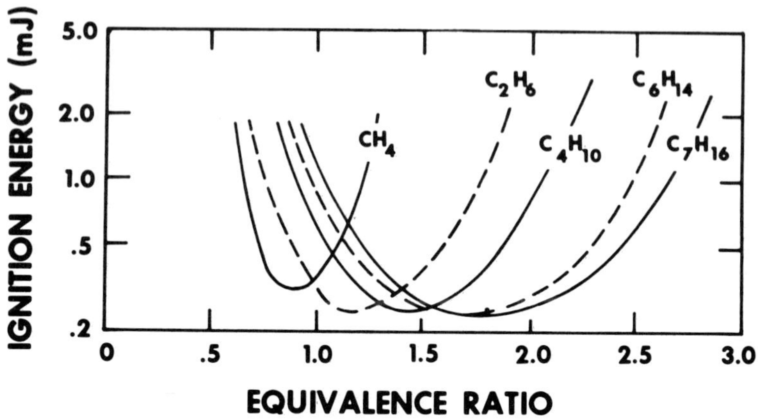

Fig. 1.20 Minimum ignition energies (from Ref. 28).

fuel vaporization is vital to ignition characteristics. Rao and Lefebvre[27] have shown that the liquid fuel droplet diameter has a powerful influence on the minimum ignition energy.

1.6 Jet Fuels

Combustion Characteristics

The most fundamental of all fuel characteristics is the heat of combustion or heating value. This empirically determined parameter represents the energy released per mass of fuel upon complete combustion when both the initial and final temperatures are nearly 25°C. The actual experiment involves a combustion bomb pressurized with pure oxygen immersed in a well-insulated water bath. The temperature rise of the water (usually only a few degrees Centigrade) is determined and the energy necessary to cause this increase for the entire system is determined. This value, which is calculated as negative for exothermic combustion reactions, is the constant-volume higher heating value.

Since the experiment is performed at 25°C, condensed water from the combustion products within the bomb provides additional energy release, which is included in the constant-volume higher heating value. The measurement can be corrected to yield the constant volume lower heating value $(\Delta h_c)_v$, which corresponds to the energy that would have been released if the water in the combustion products had remained in the vapor phase. The constant-pressure lower heating value Δh_c, which has been previously discussed in Sec. 1.3, can then be calculated,

$$\Delta h_c - (\Delta h_c)_v = (n_p - n_R)(RT/J) \tag{1.42}$$

where n_p and n_R are the number of moles of gaseous products and reactant,

Table 1.6 Heats of Combustion and Formation (from Ref. 1)

Name	Formula	Molecular Weight	$-\Delta h_c$ Constant-Pressure Lower Heating Value (cal/g)
Methane	CH_4	16.04	11,946
Ethane	C_2H_6	30.07	11,342
Propane	$C_3H_8^*$	44.09	11,072
n-butane	C_4H_{10}	58.12	10,925
Isobutane	C_4H_{10}	58.12	10,897
n-pentane	C_5H_{12}	72.15	10,744
n-hexane	C_6H_{14}	86.17	10,685
n-heptane	C_7H_{16}	100.20	10,643
n-octane	C_8H_{18}	114.22	10,611
2,2,4-trimethylpentane	C_8H_{18}	114.22	10,592
n-nonane	C_9H_{20}	128.25	10,587
n-decane	$C_{10}H_{22}$	142.28	10,567
n-tetradecane	$C_{14}H_{30}$	198.38	10,515
n-hexadecane	$C_{16}H_{34}$	226.43	10,499
n-pentatriacontane	$C_{35}H_{72}$	492.93	10,573
Ethylene	C_2H_4	28.05	11,264
Propylene	C_3H_6	42.08	10,935
Isobutene	C_4H_8	56.10	10,759
Octene	C_8H_{16}	112.21	10,556
Cyclopentane	C_5H_{10}	70.13	10,458
Cyclohexane	C_6H_{12}	84.16	10,376
Benzene	C_6H_6	78.11	9,588
Toluene	C_7H_8	92.13	9,680
α-xylene	C_8H_{10}	106.16	9,748
Methyl alcohol	CH_3OH	32.0	4,802
Ethyl alcohol	C_2H_5OH	46.0	6,447
Propyl alcohol	C_3H_7OH	60.0	7,388
Butyl alcohol	C_4H_9OH	74.1	7,936
Acetylene	C_2H_2	26.04	11,518
Hydrogen	H_2	2.016	28,651
Carbon (solid, graphite)	C	12.01	7,826
Carbon (coke) to CO_2	C	12.01	8,077
Carbon (coke) to CO	C	12.01	2,467
Carbon monoxide	CO	28.01	2,413
JP-4	$CH_{2.02}$		10,389
JP-5	$CH_{1.92}$		10,277
JP-8/Jet A/Jet A-1	$CH_{1.94}$		10,333

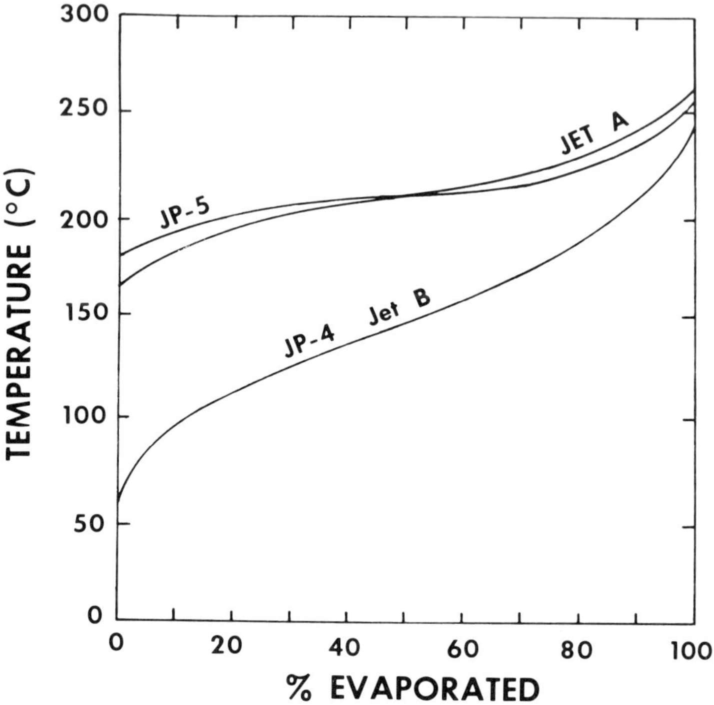

Fig. 1.21 Distillation characteristics of common jet fuels (from Ref. 29).

T the initial or final temperature (approximately 298 K), and J the mechanical equivalent of heat. Values of Δh_c for typical hydrocarbons and jet fuels are shown in Table 1.6.

While differences in the heats of combustion among hydrocarbons are relatively small, differences in volatility are substantial. Fuels can range from methane (boiling point of $-161°C$) to heavy liquid hydrocarbons containing napthalene (boiling point of 211°C). In nonaircraft turbine applications, future fuel candidates include even residual oils (which have nonvolatile components). Common aircraft turbine fuels, however, are a blend of many hydrocarbons and their volatility is usually characterized by a distillation curve as shown in Fig. 1.21.

The chemical composition of common jet fuels is extremely complex. The hundreds of hydrocarbon types present are often categorized into three groups: paraffins, olefins, and aromatics. Paraffins are the straight chain or cyclic saturated molecules like propane, butane, or cyclohexane. These are generally very clean burning fuels (low soot formation). Olefins are characterized by the presence of a carbon-carbon double bond as occurs in ethylene. These compounds are suspected of causing gum and stability problems in jet fuels. Aromatics are molecules containing unsaturated ring structures. These may be single ring (e.g., benzene) or polycyclic (e.g., napthalene) in nature. The combustion of aromatic fuels is likely to cause problems associated with carbon particle formation, flame radiation, and exhaust smoke.

A final important fuel characteristic from the handling, crash hazard, and tactical vulnerability standpoints is the flash point. This parameter is empirically determined using a controlled-temperature container partially filled with fuel. A small flame is passed over the fuel-air mixture. The minimum temperature at which some evidence of ignition is observed is defined as the flash point. It has been demonstrated that this temperature corresponds to conditions where the equilibrium vapor-air mixture above the liquid fuel is at the lean flammability limit. This characteristic is illustrated in Fig. 1.17.

Common Jet Fuels

Three jet fuel types are in wide use. JP-4 is used by the air forces of NATO, including the United States. Jet B, a fuel nearly identical to JP-4, is used by the Canadian commercial airlines. These fuels can be grossly represented as a blend of kerosene and gasoline. The high volatility of JP-4 results in a vapor pressure of about 0.17 atm (2.5 psia) at 310 K (100°F) and a flash point of approximately −25°C.

Jet A is the kerosene-based fuel used by most of the world's commercial airlines, including the United States. It has a much lower volatility than JP-4, resulting in a flash point of about 52°C. Because of the reduced probability of postcrash fires and the reduction of combat vulnerability, the NATO nation air forces are considering conversion to JP-8. This fuel is

Table 1.7 Important Jet Fuel Properties

	JP-4		Jet A (JP-8)		JP-5	
Property	Spec. Req.	Typical Value	Spec. Req.	Typical Value	Spec. Req.	Typical Value
Vapor pressure at 38°C (100°F), atm	0.13–0.2	0.18	—	0.007	—	0.003
Initial boiling point (°C)	—	60	—	169	—	182
End point (°C)	—	246	288	265	288	260
Flash point (°C)	—	−25	> 49	52	> 63	65
Aromatic content (% Vol)	< 25	12	< 20	16	< 25	16
Olefinic content (% Vol)	< 5	1	—	1	—	1
Saturates content (% Vol)	—	87	—	83	—	83
Net heat of combustion (cal/g)	> 10,222	10,388	> 10,222	10,333	> 10,166	10,277
Specific gravity	0.751–0.802	0.758	0.755–0.830	0.810	0.788–0.845	0.818
Approximate U.S. yearly consumption (10^9 gal)		3.4		13.1		0.7

nearly identical to Jet A-1, a commercial fuel similar to Jet A in all respects except freeze point ($-50°C$ vs $-40°C$ for Jet A). The combustion characteristics of JP-8, Jet A, and Jet A-1 are virtually identical.

The unique problems associated with shipboard jet fuel use cause the U.S. Navy to use a third fuel type, JP-5, which has an even higher flash point ($> 63°C$).

The physical and chemical properties of these fuels are illustrated in Table 1.7. Approximate yearly consumption figures for 1984 are also shown.

1.7 Summary

This chapter has reviewed fundamental concepts necessary for the understanding of aeropropulsion combustion. Two additional chapters will consider the practical application of this information to mainburners and afterburners. Much of this chapter has reflected the theme that the subject of combustion involves interdisciplinary study of chemistry, thermodynamics, and gasdynamics.

Key topics to the study of combustion chemistry are reaction rates, equilibrium considerations, and the mechanisms of hydrocarbon-air combustion. The Arrhenius relationship, which describes the basic dependencies of the reaction rate on pressure, temperature, and concentration, has been highlighted and its impact on combustion systems has been described. CO_2 and H_2O dissociation and the water-gas relationship are the primary equilibrium considerations. Current understanding of hydrocarbon combustion has been reviewed. This complex process can be envisioned as a sequence of events involving hydrocarbon pyrolysis and partial oxidation to H_2 and CO, chain branching reaction resulting in H_2 consumption, and CO oxidation by OH radicals generated during chain branching.

Combustion thermodynamics involves relating the energy release from fuel consumption to combustion product effects. For constant-pressure systems, the first law of thermodynamics implies the conservation of total enthalpy across the reacting system. Using this relationship, definitions and methods of calculating the flame temperature have been offered. The theoretical flame temperature, calculated assuming no dissociation, has been used to explain the effects of initial temperature, fuel-air ratio, fuel type, and extent of vitiation. Methods of more accurate flame temperature calculation, including dissociation effects, have been presented and the above-described effects illustrated.

Gasdynamics and diffusion processes affecting combustion have been described. Premixed laminar flames have been discussed and the dependence of propagation rate on temperature and especially the fuel-air ratio have been highlighted. In the case where fuel and air are not initially mixed, the rates of fuel and oxygen diffusion into the flame region control the burning rate. The key properties of diffusion flames and methods of analyzing laminar systems have been reviewed. The impact of turbulence on premixed and diffusion flames has been discussed. In the case of premixed systems, flame propagation rates are enhanced. In the case of diffusion flames, combustion zone mixing rates are increased, resulting in greater burning rates. Finally, a model of the ultimate turbulent system, the perfectly stirred reactor, has been offered. In this system, mixing rates are instantaneous relative to the chemical kinetic effects and uniform temperature and species concentration exist throughout the reactor. This perfectly stirred reactor analysis has indicated important dependencies of such a system on temperature, mixture ratio, and combustion kinetics.

Combustion parameters of importance to aeropropulsion have been reviewed and explained using fundamental information regarding the chemistry, thermodynamics, and gasdynamics. The parameters reviewed were combustion efficiency, flame stabilization, and ignition. Combustion efficiency has been defined and related to both the exhaust temperature and species concentration. Flame stabilization has been discussed relative to the definition of flammability (which applies to a quiescent system), as well as to the basic processes occurring in flameholder or primary zone regions. In the latter case, the roles of the recirculation and shear layer zones have been highlighted. Ignition has been discussed in terms of spontaneous ignition temperature, ignition delay time, and minimum ignition energy.

Finally, the important combustion characteristics of jet fuels have been defined and discussed. These include the heat of combustion, volatility and distillation characteristics, and flash point. The properties of current jet fuels, JP-4 (or Jet B), JP-8 (similar to Jet A), and JP-5 have been tabulated.

Bibliography

Barnett, H. C. and Hibbard, R. R. (eds.), "Basic Considerations in the Combustion of Hydrocarbon Fuels with Air," Propulsion Chemistry Division, Lewis Flight Propulsion Laboratory, NACA Rept. 1300, 1957.

Fenimore, C. P., *Chemistry in Premixed Flames*, Pergamon Press, New York, 1964.

Frank Kaminetskii, D. A., *Diffusion and Heat Exchange in Chemical Kinetics*, translated by N. Thon, Princeton University Press, Princeton, N.J.

Fristrom, R. M. and Westenberg, A. A., *Flame Structure*, McGraw-Hill Book Co., New York, 1965.

Gaydon, A. G. and Wolfhard, H. G., *Flames*, Chapman and Hall, London, 1960.

Laidler, K. J., "Reaction Kinetics," *Homogeneous Gas Reactions*, Vol. I, Pergamon Press, New York, 1963.

Laidler, K. J., *Chemical Kinetics*, 2nd ed., McGraw-Hill Book Co., New York, 1961.

Lewis, B. and Von Elbe, G., *Combustion Flames and Explosions of Gases*, Academic Press, New York, 1964.

Moelwyn-Hughes, E. A., *Physical Chemistry*, 2nd revised ed., MacMillan Co., New York, 1964.

Moore, W. J., *Physical Chemistry*, 3rd ed., Prentice-Hall, Englewood Cliffs, N.J., 1962.

Smith, M. L. and Stinson, K. W., *Fuels and Combustion*, McGraw-Hill Book Co., New York, 1952.

Strehlow, R. A., *Fundamentals of Combustion*, International Textbook Co., Scranton, Pa., 1967.

Swithenbank, J., *Combustion Fundamentals*, Air Force Office of Scientific Research, Washington, D.C., Feb. 1970.

Williams, F. A., *Combustion Theory*, Addison-Wesley, Reading, Mass., 1965.

"Literature of the Combustion of Petroleum," *Advances in Chemistry Series*, No. 20, American Chemical Society, Washington, D.C., 1958.

Proceedings of the International Combustion Symposia, Vols. 1–24, The Combustion Institute, Pittsburgh, Pa., 1947–1983.

References

[1]Smith, M. L. and Stinson, K. W., *Fuels and Combustion*, McGraw-Hill Book Co., New York, 1952.

[2]Edelman, R. B., Fortune, O., and Weilerstein, G., "Some Observations on Flows Described by Coupled Mixing and Kinetics," *Emissions from Continuous Combustion Systems*, edited by W. Cornelius and W. Agnew, Plenum Press, New York, 1972, pp. 55–90.

[3]Howard, J. B., Williams, G. C., and Fine, D. H., "Kinetics of Carbon Monoxide Oxidation in Postflame Gases," *14th International Symposium on Combustion*, The Combustion Institute, Pittsburgh, Pa., 1973, pp. 975–986.

[4]Mosier, S. A. and Roberts, R., "Low Power Turbopropulsion Combustor Exhaust Emissions," AFAPL-TR-73-36, Vols. 1–3, 1974.

[5]Fremont, H. A., Powell, H. N., Shaffer, A., and Siecia, S. N., *Properties of Combustion Gases*, Vols. I and II, Aircraft Gas Turbine Development Dept., General Electric Co., Cincinnati, 1955.

[6]Gordon, S. and McBride, B. J., "Computer Program for Calculation of Complex Chemical Equilibrium Compositions, Rocket Performance, Incident and Reflected Shocks, and Chapman Jouguet Detonations," NASA SP 273, 1971.

[7]Zabetakis, M. G., "Flammability Characteristics of Combustible Gases and Vapors," U.S. Dept. of Interior, Bureau of Mines, Bull. 627, 1965.

[8]Belles, F. E., "Flame Propagation in Premixed Gases," *Literature of the Combustion of Petroleum*, *Advances in Chemistry Series*, No. 20, American Chemical Society, Washington, D.C., 1958, pp. 166–186.

[9]Burke, S. P. and Schumann, T. E. W., "Diffusion Flames," *Industrial and Engineering Chemistry*, Vol. 20, 1928, pp. 998–1004.

[10]Dryer, F. L., "Fundamental Concepts on the Use of Emulsified Fuels," Paper presented at Fall Meeting of Western States Section, The Combustion Institute, Palo Alto, Calif., 1975.

[11]Brokaw, R. S. and Gerstein, M., "Diffusion Flames," *Basic Considerations in the Combustion of Hydrocarbon Fuels with Air*, NACA Rept. 1300, 1957, Chap. VII.

[12]Rao, K. V. L. and Lefebvre, A. H., "Evaporation Characteristics of Kerosene Sprays Injected into a Flowing Air Stream," *Combustion and Flame*, Vol. 26, No. 3, June 1976, pp. 303–310.

[13]Lefebvre, A. H. and Reid, R., "The Influence of Turbulence on the Structure and Propagation of Enclosed Flames," *Combustion and Flame*, Vol. 10, pp. 355–366.

[14]Damkohler, G., *Zeitschrift fuer Elektrochemie*, Vol. 46, 1949, p. 601.

[15]Shchelkin, K. I., *Soviet Physics—Technical Physics*, Vol. 13, Nos. 9–10, 1943.

[16]Karlovitz, B., Denniston, D. W., and Wells, F. E., *Journal of Chemical Physics*, Vol. 19, 1951, p. 541.

[17]Scurlock, A. C. and Grover, J. J., *Fourth Symposium (International) on Combustion*, The Combustion Institute, Pittsburgh, Pa., 1953, p. 645.

[18]Gerstein, J. and Dugger, G. L., "Turbulent Flames," *Basic Considerations in the Combustion of Hydrocarbon Fuels with Air*, NACA Rept. 1300, 1957, Chap. V.

[19]Mellor, A. M., "Gas Turbine Engine Pollution," *Pollution Formation and Destruction in Flames, Progress in Combustion Science and Energy*, Vol. 1, edited by N. A. Chigier, 1975.

[20]Gouldin, F. C., "Controlling Emissions from Gas Turbines—The Importance of Chemical Kinetics and Turbulent Mixing," *Combustion Science and Technology*, Vol. 7, 1973.

[21]Woodward, E. C., "Application of Chemical Reactor Theory to Combustion Processes," *Literature of the Combustion of Petroleum*, *Advances in Chemistry Series*, Vol. 20, American Chemical Society, Washington, D.C., 1958, pp. 22–38.

[22]Longwell, J. P., Frost, E. E., and Weiss, M. A., "Flame Stability in Bluff Body Recirculation Zones," *Industrial and Engineering Chemistry*, Vol. 45, 1953, pp. 1629–1633.

[23]"Fire Protection Research Program for Supersonic Transport," APL-TDR-64-105, 1964.

[24]Zukoski, E. E. and Marble, F. E., *Proceedings of Gas Dynamics Symposium on Aerothermochemistry*, Northwestern University, Evanston, Ill., 1956, p. 205.

[25]Altenkirch, R. A. and Mellor, A. M., "Emissions and Performance of Continuous Flow Combustors," *Fifteenth International Symposium on Combustion*, The Combustion Institute, Pittsburgh, Pa., 1974, pp. 1181–1189.

[26]Spadaccini, L. J., "Autoignition Characteristics of Hydrocarbon Fuels at Elevated Temperatures and Pressures," ASME Paper 76-GT-3, 1976.

[27]Rao, K. V. L. and Lefebvre, A. H., "Minimum Ignition Energies in Flowing Kerosene-Air Mixtures," *Combustion and Flame*, Vol. 27, No. 1, Aug. 1976, pp. 1–20.

[28]Lewis, B. and Von Elbe, G., *Combustion Flames and Explosions of Gases*, Academic Press, New York, 1961.

[29]*Data Book for Designers: Fuels, Lubricants, and Hydraulic Fluids used in Aerospace Applications*, Exxon Co., Houston, Tex., 1973.

CHAPTER 2. AFTERBURNERS

Edward E. Zukoski
California Institute of Technology, Pasadena, California

2. AFTERBURNERS

2.1 Introduction

The simple gas turbine cycle can be designed to have good performance characteristics at a particular operating or design point. However, a particular engine does not have the capability of producing a good performance for large ranges of thrust, an inflexibility that can lead to problems when the flight program for a particular vehicle is considered. For example, many airplanes require a larger thrust during takeoff and acceleration than they do at a cruise condition. Thus, if the engine is sized for takeoff and has its design point at this condition, the engine will be too large at cruise. The vehicle performance will be penalized at cruise for the poor off-design point operation of the engine components and for the larger weight of the engine. Similar problems arise when supersonic cruise vehicles are considered.

The afterburning gas turbine cycle was an early attempt to avoid some of these problems. Afterburners or augmentation devices were first added to aircraft gas turbine engines to increase their thrust during takeoff or brief periods of acceleration and supersonic flight. The devices make use of the fact that, in a gas turbine engine, the maximum gas temperature at the turbine inlet is limited by structural considerations to values less than half the adiabatic flame temperature at the stoichiometric fuel-air ratio. As a result, the gas leaving the turbine contains most of its original concentration of oxygen. This oxygen can be burned with additional fuel in a secondary combustion chamber located downstream of the turbine where temperature constraints are relaxed. The increased total temperature produced at the nozzle by this additional heat addition results in an increased exit velocity and thrust.

The advantage of using the afterburning gas turbine engine cycle is that the weight of the augmented engine is much less than the weight of a turbojet engine producing the same maximum thrust. This advantage is partially offset by the low thermal efficiency of the augmented turbojet cycle, which is characterized by values of specific fuel consumption much higher than those for the gas turbine cycle. However, when the afterburner is used for a small part of a flight, the weight reduction is more important than the increase in fuel consumption.

In the middle 1960s, an augmented gas turbine engine, the General Electric GE4, was selected as the cycle to be used on the Boeing supersonic transport. It was an afterburning turbojet engine and the afterburner was used not only during takeoff or transonic acceleration, but also during the

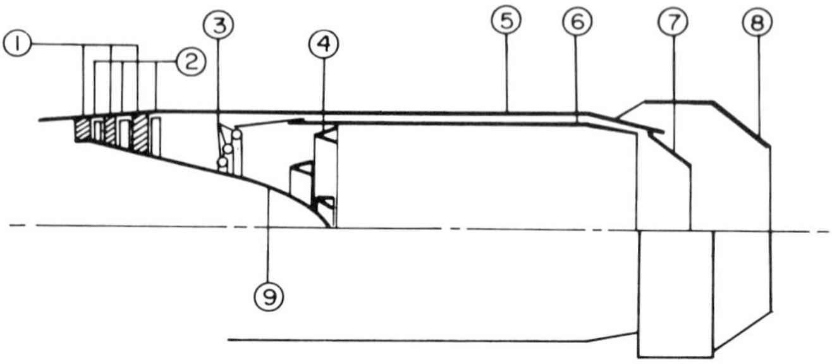

Fig. 2.1 General Electric J-79 afterburner: (1) turbine nozzles, (2) turbine blades, (3) fuel injection rings, (4) three annular V-gutter flameholders, (5) afterburner case, (6) perforated liner, (7) and (8) primary and secondary nozzle flaps, and (9) diffuser inner cone.

Mach 2.7 cruise. At these speeds, afterburning is required even during cruise to obtain a reasonable air specific thrust.

Finally, with the advent of the turbofan engines in the late 1960s and the variable-cycle engines in the 1970s, the afterburner must be viewed as one of an increasing number of devices that can be used to enhance the flexibility of the basic gas turbine cycle. The aim of these systems is to optimize engine performance over the widest possible range of operating conditions. Augmentation can be used in both fan and core streams. In some flight regimes, afterburning in the bypass airstream alone is advantageous and in others, where maximum augmentation is required, afterburning in both the bypass and core engine exhaust streams is desirable. Under some circumstances, mixing the fan and core engine exhaust streams prior to afterburning may produce a large enough performance gain to more than offset total pressure losses and increased engine weight associated with the mixing process.

An afterburner for the gas turbine engine cycle is very similar to a ramjet engine. Gas leaving the turbine is diffused, liquid fuel is added through fuel injection tubes or rings, the combustion process is initiated in the wakes of a number of flame stabilizers, and heat is added along the flame surfaces spreading from these stabilization positions. Nozzles with variable-area throats are necessary to accommodate the large total temperature changes produced by afterburning.

These elements of the afterburner are illustrated in Fig. 2.1 for the turbojet cycle and in Fig. 2.2 for the turbofan cycle. The turbojet engine is a sketch of the General Electric J79 engine and the turbofan engine is a sketch of the Pratt & Whitney F100 engine. In the latter case, afterburning is accomplished without mixing core and fan streams and the inner contour of the nozzle is shown in closed (11) and open (12) positions. In both engines, a combustion chamber liner with an aircooling passage is used to protect the outer, pressure vessel wall from heat transfer by convection and radiation.

To illustrate typical afterburner operating conditions, performance curves are shown in Fig. 2.3 for a turbofan engine, the Pratt & Whitney TF30 engine. This engine is similar to that shown in Fig. 2.2 and the core and turbine gas streams are not mixed. The specific fuel consumption (SFC) and

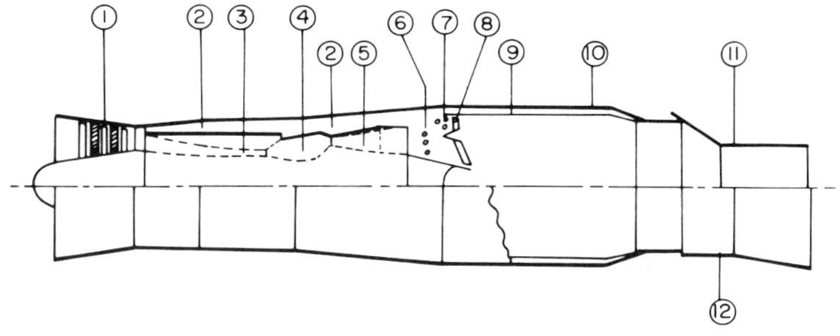

Fig. 2.2 Pratt & Whitney F100-PW-100 augmented turbofan engine: (1) three-stage fan; (2) bypass air duct, core engine compressor (3), burner (4), and turbine (5); (6) fuel injectors for core engine gas stream; (7) fuel injectors for bypass airstream; (8) flame stabilizer for afterburner; (9) perforated afterburner liner; (10) afterburner case; nozzle closed to minimum area (11) and opened to maximum area (12).

thrust F are shown as ratios of their values to their values with no afterburning and as a function of afterburner fuel-air ratio. Afterburner total pressure ratio π_{AB} and combustion efficiency η_{AB} are also given. The two curves on each plot correspond to operating altitudes of about 12 and 14.6 km and a flight Mach number of 1.4. Flow conditions at afterburner entrance for the core stream and for the two altitudes were, for the core stream: total pressure of 1.05 and 0.69 atm, total temperatures of about 900 K for both, inlet velocities of 180 and 230 m/s; for the fan stream: total temperature was about 400 K. At these conditions, thrust augmentation of about 60% can be achieved at a cost of an increase of 120% in specific fuel consumption. The performance decreases as the altitude is increased. Note that the afterburner total pressure ratio with no heat addition is about 0.94; the 6% loss represented by this ratio accounts for some diffusion loss in addition to the flameholder drag losses.

For this engine, fuel is injected through orifices with diameters of about 0.15 cm in a number of concentric rings of fuel injection tubes, which are similar to those shown in Fig. 2.2. Afterburner fuel-air ratio is increased by adding fuel first to the core flow near the interface between the core flow and fan airstream, then to the fan air, and finally to the rest of the core flow. Because the fan air contains the most oxygen and the lowest temperature, afterburning in the stream produces the largest performance gain. However, the low temperature of this stream makes vaporization of the fuel and hence afterburning most difficult. The addition of fuel in the order suggested here first produces a high temperature at the outer edge of the core stream where it can act as a pilot for the fan air combustion process. The rapid fall in combustion efficiency at 14.6 km and for low fuel-air ratios is due to the problem of burning in the cold fan streams where vaporization of the fuel is very poor.

The purpose of this chapter is to discuss the engineering information available concerning afterburner components and to indicate some of the current design practices. Some problem areas will be more thoroughly described than others. The supplementary reading list at the end of this

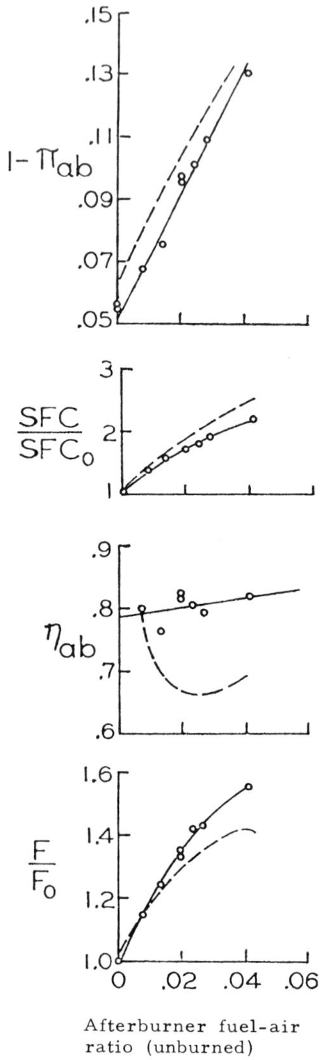

Fig. 2.3 Afterburner performance characteristics for the TF30-P-3 engine. Engine characteristics: $\pi_c = 17$, $\pi_c' = 2.1$, bypass ratio $\alpha = 1.0$, flameholder blockage 0.38, flight Mach number 1.4. Solid curves are for an altitude of 12 km and dashed curves for 14.6 km (data from Ref. 1).

chapter contains references to review articles covering a number of subjects concerning afterburners omitted here.

2.2 Diffuser

The heat that can be added to a compressible flow in a constant-area duct before choking occurs and the pressure loss accompanying this heat addition depend on the Mach number of the flow entering the burner. As the inlet Mach number decreases, the maximum heat addition increases and the total pressure loss decreases. In addition, the flame stabilization process becomes more difficult as the gas speed increases. Hence, it is desirable to have as low a Mach number as possible at the burner inlet, which leads to the use of a diffuser between the turbine exit plane and the afterburner itself.

The minimum Mach number at the burner inlet is usually fixed by the requirement that the diameter of the afterburner section of the engine not exceed that of the engine components located upstream of the afterburner. This limitation arises from the desire to minimize the drag of the engine due to frontal area and nozzle exit area and the desire to minimize the weight of the afterburner itself.

The desire to minimize weight also results in the requirement that the diffuser be kept as short as possible without producing flow separation from the inner body. Relatively large divergence angles can be used because the blockage of the flow, produced by the flameholder and fuel injection systems, reduces the tendency of the flow to separate from the diffuser cone. Finally, the general problem of producing a steady, uniform, and unseparated flow at the diffuser outlet is often complicated by the presence of a large swirl component in the gas leaving the turbine (15–20% tangential component) and the interaction of this swirling flow with the struts required to support the rear engine bearing.

In augmented turbofan engines in which the two gas streams are to be mixed before afterburning is initiated, the diffuser is usually combined with the mixer. For example, the fan and core streams can be ducted together to form a series of adjacent radial slots with fan air and core air in alternate slots. This geometry has the advantage that mixing will occur in a distance much smaller than required with the undisturbed annular geometry shown in Fig. 2.2. By keeping the cross-sectional areas of each stream almost constant, pressure losses in mixers of this type can be minimized.

2.3 Fuel Injection, Atomization, and Vaporization

The goal of the fuel injection system is to produce a specified distribution of fuel vapor in the gas stream entering the afterburner. In most engines, fuel is introduced in a staged manner so that the heat addition rate can be increased gradually from zero to the desired value. Because ignition, flame stabilization, and flame spreading are easiest to achieve when the fuel-air ratio is close to the stoichiometric value, staging is usually produced by adding fuel to successive annular stream tubes so that the mixture ratio in each tube is nearly stoichiometric. Each stream tube has its own set of fuel injectors and control system, which can be activated independently. For example, see the six injectors used in the F100 engine shown in Fig. 2.2.

The most remarkable fact concerning the fuel systems for afterburners is their simplicity. In many engine systems, fuel is supplied to a circular tube that lies with its axis perpendicular to the gas stream. Fuel is injected into the gas though small-diameter holes located in the sides of the tube such that the liquid jet enters the gas stream in a direction perpendicular to the undisturbed flow direction. The liquid jet penetrates some distance into the gas stream before its momentum is dissipated. During this penetration process, the airstream tears the jet apart and produces droplets with diameters of micrometer size. Heat transfer from the hot gas stream then vaporizes the droplets.

Given the wide range of values of the mass flows of fuel required, it is remarkable that reasonably thorough mixing of the fuel with the air can be achieved with this simple injection system. In some recent engines, efforts are being made to use simple variable-area injection ports that may possibly give better preparation of the fuel-air mixture.

The whole area of fuel penetration, atomization, and vaporization is not well understood from first principals and one of the time-consuming parts of an afterburner development program is to determine the optimum distribution of the locations for injector tubes, injector ports, and port diameters. In the following paragraphs, a very brief analysis is made of several aspects of the fuel injection problem with the aim of illustrating some of the important scaling parameters rather than of furnishing design procedures.

Penetration

The trajectory of a fluid jet injected into a high-speed airstream can be crudely analyzed by treating the boundaries of the jet and the resulting droplet stream as a solid body and applying the continuity and momentum conservation laws. The force applied to the surface of the stream tube in the direction of the flow (see Fig. 2.4) will be proportional to the dynamic pressure of the gas $\rho_\infty v_\infty^2/2$ and some cross-sectional area, say $\bar{w} \times h$ (where $\bar{w}$ is an average width and h the penetration distance). This force must be balanced by the momentum flux of the injectant stream with mass flux $\dot{m}_j$. If we assume that the fluid is accelerated to a velocity v_∞ at the exit of the control volume and that it enters with *no* momentum flux in the direction of

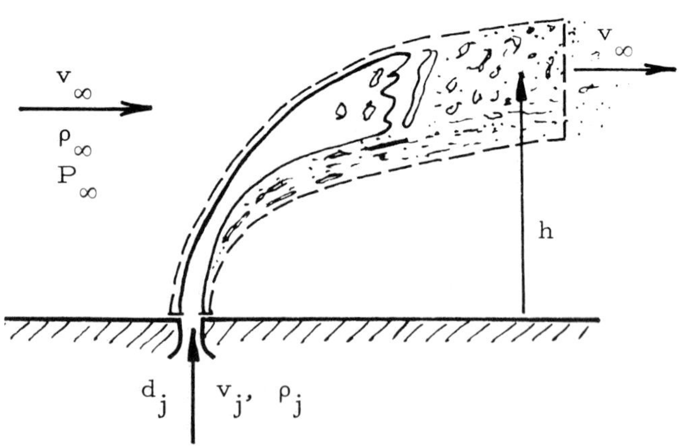

Fig. 2.4 Estimate of penetration distance.

flow, the force balance gives

$$\rho_\infty v_\infty^2(wh) \propto \dot{m}_j v_\infty$$

so that

$$\dot{m}_j/[\rho_\infty v_\infty(wh)] = \text{const} \tag{2.1}$$

The actual penetration distance is obtained by assuming $w \propto d_j$ and $\dot{m}_j \propto (\pi d_j^2/4)\rho_j v_j$; it is given by

$$\frac{h}{d_j} \propto \sqrt{\frac{\rho_j}{\rho_\infty}}\left(\frac{\rho_j v_j^2}{\rho_\infty v_\infty^2}\right)^{\frac{1}{2}}$$

Note that the denominator of Eq. (2.1) is proportional to the mass flux of air through the region fed by the injector. Thus, if the injector flow rate is changed in order to keep h fixed (as some other parameter such as flight altitude is varied), the overall fuel-air ratio in the stream tube fed by the injector will be held fixed.

Experimental work by Schetz and Padhye[2] suggests that an analysis of this type does predict a reasonable dependence of penetration distance h on the dynamic pressure ratio $\rho_j v_j^2/\rho_\infty v_\infty^2$. Penetration distances of tens of diameters can be achieved with dynamic pressure ratios between 1 and 2.

Atomization

The breakup of the injected fuel stream into small droplets depends on the dynamic pressure parameters listed above and, in addition, on the viscosities of the fuel and gas streams and on the interfacial tension of the fuel-gas system. The physical process of atomization of the fluid jet probably involves the production of waves on the fluid-gas interface, the shedding of long ligaments of fluid as the waves break, and then the breakup of the ligaments into droplets. The overall process is sufficiently complex that no model or set of scaling parameters has been generally accepted. The problem is complicated by the necessity for measuring and describing a distribution of drop sizes rather than a single drop diameter.

To give the flavor of the results obtained in experiments, the results of Ingebo and Foster[3] will be described. They worked with the volume-median droplet diameter defined as

$$D_{30} = \left(\sum_i n_i D_i^3 / \sum_i n_i\right)^{\frac{1}{3}}$$

where n_i is the number of particles with diameter D_i. They found that D_{30} depended on the injector diameter d_j and two dimensionless parameters, a Reynolds number,

$$Re \equiv \rho_l v_\infty d_j/\mu_l$$

based on the gas stream velocity v_∞, the injector diameter d_j, the liquid density and viscosity ρ_l and μ_l, and a Weber number,

$$We \equiv \sigma_j / \rho_\infty v_\infty^2 d_j$$

based on fuel surface tension σ_j and dynamic pressure of the gas. The relationship for D_{30} found by these authors was

$$D_{30}/d_j = 4(We/Re)^{\frac{1}{4}} \tag{2.2}$$

For a gas stream with a speed of about 200 m/s, temperature of 900 K, pressure of 3 atm, and a gasoline-like fuel jet, the Reynolds and Weber numbers for a millimeter diameter injector port are about 2×10^5 and 10^{-3}, respectively. Using these values in the above equation we find $D_{30} \approx 40\ \mu$m.

Equation (2.2) indicates that there is *no* dependence of atomization on the injection velocity v_j, but that there is an inverse one-quarter power dependence of D_{30} on the gas stream density and, hence, pressure. Thus, as the pressure decreases, droplet diameter will increase slowly. Finally, D_{30} is proportional to the square root of the port diameter d_j.

Evaporation

The evaporation rate of the droplets formed in the injection process is a strong function of the gas and droplet temperatures and the relative velocity of the droplets with respect to the air. For most conditions of interest here, the pressure of the vapor at the surface of the drop will be close to the equilibrium value fixed by the surface temperature of the drop. Vapor is removed from this region by diffusion and forced convection; and the heat required to produce the evaporation is transferred to the surface by conduction from within the drop and by conduction and forced convection from the gas stream. The drop radius and the temperature distribution within the drop will be functions of time. Transport of vapor and heat occur by diffusion and forced convection, both of which may be either laminar or turbulent.

The simplest situation to analyze for this complex problem is the case in which molecular transport processes for mass and heat are dominant and for which the temperature changes within the drop can be ignored. This quasisteady situation is most likely to occur for small (e.g., 10 μm diam) drops moving at a small velocity with respect to the gas stream. Under these conditions, the rate of change of mass M of a droplet is given by

$$\dot{M} = -4\pi r(\rho_v \mathscr{D}) \tag{2.3}$$

where r is the droplet radius and ρ_v and $\mathscr{D}$ the vapor density and diffusion coefficient at the particle surface. Since the particle mass is $(4/3)\pi r^3 \rho_l$,

$$\frac{r\mathrm{d}r}{\mathrm{d}t} = \frac{\rho_v \mathscr{D}}{\rho_l}$$

and the time t_e required for complete evaporation of a particle of initial radius r_0 is

$$t_e = r_0^2 / \left(\overline{\rho_v \mathscr{D} / \rho_l}\right) \tag{2.4}$$

where the denominator is the value averaged over the evaporation time of the droplet. Clearly, the time for complete evaporation will be much longer for large drops.

The term in the denominator of Eq. (2.4) is a strong function of the temperature; in order to determine its value, an energy balance for the droplet must also be used. The vapor density at the droplet surface ρ_v is only a function of temperature and, since $\mathscr{D} \propto 1/P$ holds roughly, $\rho_v \mathscr{D}$ will be proportional to $1/P$, the inverse of the local static pressure. Thus, as the pressure in the region of injection falls, the time required for evaporation of small droplets of a *fixed* temperature will decrease.

For the simple situation under investigation here, the energy balance reduces to

$$-\dot{M}L = \dot{q}$$

where L is the heat of vaporization, $(-\dot{M})$ the rate of generation of vapor, and $\dot{q}$ the heat transferred by conduction to the drop. $\dot{q}$ is given by $4\pi rk(T_\infty - T_p)$, where k is the coefficient of thermal conductivity and T_∞ and T_p the temperatures of the gas far from the drop and the temperature of the drop, respectively. Given these results, the energy balance given above reduces to

$$\frac{T_\infty - T_p}{T_\infty} = \left(\frac{\rho_v}{\rho_g}\right)\left(\frac{\rho_g \mathscr{D} C_{pg}}{k_g}\right)\left(\frac{L}{C_{pg} T_\infty}\right) \tag{2.5}$$

Note that ρ_v is an exponential function of T_p. Here the vapor density at the particle surface has been normalized by ρ_g, the density of the gas stream evaluated at T_p. The second and third terms on the right-hand side of Eq. (2.5) are almost independent of pressure and temperature. (The second term is the Lewis number.) However, the first term is a very strong function of temperature and is inversely proportional to pressure because the gas density is proportional to P/T_p. Hence, Eq. (2.5) can be written in the form

$$(T_\infty - T_p) = f\{T_p\}/P_\infty \tag{2.6}$$

where f increases exponentially with T_p.

If P_∞ is reduced, e.g., by increasing the altitude at which the engine is flying, the droplet temperature must adjust to satisfy this equality. Two regimes of interest can be defined. In the first, let the gas temperature be much greater than the particle temperature. Then a drop in P_∞ must be accompanied by a drop in T_p. Because of the strong dependence of f on T_p, a small change in T_p will be required to satisfy the equation. Hence, the temperature difference ($T_\infty - T_p$) will not change appreciably and, consequently, the conduction term and thus the evaporation rate will not change appreciably. For this example, the increase in the diffusion coefficient resulting from a reduction in the pressure is offset by a reduction in the vapor pressure (due to reduced T_p) so that the vapor diffusion and evaporation rates are held almost constant. For this case then, the evaporation time t_e of Eq. (2.4) will not be affected directly by the pressure.

In the second regime, assume that T_∞ is almost equal to T_p. Then, when P_∞ is forced to drop, the small change in T_p needed to balance Eq. (2.6) will make an appreciable difference in the temperature difference ($T_\infty - T_p$). Thus, the heat-transfer rate and consequently the evaporation rate will increase as the pressure falls.

If the simplified results presented above are combined, it is found that as the pressure in the afterburner falls, say as a result of the increase in altitude of an engine operating at a fixed Mach number, the simple injector discussed here is capable of supplying the required fuel flow to the appropriate volume of space. However, as the pressure decreases, the droplet diameter will increase. In addition, for the first vaporization regime discussed above, the evaporation rate will be independent of the pressure but will decrease rapidly as the drop radius increases. Hence, the effect of the reduction in pressure will be to increase the initial diameter of the droplets and to decrease the evaporation rate. In the second regime, the effect of the increase in the initial radius of the drops will be offset by the increase in the diffusion rates.

Also note that the evaporation rate depends on the diameter of the injector part d_j. It is clear from Eq (2.2) that D_{30} is proportional to $\sqrt{d_j}$ and consequently that the evaporation time given in Eq. (2.4) is directly proportional to d_j. Hence, more rapid evaporation could be achieved by using injection ports of a smaller diameter. However, the number and location of the injector ports would need to be changed to keep the fuel distribution unchanged. In addition, the smaller holes would be more subject to blockage by dirt and pyrolized fuel.

In real systems, the observed effect of a pressure decrease is that combustion efficiency falls off badly when the pressure is reduced below some limiting value, often on the order of 1 atm. This pressure limit is a major restriction on the operating range for an afterburner.

2.4 Ignition

The fuel-air mixture produced by the injection process has a flame propagation velocity that is much lower than the gas speed in the combustion chamber. Thus, unless sources of continuous ignition are present in the chamber, the burning gas ignited by a temporary process will be blown out of the engine as soon as the ignition is stopped. In most afterburner systems, the continuous source of ignition is the wake of a bluff body (called the flameholder) held with its axis perpendicular to the flow. Hot gas, trapped in the first few diameters of the flameholder wake, mixes with the combustible mixture flowing over the wake and acts as the source of ignition. The process of flame stabilization is the subject of the next section; here the process of the initiation of the stabilization process will be discussed. This ignition process need only start the stabilization process and may then be turned off. Furthermore, it has been found that once the stabilization process has been established in a relatively short length of flameholder, say 5 diam, the process will spread to the rest of the stabilizer system if the wakes of the stabilizers form a continuous pathway. Finally, fuel is usually added in sequence to a number of annular stream tubes to prevent pressure surges during afterburner ignition and to allow modulation of the afterburner thrust. Hence, once one region is "lit," it can act as a source of ignition for adjacent regions when fuel is added to them. Thus, the purpose of the ignition system is to establish a stabilized flame in a relatively small part of the flame stabilizer system. The bluff-body flameholder system will then furnish a path for the further spreading of the stabilized flame as additional fuel is added.

Three general systems have been used: the hot-streak technique, spark or arc ignition, and the pilot burner technique. In the hot-streak system, fuel is injected for a short period into the gas stream of the core engine just upstream of the turbine. The combustible flow formed by this process produces a very hot stream of burning gas that is positioned radially to coincide with the primary fuel-air injection stream tube for the afterburner. Combustion occurs in this stream by autoignition (because of the high temperatures present upstream of the turbine) and the flame stabilization process is initiated when the hot burning gas fills the initial wake region of the flameholders. The hot streak can be maintained for only a brief period to prevent thermal damage to the turbine.

Ignition and initiation of the flame stabilization process can also be initiated by producing a high-energy electric arc in the primary stream tube. In this case, ignition is usually produced by placing the arc in a region of the wake of the flameholder system that is particularly sheltered and that may have its own fuel supply system. Stabilization is initiated locally by the heat from the arc and spreads by the mechanism described above.

The pilot burner system is similar to the arc system and may use an arc to initiate combustion. In this system, a small can burner (similar to that used in the core-engine primary burner) is located in the primary stream tube. This system furnishes a continuous source of hot combustion products that act in a manner similar to the hot-streak system to start the stabilization process once fuel injection is started.

The energy required to initiate the flame stabilization process comes primarily from chemical reactions initiated in the premixed fuel-air stream by the ignition system. At present, the amount of energy required (e.g., joules supplied to the arc discharge or total mass of fuel injected into the hot streak) to achieve ignition cannot be calculated for a particular design. However, some trends can probably be determined by examination of experimental results concerned with the ignition of the flowing streams of fuel-air mixtures.

Studies with various sources of ignition such as arcs, hot surfaces, and hot gas streams indicate that the energy required is a strong function of the fuel-air ratio, fuel properties, local pressure and temperature, and residence time of the combustible gas in the system. The energy required is smallest for mixture ratios near the stoichiometric value and increases very rapidly as the mixture ratio is decreased below 0.5–0.7 of the stoichiometric ratio. Also, it decreases with increases in the pressure, temperature, oxygen mole fraction, and residence time of the flow in the ignition region. The exact values required are very sensitive to the particular device being studied. Thus, the calculation of the conditions required to produce ignition for a particular system cannot yet be carried out, even for very idealized experimental conditions.

Based on these results, one expects ignition in afterburner systems to be most easily achieved in gases with fuel-air ratios close to the stoichiometric value in the most sheltered regions of the stabilization system, where residence times are longest and when pressures and temperatures are highest.

Ignition is usually harder to achieve than stabilization. Because velocities and temperatures do not change a great deal at the afterburner inlet, the primary problem for afterburner ignition systems is the high-altitude relight problem. The difficulty here is associated with the low pressure in the afterburner that affects both the preparation of the fuel (by the injector system) and the ignition process directly. The operating regime for the engine is often plotted, as shown in Fig. 2.5, on an altitude vs Mach number

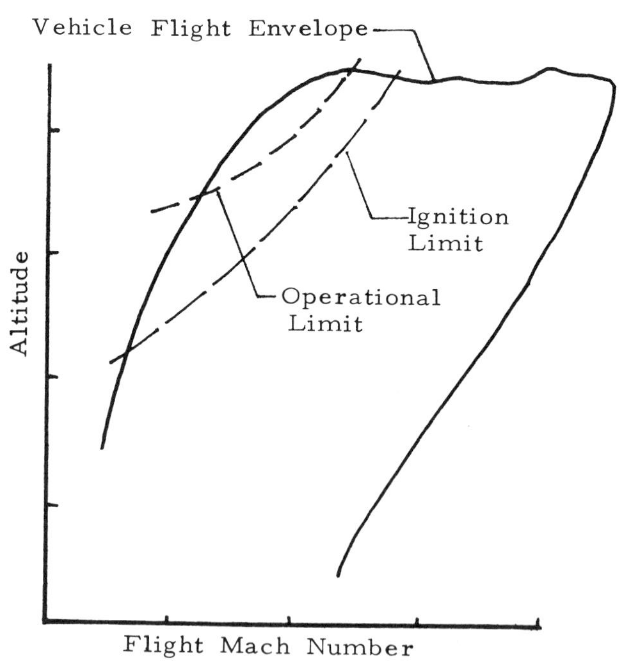

Fig. 2.5 Flight envelope for an afterburner.

map. The boundary called the ignition curve is the altitude (i.e., the pressure) limit above which ignition is no longer possible. Afterburner operation above this altitude is possible (for this example) if the afterburner is ignited at a lower altitude.

2.5 Stabilization Process

The purpose of the flame stabilization process is to establish a continuous source of ignition in a fuel-air mixture whose velocity is much greater than the turbulent flame speed for the mixture. Laminar flame speeds in mixtures of typical hydrocarbon fuels with air heated by partial combustion are in the range of a few meters per second. If the flow is turbulent, this speed may be as large as a few tens of meters per second, although the meaning of the term *turbulent flame speed* is poorly defined. These turbulent speeds are still much less than the gas speeds encountered in the flame stabilization region of afterburner systems where gas speeds of hundreds of meters per second are typically encountered. Hence, a continuous source of ignition is required to start the combustion process. Once started, the combustion wave can spread from its point of ignition across the fuel-air mixture produced by the injection system in a wave-like manner similar to the propagation of an oblique shock wave across a supersonic flow.

One of the most important parameters of the stabilization process is the state of the fuel in the fuel-air mixture. In the core flow, temperatures are high enough to insure that most of the fuel is vaporized. However, in the fan stream, temperatures can be so low that only a small fraction of the fuel will be vaporized and the stabilization mechanism will have the additional job of vaporizing the fuel used in producing a stable flame. In existing systems, this problem is usually avoided by starting the afterburning process in the core stream. The hot gas generated in the core is then used to stabilize the flame in the fan stream. The first part of the following discussion is restricted to the vaporized fuel example.

The ignition process, usually called flame stabilization in afterburner systems, is typically achieved in a mixture of fuel *vapor* and air by allowing the hot products of combustion to mix with the unburned fuel-air mixture. The steady flow of hot gas required for this process is usually obtained by setting up a recirculating flow of burned material. When a bluff-body flameholder is used (Fig. 2.6a), hot gas is generated by the recirculation of burning material in the wake of the bluff body and supplies the energy and the mass of products of combustion required to ignite the unburned material flowing past the wake. This process is described and analyzed in detail in the following sections.

The processes occurring in the wake of the bluff body also occur in the wake of a step or in a wall recess. Figures 2.6b and 2.6c show that heat transfer from the gas to walls of the cavity reduces the gas temperature in the recirculation zone and makes stabilization more difficult.

The wake region required to produce the recirculation zone described above can also be produced by injecting a secondary gas stream into the unburned flow, as shown in Figs. 2.6d and 2.6e. Again, a wake region with strong recirculation is formed by the interaction of the two streams. In these

a)

b)

c)

d)

e)

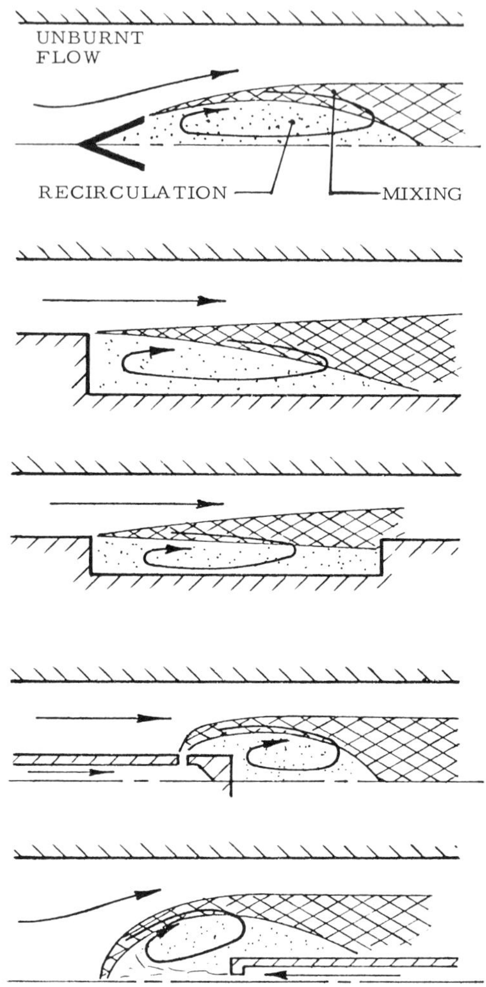

Fig. 2.6 Flame stabilization schemes for mixtures of fuel vapor and air.

examples, the fluid entering the recirculating wake is made up in part of the injected fluid. Hence, the fuel-air ratio of the recirculation zone gas can be changed by changing the fuel-air ratio of the injectant. This process can result in a powerful control over the stabilization process, as will be described later in this chapter.

The configurations shown schematically in Fig. 2.6 can be used in either axisymmetric or two-dimensional geometries. The latter configuration is the most common.

The performance of a flame stabilization system is usually presented in the form of a map on which the boundaries of a stabilization parameter are

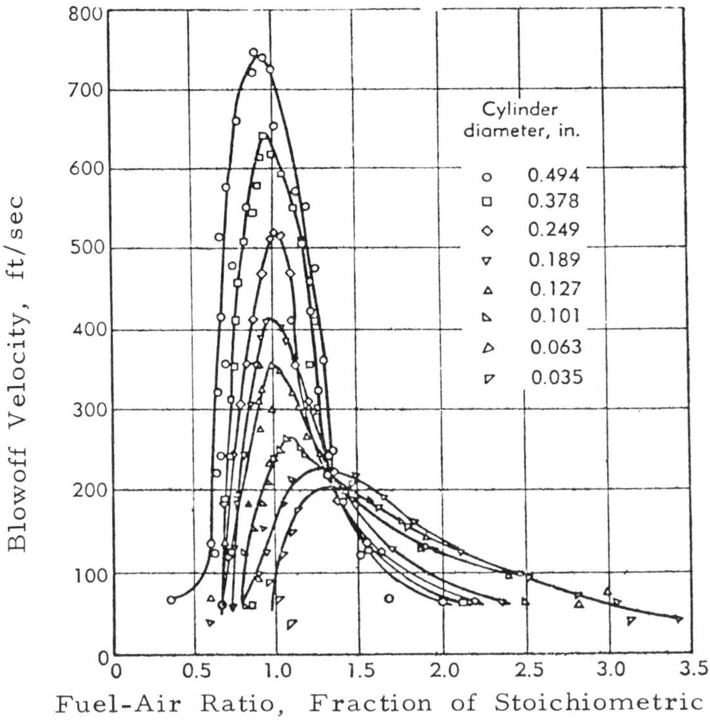

Fig. 2.7 Stability limit curves for circular cylinders, vaporized hydrocarbon fuel-air mixture at 1 atm and 60°C (data from Ref. 4).

given as a function of the fuel-air ratio. An elementary map of this type is shown in Fig. 2.7 where the velocity at which the stabilization process fails (called here the blowoff velocity) is presented as a function of the fuel-air ratio with the flameholder scale as a parameter. The experimental data of Haddock[4] were obtained in a mixture of vaporized gasoline and air and with circular cylinders used as flameholders. For each flameholder, flame stabilization was found to be possible for the range of values of the fuel-air ratio within the curves shown in Fig. 2.7. For example, stabilization was possible with the $\frac{1}{2}$-in. (1.3-cm) diam cylinder for fuel-air ratios between 0.70 and 1.15 times the stoichiometric value when the gas speed approaching the cylinder was about 600 ft/s (183 m/s). As the gas speed increased, the fuel-air ratio range for which stabilization could be achieved decreased and the peak value of about 740 ft/s (226 m/s) occurred for values close to stoichiometric. Flameholders with diameters between $\frac{1}{2}$ and $\frac{1}{8}$ in. (1.27 and 0.32 cm) had clearly turbulent wakes and exhibited similar behavior. The blowoff velocities increase regularly as the holder scale increases.

The three remaining curves are for holders with laminar wakes. The peak blowoff velocity for these examples is progressively shifted toward fuel-air ratios greater than stoichiometric as the flameholder scale is reduced. This behavior is a result of a molecular diffusion process that is usually of no importance in flameholder systems operating in gas turbine engines. However, note that even for the largest flameholder, the lean stability limit is well above 0.5 of stoichiometric and that values above 0.8 are required to stabilize flames in high-speed flows.

It is clear from these data that the velocity at which stabilization can be maintained is a strong function of the flameholder scale, fuel-air ratio, and wake condition (laminar or turbulent). Other experiments show that it also depends upon the fuel characteristics; the temperature, pressure, and oxygen content of the unburnt gas stream; and a number of geometric factors describing the flameholder-duct system. The nature of these dependencies and their origin will be discussed in later paragraphs of this section. The stability limits obtained with the other systems shown in Fig. 2.6 follow a pattern similar to that described above when mixtures of fuel vapor and air are used. However, when a large fraction of the fuel is not vaporized, the picture is quite different.

The bluff-body flame stabilization process that occurs in fuel-vapor-air mixtures will be treated in detail in the following subsection. Briefer descriptions will be given of the scaling parameters for the stabilization by secondary injection (illustrated in Fig. 2.6d) and of bluff-body stabilizers operating in unvaporized fuel-air mixtures.

Flame Stabilization by Bluff Bodies in Premixed Flow

The mechanism of flame stabilization by bluff-body flameholders is still a subject of some contention and the mechanism described here is that developed at the California Institute of Technology by F. E. Marble, E. E. Zukoski, and a number of co-workers at the Jet Propulsion Laboratory (e.g., see Refs. 5–8).

This approach is based on a picture of the wake of a two-dimensional flameholder shown in Fig. 2.8. Figure 2.8a was taken in the flame stabilization region of a circular cylinder. The line of sight is along the cylinder. Figure 2.8b is a highly simplified drawing of the velocity field for the same region. The mixing zones are illuminated in Fig. 2.8a by light from the combustion process occurring there. Chemiluminescence and hence chemical reaction is almost absent in the recirculation zone.

The wake of the flameholder can be divided into two regions: the recirculation zone that is characterized by strongly recirculating flow and the mixing zone that separates the recirculation zone from the unburned mixture. The temperature of the recirculation zone gas is typically within 5–10% of the adiabatic flame temperature corresponding to the mixture ratio of the approach stream and is independent of gas speed and the flameholder-duct geometry. Chemical reaction appears to be almost complete in this volume and the residence of gas is 5–10 times the time required by the flow at the cold side of the mixing region to pass over the recirculation zone. This long residence time insures that, if any chemical

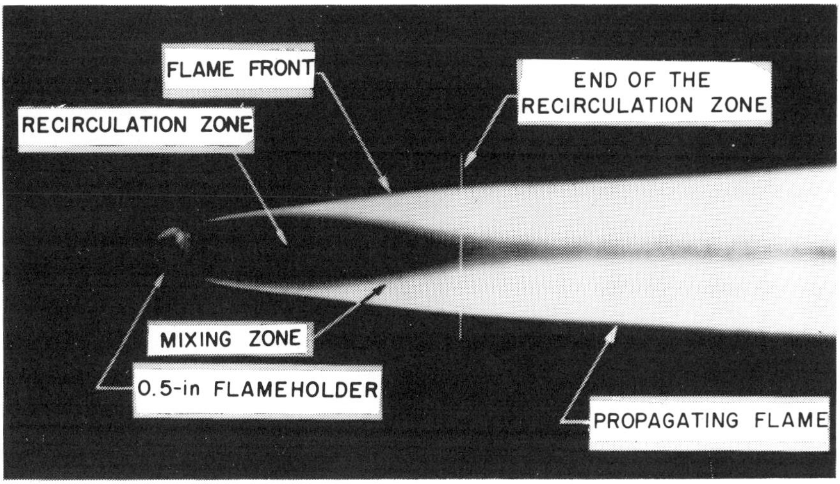

a)

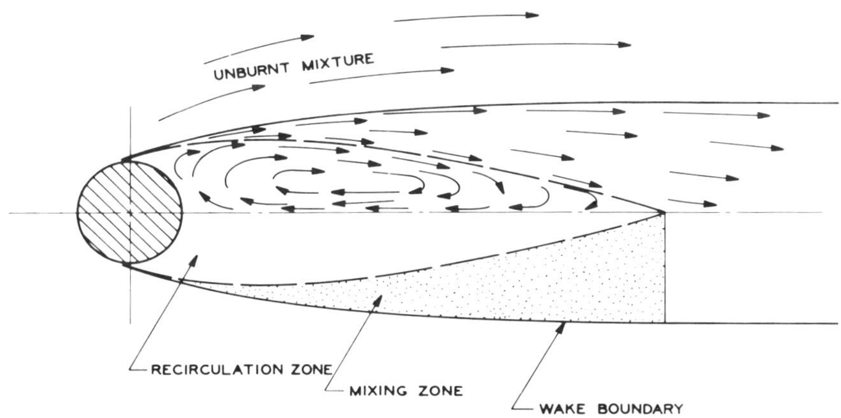

b)

Fig. 2.8 Flame stabilization region of a circular cylinder used as a bluff-body flameholder.

reaction takes place in the wake at all, it will be most nearly complete in the recirculation zone.

The mixing zones on either side of the recirculation zone are turbulent regions of very strong shear, steep temperature gradients, and vigorous chemical reaction. These regions thicken almost linearly and their junction forms the downstream end of the recirculation zone.

The process leading to the formation of a self-propagating chemical reaction takes place in the mixing zone. This region is fed by turbulent mixing processes with cool combustible gas from the approach stream and with very hot burned material from the recirculation zone. Thus, if no chemical reaction were present, the mixing zone gas would still be at a high temperature at the inner edge (or recirculation zone side) of the region. Downstream of the end of the recirculation zone, the cool, unburned gas continues to be entrained into the wake, but no more hot gas is added; hence, if no chemical reaction is present, the temperature will fall.

Chemical reaction is initiated in the mixing zone by heat and species transfer from the hot burned gases to the unburned material. The total heat and species transferred to any particle of unburned gas and its reaction rate will depend largely on the time this particle spends in contact with hot burned gas. Thus, the temperature, species concentration, and consequently, the local chemical reaction rate of the unreacted material will depend strongly on the residence time of this unburned material in the mixing zone.

For a sufficiently long residence time, a chemical reaction will be started in the mixing zone flow and will continue as this gas moves on downstream to form the wake. If this reaction is vigorous enough in the wake region, then the quenching action of the continuing entrainment of cool unburned gas will be overcome and a propagating flame will be produced.

It is reasonable to suppose that the residence time of the unburned gas in the mixing zone will be proportional to $L_e/\bar{v}$, where $\bar{v}$ is a suitably chosen average speed in the mixing zone and L_e the scale of the recirculation zone. Hence, when the velocity of the flow increases, $\bar{v}$ will increase proportionately and the residence time will decrease. When the residence time is too short, the chemical reaction in the wake will be quenched and no propagating flame will be stabilized. Thus, flame extinction or blowoff occurs due to failure of the ignition process and not as a direct result of cooling of the recirculation zone gas.

The important parameters of the system as indicated by this discussion are: $\bar{v}$, an average gas velocity in the mixing zone; τ, a time characterizing the period required to achieve ignition in the mixing zone; and L_e, the length of the recirculation zone that is effective in igniting the mixing zone. The dimensionless parameter of interest is then $\bar{v}\tau/L_e$. There is good evidence that the mixing layer flowfields are roughly similar, i.e., have similar velocity, composition, and temperature fields. Thus, if similar systems are compared with the same chemical parameters but under different fluid dynamic conditions, the stability parameter must take on a particular value at blowoff condition, i.e.,

$$(\bar{v}\tau/L_e)_{\text{blowoff}} = \beta_c$$

Because of the similarity of the turbulent mixing zones of bluff-body flameholders, the parameters L_e and $\bar{v}$ used in this expression can be replaced without loss of generality by the maximum length of the recirculating flow L and the unburned gas speed at the edge of the mixing zone V_2.

Then the blowoff condition given above can be rewritten as

$$\frac{\bar{v}\tau}{L_e} = \left(\frac{\bar{v}}{V_2}\frac{L}{L_e}\right)\left(\frac{V_2\tau}{L}\right) = \beta_c$$

When $\bar{v}/V_2$ and L/L_e can be assumed to be constants (for similar turbulent mixing regions), the stability criterion becomes

$$V_2\tau/L = \text{const}$$

at the flame extinction or blowoff condition.

A number of experiments have verified the usefulness of this relationship. However, an independent calculation of the characteristic chemical time τ has not been developed; consequently, it is necessary to experimentally determine τ from a set of flame blowoff experiments. In these experiments, V_{2c} and L are determined at the blowoff state and τ is obtained from $\tau_c \equiv L/V_{2c}$. Given this definition, the flame blowoff criteria is

$$(V_{2c}\tau_c/L)_{\text{blowoff}} = 1 \tag{2.7}$$

where the unknown constant β_c has been absorbed in the measured quantity τ_c. One of the great advantages of this approach is that all of the dependence on the chemical parameters is naturally lumped in τ_c and all of the dependence on the fluid dynamic parameters in L/V_{2c}.

When τ_c has been determined experimentally for a single duct-flameholder configuration and for the desired ranges of the chemical parameters, it can be used to predict stability limits for any other duct-flameholder configuration and the *same* range of chemical parameters. Thus, τ_c is a scaling parameter.

A number of experiments with a wide range of flameholder-duct configurations have shown that the values of τ_c do depend on a number of chemical parameters such as the fuel type, fuel-air ratio, gas temperatures, and degree of vitiation, but are substantially independent of the geometry and scale of the flameholder-duct configuration and of the gas speed so long as the flow in the mixing region is turbulent. A few typical results are summarized in Table 2.1. The top two flameholders are circular cylinders with their axis held perpendicular to the flow and the latter three are cylindric bodies with their axis parallel to the flow and several types of axisymmetric noses. The third body is the two-dimensional analog of the fourth. The large differences in the flowfield produced by these bodies is illustrated in Fig. 2.9. Note that values of the critical times evaluated at the stoichiometric ratios are about the same regardless of geometry.

In addition, for flames in which mixing length similarity does not hold, values of τ_c determined by the integration of local velocity measurements along a path in the mixing zones again confirmed the validity of this approach. For example, see the work of Broman and Zukoski.[8]

The fluid dynamic and chemical aspects of the stability criterion will be discussed separately in the following subsections.

Table 2.1 Dependence of Ignition Time τ_c Obtained at the Stoichiometric Fuel-Air Ratio on Flameholder Geometry and Gas Speed

Flameholder Geometry	D or d, in.	τ_c, 10^{-4} s
	1/8	3.09
	3/16	2.85
	1/4	2.80
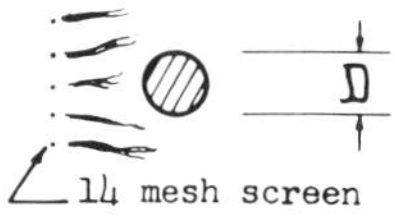	1/4	3.00
	1/4	2.38
	3/8	2.70
	1/2	2.65
	3/4	2.58
	1/4	3.46
	3/8	3.12
	1/2	3.05
	3/4	3.03
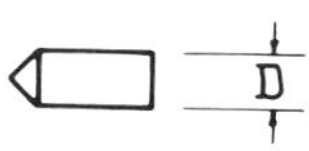	3/4	3.05
	3/4	2.70

Fluid dynamic parameters. The flame stabilization criterion given in Eq. (2.7) is not immediately useful, even when the characteristic τ_c is known as a function of the chemical parameters, because the recirculation zone length L and the gas speed in the flow over the wake V_2 are not simply related to the scale of the flameholder and the velocity far upstream. To illustrate the dependence of these parameters on other fluid dynamic variables, consider the case of a two-dimensional flameholder of height d located on the centerline of a rectangular duct of height H and subject to a flow with an upstream velocity V_1. The fluid dynamic parameters used in this model need to be connected with these parameters, which are the ones usually specified. These two sets of parameters are illustrated in Fig. 2.10. In a formal way, the stability criterion can be written as

$$\frac{V_{2c}\tau_c}{L} = 1 \quad \text{or} \quad \frac{V_{1c}\tau_c}{H} = \left(\frac{V_1}{V_2}\right)\left(\frac{L}{W}\right)\left(\frac{W}{H}\right) \tag{2.8}$$

where W is the width of the wake near the downstream end of the

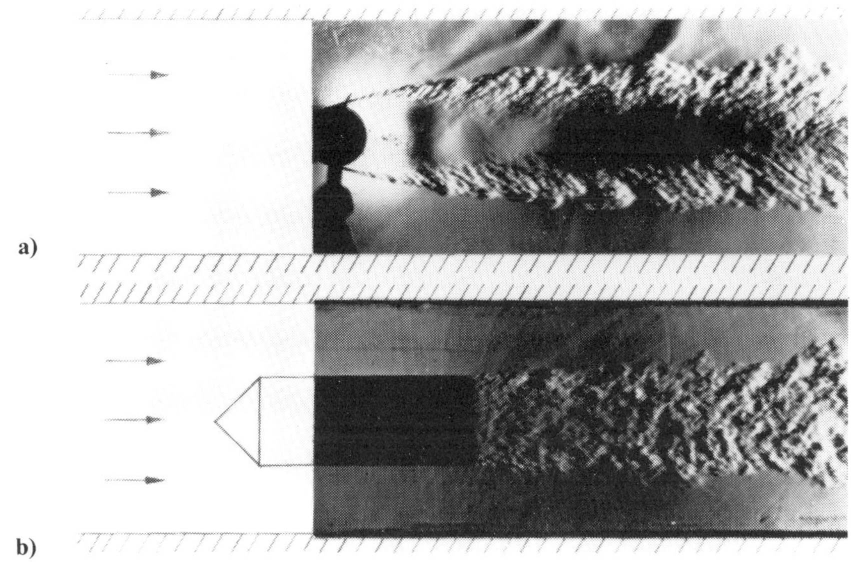

Fig. 2.9 Spark schlieren photographs of flames stabilized on: a) circular cylinder viewed from the side and b) on axisymmetric body with axis parallel to flow. [The recirculation zone, clearly outlined in a) is not visible in b).]

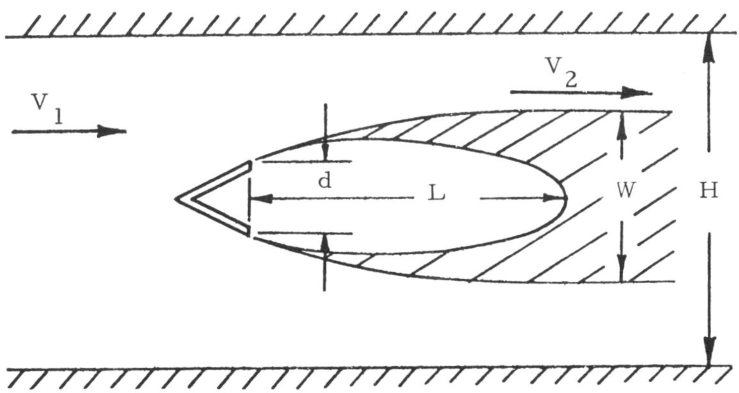

Fig. 2.10 Notation used in analysis of stabilization.

recirculation zone. Its introduction in Eq. (2.8) will lead to a useful simplification described below. Also, V_{1c} and V_{2c} are the values of velocity of the approach stream and the flow past the wake evaluated at blowoff condition.

In order to apply the criterion, V_1/V_2, L/W, and W/H as functions of the blockage ratio d/H, the flameholder geometry, and the usual parameters of compressible and viscous flows and Reynolds and Mach numbers must be known. In addition, the flameholder temperature must be considered as it affects the boundary layer on the holder and the turbulence level in the approach stream. In the following paragraphs, the effects of changing d/H (called the blockage ratio) and the flameholder geometry will be considered first, and later the effects of Reynolds number, Mach number, flameholder temperature, and turbulence will be described. (Note that in some figures D replaces d.)

(1) Blockage and flameholder geometry. Consider again the flow described schematically in Fig. 2.10. Note that the flow separates from either side of the flameholder and that the wake continues to spread downstream of the holder and asymptotically approaches a width W. (Later it will be shown that W rather than d is the most useful characteristic scale for the holder in this problem.) The flameholder and its burning wake block an appreciable part of the cross-sectional area of the duct and causes the unburned flow to accelerate to a higher velocity, V_2.

The wake width is a critical parameter since it not only affects the ratio V_2/V_1 but also fixes the recirculation zone length L. For example, the width of the mixing zone increases almost linearly with distance along its length. This spreading rate is similar to that of many other two-dimensional turbulent shear layers and has a constant width-to-length ratio of roughly $\frac{1}{8}$. At the downstream end of the zone, the two mixing layers occupy the entire width of the wake and consequently the ratio of recirculation length to wake width L/W should be close to 4. For a wide range of flameholder shapes, experimental data lie in the range

$$3.6 \leq (L/W) \leq 4$$

Data illustrating this result are shown in Fig. 2.11 for a number of circular cylinders. Despite the strong dependence of L/d on the blockage ratio, the values of L/W are almost constant for the whole range. The dependence of L/d on blockage is shown here for circular cylinders; in general, the dependence of L/d on blockage is a strong function of flameholder geometry. However, the spreading of the mixing zones and presumably, the entrainment in the mixing zones is remarkably independent of the shape of the wake region and flameholder. This simple dependence of the recirculation zone length on the wake width indicates that in this problem (as in many fluid dynamic problems involving flows over bluff bodies), the wake width rather than flameholder scale is the most useful measure of bluff-body scale.

The ratios V_2/V_1 and W/H appearing in Eq. (2.8) depend on the flameholder geometry and the blockage ratio. A simple continuity argument for incompressible flow can be used to estimate the velocity change if entrainment in the mixing layer is neglected. Conservation of mass gives

$$\rho V_1 H = \rho V_2 (H - W)$$

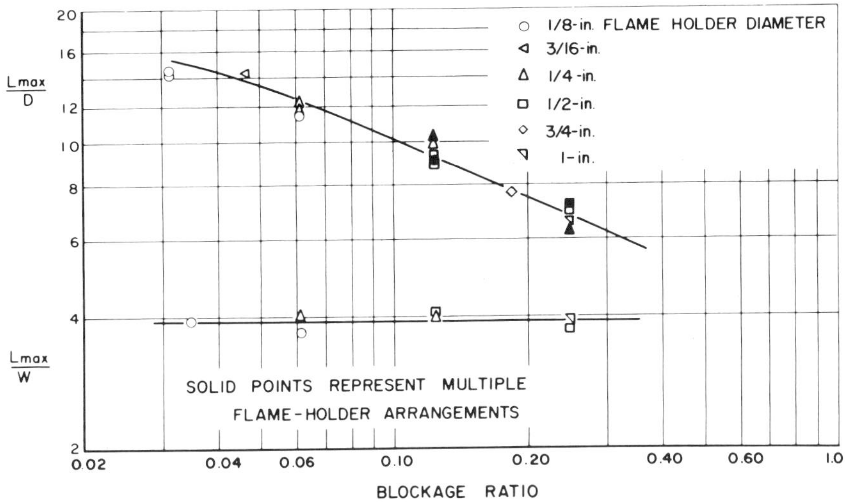

Fig. 2.11 Dependence of recirculation zone length and wake width on flameholder blockage ratio d/H.

or

$$(V_2/V_1) = 1/(1 - W/H) \tag{2.9}$$

An approximate calculation is also available that allows W/H to be calculated as a function of geometry and blockage ratio for a V-gutter or wedge flameholder geometry. Again, considering the flow shown in Fig. 2.10, examine an incompressible and inviscid flow over a wedge-shaped body of half-angle α and treat the outer boundary of the wake as a streamline that separates the unburned flow from a stagnant, constant-pressure wake. The wake spreads asymptotically to reach a width W. In the region where W is constant, the unburned flow speed reaches a value V_2 given by Eq. (2.9) and the pressure will be uniform across the wake and the unburned flow. Application of the Bernoulli equation shows that the velocity along the dividing streamline must be constant and be equal to V_2. Values of W/H can be computed from this model by a simple hodograph transformation. Calculations of this type are given by Cornell[9] for the V-gutter flameholder geometry with $0 \leq \alpha \leq 90$ deg. Values of W/H, V_2/V_1, and the parameter WV_1/HV_2 [see Eq. (2.8)] are given in Table 2.2 for wedge half-angles of 15 and 90 deg (a flat plate) and a range of blockage ratios. Note that values of W/d decrease and values of V_2/V_1 increase as the flameholder blockage is increased. The net result is that the parameter $(W/H)(V_1/V_2)$ has a rather broad maximum around blockage ratios of 0.5 for the 30 deg wedge (half-angle) and 0.3 for the 90 deg wedge or flat plate.

The wake widths and velocity ratios determined from this simple model are in reasonable agreement with values obtained experimentally for stabi-

Table 2.2 Dependence of Wake Width W, Edge Velocity V_2, and a Stability Parameter on Blockage Ratio d/H and Wedge Half-Angle α

	$\alpha = 15$ deg			$\alpha = 90$ deg		
$\frac{d}{H}$	$\frac{W}{d}$	$\frac{V_2}{V_1}$	$\left(\frac{W}{H}\right)\left(\frac{V_1}{V_2}\right)$	$\frac{W}{d}$	$\frac{V_2}{V_1}$	$\left(\frac{W}{H}\right)\left(\frac{V_1}{V_2}\right)$
0.05	2.6	1.15	0.11	4.0	1.25	0.16
0.10	1.9	1.23	0.15	3.0	1.43	0.21
0.20	1.5	1.42	0.20	2.2	1.75	0.248
0.30	1.3	1.62	0.23	1.7	2.09	0.250
0.40	1.2	1.90	0.25	1.6	2.50	0.248
0.50	1.2	2.3	0.25	1.4	3.16	0.22

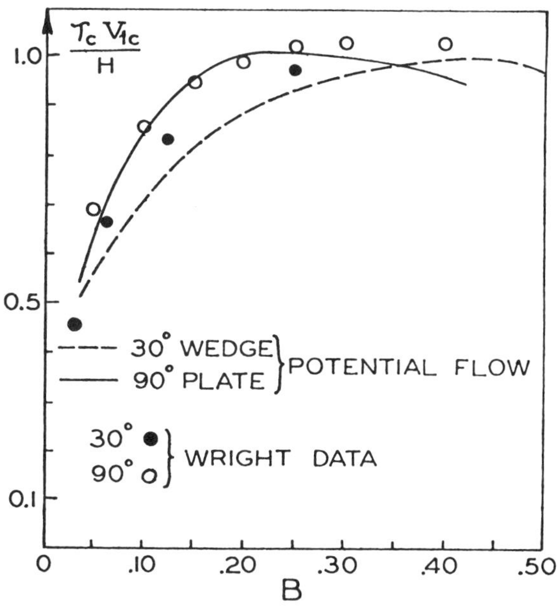

Fig. 2.12 Dependence of stability parameter on blockage ratio B and flameholder geometry for wedges with half-angles of 30 and 90 deg.

lized flames. The excellent measurements obtained by Wright[7] for flat-plate holders with $\alpha = 90$ and 30 deg are available for comparison. The differences in the values of W/H are no more than 20% and those for the ratio V_2/V_1 are less than 6%.

Given data or calculations of this type and assuming that $L/W = 4$, all of the parameters given in Eq. (2.8) can be evaluated. The experimental results of Wright and values calculated by the method described above are presented in Fig. 2.12 as a plot of $\tau_c V_{1c}/H$ vs the blockage ratio B. The agreement between the data and calculated values is good and suggests that the model discussed here is correct.

The material presented in Table 2.2 and Fig. 2.12 shows that the maximum values of the stability parameter $(V_{1c}\tau_c/H)$ are close to 1 for the 15, 30, and 90 deg half-angle wedges. This value is a reasonable estimate for all two-dimensional flameholders. For example, using the continuity argument presented above for V_1/V_2 and the rule of thumb that $L/W = 4$, Eq. (2.8) can be rewritten in terms of W/H to find the relationship for the blowoff criteria,

$$(V_{1c}\tau_c/H) = 4(W/H)(1 - W/H) \tag{2.10}$$

Note that the flameholder shape does not appear here explicitly. However, both the flameholder geometry and the blockage ratio enter in the determination of W/H. Equation (2.10) is a particularly useful and interesting result because it shows the importance of the wake width as a critical parameter (independent of flameholder-duct geometry) and because it allows a very simple starting point for the determination of the holder scale maximizing the stabilization velocity V_{1c}. The function $(W/H)(1 - W/H)$ clearly has a maximum at $W/H = \frac{1}{2}$. For small W/H, V_{1c} increases almost linearly with W. However, as W/H approaches $\frac{1}{2}$, the effect of increasing W is offset by the increase in V_2 produced by the wake-blockage effect. For $W/H > \frac{1}{2}$, the blockage effect dominates and the blowoff speed decreases.

The optimum wake width in this simple model is exactly one-half the duct height. Consequently, the maximum value of blowoff velocity that can be achieved for any flameholder geometry V_{1m} is given by

$$V_{1m}\tau_c/H = 1 \tag{2.11}$$

Note that this result is independent of the flameholder geometry and is corroborated by the experiments described above.

Flows over axisymmetric bodies can also be examined, such as the flameholder shown in the lower half of Fig. 2.8. For this geometry, the wake width W and body diameter d are equal; consequently, the recirculation zone length is $4W$ or $4d$. If this holder is placed on the axis of a circular duct of diameter D and the blockage is defined as $B = (d/D)^2$, the stabilization criterion given in Eq. (2.8) becomes

$$V_{1c}\tau_c/D = 4\sqrt{B}\,(1 - B)$$

For this example, the value of B maximizing V_1 is $\frac{1}{3}$ and

$$V_{1m}\tau_c/D = 1.54 \tag{2.12}$$

The above result holds for *any* axisymmetric flameholder-duct geometry. In the general axisymmetric case, V_{1m} is attained when the ratio of the wake width to the duct diameter W/D is $1/\sqrt{3} = 0.57$. It is interesting to note that the analysis indicates that the axisymmetric holders will allow about a

50% greater value of approach stream velocity than the two-dimensional holders.

In summary, for the two-dimensional case, the effects of blockage and flameholder geometry can be predicted from a simple model. Results obtained from the model and experimental work show that blockage effects are very important, that a maximum in blowoff velocity exists as flameholder scale is increased in a duct of fixed size, and that this maximum occurs when the flame wake width is about 50% of the duct height for all two-dimensional flameholder shapes. Similar conclusions hold for axisymmetric shapes except that for this case the optimum wake width is one-third of the duct diameter. The flameholder scale and blockage ratio required to produce the optimum wake width depend strongly on flameholder geometry.

(2) Temperature, fuel-air ratio, and vitiation. Temperature changes have little observable effects on the wake width or velocity. However, a large increase in temperature does result in a reduction in the recirculation length. For example, doubling the unburned mixture temperature can decrease recirculation zone length by 15–20%. Similar changes are observed when fuel-air mixture ratios are changed to values far from stoichiometric.

In both cases, the length of the zone decreases when the ratio of burned to unburned gas density λ increases. This trend is in agreement with trends recently observed in spreading rates of turbulent mixing regions formed between parallel flows with different velocities and densities. Experimental investigation of this simpler problem shows that if the high-speed stream is also the high-density stream, an increase in the density of the low-speed and low-density stream will increase the spreading rate. In this case, increasing spreading rates of the mixing layers will decrease the length of the recirculation zone.

For a given temperature, vitiation of the approach stream produces a reduction in the oxygen content, which results in a reduction in heat release and consequently an increase in λ. Again, the effect is qualitatively similar to an increase in the approach stream temperature.

(3) Multiple flameholder arrays. The previous discussion was restricted to a single flameholder placed on the centerline of a duct of constant height. In most practical situations, a number of flameholders must be used. When these are arranged in a single plane perpendicular to the flow direction and when they are spaced so that each holder lies on the centerline of equivalent ducts of equal height, the above analysis can be used directly to estimate the stability limits and spreading characteristics. For this case, each holder is treated as if it were in an isolated duct. For example, see Fig. 2.11 and the blowoff data of Ref. 10.

This approach gives reasonable results for laboratory-scale experiments; however, interactions do occur that may change stability limits by 5–10%. One effect results when flame stabilization fails on one holder in a multi-flameholder array slightly before the others. Failure of one stream to ignite has the effect of reducing the acceleration of the flow over the wake and hence of reducing the velocity of the flow past the wakes of the other holders. Thus, a partial blowoff can occur, leaving the remaining holders in a more stable configuration.

When the holders are spaced irregularly in either lateral dimension, or fore and aft, prediction of the wake geometry from the results of tests on isolated flameholders is no longer possible. However, these modeling ideas are still useful in a qualitative way. For example, it is expected that, if flameholders are not arranged in a plane, the disturbance produced by the flow over the downstream flameholders will pinch off the circulation zone of the upstream holder unless the spacing along the duct axis is greater than the recirculation zone length. Similarly, the burning wake of the upstream holder will increase the effective blockage of the downstream holders and hence reduce their stability limits.

(4) Reynolds number and Mach number effects. When the Reynolds number is so low that the mixing layers in the wake of the flameholder become laminar or transitional, the transport processes in the wake change drastically. Molecular diffusion becomes important and the rule of thumb for L/W is no longer applicable.

Measurements made in systems with low approach stream turbulence levels[6] have shown that the transition Reynolds number for circular cylinder flameholders is in the range $1\text{–}4 \times 10^4$. This result is for flameholders cooled to approach stream temperature and the Reynolds number is based on upstream flow properties, $R_e = \rho_1 V_1 d/\mu_1$. When the flameholder is allowed to reach temperatures hotter than the approach stream, those gas properties based on the holder temperatures should be used. Some effects of the geometrical shape of the holder on transition is expected.

Transition occurs when the separated flameholder boundary layers become turbulent very close to the separation point and upstream of the location in the mixing zone where the mixing effects or combustion can heat the gas in the separated layers. Any heating will increase the kinematic viscosity (which is roughly proportional to the temperatures to the 1.75 power) and reduce the effective Reynolds number. When the flow in the separated boundary layers remains laminar up to the point of appreciable heat addition, it remains laminar throughout the whole recirculation zone and region of flame spread. The development of a turbulent boundary layer on the flameholder upstream of separation and high levels of the approach stream turbulence will also insure that the mixing zone will be turbulent.

Transition to turbulence in the wake of a circular cylinder used as a flameholder is shown in Fig. 2.13. The schlieren photographs are taken along a line of sight looking down on the plane containing the undisturbed velocity vector and the flameholder axis. At the lowest Reynolds number (Fig. 2.13a), large-scale vortices are present, which appear as vertical lines in the left picture. However, at the highest Reynolds number, these regular features are hidden by small-scale disturbances assumed to be evidence of turbulent flow. The Reynolds numbers examined here differ by a factor of less than two and the change in the appearance is quite striking.

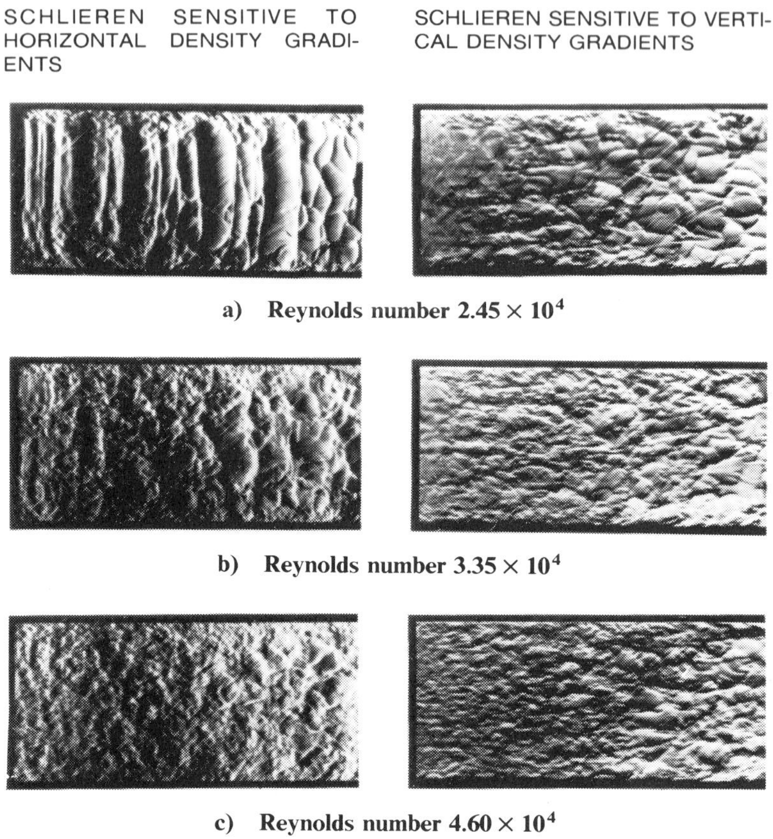

Fig. 2.13 Transition to turbulence in recirculation zone of cooled circular cylinder.

Because the Reynolds number is directly proportional to the pressure level in the engine (through the density dependence), the Reynolds number will *decrease* as the altitude increases. Hence, the Reynolds number should be evaluated at high altitudes to insure that the transition to laminar flow described here does not occur.

When appreciable heat is to be added in a burner with a constant cross-sectional area, inlet Mach numbers of about 0.15–0.25 must be used to prevent choking due to heat addition. Thus, compressibility effects usually are not important near the flameholder and recirculation zone.

However, note that the near-optimum one-half wake width (for the two-dimensional case), the flow area is reduced by a factor of two. This results in sonic speed past the recirculation zone for approach stream Mach numbers as low as $M_1 = 0.3$. Measurements made by Wright[7] bear out this prediction and further show that the W/H correlation discussed above fails when the Mach number past the recirculation zone M_2 is greater than 0.8.

(5) Freestream turbulence. The effects of freestream turbulence can be described only in a qualitative manner. As the intensity of turbulence increases, the recirculation zone shortens. However, the stabilization criterion in its most direct form, $V_{2c}\tau_c/L = 1$, remains valid even when the recirculation zone length is reduced by factors of two. Thus, turbulence appears to effect the rate of the spread of the mixing layers without changing the mechanism of stabilization.

Chemical parameters. The dependence of τ_c on chemical parameters is very strong and unfortunately much less well understood than the influence of the fluid dynamic parameters. Hence, the chief use of the scaling scheme discussed here is to predict the effect of changes in the fluid dynamic parameters for fixed chemical parameters. However, it is still interesting to list the important chemical parameters and to indicate the nature of their effects. The principal parameters are: fuel properties, fuel-air mixture ratio,

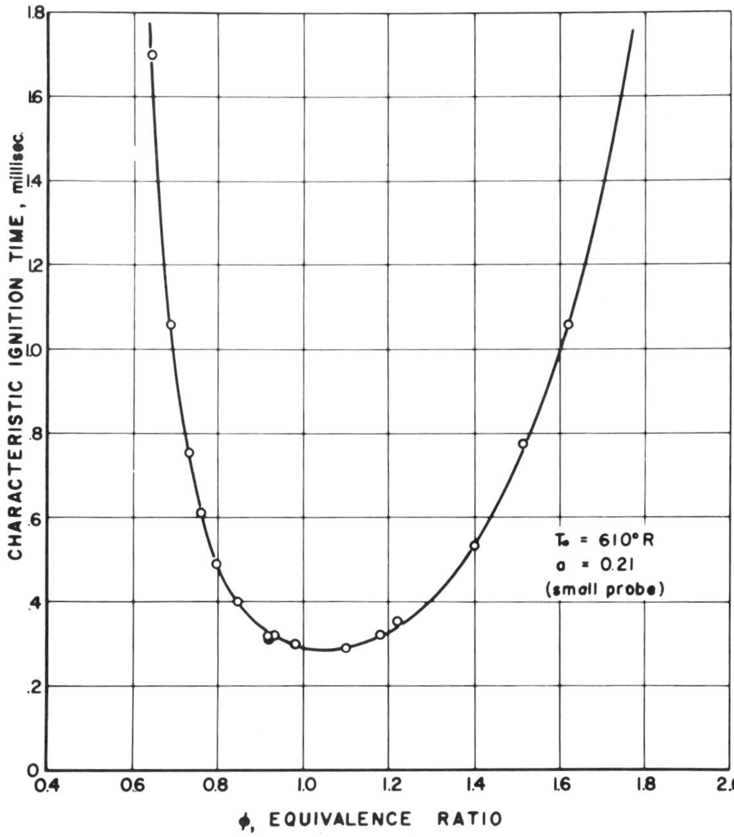

Fig. 2.14 Variation of characteristic ignition time τ_c with equivalence ratio.

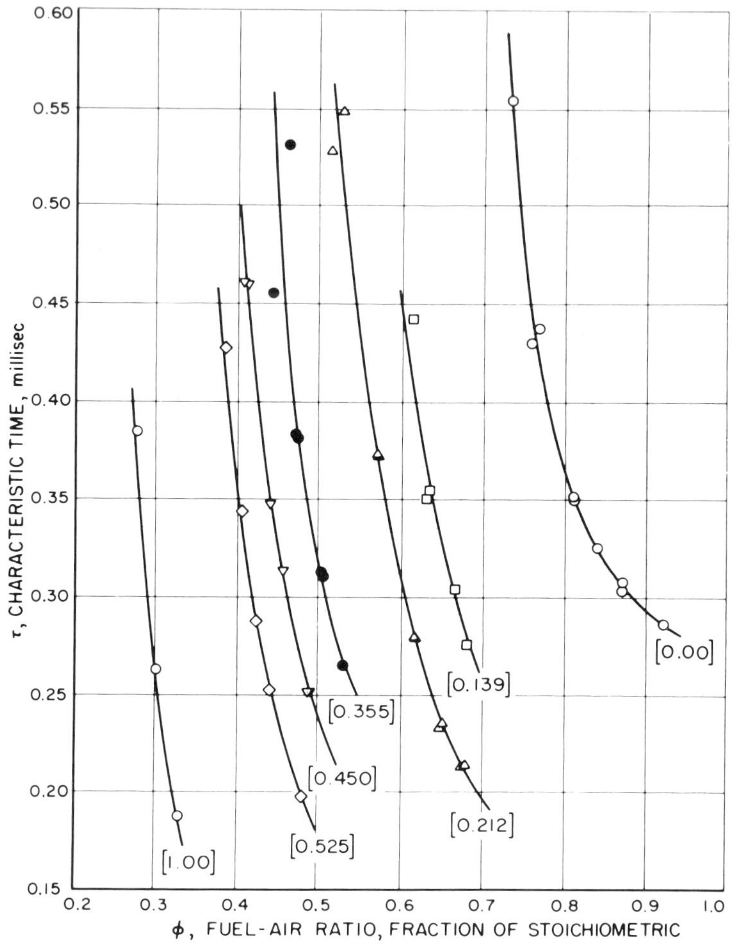

Fig. 2.15 Effect of fuel properties on characteristic time (parameter is mass fraction of hydrogen in the fuel).

approach stream temperature and oxygen concentration, and approach stream pressure.

The dependence of the critical time on fuel-air ratio and fuel type is illustrated in Figs. 2.14 and 2.15. In both, τ_c is plotted as a function of the equivalence ratio (the fuel-air ratio divided by the stoichiometric fuel-air ratio). In Fig. 2.14, values are presented for a hydrocarbon fuel vapor with a molecular weight of about 100. A number of flameholder geometries were used to obtain these data. In Fig. 2.15, a number of fuels were made up of this hydrocarbon plus various mass fractions of hydrogen. The values of τ_c decrease dramatically as the fraction of hydrogen increases. Also note that

τ_c increases very rapidly for both high and low values of the equivalence ratio.

The characteristic time is also a sensitive function of approach stream temperature and oxygen concentration. In general, τ_c decreases rapidly as the temperatures increase, even when this increase is due to vitiation (i.e., preburning at a fuel-air ratio below stoichiometric). However, for a given temperature, τ_c increases as the oxygen mass fraction decreases due to increasing vitiation.

The critical time increases as pressure decreases roughly as $\tau_c \propto 1/P$ for hydrocarbon fuels of high molecular weight (e.g., see Ref. 11). A similar result was obtained in small-scale experiments carried out with hydrogen.

In attempting to obtain an independent estimate of values of τ_c or some other experimentally determined quantity proportional to τ_c, the dependence of a time based on the ratio of the laminar flame thickness δ to laminar flame speed S has been examined. In the single case for which comparable data were available, values of $\delta/S\tau_c$ for methane were computed for stoichiometric fuel-air ratio of one and approach stream temperature between 300 and 400 K. The ratio had a value close to one for the whole range of temperatures examined. Similarly, values of the ratios of δ/S and τ_c for hydrocarbon and hydrogen fuels are about 10. Thus, δ/S may be a useful predictor for the dependence of τ_c on various chemical parameters.

Finally, there is a persistent attempt to relate the characteristic stabilization time to a thermal ignition time (e.g., Ref. 12) or a global reaction rate (e.g., Refs. 13 and 14). These efforts often lead to a representation for τ similar to the reciprocal of a reaction rate

$$\tau_c \propto T^m P^{-n} e^{+(A/RT_f)}/\phi$$

where m and n are numbers of the order of one or two, A an activation energy determined empirically, R the universal gas constant, T_f an ignition temperature or the gas temperature in the recirculation zone, and ϕ an equivalence ratio of less than 1, which is the stoichiometric value. In the works of Solokhin and Mironenko[13,14] where a more complex expression is used, the effective values of the parameters are

$$n = 1, \; m = 2.5, \; A/R = 2 \times 10^4 \text{ K}$$

Although the dependence on temperature and pressure by this approach is plausible, the use of global reaction rates and the application of this approach to processes involving chemical reactions in turbulent mixing regions does not have a sound physical basis and should be viewed as a sophisticated form of curve fitting.

Because of the lack of understanding of the chemical parameters, the stabilization criterion is useful only when τ_c values have been determined for the range of chemical parameters expected in practice. However, small-scale experiments can be used to make the required determinations and some physical feel is given by the flame speed correlation suggested above.

Alternate schemes. One popular alternate scheme for scaling stabilization phenomena is based on the arbitrary use of a dimensional parameter group of the form

$$V_{1c}/P^a d^b T^c = F\{f\}$$

to correlate a body of experimental results. Here f is the fuel-air ratio, d the flameholder scale, P and T the pressure and temperature of the approach flow, and F a function of the fuel-air ratio. (For example, see Ref. 15.) The values for the exponents selected by various authors to correlate their data have ranges of $0.8 \le a \le 2$, $\frac{1}{2} \le b \le 1$, and $\frac{1}{2} \le c \le 2.5$ and F must be determined experimentally.

At the beginning of the study of bluff-body flame stabilizers, there was great confusion concerning the exponent for d that arose because the influence on the recirculation zone region of flameholder and duct geometry, described above, was not fully appreciated. For example, when a circular cylinder held with its axis perpendicular to the gas stream is used as a flameholder, the duct walls have a large effect on the flameholder wake width, even when the ratio of holder diameter to duct height is as small as $\frac{1}{50}$. In the range $\frac{1}{20} \le d/H \le \frac{1}{4}$ and in a duct of fixed size, the wake width and hence the recirculation zone length scale approximately as $L \propto W \propto \sqrt{d}$ rather than as $L \propto d$. This square root dependence leads to a value of the exponent b of $\frac{1}{2}$. Similarly, if cylinders with their axis parallel to the flow are used as flameholders, the wake width grows linearly when d is increased in a duct of fixed size. There is an effect of blockage in this example as well, since the ratio V_2/V_1 will increase with d due to blockage changes but this increase depends on $[1-(d/D)^2]^{-1}$ and hence is hard to detect when $d/D < \frac{1}{4}$. Hence, in this example the exponent b would be close to 1. Further confusion arose because data in the laminar and turbulent regimes were used together to determine these exponents.

A value of the pressure exponent near one is a typical choice and, for the temperature, values still range between 0.5 and 2.5. Thus, a scaling parameter of the form V_{1c}/Pd is often used for fixed inlet temperature. When d is changed by scaling the entire flameholder-duct system, this parameter gives an excellent correlation—as would be expected from the previous analysis. A correlation of this type is given by Hottel et al.[15] However, note that this correlation is useful only in general if d characterizes the scale of the system; if d is changed and the duct height is held fixed, the correlation will fail because of the effect of changes in blockage on recirculation zone length and the velocity ratio V_2/V_1.

Flame Stabilization by Jets in a Homogeneous Stream

The process of flame stabilization in the wake of a gas jet is similar to the processes, described above, occurring in the wake of a bluff body. In either example, a region of strong recirculating flow is created by a physical obstacle or in the wake of interacting jets in which the hot products of combustion can be trapped. This hot gas then acts as a steady source of

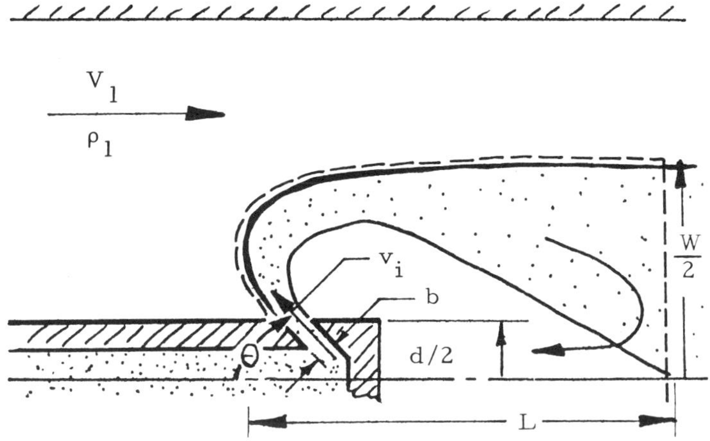

Fig. 2.16 Flame stabilization by a jet.

ignition for the oncoming stream. In the case of flame stabilization by gas jets, two new features are added: (1) the size of the recirculation zone can be changed by changing the rate of gas injection, and (2) the fuel-air ratio and hence the temperature of the gas in the recirculation zone can be changed by changing the fuel-air ratio of the injectant. Because the stabilization process is very strongly affected by changes in the temperature of the recirculation zone gas, having an independent control on this parameter will allow stabilization at fuel-air ratios far below those that could be achieved with bluff-body holders producing the same sized recirculation zone. The advantages of these two features are offset in part by the performance losses and the mechanical problems associated with supplying the gas flow required for flame stabilization and, if a fuel-air mixture is to be used, the problems involved with the production of the vaporized fuel and the preparation of a homogeneous mixture prior to injection.

A crude scaling law for one example of this type of system will be developed here, along with a few experimental results to illustrate the general features of the stabilization process.

The flowfield produced by axisymmetric injection of a jet into a cross flow is shown in Fig. 2.16. The jet is injected through an annular slot of width b in the wall of a center body of diameter d and at an angle $(\pi - \theta)$ with respect to the oncoming flow. In this model, a momentum balance is made on the injectant and the approximation that the drag of the effective body produced by injection (see dotted contour in Fig. 2.16) is balanced by the momentum change of the injectant. Thus,

$$\frac{1}{2}\rho_1 V_1^2 C_d\left(\frac{\pi W^2}{4}\right) = (\rho_i v_i \pi \mathrm{d} b) v_i (1 + \cos\theta)$$

where the drag is characterized by a drag coefficient C_d having a value around one, and it is assumed that the injected mass flow $(\rho_i v_i \pi \mathrm{d} b)$ enters and leaves the control volume with its velocity of injection v_i. (The assumption that the exit velocity is v_i and not a value closer to V_1 is certainly questionable.) Given this crude balance,

$$\frac{W}{b} = \left[\left(\frac{d}{b}\right)\left(\frac{8(1+\cos\theta)}{C_d}\right)q\right]^{\frac{1}{2}}$$

where $q = (\frac{1}{2}\rho_i v_i^2)/(\frac{1}{2}\rho_1 V_1^2)$.

When W is much larger than the thickness of the boundary layer on the center body and when W is much smaller than the duct diameter (so that blockage effects can be ignored), C_d should be constant. Consequently, the above equation indicates that W will scale as $q^{\frac{1}{2}}$ and when ρ_i/ρ_1 and the geometric parameters are constant, this equation reduces to

$$W \propto \rho_i v_i \propto \dot{m}_i$$

where $\dot{m}_i$ is the mass flow of the injectant.

Experiments (e.g., Ref. 16) have shown that the length of the recirculation zone formed by this injection process is between one and two times the width of the region and that the value of L/W is independent of q but does depend on the injection angle θ. Hence, $L \propto \dot{m}_i$ will hold for this system and $L \propto \sqrt{q}$ for the more general case. Combining the latter results with the scaling law for W and the blowoff criterion, $\tau_c V_{1c}/L = 1$, results in an equation for the blowoff velocity,

$$V_{1c} \propto L/\tau_c \propto (L/W)(W/\tau_c)$$

or

$$V_{1c} = \left[\left(\frac{L}{W}\right)^2 \frac{8(1+\cos\theta)}{C_d}\right]^{\frac{1}{4}}\left[\frac{v_i\sqrt{b}\,d}{\tau_c}\right]^{\frac{1}{2}}$$

when the densities of the injectant and the approaching stream are equal. The terms in the first square bracket on the right-hand side of this equation depend on the angle of injection and those in the second on the velocity and area of the injector. If the geometry and chemical parameters are held fixed, the blowoff velocity is proportional to the square root of the injector velocity and the injector mass flow rate.

As an example, the experiments of Kosterin et al.[16] indicate that flame stabilization in a stream with approach stream speeds of about 100 m/s and fuel-air ratio near $\frac{1}{2}$ of stoichiometric could be achieved with values of q near 50. Mass flows in the injector were less than 1% of the approach stream flow for this example. In addition, the data are roughly correlated by $V_{1c} \propto \sqrt{q}$, which agrees with the above analysis.

The gas entering the recirculation zone is made up from the approach stream as well as the injectant stream and the ratio of these two mass flow rates has been found to be independent of q and strongly dependent on the injectant angle. For example, Kosterin et al.[16] find that the ratio of approach stream to injectant entrainment rates E_r is about 6.5 at $\theta = 135$

Table 2.3 Effect of Injectant Equivalence Ratio on Equivalence Ratio of Mainstream at Blowoff

ϕ_i	ϕ_{1c}	ϕ_e	Flameholder characteristics, mm
0	0.60	0.52	$b = 0.5$
			$d = 15$
0.54	0.54	0.54	$q = 45$
1.0	0.43	0.51	Kerosene fuel
1.2	0.40	0.51	$\theta = 135$ deg

deg, 4.0 at 90 deg, and 2.5 at 180 and 70 deg. Thus, the mixture ratio and hence temperature of the recirculation zone can be strongly influenced by the injectant.

The magnitude of this effect is shown in Table 2.3 where equivalence ratio (the fuel-air ratio, fraction of stoichiometric), in the approaching stream at the blowoff condition ϕ_{1c} is shown as a function of the equivalence ratio in the injectant stream ϕ_i when the gas speed of the approach stream was 100 m/s. In this example, the lean blowoff limit of the approach stream was reduced from 60% of the stoichiometric fuel-air ratio to 40% when the fuel-air ratio of the injectant fluid was increased from 0 to 120% of stoichiometric. The quantity ϕ_e is the calculated value of the equivalence ratio in the recirculation zone based on a measured value of entrainment ratio E_r. Note that ϕ_e is almost constant. Hence, the recirculation zone temperature, which is presumably also almost constant, is believed to be a critical feature in the ignition process.

Note that using injectant angles with smaller values of E_r will increase the sensitivity of the equivalence ratio at the blowoff condition to changes in ϕ_i, but at the same time will change the relationship between wake width, recirculation zone length, and injectant parameter q.

Large-scale tests of a complete afterburner system using two-dimensional arrays of jets of the type discussed here as flame stabilizers are reported in Ref. 17. Good stabilization characteristics and afterburner combustion efficiency were achieved with a total injectant flow rate of 2–4% of the total flow to the afterburner. Total pressure losses associated with jet flameholder systems were found to be 3–4% lower than corresponding values for bluff-body systems when the augmentation was zero. The possibility that gains can be made in reducing the nonafterburning total pressure loss is another reason for pursuing the investigation of this system.

Flame Stabilization in a Heterogeneous Fuel-Air Mixture

In almost any afterburner configuration, the weight savings to be made by reducing all of the length scales in the system will insure that the fuel injection system will be located so close to the flame stabilizers that some of the liquid fuel will arrive at the plane of the flameholders in an unvaporized state. When afterburners are to be used in the airstream of a fan engine, the low temperatures of these streams will greatly increase the fraction of fuel that is not evaporated. For example, in low-pressure-ratio fan engines operating at high altitudes, present-day jet fuels will be almost completely in the liquid state. In such streams, the flame stabilization system must produce some vaporization of the fuel in addition to acting as a continuous source of ignition.

In addition to these low-temperature problems, special requirements are placed on the fan stream augmentation system (often called a duct burner) by the operating characteristics of the fan. The fan is typically a low-pressure-ratio device and a relatively weak pressure disturbance propagating upstream from the augmentor can push the fan into a strong surge or stall. Hence, ignition of the fan stream augmentor must be achieved at a very low overall fuel-air ratio so that the sudden increase in total temperature (due to the start of afterburning) will not interact with the choked nozzle to produce a pressure pulse which will cause the fan to stall.

Common duct system. Under conditions such that the fuel in the fan stream is poorly vaporized, the simplest system is the engine in which afterburning takes place in a common duct with fuel injection system modulated so that combustion starts in the afterburner in the core stream where high temperatures insure good vaporization. Some of the hot gas produced by afterburning in the homogeneous core stream can then serve to support the flame stabilization process in the fan airstream and to produce vaporization in regions where the fan and core streams mix. The fan engine described in the introduction of this chapter (e.g., see Fig. 2.2) uses this system.

If fan air is to be burned in a separate duct where this support is not available, the flameholder must operate alone. In the following paragraphs, a qualitative picture of several flame stabilization schemes is given for this second and most extreme example. Few experimental data are available and because the process is complex, systematic experimental or theoretical treatment of this important problem is lacking.

Bluff-body flameholders. There is evidence that the picture of the bluff-body stabilization process described above applies in most respects to bluff-body stabilization in a heterogeneous flow. The principal difference is that in heterogeneous flows fuel vaporization must take place during the stabilization process itself. This can occur in two ways: (1) when liquid fuel drops impinge on the hot flameholder and (2) when fuel drops enter the mixing zone. In either case, the fuel-air ratio in the wake will be fixed by local conditions in a manner similar to that described above for the gas jets when a fuel-air mixture is used as the injectant. In this case, as there, the fuel-air ratio of the gas entering the wake appears to be the dominant factor and maximum blowoff velocity occurs for the injection conditions that supply the recirculation zone with a stoichiometric fuel-air mixture. Proper design and control of the fuel injection system can be used to get the best conditions for stabilization regardless of the overall fuel flow requirements.

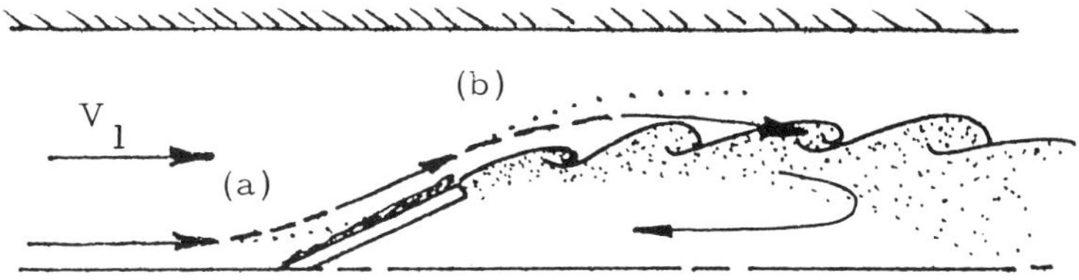

Fig. 2.17 Capture of fuel droplets.

The process of fuel vaporization by the flameholder involves a number of steps. First, the fuel droplets must be captured by the flameholder. Capture results because the droplets cannot exactly follow the gas streamlines; as the gas flows over the flameholder, the droplets will first be accelerated toward the holder (region a of Fig. 2.17) and then away from the mixing zone at region b. The drift rate across streamlines will depend strongly on particle size, the flameholder shape, and the velocity, density, and viscosity of the gas. After the fluid is captured on the holder, its residence time there will be fixed by a balance between the shear forces produced by the gas stream and the shear and surface tension forces acting between the liquid film and the flameholder. If too thick a film is formed, ablation of liquid drops from the downstream edge of the holder will result. If the vaporization rates are high, the holder may operate in a dry state. Heat transferred between the recirculation zone gas and the liquid film will depend on the usual convective parameters, and hence will increase with the speed, temperature, and pressure of the gas in the recirculation zone, and again will depend on the flameholder geometry.

Given the capture and heat-transfer processes, the properties of the fuel will then fix the rate of production of the fuel vapor at the flameholder, and the entrainment of vapor and air in the mixing zones will fix the fuel-air ratio of the recirculation zone gas. Additional vaporization of fuel droplets entrained in the mixing and recirculation zones will further increase the fuel-air ratio.

When the capture, heat-transfer, and evaporation rates are high, the fuel-air ratio in the recirculation zone can be much higher than in the stream approaching the flameholder. This difference would make possible ignition and stabilization when the overall fuel-air ratio is far below stoichiometric and hence would be advantageous during startup of the augmentation process. However, when the overall fuel-air ratio is increased toward the stoichiometric value (as it must be to achieve maximum augmentation), the fuel-air ratio in the recirculation zone could increase to values far above stoichiometric and a flame blowoff on the fuel rich side would occur. Hence, having the fuel-air ratio in the wake larger than that in the flow is not always advantageous.

It is clear from this brief qualitative description that the fuel injection system and the fuel capture and heat-transfer rates of the flameholder must be carefully controlled over a wide range of operating conditions if the simple bluff-body flameholder is to be used successfully. In general, such control is not possible with existing injection systems over wide operating ranges.

Flame stabilization by jets. A second system that has shown promise in heterogeneous systems is the aerodynamic flameholder or gas-jet holder described in a previous section. For duct burners, this system is perhaps the best of the three discussed here as far as its flame stabilization properties are concerned. However, losses and mechanical problems associated with the system may make its use impractical. (For example, see Ref. 17.)

Pilot burner. A third system suggested for use in heterogeneous fuel-air mixtures is a piloted burner. A small part of the afterburner flow, say 5–10%, is burned in a can-like pilot burner (or a number of burners) at the stoichiometric fuel-air ratio. The hot gas from this source is then used to support the stabilization by a system of conventional flameholders. The pilot burners require separate fuel injection and control systems to maintain fuel-air ratios different from those in the main flow. Systems of this type can be ignited at very low overall fuel-air ratios (e.g., Ref. 17) and low pressure levels. However, larger total pressure losses are produced by this system both with and without augmentation.

In summary, a number of schemes are available to produce flame stabilization in heterogeneous flows under conditions suitable for fan engine applications. Although the common burner scheme is the most well developed, the aerodynamic and pilot burner schemes offer advantages that are worth further exploration.

Discussion

The model used above to describe flame stabilization in homogeneous fuel-air streams by bluff bodies has the advantage of cleanly separating aerodynamic and chemical features of the process. The influence of various aerodynamic parameters is well understood from a qualitative point of view and many features can be treated in a quantitative manner. In particular, the dependence of stability limits on the geometry and scale of the flameholder-duct system is now clear.

The influence of the various chemical parameters of the problem are much less well understood. The use of an ignition time delay as suggested by Mullins,[12] Solokhin and Mironenko,[14] and Kosterin et al.[16] to describe ignition in a turbulent mixing zone is not correct, in the opinion of the author.

In one typical version of this approach (by Solokhin and Mironenko[14]), the ignition time is calculated from a global model of the chemical reaction rate and the chemical concentrations and temperature used in the calculation are taken from a mathematical model of the mixing zone that is based on time-averaged measurements of these parameters. There are three problems here. Global models for reaction rates have been used in calculations of laminar flame speeds and have led to qualitatively useful results. However, in order to obtain quantitatively accurate predictions, it has been found necessary to consider detailed chemical analyses that usually involve a large number of reaction steps, reaction rates, and activation energies.

A second and perhaps more serious problem arises from the treatment of the mixing layer. The model used by Solokhin pictures the layer as a region in which the temperature and concentrations change smoothly from values corresponding to the unburned mixture on one side to values corresponding to the products of combustion on the other. This is the result obtained experimentally with instrumentation producing time-averaged values. However, recent experimental developments (e.g., Brown and Roshko[18]) suggest that the conditions in the shear or mixing layer are quite different. Experimental results indicate that large-scale structures predominate in the mixing layer and that consequently gas in the layer, at a given instant and at a given point, has a high probability of being either completely burned or unburned. The probability of finding gas with a temperature or concentration of an intermediate value is small even at the center of the layer. In this picture of the flow, chemical reactions will start at boundaries between fully burned and unburned masses of gas and *not* in a uniform mixture of burned and unburned material. Thus, if the new model of the mixing layer is correct, the use of the time-averaged values of temperature and composition in the calculation of chemical reaction rates is inappropriate.

Finally, flame stabilization involves more than the simple ignition process. In order to stabilize a flame, the gas ignited in the mixing layer must continue to burn after it moves past the downstream end of the recirculation zone. Thus, the heat release rate in the mixing zone gas must be high enough to overcome the quenching effects of the entrainment of unburned gas in the region downstream of the recirculation zone. A simple ignition model is probably not sufficient to describe this process.

2.6 Flame Spread in Premixed and Homogeneous Fuel-Air Mixtures

In this section, the process of heat addition after flame stabilization has been achieved is discussed. Processes occurring in a premixed and homogeneous fuel-air mixture are considered. The term homogeneous is used here to denote mixtures of fuel vapor and air, as contrasted with heterogeneous mixtures by which is meant mixtures of fuel droplets and perhaps some fuel vapor with air. The combustion in the latter fuel-air system is important in duct-burner systems for fans, but will not be discussed here. (See the supplementary reading list at the end of this chapter.)

Given the flame stabilized in a duct, it should be possible to calculate the distance downstream of the stabilizer required to achieve a selected value of the heat release or combustion efficiency. The parameters that may influence the required length are: (1) *flow properties* (such as the pressure, temperature, and oxygen concentration; the fuel-air ratio and fuel properties; and the Mach number, velocity, and turbulence level of the unburned stream); and (2) *duct parameters* (such as duct height, flameholder geometry and blockage, and cross-sectional area changes with axial distance).

Unfortunately, at the present time, the dependence of the heat release rate in a combustion changer of fixed size on any of these parameters from basic principles cannot be predicted. Indeed, the understanding is so poor that the appropriate dimensionless parameters have not been identified or agreed upon. However, several fluid dynamic parameters (the Reynolds number and Mach number) are used to characterize the flow.

Although a great deal of experience is available that can serve as a guide for a new combustion chamber design, a large and expensive development effort is usually required to produce a satisfactory configuration. Conventional wisdom is in agreement, for example, that an increase in pressure, temperature, oxygen concentration, and turbulence level will increase the heat release rate and reduce the combustion chamber length required for high combustion efficiency. However, quantitative measures of the effects to be expected, given a particular change in a parameter, are not available. The reason for this is that the combustion process is turbulent and occurs in a region of strong shear and large axial pressure gradient.

General features of the flowfield of a typical flame are shown in Fig. 2.18. In this example the flame is stabilized by a bluff body placed on the centerline of a constant-area duct. The downstream end of the recirculation zone is about one tunnel height H downstream of the flameholder. The fuel is a hydrocarbon that produces a highly visible flame and the outer boundaries, shown in Fig. 2.18a, are based on time-exposure photographs taken in the light of the flame itself.

Temperature profiles shown in the lower half of Fig. 2.18a at a number of positions exhibit a sharp initial rise from the cold-gas temperature, which is followed by a very small and more gradual further increase. The maximum values, reached at the centerline of the duct, are close to the adiabatic flame temperature. The two boundaries based on the positions at which the initial temperature rise starts and stops are also shown in the lower half of Fig. 2.18b and they contain the region of strong chemiluminescence (shown here as a dotted region).

The temperature values shown here are averaged over long periods of time. Time-resolved temperature measurements suggest that the rapid initial rise is produced by averaging in time over temperatures that fluctuate rapidly between values close to the unburned and burned gas temperatures.

Spark schlieren photographs, similar to the sketch at the top half of Fig. 2.18b, support this picture. They indicate that the edge of the flame contains distinct vortex-like structures that produce a strongly corrugated surface (see Figs. 2.9 and 2.13). Hence, a probe located in this region would alternately observe hot and cold fluid. The scale of these structures grows slowly with the increasing distance from the flameholder and typically occupies between 30–40% of the width of the flame W. That is, regions of strong temperature gradients penetrate far into the flame front.

Concentration profiles for the products of combustion are similar to the temperature profiles: regions of the strong chemiluminescence produced by combustion (the dotted region in Fig. 2.18b) and of strong ionization, concentration, and temperature gradients almost exactly coincide. This result suggests that the sharp boundaries shown in the schlieren photo-

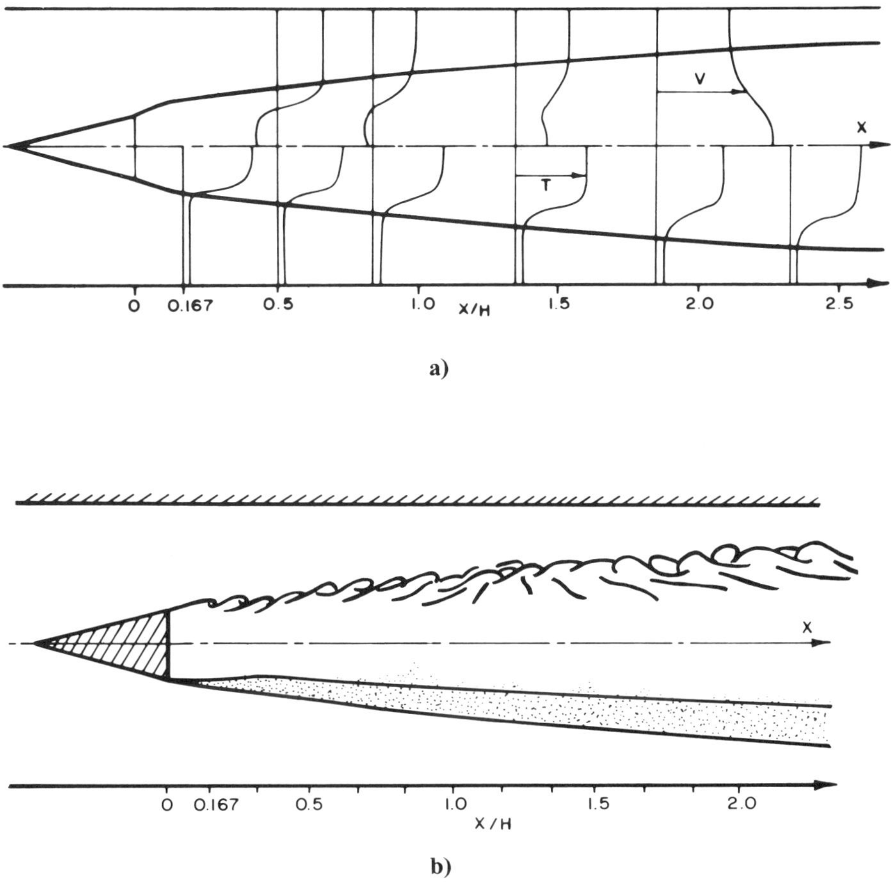

Fig. 2.18 Flame stabilized on a two-dimensional wedge flameholder: a) temperature and velocity profiles; b) schlieren and chemiluminescence boundaries.

graphs are flame fronts and that their positions are restricted to the region of sharp temperature rise.

The velocity profiles (shown in Fig. 2.18a) are also averaged over long periods of time and show less steep gradients than the temperature profiles. Near the flameholder and in the recirculation zone, the time-averaged velocity is reversed. The centerline velocity increases rapidly for positions farther downstream and exceed the unburned gas speed at positions farther downstream than $1\frac{1}{2}$ duct heights. This acceleration of the burned gas is a result of the action of the axial pressure gradient, produced by heat addition, on the high-density unburned gas and the lower-density burned material. The axial gradient, which is uniform across the duct, causes the lower-density fluid to accelerate more rapidly than the high-density stream and thus produces the hat-shaped profile.

In summary, the flame front appears to be made up of thin regions of chemical reaction that are rolled up into vortex-like structures. The size of these structures grows slowly as the axial distance increases and they occupy between 30–40% of the "flame" width. Strong chemical reaction and large heat release occur in the shear layers forming the boundaries between streams with large density and velocity differences.

The vortex-like structures lie in the region with a strong average velocity gradient. This suggests that they are related to the large-scale structures observed in two-dimensional shear layers. However, the vortex pairing observed without combustion has not been observed in spreading flames.

This picture of the spreading flame suggests that the rate of consumption of unburned fluid in the spreading flame is fixed by an entrainment process, rather than by a simple flame propagation process that might be expected to depend at least weakly on molecular transport properties. Much of the experimental data presented later support this view. The primary problem concerned with the prediction of flame spreading rates is the determination of this entrainment process. At the present time, no satisfactory physical model has been developed to describe it.

The remainder of this section reviews experimental information concerning flame spreading rates and discusses the implications of these data for turbulent entrainment rates or flame speed. Several simple models are described that allow a reasonable description of the dependence of some of the fluid dynamic parameters on the heat addition from the flame. However, even these restricted models remain incomplete because the entrainment rate of the turbulent flame cannot yet be prescribed.

Spreading Rates of Turbulent Flames

The quantity that the afterburner designer needs to know is the manner in which the combustion efficiency of a burner varies with the parameters described in the previous paragraphs. Unfortunately, combustion efficiency is difficult to measure accurately and has not been the subject of detailed investigations under conditions in which the effects of changing combustion chamber parameters were clearly isolated. Instead, a number of investigations have been made of the spreading rate of the flame front and then conclusions regarding the more applied problem of combustion efficiency have been drawn based on these results. See, for example, Williams et al.,[19] Wright and Zukoski,[20] and Solntsev.[21]

There are a number of problems with this approach. First, the flame is observed to have a finite thickness, which may be as great as 20% of the duct height (e.g., see Fig. 2.18). The spreading rate and conclusions drawn from it will depend strongly upon which surface is defined as the flame front. Further, since the combustion process presumably takes place within this thickness, any interpretation of combustion efficiency based on a single boundary will be suspect.

Second, there is a strong interaction of the flow with the heat addition process. The strong axial pressure gradient produced by heat addition produces acceleration in the unburned fluid that, in turn, produces an

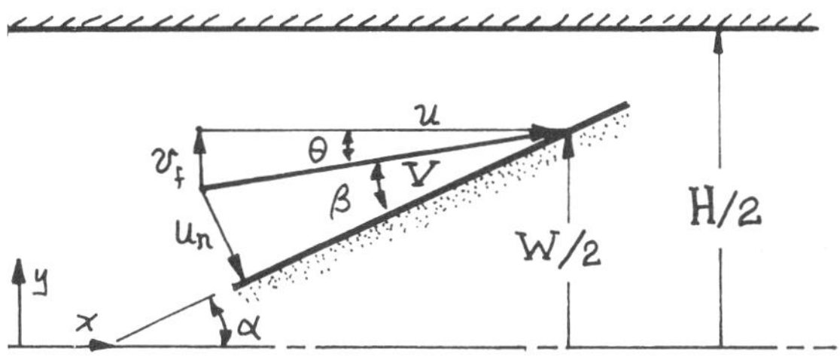

Fig. 2.19 Calculation of streamline angle at flame front.

appreciable curvature of streamlines near the flame front. The acceleration and curvature can have important effects on the interpretation of the flame width data.

This process is illustrated by the two-dimensional flow shown in Fig. 2.19. The problem is simplified by assuming that the unburned flow is isentropic, incompressible, and has a velocity vector which is almost axial and almost uniform. Under these assumptions the continuity and axial momentum equations can be written for the unburned flow as

$$\frac{\partial u}{\partial x}+\frac{\partial v}{\partial y}=0 \quad \text{and} \quad \rho u\frac{\partial u}{\partial x}+\frac{\partial P}{\partial x}=0$$

Combining these and making use of the approximation that ρ, u, $\partial u/\partial x$, and P are independent of vertical position y leads to the small angle that the velocity vector V makes with respect to the axis at the flame front,

$$\theta \approx \frac{v_f}{V}=\left(\frac{V_1}{V}\right)^2\left(\frac{H-W}{4}\right)\left(\frac{\partial C_p}{\partial x}\right) \tag{2.13}$$

where V_1 is the gas speed far upstream of the flameholder, v_f the y component of the velocity at the flame front, and C_p the pressure coefficient defined as

$$C_p=(P_1-P)/\tfrac{1}{2}\rho_1 V_1^2$$

The ratio V/V_1 is different from 1 because of the velocity change produced by acceleration of the flow around the flameholder and then by heat addition.

The angle α is one made between the flame front at the point of interest and the axis of the duct. It can be obtained from the shape of the flame front and is just

$$\alpha=\frac{1}{2}\frac{\mathrm{d}W}{\mathrm{d}x} \tag{2.14}$$

The velocity of the gas normal to the flame front u_n, sometimes called the turbulent entrainment velocity or flame speed, is given by $V \sin \beta$. β can be obtained from the difference between angles θ and α, that is,

$$u_n/V = \sin(\alpha - \theta) \tag{2.15}$$

Thus, to calculate u_n consistent with any choice of the flame front, V/V_1 and both angles α and θ must be known. In many situations in which turbulent flame spreading is considered, the ratio V/V_1 can be as large as 2 to 3 and the angles α and θ are often nearly equal. Hence, neither of these effects can be ignored a priori.

This section outlines the dependence of the wake width on a number of parameters that describe the combustion chamber and the implications of these results with regard to the combustion efficiency and turbulent entrainment velocity.

Flame spreading. The spreading rate of the flame is a very strong function of the condition of the flame. When the flame is laminar, the flame width is a strong function of laminar flame speed and turbulence level in the approach stream.[22,23] However, at higher speeds and Reynolds numbers, the flame becomes turbulent under the same conditions that the mixing layers become turbulent and the dependence of the flame shape upon the molecular transport processes becomes negligible.

This transition is illustrated by the schlieren photographs of Fig. 2.13 and by the data shown in Table 2.4, taken from Thurston.[24] He examined flame spreading in a rectangular duct (about 15 × 7.5 cm in cross section) and used several cooled circular cylinders as flameholders. The data of this table show the wake width and velocity ratios measured near the downstream end of the recirculation zone (subscript 2) at about 4 cm and at a station 37 cm downstream of the holder (subscript 37). The outer edge of the flame defined in schlieren photographs and averaged through its bumpy surface

Table 2.4 Variation with Approach Speed of Velocity Ratio and Wake Width at the Downstream End of Recirculation Zone (sub 2) and 37 cm (sub 37) from Flameholder[a]

V_1 (m/s)	$\frac{V_2}{V_1}$	$\frac{W_2}{H}$	$\frac{V_{37}}{V_1}$	$\frac{W_{37}}{H}$
30	1.09	0.23	1.55	0.40
60	1.06	0.22	1.34	0.34
80	1.06	0.21	1.31	0.33
100	1.05	0.21	1.29	0.33
120	1.04	0.21	1.27	0.33
140	1.04	0.21	1.27	0.33

[a] Duct about 15 × 7.5 cm with 0.32 cm diam cylinder spanning the 7.5 cm dimension; stoichiometric fuel-air ratio.

was used to determine the wake width. Total and static pressure measurements were made as a function of axial position and the velocity in the unburned flow was determined from the pressure measurements.

Note in Table 2.4 that as the approach speed V_1 is increased from 30 m/s (at the beginning of the transition to turbulence) to 60 m/s, the wake width at the 37 cm station decreases by about 15%. A similar doubling of speeds to 120 m/s produces a much smaller change. Similarly, at $x = 37$ cm, the ratio of the velocity to the approach stream speed decreases rapidly at first and then approaches a constant value. Further, of the 6% velocity reduction occurring as the approach speed is changed from 60 to 120 m/s, 2% is evidently caused by changes occurring in the neighborhood of the recirculation zone, e.g., see the V_2/V_1 column. The small changes in the wake width and velocity ratio observed here for the turbulent flow condition are typical of measurements obtained in the turbulent regime for holder sizes in the range of 0.32–2.54 cm. A similar transition was also reported in Ref. 19.

The spreading rates of flames stabilized on bluff-body flameholders operating in constant-area ducts and in the turbulent regime have been determined by Wright and Zukoski[20] as a function of the approach stream speed, fuel-air ratio, temperature, fuel type, and flameholder-duct geometry. Fortunately for the designer of afterburners, the observed spreading rates have been found to be almost independent of these parameters. Typical results are shown in Fig. 2.20 taken from Ref. 20.

In Fig. 2.20a, the dependence of the wake width on the fuel-air ratio is shown. In this range of equivalence ratios, the laminar flame speed for a typical hydrocarbon fuel has a maximum near an equivalence ratio of 1.10 and would decrease to about 60% of the maximum value for $\phi \approx 0.75$. Similarly for the temperature increase shown in Fig. 2.20b, the laminar flame speed would increase by factors of about two or more. Finally, the laminar flame speed for hydrogen-air mixtures is of the order of 10 times that of the hydrocarbon fuel used in these tests. Despite these changes in parameters, which produce large changes in laminar flame speed, little change is observed in the flame geometry. The lack of change in Fig. 2.20c is particularly interesting.

Similar results concerning the very weak dependence of the flame geometry on the approach stream speed and fuel-air ratio are reported by Williams et al.[19] and also by Solntsev[21] who carried out experiments in much larger-scale apparatus. The former measurements were made from schlieren photographs and Solntsev used photographs, as well as temperature, oxygen concentration, and ion density profiles to determine the widths. Thus, it is clear that the measurements described above are general and that the laminar flame speed and presumably other molecular transport processes are *not* important in fixing the geometry of the spreading turbulent flame.

The dependence of flame geometry on the turbulence level of the unburned mixture is less clear. The experiments of Williams et al. showed a very weak dependence in small-scale experiments, whereas Solntsev found a somewhat larger dependence. The latter suggested that the entrainment rate of the turbulent flame is proportional to the turbulence level in the unburned flow.

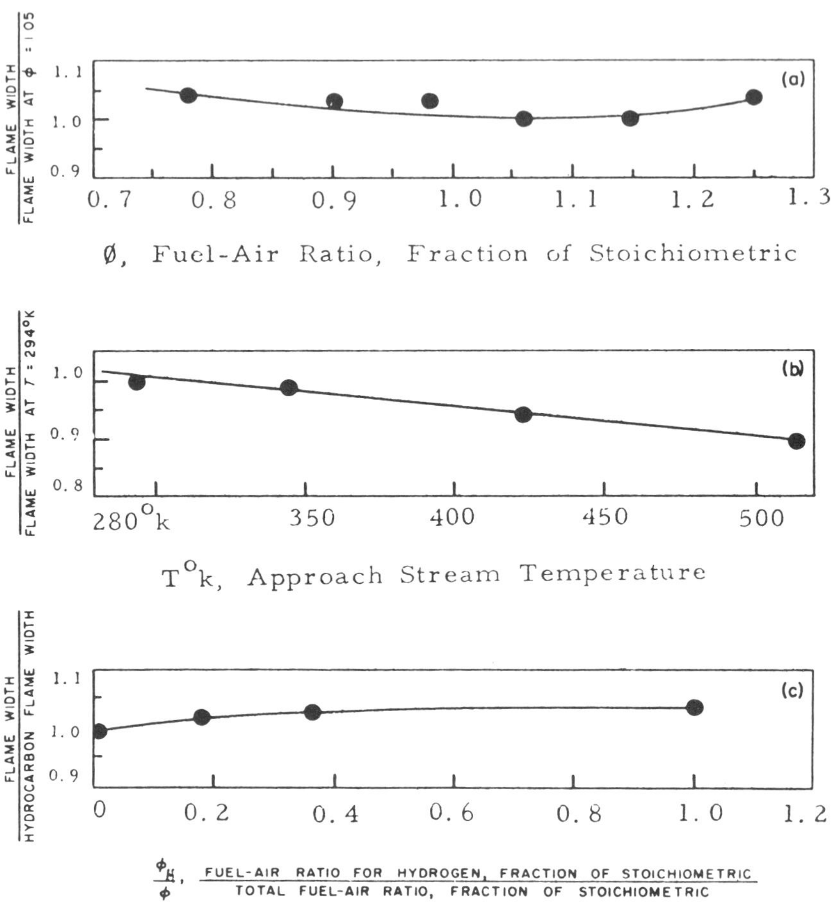

Fig. 2.20 Dependence of flame width on fuel-air ratio, approach stream temperatures, and fuel type (holder is a $\frac{1}{8}$ in. circular cylinder and width is measured 15 in. from holder).

Experiments carried out by Wright[7] in which high subsonic speeds were observed in the flow past the wake indicate that the Mach number does not have a large effect on flame geometry as long as the local Mach number is below 0.8.

No experiments dealing directly with the pressure dependence of the flame spreading phenomena have been reported. A number of experiments have been made of complete afterburner systems in which the pressure effects were examined and were found to have strong effects on combustion efficiency when the pressure fell below a limiting value.[15] However, since the complete system was involved, it is not clear which process (injection, flame stabilization, or flame spreading) was responsible for the drop in efficiency

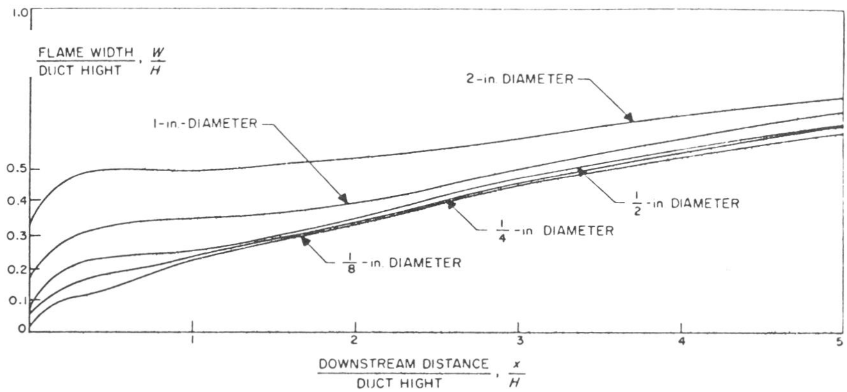

Fig. 2.21 Flame shapes for flameholders of several sizes in a duct of fixed size, $V = 300$ ft/s, $\phi = 1.0$, $H = 15$ cm.

as the pressure was reduced. The Reynolds number is reduced when the pressure falls and the velocity is held fixed. Such a reduction could cause a transition from turbulent to laminar flow, which would have a very adverse effect on flame spreading and combustion efficiency as well as on the flame stabilization process. In addition, Hottel et al.[15] suggest that a pressure reduction would also reduce the turbulence level of the approach stream and that this will result in a reduction of the flame spreading rate.

The dependence of flame geometry on flameholder-duct geometry is weak and complicated. The wake width at a given distance downstream from the flameholder in a duct of fixed size does increase slowly as the flameholder scale is increased. However, at distances greater than several recirculation zone lengths of the larger holder, wake widths are almost independent of holder scale. Thus, an increase in blockage for a flameholder of fixed size will cause a slight decrease in the flame spreading rate.

Data illustrating this result are shown in Fig. 2.21 and are taken from Ref. 20. Flame boundaries are shown for flames stabilized in a rectangular duct (15 cm high by 7.5 cm wide) for five cooled circular cylinders with diameters of 0.32, 0.63, 1.27, 2.54, and 5 cm. The outer edge of the flame, determined from schlieren photographs and normalized by the duct height H of 15 cm, is given as a function of the axial position, which is also normalized by the duct height. The large initial differences are due to differences in the width of the recirculation zone, discussed earlier. However, farther downstream, the boundaries begin to merge. At $x/H = 5$, the differences in the wake width are small despite the change in the blockage ratio from $1/48$ to $\frac{1}{3}$.

If the flame geometry of a single flameholder in a duct is compared with the flames produced by two holders of the same scale located in the same duct, each flame width of the latter configuration will be slightly smaller than that of the single holder. However, the fraction burned will be greatly

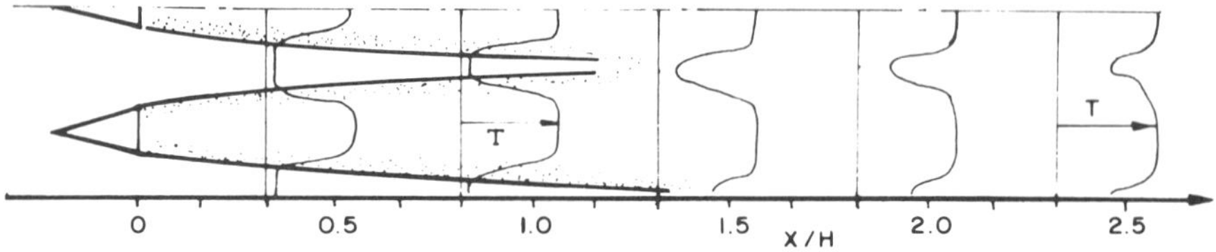

Fig. 2.22 Flame edge and temperature profiles for a three-flameholder array (from Ref. 21).

Table 2.5 Comparison of Flame Widths for 30 deg Half-Angle Wedge Holders in a Duct of 300 mm Height

x (cm)	$\frac{x}{H}$	$\frac{W_1}{H}$	$\frac{W_3}{\frac{1}{3}H}$	$\frac{x}{\frac{1}{3}H}$
100	0.33	0.31	0.68	1.0
250	0.83	0.48	0.89	2.5
400	1.33	0.60	1.00	4.0
850	2.83	0.71	1.00	8.5

Subscript 1: one single 70 mm holder; subscript 3: three 30 mm holders.

increased at a given distance downstream of the holder. Hence, if stabilization problems are not a limiting constraint, increasing the number of holders will always improve combustion efficiency.

This situation is illustrated by the data of Solntsev[21] shown in Figs. 2.18 and 2.22 and in Table 2.5. The sketches show flame boundaries and temperature profiles for two flameholder configurations placed in the same 300 cm high duct. In Fig. 2.18, the holder is a 30 deg half-angle wedge with a 70 mm base height; in Fig. 2.22, three 30 deg half-angle wedges are used. Wake widths are presented as a fraction of the duct height and for a number of axial positions in Table 2.5. The duct height used in presenting the three-flameholder data is divided by three so that each holder is charged with its equivalent duct height. The flame fronts for the three-holder configurations begin to merge near $x/H = 1.3$ and combustion is complete before the $x/H = 2.8$ station at which the flame width of the single holder configuration is less than 75% of the duct height. If the wakes widths and axial positions are normalized by the height of the duct occupied by each holder, i.e., by 300 mm for the single holder and 100 mm for the three-holder configuration, the systems look more similar. Thus, the flames occupy about 65% of their ducts at a station one equivalent duct height downstream of the holder.

Entrainment rates. The entrainment rate of the flame can be determined from knowledge of the flame geometry and the pressure and

Table 2.6 Parameters Used in the Calculation of Entrainment Velocity u_n for Stoichiometric Fuel-Air Ratio[a]

D (cm)	V_1 (m/s)	C_{p37}	$\partial C_p/\partial x$ (1/cm)	$(H-W)/2$ (cm)
0.32	30	1.40	0.040	4.6
0.32	60	0.80	0.020	5.0
0.32	120	0.62	0.016	5.1
0.32	140	0.62	0.016	5.1
0.64	60	0.87	0.015	4.9
0.64	120	0.70	0.013	5.0
1.27	60	1.10	0.014	4.6
1.27	120	0.96	0.010	4.6

$\partial W/\partial x$	α	θ	β, u_n/V	u_n (m/s)
0.140	0.070	0.040	0.030	1.4
0.086	0.043	0.028	0.015	1.2
0.090	0.045	0.025	0.020	3.1
0.090	0.045	0.025	0.020	3.6
0.076	0.038	0.019	0.019	1.6
0.062	0.031	0.011	0.020	3.1
0.066	0.033	0.015	0.018	1.6
0.065	0.033	0.012	0.021	3.5

[a] Data from Thurston.[24] (Note: no boundary-layer correction has been made for β.)

velocity field produced by the flame. However, the entrainment rates determined by this method are very strongly dependent on the definition used for selection of the flame front position and on a number of corrections. A set of calculations made from the data of Thurston[24] illustrates the process.

Consider Fig. 2.19 and the data presented in Table 2.6. Values of experimental parameters and the angles α, β, and θ calculated from Eqs. (2.13) and (2.14) are presented in the table for three flameholders of diameter D and a range of approach stream speeds V_1. Calculated values of the entrainment speed are given in the last column.

The entrainment rate per unit area of the flame is ρu_n, which in Eq. (2.15) was given by

$$\text{Entrainment rate} = \rho u_n = \rho V \sin \beta$$

and when β is small

$$\text{Entrainment rate} = \rho V \cdot \beta$$

One interesting result shown in Table 2.6 is that the entrainment angle β is almost independent of approach stream speed and flameholder blockage when the flame is turbulent. The values of β lie around 0.02 with a scatter of at least 0.005 or 25%. Some of the data used to calculate values of $\beta = \alpha - \theta$ are also shown in the table. Examination of the quantities shows that β is constant despite substantial changes in these quantities resulting from changes in flameholder diameter D. Note also that β changes only slightly when the speed of the approaching stream V_1 is increased by a factor of two.

The large scatter is due to the rather arbitrary definition of the flame width W and uncertainties in the estimation of the slopes for W and C_p that appear in calculation of α and θ. Notice that in most cases the velocity vector V is turned away from the axis through an angle θ that is roughly half of α, the angle between the spreading flame front and the axis of the duct. Because θ is greater than the angle β for many of the conditions presented here, large errors in β must be expected from this source. In addition, the boundary layers growing on the walls of the combustion chamber will also produce an acceleration of the flow by reducing the effective cross-sectional area of the duct and hence must be taken into account in calculating the angle θ. For the data presented in Table 2.5, the growth of the displacement thickness on the top wall of the duct would decrease θ and hence increase β by about 0.002 rad. Side wall boundary layers will have a similar effect. Hence, the entrainment velocity for those experiments is about 0.025 ± 0.007 of the local unburned gas speed.

The entrainment rate can also be obtained directly from the flame width and velocity data by a different mass balance technique. The idea here is to measure the mass flow of the unburned gas outside the flame as a function of axial position. The rate of change of this flow rate $\dot{m}_c$ with the axial position can then be used to determine the entrainment rate. When α is small, the entrainment rate per unit area at one side of the flame is $(-\frac{1}{2})(\mathrm{d}\dot{m}_c/\mathrm{d}x)$ and can be expressed in terms of the total mass flow $\dot{m}_1 = \rho_1 V_1 H$ as

$$\beta = \frac{u_n}{V} = -\frac{H}{2}\left[\frac{\mathrm{d}}{\mathrm{d}x}\left(\frac{\dot{m}_c}{\dot{m}_1}\right)\right]\left(\frac{V_1}{V}\right)$$

Measurements of $\dot{m}_c$ of the type required were carried out by Thurston[24] for one of the experiments reported in Table 2.5. An estimate of β based on the above equation and Thurston's data was 0.027 for the 0.32 cm flameholder. The agreement of the two methods is close considering all the uncertainties.

The entrainment rate of a turbulent flame will certainly depend on a number of parameters not changed in the experiments described here and hence this value for β cannot be viewed as having any general applicability. However, the small rate of entrainment is of general interest and probably represents the lower bound of values for entrainment in a low-turbulence experiment.

One-Dimensional Heat Release

As a first crude approximation for the processes occurring in the combustion chamber of an afterburner, it is convenient to investigate the process of heat addition in simple, one-dimensional channel flow. The calculation will be carried out with the assumption that the heat is added uniformally across the channel.

Differential relations. The aim of the calculations is to illustrate the variation of the variables as heat is added; of particular interest are the dependence of the stagnation pressure and Mach number on the stagnation enthalpy or temperature. The process when the assumption of constant-channel area is made will be considered first and later the expressions including the variation of area will be derived.

Referring to Fig. 2.23, a quantity of heat $\mathrm{d}H$ is added between two control planes separated by a distance $\mathrm{d}x$. It is assumed that no change in the initial conditions (u, ρ, p, T, M) is invoked by this addition of heat. Therefore, the flow process may be described by the laws of continuity, momentum, and energy transport

$$\mathrm{d}(\rho u) = 0$$

$$(\rho u)\,\mathrm{d}u + \mathrm{d}p = 0$$

$$C_p\,\mathrm{d}T + u\,\mathrm{d}u = \mathrm{d}H = C_p\,\mathrm{d}T_t \tag{2.16}$$

and the equation of state for a perfect gas

$$p = \rho RT$$

The change of gas velocity accompanying an addition of heat may be found

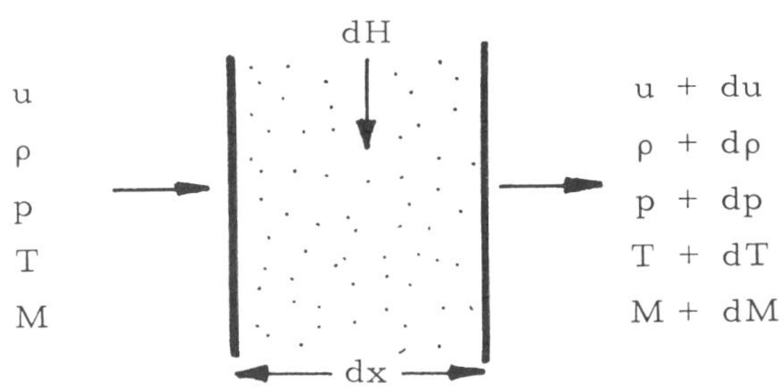

Fig. 2.23 One-dimensional flow of gas with infinitesimal heat addition.

by eliminating the temperature variation from the energy equation, using the equation of state

$$u\,\mathrm{d}u + C_pT\left(\frac{\mathrm{d}p}{p} + \frac{\mathrm{d}\rho}{\rho}\right) = \mathrm{d}H \tag{2.17}$$

Now computing $\mathrm{d}p/p = -u\,\mathrm{d}u/RT$ and $\mathrm{d}\rho/\rho = -\mathrm{d}u/u$ from the momentum and continuity equations, respectively, it follows, upon collecting and rearranging terms, that

$$\frac{\mathrm{d}u}{u} = -\left(\frac{1}{M^2 - 1}\right)\frac{\mathrm{d}H}{C_pT} \tag{2.18}$$

The result indicates that the variation of the gas velocity with heat addition depends critically upon the Mach number of the flow at the point of heat addition. The velocity increases with heat addition for subsonic flow and decreases with heat addition for supersonic flow. This distinction between the behavior of flow at subsonic and supersonic velocities is somewhat reminiscent of that occurring in the behavior of the velocity during a contraction of the channel cross section in isentropic flow.

The pressure variation may be computed directly from the momentum relation as

$$u\,\mathrm{d}u = -\frac{1}{\gamma}a^2\frac{\mathrm{d}p}{p} \tag{2.19}$$

and employing Eq. (2.18) for the velocity variation,

$$\frac{\mathrm{d}p}{p} = +\left(\frac{\gamma M^2}{M^2 - 1}\right)\left(\frac{\mathrm{d}H}{C_pT}\right) \tag{2.20}$$

The addition of heat causes a drop in the gas pressure for subsonic velocities, which becomes progressively more severe as the Mach number approaches unity.

The static temperature is found by writing the logarithmic derivative of the equation of state as

$$\frac{\mathrm{d}T}{T} = \frac{\mathrm{d}p}{p} - \frac{\mathrm{d}\rho}{\rho}$$

Substituting for the terms on the right from the continuity and momentum equations, this gives

$$\frac{\mathrm{d}T}{T} = 1 - \left(\frac{u^2}{RT}\right)\frac{\mathrm{d}u}{u}$$

or, substituting from Eq. (2.18) for the velocity variation,

$$\frac{\mathrm{d}T}{T} = \left(\frac{\gamma M^2 - 1}{M^2 - 1}\right)\left(\frac{\mathrm{d}H}{C_p T}\right) \tag{2.21}$$

The factor $(\gamma M^2 - 1)/(M^2 - 1)$ exhibits two changes of sign because the numerator and denominator pass through zero at $M^2 = 1/\gamma$ and $M^2 = 1$, respectively. So long as $M^2 < 1/\gamma$, the temperature rises with the addition of heat, which is the trend naturally to be expected. However, as the Mach number increases, but remains in the range $1/\gamma < M^2 < 1$, the temperature decreases as heat is added to the gas, a result that is not at all obvious physically and requires a bit closer investigation. Finally, when $M > 1$, the static temperature again increases with $\mathrm{d}H$, as would be expected.

The energy equation (2.16) indicates the proportion of a heat increment $\mathrm{d}H$ appearing as gas enthalpy $C_p \mathrm{d}T$ and appearing as kinetic energy of mean motion $u\,\mathrm{d}u$. Expressing the differentials $\mathrm{d}u$ and $\mathrm{d}T$ in terms of $\mathrm{d}H$ and simplifying the result, it is found that

$$\underbrace{\left[\frac{(\gamma - 1)M^2}{1 - M^2}\right]}_{\text{increment of kinetic energy}}\frac{\mathrm{d}H}{C_p T} + \underbrace{\left(\frac{1 - \gamma M^2}{1 - M^2}\right)}_{\text{increment of gas enthalpy}}\frac{\mathrm{d}H}{C_p T} = \frac{\mathrm{d}H}{C_p T} \tag{2.22}$$

so that a portion $[(\gamma - 1)M^2]/(1 - M^2)$ of the added energy $\mathrm{d}H$ is devoted to increasing the kinetic energy of the gas, while a portion $(1 - \gamma M^2)/(1 - M^2)$ appears as the enthalpy of the gas. Now clearly, as the Mach number increases, the amount of heat required to supply the kinetic energy increases until, at $M^2 = 1/\gamma$, the entire heat addition is required to supply the kinetic energy alone, with the result that the gas enthalpy cannot change. Further increase of the Mach number increases the kinetic energy requirement even more, so that a portion of it must be furnished by the gas enthalpy itself. As a consequence, the gas temperature decreases.

The situation is clear for supersonic flow. Here, the gas velocity decreases as heat is added, with the result that all of the heat added, plus that resulting from kinetic energy reduction, is available to increase the gas enthalpy. The enthalpy increase is

$$\left[1 + \frac{(\gamma - 1)M^2}{M^2 - 1}\right]\mathrm{d}H$$

and hence is in excess of the amount supplied. Note that the ratio of the quantity of heat passing to kinetic energy, to that passing to gas enthalpy, is given by

$$\frac{\text{Kinetic energy increment}}{\text{Gas enthalpy increment}} = \left[\frac{(\gamma - 1)M^2}{1 - \gamma M^2}\right] \tag{2.23}$$

and hence is a function of the Mach number alone.

It is a simple matter now to compute the change of Mach number with heat addition; logarithmic differentiation indicates that

$$\frac{\mathrm{d}M}{M} = \frac{\mathrm{d}u}{u} - \frac{1}{2}\frac{\mathrm{d}T}{T}$$

Combining the known values of $\mathrm{d}u/u$ and $\mathrm{d}T/T$ from Eqs. (2.18) and (2.21) the formula for $\mathrm{d}M/M$ is simply

$$\frac{\mathrm{d}M}{M} = \frac{1}{2}\left(\frac{1+\gamma M^2}{1-M^2}\right)\frac{\mathrm{d}H}{C_pT} \tag{2.24}$$

It is clear from Eq. (2.24) that heat addition always brings the flow toward a Mach number of unity, that is, heat addition results in a Mach number increase for subsonic flow and a Mach number decrease for supersonic flow.

The variation of stagnation pressure can now be simply determined by taking the logarithmic derivative of

$$p_t = p\left(1 + \frac{\gamma-1}{2}M^2\right)^{\gamma/(\gamma-1)}$$

and substituting for the appropriate results given above. Then it is found that

$$\frac{\mathrm{d}p_t}{p_t} = \left[\frac{(1-\gamma/2)M^2}{1+[(\gamma-1)/2]M^2}\right]\frac{\mathrm{d}H}{C_pT}$$

or, more simply,

$$\frac{\mathrm{d}p_t}{p_t} = -\left(\frac{\gamma}{2}M^2\right)\left(\frac{\mathrm{d}H}{C_pT_t}\right) \tag{2.25}$$

It is evident that the stagnation pressure drops with the heat addition and that the rate depends very strongly on the Mach number at which the heat is added.

The results derived in the previous paragraphs concern heat addition in a constant-area channel. The effect of simultaneous changes in area and total gas enthalpy are also of interest and may be simply derived on the basis of the following considerations. For example, consider the Mach number to be a function of both area and stagnation temperature. Then,

$$\mathrm{d}M = \left(\frac{\partial M}{\partial T_t}\right)_A \mathrm{d}T_t + \left(\frac{\partial M}{\partial A}\right)_{T_t}\mathrm{d}A$$

However, the quantity $(\partial M/\partial T_t)$ is obtained from Eq. (2.24) since the derivative given in the above equation was obtained with the area held fixed.

Thus,

$$\left(\frac{\partial M}{\partial T_t}\right)_A = \left(\frac{M}{T}\right)\left(\frac{1}{2}\right)\left(\frac{1+\gamma M^2}{1-M^2}\right)$$

Similarly, the quantity $(\partial M/\partial A)_{T_t}$ was previously obtained during study of isentropic channel flow, and is given by

$$\left(\frac{\partial M}{\partial A}\right)_{T_t} = \left(\frac{M}{A}\right)\left(-\frac{1-[(\gamma-1)/2]M^2}{1-M^2}\right)$$

Therefore,

$$\frac{\mathrm{d}M}{M} = \left(\frac{1}{2}\frac{(1+\gamma M^2)}{(1-M^2)}\right)\left(\frac{\mathrm{d}H}{C_pT}\right) + \left(-\frac{1+(\gamma-1)M^2}{1-M^2}\right)\frac{\mathrm{d}A}{A}$$

The results for other parameters of interest are

$$\frac{\mathrm{d}u}{u} = \left(\frac{1}{1-M^2}\right)\frac{\mathrm{d}H}{C_pT} + \left(-\frac{1}{1-M^2}\right)\frac{\mathrm{d}A}{A}$$

$$\frac{\mathrm{d}p}{p} = \left(\frac{-\gamma M^2}{1-M^2}\right)\frac{\mathrm{d}H}{C_pT} + \left(\frac{\gamma M^2}{1-M^2}\right)\frac{\mathrm{d}A}{A}$$

$$\frac{\mathrm{d}T}{T} = \left(\frac{1-\gamma M^2}{1-M^2}\right)\frac{\mathrm{d}H}{C_pT} + \left(\frac{(\gamma-1)M^2}{1-M^2}\right)\frac{\mathrm{d}A}{A}$$

$$\frac{\mathrm{d}\rho}{\rho} = \left(\frac{-1}{1-M^2}\right)\frac{\mathrm{d}H}{C_pT} + \left(\frac{M^2}{1-M^2}\right)\frac{\mathrm{d}A}{A}$$

$$\frac{\mathrm{d}p_t}{p_t} = \left(\frac{-(\gamma/2)M^2}{1+[(\gamma-1)/2]M^2}\right)\frac{\mathrm{d}H}{C_pT} = \left(-\frac{\gamma}{2}M^2\right)\frac{\mathrm{d}H}{C_pT_t}$$

$$\frac{\mathrm{d}T_t}{T_t} = \left(\frac{1}{1+[(\gamma-1)/2]M^2}\right)\frac{\mathrm{d}H}{C_pT} = \frac{\mathrm{d}H}{C_pT_t} \tag{2.26}$$

Use of these equations makes possible a determination of the local rate of the changes in the various parameters when both heat addition $\mathrm{d}H$ and area change $\mathrm{d}A$ are made. Note, although, that $\mathrm{d}M/M$, $\mathrm{d}p/p$, $\mathrm{d}T/T$, $\mathrm{d}\rho/\rho$ may be held constant by appropriate matching of the variation of $\mathrm{d}H/C_pT$ and $\mathrm{d}A/A$. For example, if $\mathrm{d}H/C_pT = \mathrm{d}A/A$, then the static pressure and velocity remain constant. However, the ratio $\mathrm{d}p_t/p_t$ is always finite and negative.

The equations cannot be used in general to obtain algebraic solutions for the flowfield in a duct of specified area variation and heat addition. However, they can be used in numerical calculations.

Integrated relations. The equations for heat addition in *constant-area channel flows* can be integrated to give the changes in the variables of interest occurring when heat is added or the equations for conservation of energy, momentum, and mass can be applied directly across the heat addition zone. The results are

$$\frac{T_{t_2}}{T_{t_1}} = \left(1 + \frac{\Delta H}{C_p T_{t_1}}\right) = \left(\frac{1+\gamma M_1^2}{1+\gamma M_2^2}\right)^2 \left(\frac{1+[(\gamma-1)/2]\,M_2^2}{1+[(\gamma-1)/2]\,M_1^2}\right)\left(\frac{M_2^2}{M_1^2}\right)$$

$$\frac{p_{t_2}}{p_{t_1}} = \left(\frac{1+\gamma M_1^2}{1+\gamma M_2^2}\right)\left(\frac{1+[(\gamma-1)/2]\,M_2^2}{1+[(\gamma-1)/2]\,M_1^2}\right)^{\gamma/(\gamma-1)}$$

$$\frac{u_2}{u_1} = \frac{\rho_1}{\rho_2} = \left(\frac{1+\gamma M_1^2}{1+\gamma M_2^2}\right)\left(\frac{M_2}{M_1}\right)^2$$

$$\frac{p_2}{p_1} = \left(\frac{1+\gamma M_1^2}{1+\gamma M_2^2}\right)$$

$$\frac{T_2}{T_1} = \left(\frac{1+\gamma M_1^2}{1+\gamma M_2^2}\right)^2\left(\frac{M_2}{M_1}\right)^2 \tag{2.27}$$

These equations are given in terms of M_1 and M_2, whereas in most problems of interest, the heat addition $\Delta H/C_p T_{t_1}$ and M_1 will be the specified quantities. Since elimination of M_2 from the above equations is algebraically complicated, the results will be expressed numerically in order to obtain a directly useful form.

Pick state 1 as the initial condition corresponding to a Mach number M at the start of heat addition in a constant-area duct. Define a second state, denoted by a super *, at which heat addition has been sufficient to drive the Mach number to 1. This state is to be used as a reference condition in a manner analogous to the A^* state (of Sec. 2.19 of *Aerothermodynamics of Gas Turbine and Rocket Propulsion*) of one-dimensional, isentropic, channel flow.

In the above equations, P_{t_1}, for example, becomes P_t, P_{t_2} becomes P_t^*, and M_1 becomes M and M_2 becomes 1.0. Values of p_t/p_t^*, p/p^*, T/T^*, and T_t/T_t^* are given in Table 2.7 as a function of the initial Mach number M and for $\gamma = 1.40$. The variation with γ is small in subsonic regions, but becomes important when $M > 1.5$. Also note the rapid increase in p_t/p_t^* for supersonic speeds.

Table 2.7 or a plot may be used to obtain solutions of problems involving uniform heat addition in a channel of constant cross section. Note that the

Table 2.7 Frictionless, Constant-Area Flow with Change in Stagnation Temperature (perfect gas, $\gamma = 1.4$ exactly)

M	T_t/T_t^*	T/T^*	p/p^*	p_t/p_t^*
0.02	0.00192	0.00230	2.3987	1.2675
0.04	0.00765	0.00917	2.3946	1.2665
0.06	0.01712	0.02053	2.3880	1.2647
0.08	0.03021	0.03621	2.3787	1.2623
0.10	0.04678	0.05602	2.3669	1.2591
0.12	0.06661	0.07970	2.3526	1.2554
0.14	0.08947	0.10695	2.3359	1.2510
0.16	0.11511	0.13843	2.3170	1.2461
0.18	0.14324	0.17078	2.2959	1.2406
0.20	0.17355	0.20661	2.2727	1.2346
0.22	0.20574	0.24452	2.2477	1.2281
0.24	0.23948	0.28411	2.2209	1.2213
0.26	0.27446	0.32496	2.1925	1.2140
0.28	0.31035	0.36667	2.1626	1.2064
0.30	0.34686	0.40887	2.1314	1.1985
0.32	0.38369	0.45119	2.0991	1.1904
0.34	0.42057	0.49327	2.0647	1.1821
0.36	0.45723	0.53482	2.0314	1.1737
0.38	0.49346	0.57553	1.9964	1.1652
0.40	0.52903	0.61515	1.9608	1.1566
0.42	0.56376	0.65345	1.9247	1.1480
0.44	0.59748	0.69025	1.8882	1.1394
0.46	0.63007	0.72538	1.8515	1.1308
0.48	0.66139	0.75871	1.8147	1.1224
0.50	0.69136	0.79012	1.7778	1.1140
0.52	0.71990	0.81955	1.7410	1.1059
0.54	0.74695	0.84695	1.7043	1.0979
0.56	0.77248	0.87227	1.6678	1.09010
0.58	0.79647	0.89552	1.6316	1.08255
0.60	0.81892	0.91670	1.5957	1.07525
0.62	0.83982	0.93585	1.5603	1.06821
0.64	0.86920	0.95298	1.5253	1.06146
0.66	0.87709	0.96816	1.4908	1.05502
0.68	0.89350	0.98144	1.4569	1.04890
0.70	0.90850	0.99289	1.4235	1.04310
0.72	0.92212	1.00260	1.3907	1.03764
0.74	0.93442	1.01062	1.3585	1.03253
0.76	0.94546	1.01706	1.3270	1.02776
0.78	0.95528	1.02198	1.2961	1.02337
0.80	0.96394	1.02548	1.2658	1.01934
0.82	0.97152	1.02763	1.2362	1.01569
0.84	0.97807	1.02853	1.2073	1.01240
0.86	0.98363	1.02826	1.1791	1.00951
0.88	0.98828	1.02690	1.1515	1.00698
0.90	0.99207	1.02451	1.1246	1.00485
0.92	0.99506	1.02120	1.09842	1.00310
0.94	0.99729	1.01702	1.07285	1.00174
0.96	0.99883	1.01205	1.04792	1.00077
0.98	0.99972	1.00636	1.02364	1.00019
1.00	1.00000	1.00000	1.00000	1.00000
1.10	0.99392	0.96031	0.89086	1.00486
1.20	0.97872	0.91185	0.79576	1.01941
1.30	0.95798	0.85917	0.71301	1.04365
1.40	0.93425	0.80540	0.64102	1.07765
1.50	0.90928	0.75250	0.57831	1.1215
2.00	0.79339	0.52893	0.36364	1.5031
2.50	0.71005	0.37870	0.24616	2.2218
3.00	0.65398	0.28028	0.17647	3.4244
3.50	0.61580	0.21419	0.13223	5.3280
4.00	0.58909	0.16831	0.10256	8.2268
4.50	0.56983	0.13540	0.08277	12.502
5.00	0.55555	0.11111	0.06667	18.634

starred quantities are functions of the initial conditions, e.g., $T_t^* = T_t^*\{M_0, T_{t_0}, \gamma\}$, and hence vary in magnitude as M_0 and T_{t_0} are changed. The tabulated values can be used in the following manner. Assume the conditions at the inlet and total temperature ratio across the burner τ_b are given. Then,

$$\tau_b = \frac{T_{t_3}}{T_{t_2}} = \frac{T_{t_3}}{T_{t_3}^*} \frac{T_{t_3}^*}{T_{t_2}^*} \frac{T_{t_2}^*}{T_{t_2}}$$

Since a constant-area burner with constant mass flow is used,

$$T_{t_3}^* = T_{t_2}^* = T_t^*$$

and

$$\frac{T_{t_3}}{T_t^*} = \tau_b \frac{T_{t_2}}{T_t^*}$$

But if M_2 is specified, T_{t_2}/T_t^* can be found from the table and consequently T_{t_3}/T_t^* can be evaluated. Given this result, the other properties at state 3 can be determined. Consider a few examples.

Example 2.1

Letting $M_1 = 0.2$, how much heat can be added to the flow if heat addition is limited by choking the flow? Since the final Mach number is 1, $T_{t_{max}} = T_t^*$, and from Table 2.4,

$$T_t^*/T_t = (1/0.17355) = 5.8$$

Thus, the total temperature of the flow may be increased by a factor of 5.8. The total pressure ratio across the area of heat addition is

$$(p_t^*/p_t) = (1/1.2346) = 0.81$$

Example 2.2

To determine the Mach number change and stagnation pressure ratio across a combustion chamber when $M_2 = 0.2$ and $T_{t_3}/T_{t_2} = 4$, let

$$\frac{T_{t_3}}{T_{t_2}} = \left(\frac{T_{t_3}}{T_t^*}\right)\left(\frac{T_t^*}{T_{t_2}}\right) = 4$$

and

$$\frac{T_{t_3}}{T_t^*} = \frac{T_{t_2}}{T_t^*} \times 4 = 0.17355 \times 4 = 0.695$$

The Mach number corresponding to this value of T_t/T_t^* is 0.50. Consequently, $M_3 = 0.50$, and

$$\frac{p_{t_3}}{p_{t_2}} = \left(\frac{p_{t_3}}{p_t^*}\right)\left(\frac{p_t^*}{p_{t_2}}\right) = \frac{1.114}{1.235} = 0.90$$

Example 2.3

The total temperature of a stream is to be increased by 50% and the total pressure loss compared if the heat is added at Mach 3.0 and 0.3. For the high Mach number case,

$$T_{t_3}/T_{t_2} = 1.5$$

therefore,

$$T_{t_3}/T_t^* = 1.5 \times 0.654 = 0.98 \quad \text{and} \quad M_3 = 1.15$$

Then,

$$\frac{p_{t_3}}{p_{t_2}} = \frac{p_{t_3}/p_t^*}{p_{t_2}/p_t^*} = \frac{1.01}{3.42} = 0.296$$

For the low Mach number case,

$$M_3 = 0.40 \quad \text{and} \quad \frac{p_{t_3}}{p_{t_2}} = \frac{1.16}{1.20} = 0.97$$

Obviously, supersonic heat addition causes a much greater reduction in total pressure.

Thermal choking. If the heat addition for example 2.1 had been greater than $(4.8T_{t_1})$, the tables give no solution. This is a result of the fact that as heat is added to a subsonic flow of constant cross-sectional area, the Mach number approaches unity. If more heat is added, the Mach number at the channel exit will remain at the value unity, but the upstream boundary condition must change so that the ratio $[T_{t_1}/T_t^*\{M_1\}]$ corresponds to the actual head addition. For example, if heat addition in example 2.1 had been $(5.32)T_{t_1}$, then $T_t^* = T_{t_1} + (5.32)T_{t_1}$, or $T_{t_1}/T_t^* = 1/6.32 = 0.158$, and the value of M_1 would drop from 0.2 to 0.19.

Two-Dimensional Heat Addition Calculations

The effect on the flow parameters of the heat addition from a spreading flame in a constant-cross-section duct will be calculated in this section. Although the calculations are idealized, they give a more realistic picture of

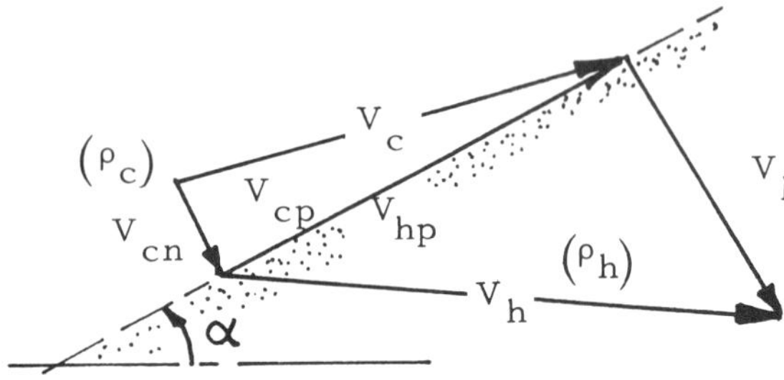

Fig. 2.24 Velocity change across an infinitely thin flame.

heat addition than the one-dimensional calculation described above. After a brief description of the flow across a flame sheet, the heat addition at a thin flame spreading across a constant-area duct will be analyzed, along with the effects of the flameholder and the compressibility. Finally, problems involved with the calculation of the flame geometry are outlined briefly.

Flow across a flame sheet. Before examining the geometric spreading of a flame in a duct, it is interesting to examine the processes occurring at a flame front. Consider Fig. 2.24. Unburned gas with velocity V_c approaches the flame front, making a small angle α with respect to the axis of the combustion chamber. After passing through the flame, the velocity is V_h and the density ρ_h. Subscripts n and p denote components of velocity normal and parallel to the flame. The continuity equation and the two momentum equations can be applied to determine the relationship of the vector components and the static pressure. Consider an incompressible flow and replace the energy equation with the statement that the density ratio $\lambda = \rho_h/\rho_c$ is a given quantity. The three equations are

$$\rho_c V_{cn} = \rho_h V_{hn}$$

$$\rho_c V_{cn}^2 + P_c = \rho_h V_{hn}^2 + P_h$$

$$(\rho_c V_{cn}) V_{cp} = (\rho_h V_{hn}) V_{hp}$$

These equations can be manipulated to give

$$V_{hn}/V_{cn} = \rho_c/\rho_h = 1/\lambda$$

$$V_{cp} = V_{hp}$$

$$(P_c - P_h)/\rho_c V_c^2 = (V_{cn}/V_c)^2[(1/\lambda) - 1] \tag{2.28}$$

The total pressure loss can be written as

$$\frac{P_{ct} - P_{ht}}{\frac{1}{2}\rho_c V_c^2} = \left(\frac{V_p}{V_c}\right)^2 (1 - \lambda) + \left(\frac{V_{cn}}{V_c}\right)^2 \left(\frac{1}{\lambda} - 1\right) \tag{2.29}$$

The major problem arising in computing flame spreading is that the values of the normal velocity of the cold stream cannot be predicted accurately. However, some idea of the order of magnitude of the static pressure difference can be obtained by selecting a value for λ, say 0.25, and using the value of 0.025 for V_{cn}/V_c. (The latter value is that determined from experimental data in the above subsection on spreading rates, where the velocity ratio was called β or u_n/V.) With these assumptions, the pressure jump is less than 2×10^{-3} of the dynamic pressure in the cold gas. If a laminar flame speed were used, V_{cn}/V_c would be even smaller for conditions of interest. Hence, it is justified to ignore the effects of the pressure rise on the flowfield and to assume that the static pressure across the entire duct is constant for the afterburner flowfield. [However, if one is interested in laminar flame shapes, $(V_{cn}/V_c)^2$ may be made as large as necessary and the pressure difference clearly cannot be ignored.]

In addition, when $V_{cn} \ll V_c$, the magnitudes of the velocity vectors V_c and V_h will be approximately equal to the parallel velocity component and hence they will be equal. That is, $V_c \approx V_{cp} \equiv V_{ch} \approx V_h$. The total pressure change [see Eq. (2.29)] reduces to $(1 - \lambda)$ or $(1 - \rho_h/\rho_c)$ and is produced by the change in the density due to heat addition. Hence, the velocity and pressure jumps that occur at the flame front are often negligibly small. Given these results, the velocity profile chosen for the following study and the assumption that pressure depends only on axial position are reasonable.

Idealized flame spread. The effects of adding heat to a one-dimensional compressible flow in a constant-area duct have been discussed above. Here, the effect on the flow of heat addition by a flame spreading from the center of a constant-area duct will be investigated.

To simplify the analysis, first consider a flame developing in a two-dimensional flow of an incompressible gas. Because the entrainment rate of a turbulent flame is not known, the pressure drop, velocity profile, and width between flame fronts must be described as functions of the fraction of the gas that has been burned. Even though the distance required to obtain complete combustion cannot be obtained from this analysis, the total pressure loss, flow acceleration, and a rough idea of the resulting velocity profile can be obtained. The approach used here is taken from that of Tsien[25] whose work was based on profiles suggested by the experiments and calculations of Williams et al.[19]

Tsien proceeded by using an integral approach. He specified the shape of the velocity profile in terms of the width of the burned region W and two velocity parameters. These unknown parameters and a pressure parameter were found as functions of the fraction of the mass flow that is burned at

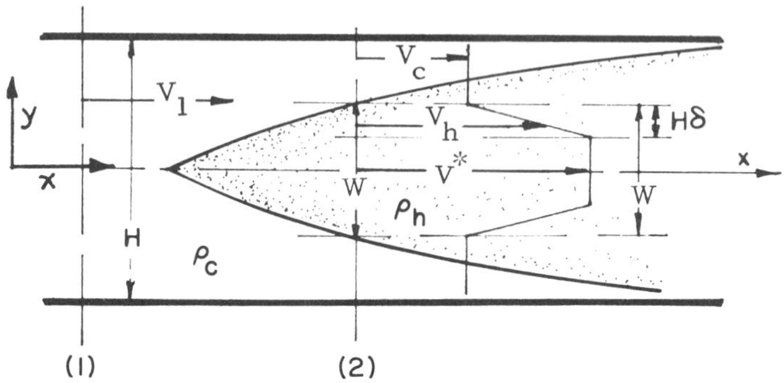

Fig. 2.25 Notation for ideal flame spread calculation.

any station by use of integrated forms of the continuity equation and the Bernoulli relationship applied to the burned and unburned streams. A similar approach is used here, but a momentum equation is applied in place of the Bernoulli equations for the burned gas and the velocity distribution is slightly more complex. Since the burned gas flow is rotational, this appears to be a more satisfactory approximation. The notation and velocity distribution are identified in Fig. 2.25.

A trapezoidal velocity profile is shown in this figure. The axial velocity V_c in the unburned gas is constant at any x station; the velocity in the burned gas rises linearly from V_c to V^* over a distance $H\delta$ at the edge of the flame and is constant with a value V^* over the central region of the burned gas. The static pressure P is assumed to be a function of axial position x alone and the heat addition at the flame is taken into account by specifying that the burned stream density ρ_h is constant and is given by $\lambda\rho_c$. The y component of velocity is ignored.

Continuity and momentum equations are

$$\rho_c V_1 H = 2\left(\int_0^{H/2} (\rho V \mathrm{d}y)\right) = \dot{m}_1 \tag{2.30}$$

$$(P_1 - P)H = 2\left(\int_0^{H/2} \rho V^2 \mathrm{d}y\right) - \dot{m}_1 V_1 \tag{2.31}$$

Here subscript 1 refers to conditions for upstream of the flame. The Bernoulli equation for the unburned stream is

$$\tfrac{1}{2}\rho_c V_1^2 + P_1 = \tfrac{1}{2}\rho_c V^2 + P \tag{2.32}$$

The fraction burned is

$$f \equiv \left(2\int_0^{W/2} \rho V \mathrm{d}y\right) \Big/ \dot{m}_1 \tag{2.33}$$

where $\dot{m}$ is total mass flow rate, $\rho_1 V_1 H$. The trapezoidal velocity profile is

$$V = V_c,\ W/2 \le y \le H/2$$

$$V = \left[V_c + (V^* - V_c)\left(\frac{W}{2} - y\right)\Big/ H\delta\right]$$

$$(W/2 - H\delta) \le y \le (W/2)$$

$$V = V^*,\ 0 \le y \le (W/2 - H\delta) \tag{2.34}$$

The unknowns are f, W, V_c, V^*, and P. Since there are four equations, it is possible to solve for four of them, say V_c, V^*, f, and P, as a function of the fifth, the width W. The parameter δ is treated as a fixed fraction of duct height H and must be specified. It is included here so that the effect of the velocity profile shape on the other parameters can be examined.

The solution of the four algebraic equations resulting from the substitution of the velocity profiles assumed here in Eqs. (2.30–2.33) is straightforward and can be expressed in terms of dimensionless parameters defined as

$$\psi \equiv (P_1 - P)/\tfrac{1}{2}\rho_c V_1^2$$

$$\bar{u} \equiv V_c/V_1$$

$$\bar{u}^* \equiv V^*/V_1$$

$$\eta = W/H$$

$$\lambda = \rho_h/\rho_c$$

$$f = \text{fraction burned} \tag{2.35}$$

The solutions are

$$\psi = (\bar{u}^2 - 1)$$

$$f = [1 - \bar{u}(1 - \eta)]$$

and a pair of equations for $\bar{u}$ and $\bar{u}^*$,

$$\bar{u}(1 + \lambda\delta - \eta) + \bar{u}^*(\eta - \delta)\lambda - 1 = 0$$

$$\bar{u}^2\left(\tfrac{1}{2} + \tfrac{2}{3}\lambda\delta - \eta\right) + \bar{u}^*\left[\eta - \left(\tfrac{4}{3}\right)\delta\right]\lambda$$

$$+ \left(\tfrac{2}{3}\right)\bar{u}^*\bar{u}(\delta\lambda) - \tfrac{1}{2} = 0$$

Numerical solutions are found by specifying values for δ and λ and calculating values of $\bar{u}$, $\bar{u}^*$, and f from the above equations for various

Table 2.8 Dependence of Flame Spread Parameters on Dimensionless Flame Width η for $\lambda = 0.25$ and $\delta = 0.1$

η	f	ψ	$\bar{u}$	$\bar{u}^*$
0.2	0.07	0.36	1.17	1.55
0.3	0.12	0.59	1.26	1.69
0.4	0.18	0.89	1.37	1.88
0.5	0.25	1.3	1.50	2.11
0.6	0.34	1.7	1.65	2.38
0.7	0.45	2.4	1.83	2.70
0.8	0.59	3.2	2.05	3.08
0.9	0.76	4.4	2.31	3.55
1.0	1.0	6.1	2.66	4.15

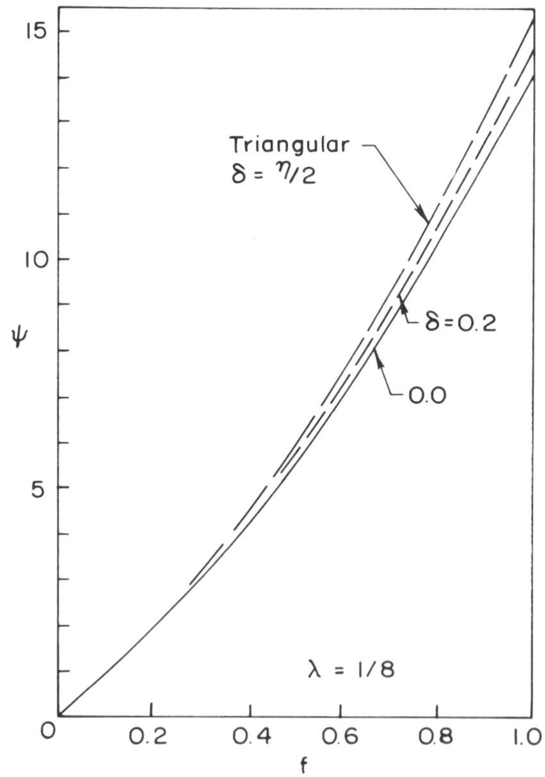

Fig. 2.26 Dependence of normalized pressure ψ on fraction burned and hot steam profile parameter.

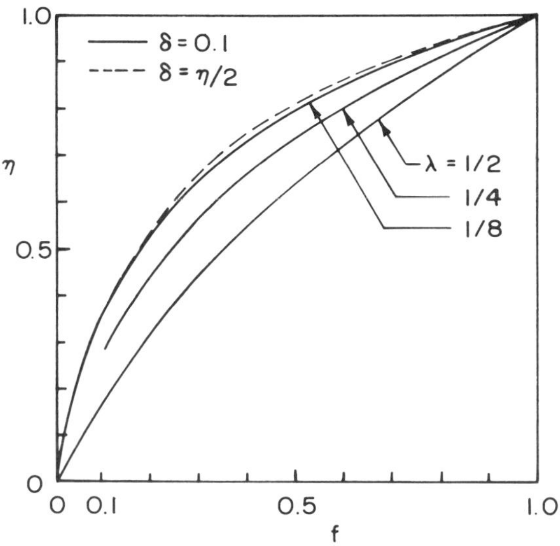

Fig. 2.27 Dependence of normalized width of flame on fraction burned with density ratio as a parameter.

values of η. Table 2.8 presents numerical examples to illustrate the variation of the parameters for the conditions that $\rho_h/\rho_c = 0.25$ and $\delta = 0.1$.

Note that the fraction burned does not reach 50% until the flame width is about 75% of the duct height and that both the cold and hot streams have large velocity changes. The velocity increase for the simple one-dimensional case is just $1/\lambda$ or 4.0 for the incompressible when $\delta = 0$. The total pressure change along the central streamline for this incompressible case is

$$\Delta P_t / \tfrac{1}{2}\rho_c V_1^2 = -\left(1 + \psi - \lambda(\bar{u}^*)^2\right)$$

The dependence of the parameters on f is shown in Figs. 2.26–2.30 with λ and δ as parameters. Consider first the effect of changing the profile parameter δ. Remember that $\delta = 0$ corresponds to a square velocity profile with all the burned fluid moving with speed V^* and that the $\delta = \eta/2$ corresponds to a triangular profile with a linear change from velocity V_c at the flame front to V^* on the axis.

The dependence of the pressure drop on fraction burned is shown in Fig. 2.26 with δ as a parameter. The pressure drop is almost independent of δ for this example and similar results were obtained for values of the density ratio λ between $\frac{1}{8}$ and $\frac{1}{2}$. A similar result is obtained for the cold stream velocity $\bar{u}$ and wake width η; the examples shown in Figs. 2.27 and 2.28 for $\delta = 0.1$ are representative for all δ values. The velocity profile shape factor δ was found to have very little effect on these parameters, so it can be concluded that the predicted relationships between burned gas wake width ηH, the cold stream velocity V_c/V_1, and the fraction burned f will be reasonably accurate regardless of the real velocity profile in the burned

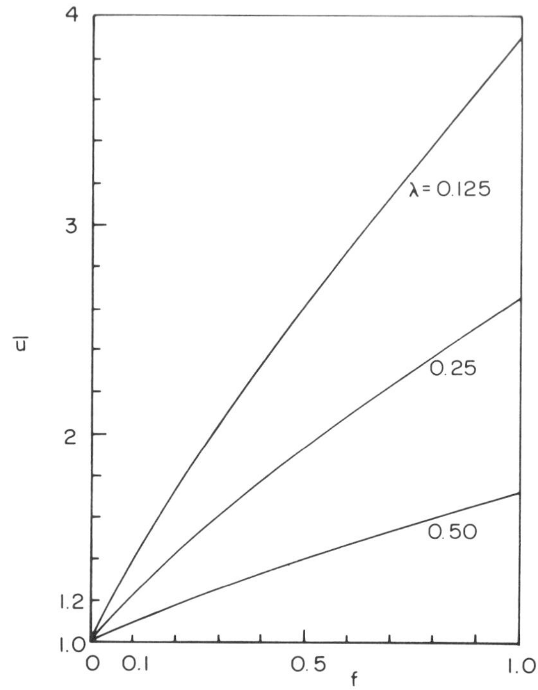

Fig. 2.28 Dependence of cold stream velocity ratio on fraction burned with density ratio as a parameter.

stream. Of course, the choice of profile parameter will have a large effect on the value of the maximum velocity in the hot stream. An example of this dependence is shown in Fig. 2.29 for $\lambda = \frac{1}{8}$ and in Fig. 2.30 for all λs and small δ values.

The strong dependence of ψ and $\bar{u}$ on density ratio is to be expected because the density change fixes the fluid acceleration. The weak dependence of $\eta\{f\}$ on λ is more surprising. An appreciable change in λ (see Fig. 2.27) would produce only a few percent change in the fraction burned. For example, when λ is changed from $\frac{1}{8}$ to $\frac{1}{6}$, values of f corresponding to $\eta = 0.4$ change from 0.12 to 0.15 and values of f corresponding to $\eta = 0.7$ change from 0.37 to 0.39.

Comparison of Figs. 2.28 and 2.29 for $\lambda = 8$ and $\delta = 0.1$ shows the rapid acceleration of the hot gas as compared with the unburned stream. The gas speed on the centerline is always greater than the cold flow.

Effect of flameholder. This type of calculation can be easily extended to include the effects of the flameholder and recirculation zone on the flame spreading calculation. For example, consider the flow shown schematically in Fig. 2.31. The simple triangular velocity profile will be used with the

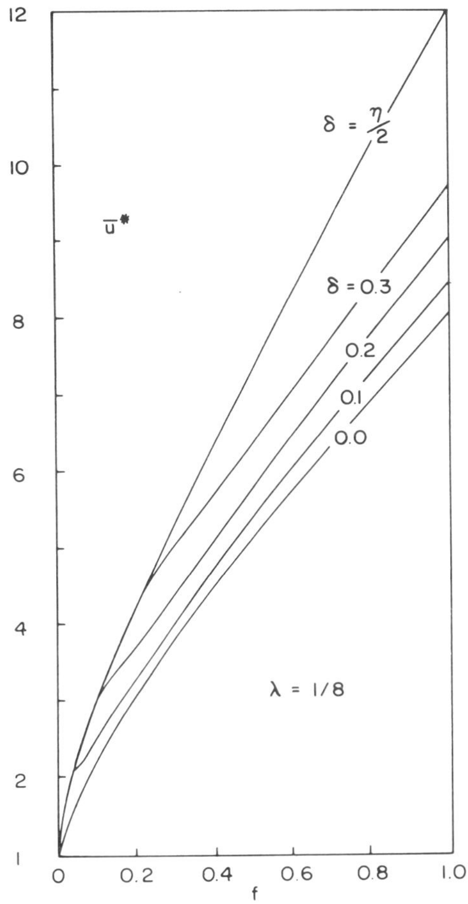

Fig. 2.29 Dependence of hot stream central velocity on fraction burned and velocity profile parameter.

additional assumption that the velocity on the centerline of the flow and at the downstream end of the recirculation zone (station 2) is zero. This approximation is a reasonable one since this point corresponds to the rear stagnation point of the recirculation zone.

Conditions at station 2 are specified when the wake width there, $\eta_2 H$, and density ratio λ are chosen. Combining the continuity and Bernoulli equations for the unburned flow results in

$$\bar{u}_2 = 1/[1 - \eta_2(1 - \lambda/2)]$$

$$f_2 = \left(\tfrac{1}{2}\lambda\right)\eta_2\bar{u}_2$$

$$(P_1 - P_2)/\tfrac{1}{2}\rho_c V_1^2 = (\bar{u}_2)^2 - 1 = \psi_2$$

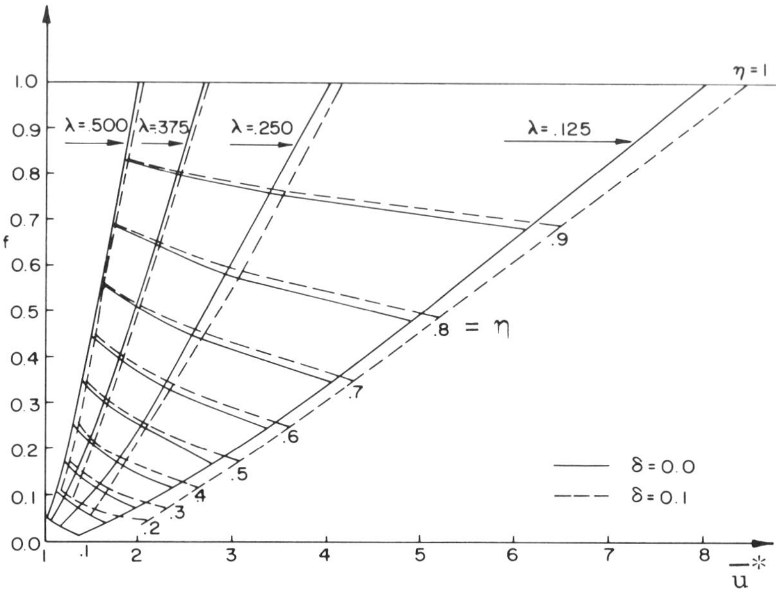

Fig. 2.30 Dependence of hot stream central velocity on fraction burned with density ratio λ and velocity profile factor δ as parameters.

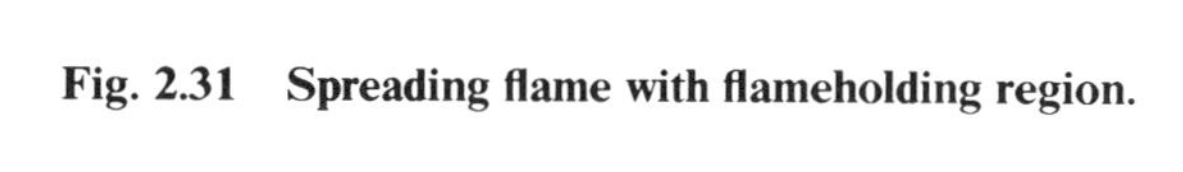

Fig. 2.31 Spreading flame with flameholding region.

Conditions at an arbitrary point downstream of station 2 can be calculated by an analysis identical to that described above. Tabulated values are shown in Table 2.9 for $\lambda = 0.25$ and as a function of η_2, the flame width at station 2. Here f_2, $\bar{u}_2$, and ψ_2 are the fraction burned, cold stream velocity ratio, and normalized pressure difference at station 2, respectively, and $\bar{u}_3$ and $\bar{u}_3^*$ are the cold stream and centerline values of the gas speeds at station 3 where the fraction burned is 1. When $\eta_2 \equiv 0$, this solution reduces to that described above for the triangular velocity profile. Values of $\bar{u}$ and η are presented in Figs. 2.32 and 2.33 as functions of the fraction burned with $\lambda = 0.25$.

Table 2.9 Dependence of Conditions at Station 3, Where $f = 1$ on Recirculation Zone Width η_2 for $\delta = \eta/2$ and $\lambda = \frac{1}{4}$

η_2	f_2	$\bar{u}_2$	ψ_2	$\bar{u}_3$	$\bar{u}_3^*$	ψ_3
0.0	0.0	1.00	0.0	2.69	5.30	6.28
0.2	0.04	1.21	0.46	2.70	5.28	6.32
0.4	0.08	1.54	1.37	2.75	5.24	6.64
0.5	0.11	1.78	2.17	2.82	5.18	6.99

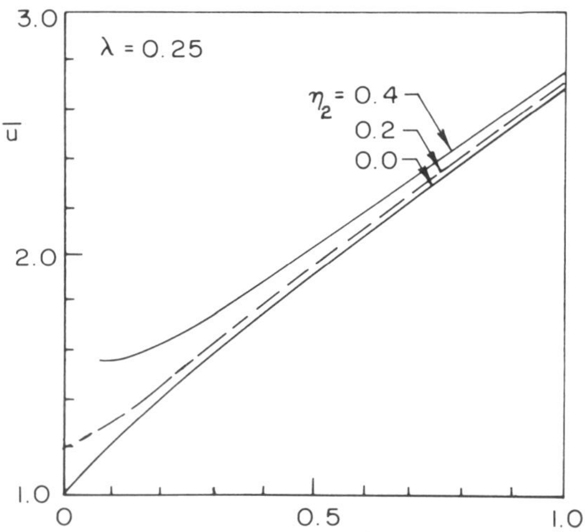

Fig. 2.32 Dependence of $\bar{u} \equiv V_c/V_1$ on fraction burned f and flame width at recirculation zone.

At low values of f, the present solutions differ appreciably from the simple $\eta_2 = 0$ case. However, for values of the fraction burned greater than about $\frac{1}{3}$, the solutions are insensitive to η_2. Hence, the overall pressure drop and acceleration of the burned stream at station 3, where the flow is completely burned, can be estimated with reasonable accuracy from the simple $\eta_2 = 0$ case. Changing η_2 in this example corresponds to changing the wake width and, hence, to increasing the flameholder size in a duct of constant height. This calculation suggests that flameholder size or blockage effects will not be very strong when η is greater than $(1.5\ \eta_2)$. Also note in Table 2.9 that, despite very large differences in the pressure coefficient ψ_2 at station 2, differences in values of ψ_3 are no more than 10%. The dependence in $\eta\{f\}$ on η_2 is even weaker.

This analysis indicates that flameholder blockage effects will not be large.

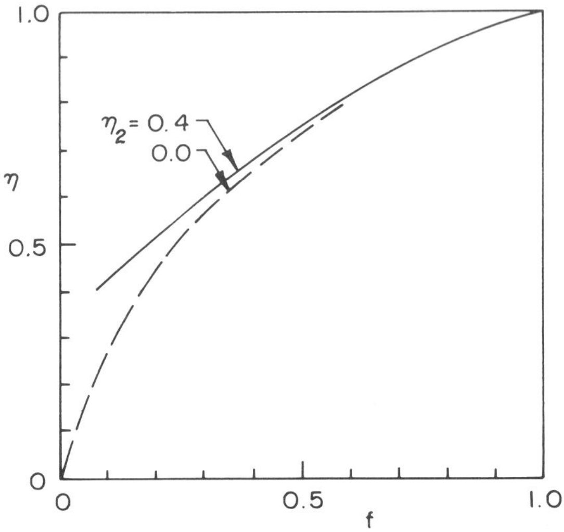

Fig. 2.33 Dependence of flame width η on fraction burned f and flame width at recirculation zone η_2.

Effect of compressibility. Tsien[25] carried out the calculation for the triangular velocity profile discussed above. He also extended the calculation to include compressible flows in which density changes in the unburned flow must be considered. The assumption is made that the density $\rho_h\{x\}$ in the burned region at position x is uniform and equal to $\lambda(\rho_c\{x\})$, where λ is a constant.

Values of the parameters for the spreading flame discussed above were calculated for the triangular velocity profile. The dependence of the width of the burned region η on the fraction burned f is almost identical to that obtained here for the incompressible case as long as no choking occurs.

For large enough values of the approach stream Mach number M_1, Tsien's solutions do not reach $\eta = f = 1$, but reverse themselves. These solutions are inadmissible and it is assumed that the smallest value of M_1 at which this behavior is observed corresponds to the "choking" limit. Values of the critical inlet Mach number for the "choked" flow, M_{1c}, were calculated by this technique and are compared in Table 2.10 with values found in the above subsection for the inlet Mach number required to choke the flow for one-dimensional heat addition.

The density ratio λ in the two-dimensional treatment presented here is the density ratio across the flame (or the temperature ratio since the static pressure change across the flame front is small). Critical Mach numbers are given in Table 2.10 for both the static and total temperature ratios across the one-dimensional heat addition region described earlier. There is little difference between these two nor between the one- and two-dimensional

Table 2.10 Comparison of Critical Mach Numbers Calculated by One- and Two-Dimensional Techniques[a]

$1/\lambda$	$(M_{1c})_{2D}$ ($\lambda \equiv T_1/T_2$)	$(M_{1c})_{1D}$ ($\lambda \equiv T_1/T_2$)	$(M_{1c})_{1D}$ ($\lambda \equiv T_{t_1}/T_{t_2}$)
2	0.36	0.34	0.38
2.67	0.29	0.29	0.32
4	0.22	0.22	0.25
8	0.13	0.15	0.17

[a]Air with $\gamma = 1.40$.

results. Hence, the one-dimensional calculation gives as good an approximation of choking as the more complex two-dimensional one.

A number of other calculations of this type have been made (e.g., Fabri et al.[26]) and they also lead to predicted values of the critical Mach numbers that are close to the one-dimensional values.

Another parameter of interest is the total pressure loss associated with the heat addition. Calculations based on the two-dimensional incompressible model discussed above predict mass-averaged values of the total pressure ratio across the burner that are a few percent higher than the one-dimensional values when the triangular velocity profile was used in the calculation. Again, the simple one-dimensional heat addition calculation appears to give satisfactory estimates.

Burning rate. The ideas developed in several of the preceding paragraphs can be used to illustrate the problems associated with calculating the position of the flame front in a duct. Consider the flow shown in Fig. 2.24 to be a small part of the spreading flame shown in Fig. 2.25. Let the mass flux in the burned flow be $\dot{m}_b$ and total mass flux be $m_1 = \rho_c V_1 H$. Then, the continuity of mass requires that

$$\frac{d(\dot{m}_b)}{dx} = 2\left(\frac{\rho_c V_{cn}}{\cos\alpha}\right)$$

where the $(1/\cos\alpha)$ term enters because the flame front length corresponding to the length dx is $dx/\cos\alpha$ and the factor 2 because there are two flame fronts. The fraction burned f has been defined as $\dot{m}_b/\dot{m}_1$ [see Eq. (2.35)] and thus the above equation becomes

$$\frac{df}{dx} = \left(\frac{V_{cn}}{V_1}\right)\left(\frac{2}{\cos\alpha}\right)\left(\frac{1}{H}\right) \tag{2.36}$$

In order to proceed further in a simple way, some other approximations must be used. One reasonable one is that α is small enough so that $\cos\alpha$ is approximately 1. With this approximation, the effects of flame shape on f are removed from the calculation.

In addition, information concerning the normal velocity component must be supplied. If the flow is laminar (an unlikely event for afterburners), then V_{cn} = laminar flame speed = S is a reasonable supposition. Then

$$\frac{\mathrm{d}f}{\mathrm{d}x} = \frac{2S}{V_1 H}$$

and integration with the assumption that S is constant gives

$$f = (2S/V_1 H)x\{f\}$$

Defining the position at which the combustion is complete, i.e., at $f = 1$, as $x\{1\}$ results in

$$x\{1\}/H = V_1/2S$$

Thus, as the initial velocity V_1 increases or the flame speed S decreases, the length required for complete combustion will increase. Also, the fraction burned at any station x will decrease under these same circumstances.

If the flow is turbulent, other approaches to describe V_{cn} must be developed. As an illustrative example, consider the combustion region at the edge of the flame as a turbulent mixing layer with unburned gas of density ρ_c and velocity V_c on one side and hot combustion products of density ρ_h and velocity V_h on the other. Entrainment rates in shear layers have been the subject of considerable study and the conventional wisdom is that entrainment rates scale as the velocity difference or as $(V_h - V_c)$ in this example. Studies by Brown[27] and Brown and Roshko[18] of shear layers with large density differences (e.g., large differences in ρ_h and ρ_c) have led to an appreciation of the importance of large-scale structures in the mixing region and the influence of the density ratio in fixing the entrainment rates. Brown suggests that the entrainment velocity from the cold stream could be expressed as

$$V_{cn}/V_c \approx 0.18\left[(\bar{u}^*/\bar{u}) - 1\right]\left[1 + (1/\lambda)^{1/2}\right]$$

Direct use of Brown's suggestion leads to entrainment rates that are within a factor of two to three of those presented above in Table 2.10 for mixing layer experiments in which the density and velocity ratios are similar to those in the combustion experiments discussed above.

The axial pressure gradients and the complicated flow at the duct centerline are certainly not anticipated by Brown's model and may be responsible for the differences. Errors in determining the experimental values for entrainment rates may also be responsible.

Finally, part of this difference may also be associated with the definition used to determine the flame front and with the effects of combustion on the mixing process. The similarities of the processes of entrainment with and without combustion appear to be large and indicate that approaching the determination of turbulent flame speeds or entrainment rates from this point of view will be useful.

Despite differences, it is interesting to calculate spreading rates when entrainment rates are given by an ad hoc expression of the form

$$\beta = V_{cn}/V_c = (0.03)[(\bar{u}^*/\bar{u}) - 1] \tag{2.37}$$

where

$$\bar{u}^* = (V_h^*/V_1) \quad \text{and} \quad \bar{u} = (V_c/V_1)$$

The dependence on density is ignored here and the constant 0.03 was picked to give values of β of 0.025 that were observed experimentally for $\lambda = \frac{1}{8}$. If this suggestion is followed, Eq. (2.36) can be usefully rewritten as

$$\frac{df}{dx} = 2\left(\frac{V_{cn}}{V_c}\right)\left(\frac{V_c}{V_1}\right)\left(\frac{1}{H}\right) = 2\left(\frac{V_{cn}}{V_c}\right)\frac{\bar{u}}{H}$$

where $\bar{u} = (V_c/V_1)$ is the function of f and the density parameter λ determined in preceding paragraphs. Integration gives

$$\frac{2x\{f\}}{H} = \int_0^f \frac{df}{(\beta\{\bar{u}^*/\bar{u}\})\bar{u}\{f,\lambda\}} \tag{2.38}$$

This form is convenient since the ratio $\bar{u}^*/\bar{u}$ appearing in the expression for β is a very weak function of f.

The results of carrying out the integration are presented in Table 2.11, where the changes in several parameters are shown as $\eta = W/H$ increases from 0.2 to 1.0 for $\lambda = \frac{1}{8}$ and $\delta = 0.10$. The orders of magnitude for values of x determined here agree roughly with experimental values for $0 \leq \eta \leq 0.4$. Solutions of this type show a strong dependence on the density ratio. This dependence arises because both $\bar{u}^*/\bar{u}$ and $\bar{u}$ decrease as the density ratio

Table 2.11 Variation of Spreading Parameters with Dimensionless Flame Width η for $\lambda = \frac{1}{8}$ and $\delta = 0.1$

η	f	$\bar{u}^*/\bar{u}$	β	x/H
0.2	0.04	1.73	0.022	0.9
0.4	0.12	1.80	0.023	2.2
0.6	0.25	1.93	0.024	2.7
0.8	0.49	2.05	0.027	5.4
1.0	1.00	2.16	0.029	7.8

increases. It can be seen from Eqs. (2.37) and (2.38) that these changes cause $x\{f\}/H$ and $x\{W/H\}/H$ to increase. However, experimental results indicate that, for the range of values of W/H and f examined experimentally, the dependence of the flame geometry on the density ratio is very small.

This discrepancy can be eliminated by replacing the constant 0.03 appearing in the definition for β [Eq. (2.37)] by a function of λ. However, currently there is no rational explanation for this dependence, although the work of Brown[27] indicates that β does depend on λ in a mixing layer.

The preceding calculation is indicative of the types of simple approaches used in calculating the geometric features of turbulent flames. Calculations of this type cannot be taken very seriously until the dependence of V_{cn} or β on the density ratio, turbulence level, flame width, etc., has been established.

Other more ambitious schemes have been developed to calculate the complete flowfield. For example, the calculations of Spalding[28] are based on a general model for computing turbulent flows. A set of differential equations is developed in an ad hoc manner from which turbulent transport parameters can be determined as part of the general solution of the equations of motion. Combustion phenomena are included by assuming infinitely fast reaction rates when the premixed fuel-air gas is mixed with the burned gas at the flame front. The results of one set of calculations made by Spalding agree with the picture reported here of the dependence of flame spreading rates on approach stream parameters. However, the infinite reaction rate assumption and the ad hoc nature of the treatment of the turbulent transport phenomena suggest that great care must be exercised in using the results of this type of calculation outside of the range of the parameters which fixed the constants in the model and which were used in checking the accuracy of the results.

Combustion efficiency in full-scale tests. The previous discussion concerned flame spreading rather than combustion efficiency. A large number of experiments have been carried out on full-scale systems in which the combustion efficiency has been measured as a function of parameters such as the inlet temperature, pressure, velocity, fuel-air ratio, the flameholder geometry, and, in at least one case, the combustion chamber length. Typical results are presented in Refs. 29–31. Because both the injection and vaporization processes and the stabilization and spreading processes are affected by changing the inlet conditions, the process affected by the changes cannot be distinguished.

However, Useller[30] has measured the effects of changing the combustor length while holding the other parameters constant. Some of his results are shown in Fig. 2.34, where combustion efficiency η_{AB} is plotted as a function of combustion chamber length downstream of the plane of the flameholder. The flameholders were 1.5 in. high V-gutters arranged in two concentric rings and with two crossed gutters on diameters serving as supports. Each flameholder occupied a part of duct with an equivalent duct height of very roughly 10 cm. The actual duct diameter was 62 cm and the inlet velocity

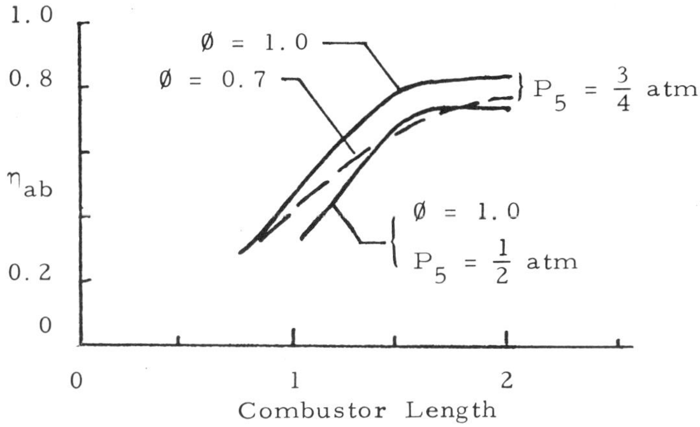

Fig. 2.34 Dependence of combustion efficiency on combustion chamber length (data from Ref. 30).

was 170 m/s. Note that all three curves of Figure 2.34 have a well-defined knee for chamber lengths of about 150 cm and that a further increase in combustor length has little effect on efficiency. Thus, the length-to-height ratios of roughly 15 are required to approach the maximum efficiency, a ratio that is unchanged as the pressure is dropped. This length ratio is slightly larger than one would expect from the smaller-scale flame spreading experiments discussed above. However, the complex flameholder shape makes direct comparison impossible. As noted above, the effects of pressure and fuel-air ratio may not be due to the influence of these parameters on the spreading process.

Summary. Simple one- and two-dimensional calculations are available that allow reasonably accurate predictions for the changes in the static and total pressure and the velocity across a region of heat addition in a constant-area duct. The inlet conditions leading to "choking" can also be predicted with reasonable accuracy. However, predictions cannot be made of the combustion chamber length required to produce a given level of combustion efficiency because the entrainment rate at a turbulent flame front is presently not understood.

A number of full-scale tests are available. However, because all components are subject to the same changes in operating parameters, it is not possible to determine whether or not changes in the spreading process produce the observed performance changes.

2.7 Nozzle and Fuel Control Systems

The afterburner control system has the primary function of controlling the flow of fuel to the afterburner and of keeping the pressure level in the afterburner at the desired value by controlling the nozzle throat area. The control system is also responsible for carrying out the ignition process and supplying the various zones of the afterburner injection system with fuel in the proper sequence and at the proper rate during the starting sequence and, later, during steady operation and termination of thrust. The control must be able to receive data from the core engine control system from which it can determine the airflow rate to the afterburner and then set the required fuel flow rate to each injector zone. Although inlet temperature and velocity do not change a great deal, pressure levels may vary by factors of 10 and thus a wide range of fuel injection rates must be accommodated.

In addition to this control function, the control system must carry out a number of other functions. For example, the fuel lines and injector rings within the afterburner are heated to high temperatures when the afterburner is turned off by convective heat transfer from the high-temperature flow from the engine. If any jet fuel is left in these tubes, the high wall temperatures may cause the fuel to decompose and produce tar-like deposits. These deposits can clog the tubes and the small-diameter injector ports in the tube walls. This problem is usually avoided by venting the lines to a region at lower pressure than the combustion chamber when afterburner thrust is terminated. The resultant flow of gas cleans out the lines and leaves them empty. However, during the afterburner startup process, these lines must be filled again before the ignition process is initiated. The problem here is that in order to reduce the pressure pulse associated with ignition, the initial fuel flow rate is usually set at a value far below that required to fill the injector rings and supply lines in a reasonable length of time, i.e., in several seconds. Thus, the control system must be designed to fill the lines at a high flow rate and then to stop the flow so that no fuel is allowed to spill into the combustion chamber where the high temperatures in the core stream could cause autoignition.

Another important function of the afterburner control system is to maintain conditions at the turbine and fan outlets in an unchanged state so that the operation of these components will not be affected by the afterburner. This control function is particularly difficult to perform during the starting and stopping transients. To understand this problem, examine the mass flux equation for the afterburner nozzle under steady-state conditions and for a simple turbojet cycle. The mass flux at the throat of the nozzle throat can be written in terms of the total pressure and temperature P_{t_7} and T_{t_7}, the nozzle throat area A_n, the gas constant R, and specific heat ratio γ (e.g., see Secs. 2.19 and 5.11 of *Aerothermodynamics of Gas Turbine and Rocket Propulsion*). The result is

$$\dot{m}_7 = \frac{P_{t_7}}{\sqrt{T_{t_7}}} A_n \left[\sqrt{\frac{\gamma}{R}} \left(\frac{2}{\gamma + 1} \right)^{(\gamma+1)/2(\gamma-1)} \right] \tag{2.39}$$

Designating the fuel-air ratio f_{AB}, the mass flow at the inlet (station 5) can be related to that at the nozzle (station 7) by

$$\dot{m}_7 = \dot{m}_5 + \dot{m}_{fuel} = (1 + f_{AB})(\dot{m}_5) \tag{2.40}$$

Rewriting Eq. (2.39) in terms of the conditions at station 5 and using the definitions $\pi_{AB} = P_{t_7}/P_{t_5}$ and $\tau_{AB} = T_{t_7}/T_{t_5}$, Eq. (2.39) can be stated as

$$\dot{m}_7 = \left(\frac{\pi_{AB}}{\sqrt{\tau_{AB}}} A_n\right)\left\{\frac{P_{t_5}}{\sqrt{T_{t_5}}}\left[\sqrt{\frac{\gamma}{R}}\left(\frac{2}{\gamma+1}\right)^{(\gamma+1)/2(\gamma-1)}\right]\right\} \tag{2.41}$$

Let the condition for which the afterburner fuel-air ratio is zero be designated by a subzero suffix. Then, if conditions at station 5 are held fixed by the control system, $\dot{m}_5$ will be a constant and it can be shown from Eqs. (2.40) and (2.41) that

$$\frac{\dot{m}_7}{\dot{m}_{7_0}} = \frac{\dot{m}_7}{\dot{m}_5} = (1 + f_{AB}) = \frac{\left(\pi_{AB}/\sqrt{\tau_{AB}}\right) A_n}{\left(\pi_{AB_0}/\sqrt{\tau_{AB_0}}\right) A_{n_0}}$$

Consequently,

$$\frac{A_n}{A_{n_0}} \approx \left(\frac{\pi_{AB_0}}{\pi_{AB}}\right)\left(\sqrt{\tau_{AB}}\right)(1 + f_{AB}) \tag{2.42}$$

where $\tau_{AB_0} \equiv 1.0$ since the fuel flow and consequently the heat addition are zero when the afterburner is turned off. However, π_{AB_0} is not 1 since flameholder drag and wall losses will always reduce the total pressure. The nozzle throat area required to keep conditions at station 5 unchanged, when f_{AB}, π_{AB}, and τ_{AB} are changed due to afterburning, can be obtained from Eq. (2.42). This equation is only approximate because changes in γ and R due to heat addition, nonuniform velocity, etc., have been ignored.

A similar analysis can be used for a fan engine, but account must be taken of the differing conditions in the fan and core streams. To illustrate the changes required in nozzle throat area, data[29] are shown in Table 2.12 for a TF30-P-3 afterburner with an inlet pressure of about 1 atm. Equation (2.42) is applied to this fan cycle and averaged values of π_{AB} and τ_{AB} are used to evaluate the area changes. For an afterburner fuel-air ratio of 0.005 (typical of values used for ignition), the nozzle area must still be increased by about 13% and for $f = 0.04$ the increase is nearly 80%. The change is almost totally the result of large changes in τ_{AB}. Nozzle area changes of this magnitude must be effected within 5–10 s to keep the duration of the starting transient within reasonable limits. The nozzle actuation rate is usually the slowest and hence the limiting rate in the control process.

This example shows that large changes in the nozzle area must be controlled in order to maintain unchanged flow conditions at the turbine and fan exit stations. High precision is also required. For example, it is clear from Eq. (2.39) that under steady conditions, a 10% error in area A_n must be balanced by a 10% error of opposite sign in P_{t_7} when $\dot{m}_7$ and T_{t_7} are held fixed. A 10% increase in P_{t_7} and P_{t_5} would often be sufficient to stall

Table 2.12 Variation of Nozzle Throat Area A_{nt} and Afterburner Total Pressure and Temperature Ratios with Afterburner Fuel-Air Ratio f_{AB}

f_{AB}	τ_{AB}	$(\pi_{AB}/\pi_{AB})_0$	$(A_{nt}/A_{nt})_0$
0.0	1.0	1.0	1.0
0.01	1.5	0.98	1.26
0.02	1.90	0.96	1.46
0.03	2.30	0.94	1.57
0.04	2.50	0.93	1.77

Data from the TF30-P-3 afterburner (Ref. 29).

the fan stages and later the core compressor of the TF30 engine examined here. Hence, control of the exit area must be precise during both the starting transients and steady-state operation of the afterburner.

The afterburner control system can be organized in a number of ways concerning the primary functions of the fuel flow rate and nozzle throat area. One common technique is the double open-loop system. In this system, movement of the power lever by the pilot produces a demand in the control unit for a fuel flow rate and nozzle exit area. The control system changes the fuel flow rates and nozzle area according to specified schedules that must be very accurately matched and timed to avoid an excursion into burner pressure levels far from the desired values. As the term "open loop" implies, no feedback control is used, thus eliminating the dynamic interaction between the two control functions and any chance for control system instabilities.

A large number of alternate schemes are possible, including systems involving feedback. A number of existing systems are based on hydraulic control systems and digital systems are currently being studied. The use of closed-loop systems have been delayed by the lack of fast-response transducers that can perform well in the high-temperature environment of the afterburner.

In the selection of a control system, the designer must choose a device from a range bounded on one side by the very complicated systems that attempt to get the highest possible performance with the shortest delay times in all engine operating modes, and on the other by very simple systems with reduced average performance and longer delay times. See Ref. 31 for a detailed discussion of control systems and control system hardware.

A detailed design of the afterburner or any variable-area jet engine exhaust nozzle for supersonic flight is also beyond the scope of this chapter. A general discussion of this subject is given in Refs. 32 and 33. However, a few general remarks are made here concerning several important nozzle design problems.

The afterburner is often used during takeoff and at supersonic speeds. In the first regime, a high subsonic or sonic exit velocity may be desired and, in the second, a supersonic exhaust velocity will be necessary. Thus, the exhaust nozzle must be capable of operating both as a convergent and a convergent-divergent nozzle. This can be achieved in a simple design by several techniques. Consider first the system illustrated in Fig. 2.35 and described in detail in the caption. The nozzle consists of two basically coaxial convergent nozzles designed to have variable exit areas. The primary nozzle (e.g., No. 3) is made up of a large number of overlapping leaves hinged at their upstream end and sealed against each other to prevent leaks. Exit area variations by factors of 1.5–2 can be accomplished with this type of nozzle. The secondary nozzle (Nos. 5, 10, or 16) is a variable-area design of the same type. In the example shown here, it has a contoured inner surface. In other designs, it may be the same type as the primary nozzle.

The primary nozzle is responsible for fixing the throat area that controls the primary or afterburner mass flow rate and the secondary nozzle is used to fix the expansion ratio for the nozzle. The primary flow separates from the downstream lip of the primary nozzle and usually reattaches to the secondary nozzle. Reattachment is not necessary in the subsonic case (Fig. 2.35b), but is necessary in the supersonic case (Fig. 2.35a) if the secondary nozzle is to be effective in fixing the exit area.

A supersonic nozzle configuration with maximum primary throat area is shown in Fig. 2.35a and a sonic jet with minimum throat area is shown in Fig. 2.35b. The primary nozzle area changes between examples a and b by a factor of about two and the expansion ratio for example a is about 1.5.

In both Figs. 2.35a and 2.35b, a secondary gas stream (1) flows between the outer and inner cases of the engine. This flow serves several purposes in the nozzle. First, the secondary gas is usually much cooler than the primary and hence this stream can act as a film coolant for the secondary nozzle. Second, the secondary flow occupies a coaxial region about the primary flow and hence can be used to determine and control the exit area of the primary flow. For example, if the secondary flow of Fig. 2.35b were turned off, the effective exit area of the primary stream would be increased. Finally, the entrainment of fluid by the shear layer (e.g., see No. 6) removes fluid from the cavity between the secondary and primary nozzles. If no secondary flow were supplied to compensate for this effect, a strong recirculating flow would be set up to supply this entrainment material. This flow would reduce the nozzle thrust coefficient, increase heat-transfer rates, and contribute to the unsteadiness of the exhaust flow.

The motion required by the secondary nozzle flaps (Nos. 5 and 10) to accommodate the area changes necessary to produce optimum expansion ratios are large when both afterburning and supersonic flight speeds are contemplated. Some of this motion can be avoided by use of the "blow-in" doors shown in Fig. 2.35c as No. 14. When doors of this type are opened, a tertiary flow of ambient gas can enter the nozzle and form a stream coaxial with the secondary and primary flows. Control of the secondary and tertiary streams can again change the throat area of the primary stream and hence

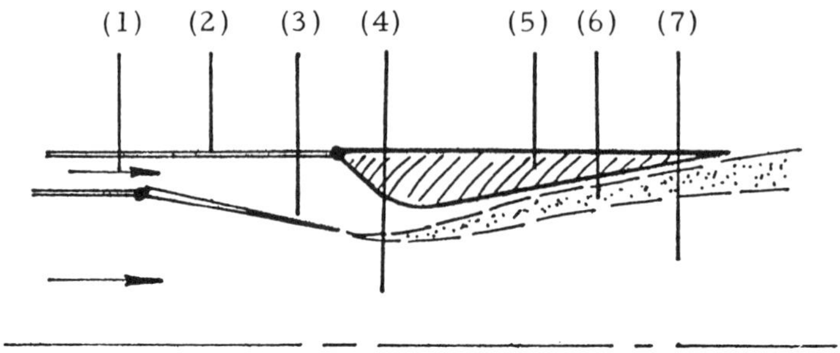

a) Supersonic nozzle configuration with afterburning: (1) secondary flow, (2) outer case of engine, (3) movable primary nozzle shown at maximum area, (4) primary flow, effective throat, (5) movable secondary nozzle shown at maximum exit area, (6) mixing layer between primary and secondary streams, and (7) supersonic primary flow.

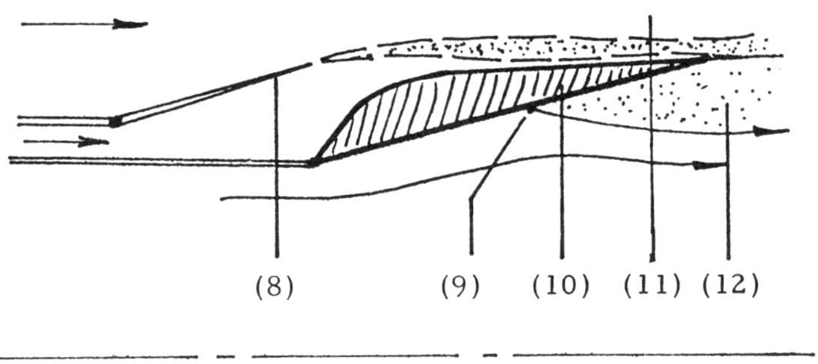

b) Subsonic nozzle configuration with no afterburning: (8) primary nozzle at minimum area, (9) separation point of external flow, (10) secondary nozzle at minimum area, (11) sonic primary stream, and (12) region of separated flow in external flow.

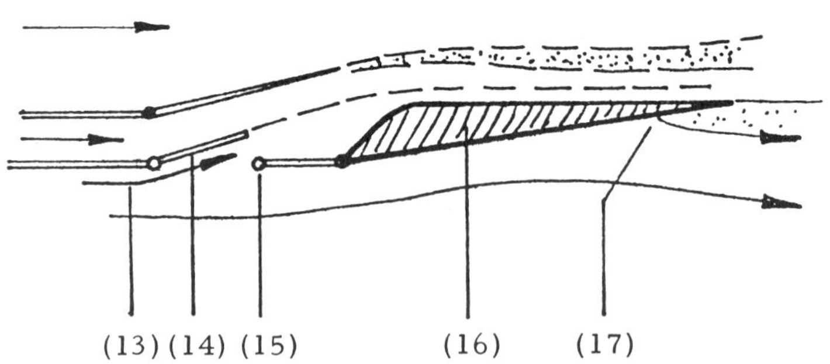

c) Subsonic nozzle configuration, not afterburning, and blow-in door in use: (13) tertiary flow of ambient gas into nozzle, (14) blow-in door and inflow configuration, (15) reversible hinge/latch, (16) movable secondary nozzle, and (17) separation point of external flow.

Fig. 2.35 Ejector nozzle configuration.

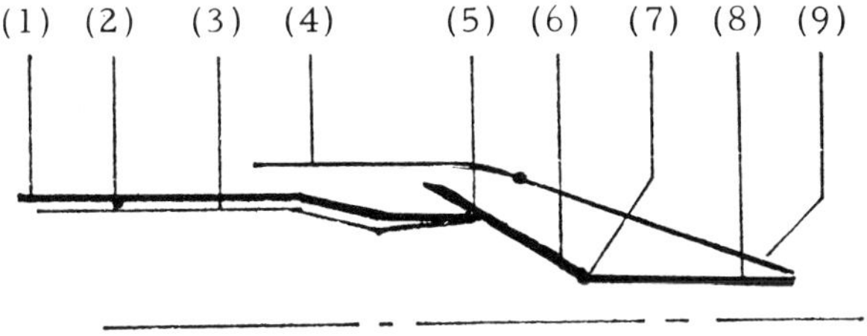

a) Subsonic nozzle configuration: (1) engine case, (2) cooling passage, (3) perforated cooling liner, (4) fixed outer cowl, (5) hinge for nozzle contraction, (6) nozzle contraction member, (7) hinge for nozzle expansion, (8) nozzle expansion member, and (9) movable outer cowl.

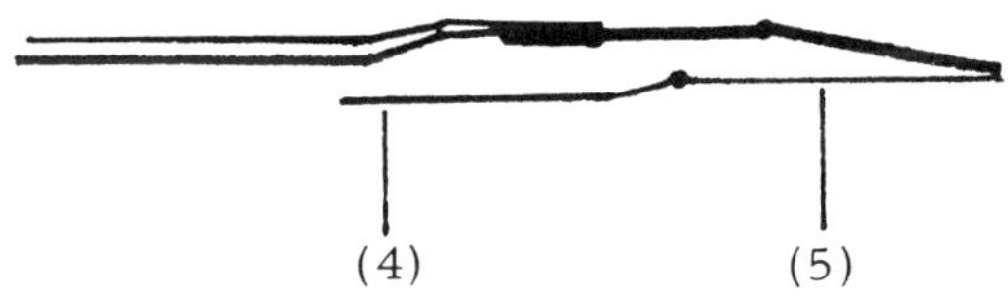

b) Supersonic nozzle configuration.

Fig. 2.36 Nozzle for Pratt & Whitney F100-PW-100 augmented turbofan engine.

reduce the motion required of the secondary nozzle flaps. The "blow-in" doors can also be used as blow-out doors, i.e., can be used as part of the control system to dump some of the secondary flow around the nozzle.

The secondary flow is usually made up of air used to cool the afterburner walls and it may also be drawn from the combustion chamber inlet flow or as far forward as the engine inlet. This nozzle is called an ejector nozzle because the primary stream can be used to pump a low-total-pressure secondary stream, just as the primary stream of an ejector pump is used to actuate a secondary flow.

A second nozzle type, shown in Fig. 2.36, does not use the ejector to minimize the motion of the secondary nozzle. In this case, a single flap (6 and 8) with two hinged joints (5 and 7) forms the nozzle contour. In Fig. 2.36a, a subsonic outlet is formed by maximum inward rotation of both members; in Fig. 2.36b, the outward rotation of both produces a larger throat area and a diverging nozzle at the exit. The nozzle is again constructed of overlapping leaves that must be sealed against each other. The throat area for this example changes by a factor of about two and the exit area ratio for the supersonic nozzle is about 1.5.

Another major area of concern for nozzle designers is the external drag of the nozzle and aft end of the propulsion pod or vehicle. Large changes in the nozzle exit area are required to accommodate the afterburner operation; consequently, it is difficult to make an aerodynamic design for the aft end of the engine that will not result in excessive drag during either afterburning or nonafterburning operation. For example, if the external surface is fixed to accommodate the maximum nozzle exit area (see Fig. 2.35a), a large base drag will occur because of the flow separation on the "boat tail" of the nozzle when a smaller exit area is actually used (see Fig. 2.35b). The flow separation shown in Fig. 2.35b produces a significant increase in drag. Use of "blow-in" doors of Fig. 2.35c reduce the required motion of the secondary nozzle and hence decrease the boat-tail angle and the size and importance of the separated region (compare Figs. 2.35b and 2.35c). A similar problem arises with the nozzle shown in Fig. 2.36. Drag from this and other sources associated with the installation of the engines in vehicles or pods is an important problem for the nozzle designer.

Ejector nozzles are used on a number of engines in current operational use. Examples are the General Electric TF30-P-3 augmented turbofan engine used in the U.S. Air Force F111 airplane and the Olympus engine used in the Concord. In the latter case, the secondary nozzle flaps are also used as part of the thrust reverser system.[34]

The system illustrated in Fig. 2.36 is used on the Pratt & Whitney F100-PW-100 augmented turbofan engine used in the U.S. Air Force F-15 and F-16 fighter airplanes.

2.8 Complete Afterburner Systems

In this section the design details and performance for a typical afterburner system will be described. The afterburner selected for examination, the standard TF30-P-3 afterburner, is shown in Figs. 2.37 and 2.38. (The material discussed here is taken directly from Ref. 29.) This is the same system that was described briefly in the introduction. Conditions in the fan stream at the afterburner inlet were a velocity of about 95 m/s, a temperature of about 390 K, with conditions in the core or gas generator stream of about 220 m/s and 875 K, respectively. The bypass ratio was about 1 for these tests.

The flameholder system consists of three rings of V-gutters and a number of short radial segments of V-gutters. The gutters all have an included angle of 45 deg; the outer gutter ring and the radial gutters attached to it have a height of 5.1 cm and the inner two gutter rings have a height of 3.6 cm. The outer ring is held in place by a rod assembly, shown in Fig. 2.38, and the ring supports 18 radial V-gutters (which have their axis lying along a radius). The inner two rings are supported by six V-gutters, which are attached to the diffuser cone and are visible in Fig. 2.37. The projected area of all the flameholders was about 38% of the duct cross-sectional area, although the holders do not lie in a single plane. The combustion chamber is made up of an outer shell (the pressure vessel) and an inner shell (part of the cooling system). The inner shell is perforated at its upstream end and louvered at its downstream end. The narrow space between the shells acts as a cooling passage; the pressure at the upstream end of this passage is high enough compared to that in the combustion chamber to produce a flow of cool fan air through the passage and out into the combustion chamber through the holes and louvers. Flow in the passage cools by forced convection, and flow through the holes and louvers cools by a film-cooling effect. The holes in the upstream section have also been designed to produce

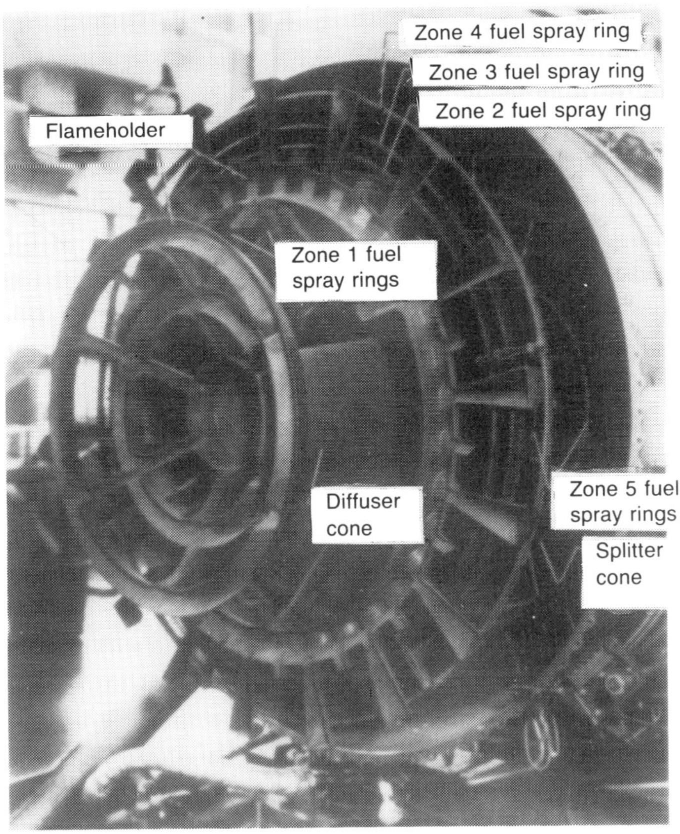

Fig. 2.37 Afterburner flameholders and fuel spray rings (from Ref. 29).

maximum damping of the acoustic disturbances at frequencies for which combustion instabilities have been observed in the chamber.

Fuel is injected into the flow through seven fuel spray rings arranged in five zones (see Figs. 2.37 and 2.38). Modulation of the fuel injection system is accomplished by an integrated afterburner/nozzle control system that simultaneously opens the nozzle and increases the fuel flow. These two operations are controlled by an open-loop hydraulic control mechanism that has no feedback information concerning the afterburner ignition or pressure level. In the present engine, fuel is first introduced to zone 1. Under low-pressure (i.e., high-altitude) operation, fuel is injected through one of the rings and, when the pressure level is above 124 kPa, the second ring is also used. These fuel spray rings lie close to the interface between the fan and the core gas streams and the fuel is primarily injected into the core or gas generator stream. If maximum power is required, the control system will start the fuel flowing to the other four zones at intervals of about 1.5 s.

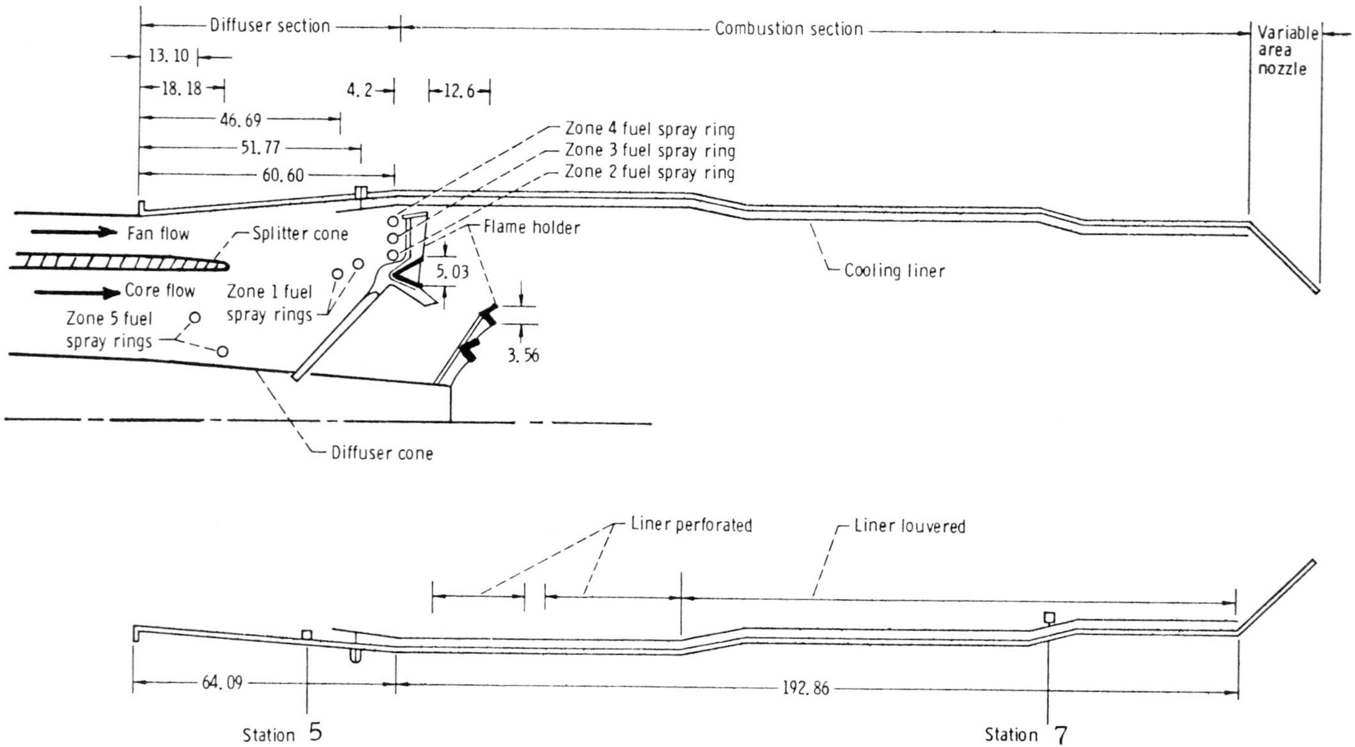

Fig. 2.38 **Afterburner for TF30-P-3 augmented turbofan engine [all dimensions are in centimeters (from Ref. 29)].**

Zones 2–4 are single-ring injectors that feed fuel to the fan stream. They are placed close to the radial flameholders to insure a supply of liquid fuel to the flameholders, which is then evaporated and fed to the recirculation in the manner suggested in Sec. 2.5.

Zone 5 is a double ring that supplies the core engine exhaust stream. These rings are placed about 65 cm upstream of the flameholders to give the fuel time (about 3 m/s) to vaporize before reaching the flame stabilization region.

Each of the three flameholder rings occupies an equivalent duct with a height of about 15 cm and a length of about 175 cm. Thus, the length-to-height ratio is very roughly 12, but this interpretation is complicated by the presence of numerous radial gutters. Thus, in the outer region (which has many radial holders), this ratio is much larger.

The transient processes that occur during a rapid advance of the throttle from a military power setting to a maximum power setting are shown in Fig. 2.39, which is taken directly from Fig. 9 of Ref. 29. The position of the power level controlled by the pilot is shown in Fig. 2.39a as a function of the time. It is moved forward from a military power level at time zero to maximum power at about 1.4 s. The nozzle control system opens the nozzle throat area according to the schedule shown in Fig. 2.39b.

The total fuel flow rate to the afterburner system is shown in Fig. 2.39c. Fuel flow starts at about 0.3 s after the power lever is moved to maximum power and ignition of the zone 1 system occurs at about 1.6 s. The time for ignition is deduced from the first rapid increase in static pressure at the nozzle exit (P_7) shown in Fig. 2.39d. As the fuel injection into each zone is started, the overall injection rate increases (Fig. 2.39c). The particularly large excursion in flow that peaks at about 2.6 s is an artifact of the measuring system and does not represent an increase in fuel delivered to the combustion chamber. It is caused by a change in fuel pumps from the afterburner hydraulic pump, used at very low fuel flow rates, to the main afterburner fuel pump. Note that the maximum fuel flow of the afterburner system is about 6000 kg/h, whereas that of the core engine is only about 1200 kg/h at the condition discussed here. The ignition of each zone is indicated by the pressure cusps appearing in the nozzle inlet pressure (Fig. 2.39d). The schedule for nozzle area changes (Fig. 2.39b) lags these ignition events by about 0.2 s, a mismatch producing the observed pressure disturbances.

For this particular throttle excursion, the only pressure pulse that can cause any trouble at the fan or compressor is the first pulse starting at 1.6 s. The large rise in nozzle inlet pressure (Fig. 2.39d) also appears at the fan exit station (Fig. 2.39e). This excursion increases the fan exit pressure above the steady-state stall line for the fan. In this example, the transient nature of the disturbance kept the fan from stalling. However, the disturbance was observed throughout the complete compressor system as a 5–10% excursion in pressure.

The steady-state characteristics for this afterburner are shown in Fig. 2.40 as a function of the afterburner fuel-air ratio and with the inlet pressure as a parameter. Afterburner total temperature τ_{AB} and pressure π_{AB} ratios are

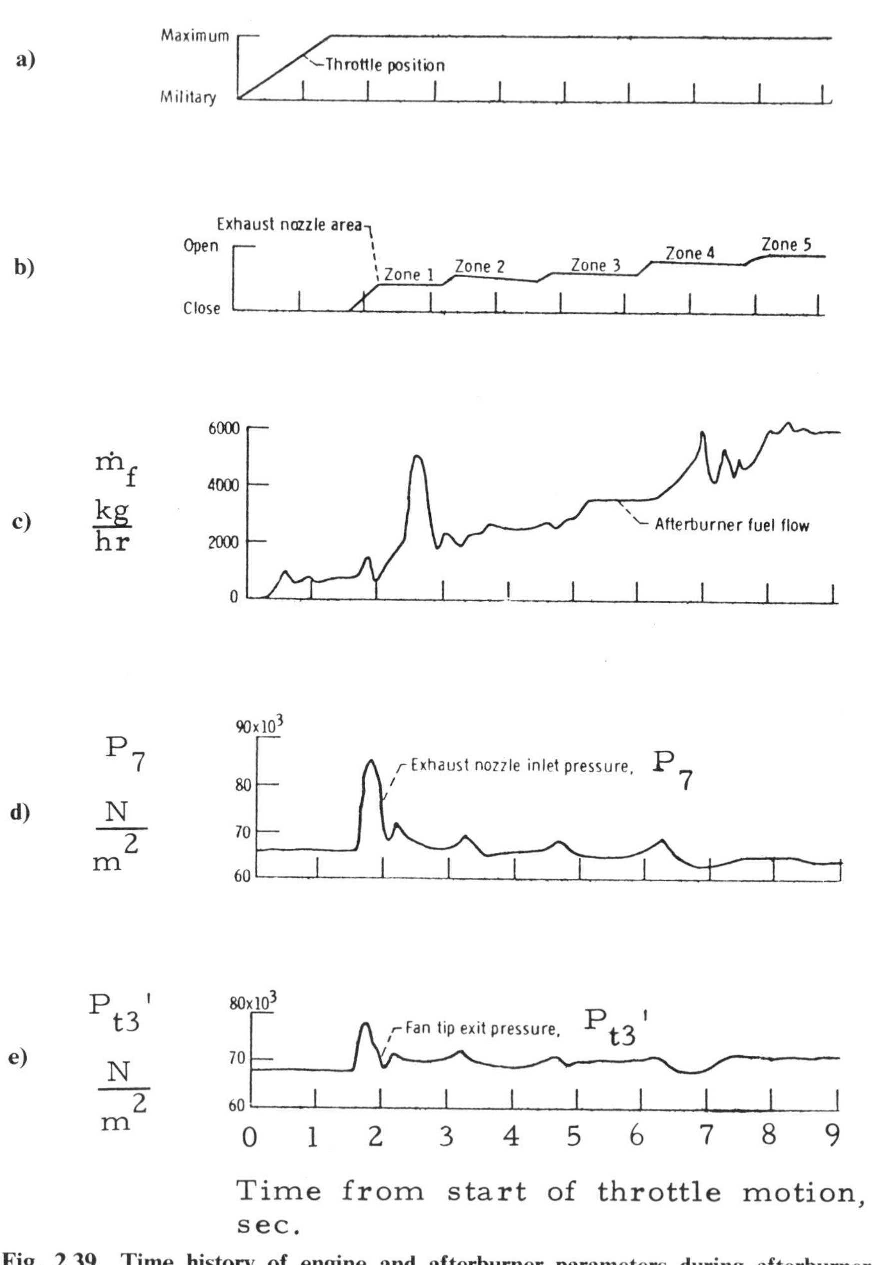

Fig. 2.39 Time history of engine and afterburner parameters during afterburner starting transient (data from Ref. 29).

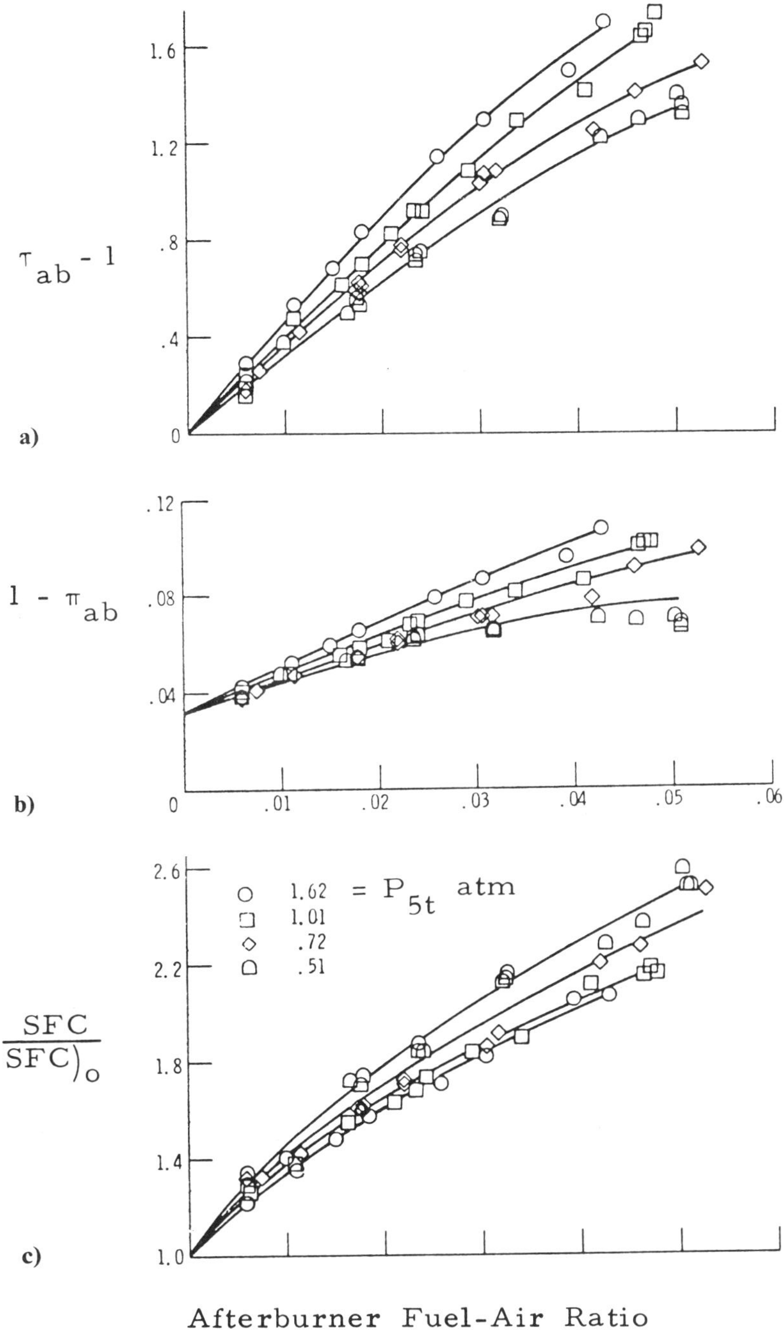

Fig. 2.40 Effect of fuel-air ratio and pressure on steady-state performance parameters (from Ref. 29).

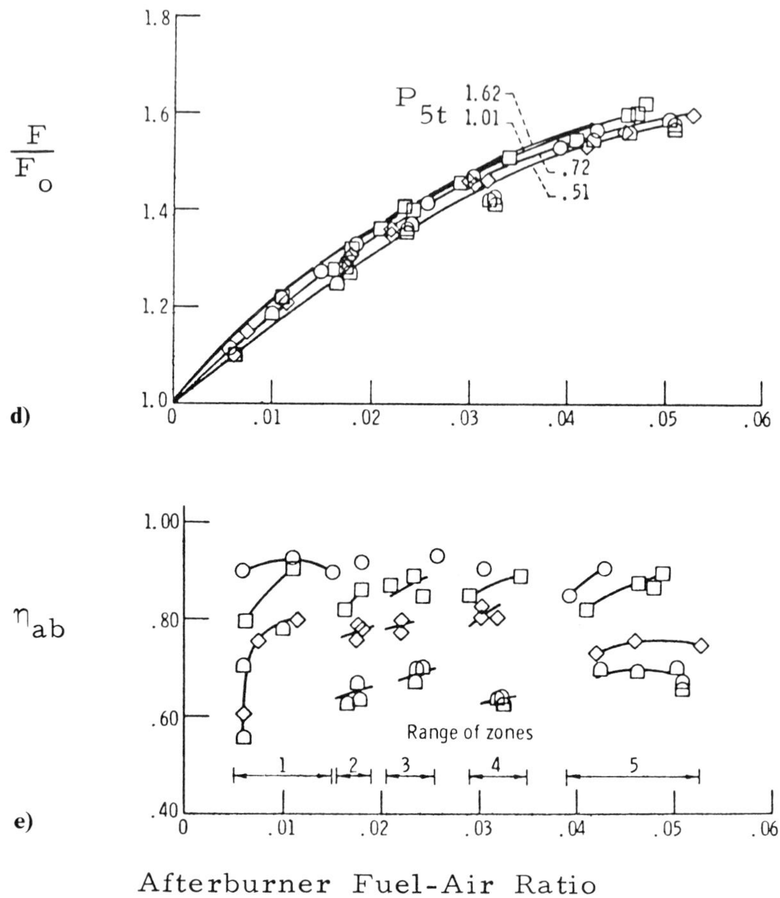

Fig. 2.40 (cont.) Effect of fuel-air ratio and pressure on steady-state performance parameters (from Ref. 29).

shown in Figs. 2.40a and 2.40b, and the specific fuel consumption (SFC) and thrust F are shown as ratios of their values with and without afterburning in Figs. 2.40c and 2.40d. Finally, the combustion efficiency is shown in Fig. 2.40e. The values presented here are based on measurements of the static pressure at stations 8 and 9 of Fig. 2.36, the total temperatures at station 8, and the measured thrust of the engine. Hence, the values are reasonable estimates, but they are not the result of detailed measurements made in the flow.

Figure 2.40 also shows the relationship between the five zones and the afterburner fuel-air ratios that can be achieved when each is activated. Thus, when zones 1–3 are operating, fuel-air ratios of 0.020–0.026 can be used. However, the value $\eta_{AB} = 0.80$ obtained for an inlet total pressure of 0.72

atm and a fuel-air ratio of 0.022 is the *overall* efficiency and is not associated with any particular zone. For inlet pressures greater than 1 atm, the combustion efficiency is uniform and greater than 85%.

Note that all of the steady-state parameters except the thrust drop off rapidly when the inlet pressure falls below about 1 atm.

It is comforting to find that relatively effective augmentation systems can be developed despite the lack of understanding of many very important features in the fuel preparation, stabilization, and flame spreading processes. However, the cost of developing these systems is greatly increased by such ignorance.

2.9 Combustion Instabilities

One of the principal problems associated with the development of high-performance augmentation systems is the suppression of the large-amplitude pressure oscillations often accompanying the combustion process. Disturbances of this type are called combustion instabilities and most appear to be driven by a coupling between an acoustic oscillation at one of the many resonant modes of the system and some step in the heat addition process.

The range of frequencies is very wide. Longitudinal modes with frequencies as low as 50–100 Hz have been identified and frequencies for radial or tangential modes can be as high as 5000 Hz. The frequency spectra of strong oscillations often exhibit 5–10 harmonics with amplitudes within 10 dB of the fundamental frequency. Also, peak-to-peak amplitudes of oscillations as large as 50% of the average pressure are observed and many systems regularly operate with oscillations with an amplitude of 5–10% of the average pressure.

In a typical situation, the amplitude of the oscillation at low fuel-air ratios is small and grows slowly as the fuel-air ratio is increased. Finally, near some critical fuel-air ratio, the amplitude grows very rapidly to a much larger value. The fuel-air ratio at the critical threshold and the rapidity with which the amplitude of the disturbance grows after the threshold has been passed often depend on the combustor inlet velocity, pressure, and temperature.

The oscillations reduce the life of combustion chamber apparatus or force an increase in weight to obtain a given life expectancy. The life of the components is decreased because of the fatigue produced by high-frequency oscillating forces and by the increased heat transfer that often accompanies the oscillations.

In analyzing problems of this type, it is helpful to consider the energy stored in a mode of oscillation of the gas in the afterburner system. This energy can be increased by a number of driving mechanisms and it can be decreased by a number of damping or attenuation mechanisms. In order to tune the driving processes to the natural resonant frequencies for acoustic disturbances, a feedback or coupling mechanism is required. A steady amplitude is reached when the attenuation and driving processes are in balance. In the following discussion, several of these damping, driving, and coupling processes will be described, but no attempt will be made to give an overall description of a stability calculation.

Damping

There are a number of damping terms that arise at the inlet or inlets and outlet to the cavity. For example, when a traveling acoustic wave impinges on a nozzle, reflected and transmitted waves are generated. The energy carried through the nozzle by the transmitted wave is lost from the cavity and represents a loss. Similar losses can occur at the inlet. However, interactions at the inlet to the afterburner occur at interfaces defined by the turbine blades (in the case of the core stream) and by compressor blades (in the case of the fan stream). Interaction between the gas flowing through these components and the acoustic disturbances can produce either a loss or gain for the disturbance.

Additional sources of acoustic damping arise due to viscous interactions involving the boundary layers on the walls and the drag forces on bluff bodies such as flame stabilizers, fuel injection rings, and support struts.

In addition, resonators built into the combustion chamber walls can be used to greatly increase the damping rates. The cooling liner shown in Fig. 2.38 is an example of the application of this technique and it will be described briefly here because this technique is commonly used to control high-frequency disturbances. The liner has been perforated in the region close to the flame stabilizers. When there is no mean flow across these holes, a pressure oscillation in the combustor and close to the liner will produce a pressure difference across the holes that will drive gas from the combustion chamber into the cooling passage when the combustor pressure is high and suck gas back into the combustion chamber when the pressure is low. This process feeds energy from the acoustic mode into the kinetic energy of the gas that flows through the openings and is subsequently dissipated. This process removes energy from the acoustic mode and hence represents a damping of the oscillation.

The rate at which energy is removed from a particular mode of oscillation can be maximized by selecting the proper hole size and volume for the cooling passage associated with each hole, so that the natural frequency for the cavity is close to that of the acoustic mode. For example, Helmholtz showed that a cavity with a narrow neck had a natural frequency that could be analyzed approximately by treating the compressible gas in the cavity as the spring and the mass of gas in the narrow throat as the mass in a spring-mass system. The natural frequency for the cavity is approximately $a\sqrt{A/L_eV_e}$ where a is the speed of sound, A the cross-sectional area of the neck, L_e the effective length of the narrow neck (which in this thin-wall case is roughly the hole diameter), and V_e the volume of the cavity associated with each hole. By proper selection of A, L_e, and V_e the cavity can be designed so that its natural frequency is close to the frequency of the combustion instability and thus maximize the energy loss rate or the damping. Damping can be further increased by allowing fluid to flow from the liner cooling passage into the combustion chamber. Fortunately, this flow direction is in agreement with that required to produce the best cooling effect.

Perforated cooling liners are often used successfully to suppress high-frequency oscillations. However, the effectiveness decreases for frequencies below 1000 Hz and devices of this type are almost impossible to implement for the lowest frequencies encountered in augmented turbofan systems because of constraints on the allowable volume of the cavity.

A number of numerical computer codes have been developed to analyze acoustic fields inside arbitrarily shaped cavities. Resonant frequencies and damping due to wall friction effects can be calculated by these codes. However, the effects of regions with large temperature differences and high Mach numbers have not yet been included in the codes. A discussion of calculations of this type is given in Ref. 35, which deals primarily with combustion instabilities in solid- and liquid-propellant rockets.

The physical principals required to formulate the description of the acoustic field in a cavity, which contains a gas with large temperature and mean flow variations, are well understood and the properties of the acoustic field can be calculated as accurately as required (given sufficient effort and funds). Most of the damping characteristics can also be treated with reasonable accuracy for simple modes and combustor configurations, the physical processes of which are well understood. The same statement cannot be made for the driving and coupling mechanisms discussed in the following paragraphs.

Driving Mechanisms

Consider a one-dimensional compressible gas stream flowing in a constant area duct and let heat be added uniformly at a rate $\dot{Q}$ over some finite length of the duct. The heat addition process will produce an acceleration of the gas and a drop in the pressure according to the relationships derived in Sec. 2.6. However, as long as $\dot{Q}$ is independent of time, no acoustic field will be produced.

When $\dot{Q}$ has a periodic component, say $\dot{Q} = (\dot{Q}_0 + \varepsilon \sin \omega t)$ and $\varepsilon \ll \dot{Q}_0$, a pair of traveling acoustic waves will be generated, one propagating upstream and the other downstream. The strength of the waves will depend on ε, $\dot{Q}_0$, and the Mach number at the heat addition station. If these waves are in phase with an acoustic disturbance already present in the duct, the fluctuating heat addition will drive the disturbance, i.e., will put energy into the acoustic mode.

The fluctuating heat input rate also produces a fluctuating entropy or total temperature in the gas flowing through the heater that is convected downstream with the flow. Marble and Candel[36] have pointed out that acoustic disturbances are created when a flow containing periodic entropy variations of this type passes through a region containing pressure gradients. Thus, if a nozzle is placed at the downstream end of the duct discussed above, acoustic disturbances will be generated at the nozzle by the entropy fluctuations produced farther upstream by the periodic variations in heat input. In the present example, the nozzle will generate a second pair of traveling waves, one moving upstream and the other downstream. The nozzle will also act as a partial reflector of the traveling waves produced directly in the heat addition process and the reflected wave will also move back upstream. The acoustic disturbances travel downstream to the nozzle at a velocity that is the sum of the local speed of sound a and the flow speed u [i.e., $(u + a)$], whereas the entropy fluctuations are carried by the flowing gas and travel at a speed u. Hence, the phase difference in the two traveling waves moving upstream from the nozzle will depend on the distance between the heat addition region and the nozzle and also on the Mach number of the flow. If the phase difference is close to a multiple of 2π, the waves originating at the nozzle will reinforce each other.

In summary, any fluctuation in the heat input rate to a gas flow will directly produce acoustic disturbances and variations in the entropy of the flowing gas. When these entropy variations are convected through a pressure gradient, a second acoustic disturbance is produced. The interaction of the traveling waves produced in this process will depend on the geometry and Mach number of the flow.

Coupling

The most difficult step in the stability calculations or afterburner systems is the description of the process by which acoustic disturbances interact with the combustion process to produce fluctuations in the heat input rate. This coupling process must be understood before any meaningful stability calculations can be made.

A large number of coupling mechanisms are present in the afterburner system. These may be conveniently grouped under the heading of fuel preparation, flame stabilization, and flame spreading mechanisms.

Acoustic disturbances can interact with each step in the fuel preparation process. Thus, the rates of fuel injection, atomization, and vaporization processes can each be affected by local pressure and velocity fluctuations. For example, the rate of fuel flow at the injector orifice is proportional to the square root of the pressure drop across the orifice and hence will be affected by fluctuations in the gas pressure at the injector due to an acoustic disturbance in the duct. The fluctuations in mass flow and pressure at the upstream side of the orifice produced by this interaction must be treated as matching conditions that will specify the nonsteady operation of the whole fuel feed system (a system that will also have a set of resonant frequencies). For simplicity, consider a system with perfect atomization and vaporization. The resulting fluctuation in fuel injection rate at the injector will produce a fluctuation in the fuel-air ratio at the flame front after a time delay, which is fixed by the time required for the fuel to be carried from the injector to the flame front. The variation in fuel-air ratio will produce a fluctuation in the heat release rate at the flame front and when the acoustic disturbance in the duct is in phase with this variation of heat release, it will be driven by this process. If the two disturbances are sufficiently out of phase, the disturbance in the duct will be damped. Analysis of the reaction of feed systems to pressure oscillations of the type described here are presented in detail in Ref. 35.

Several coupling mechanisms associated with the flame stabilization region have been suggested, one of which is described here in some detail to illustrate this class of coupling processes. This discussion is drawn from the work of Marble and Rogers[37] and Barker.[38] Consider the two-dimensional flow shown in Fig. 2.41 and assume that a strong transverse oscillation is present downstream of the flameholder. The smallest resonant frequency for a transverse mode is that corresponding to a plane wave moving vertically and having a pressure node at the axis and a maximum pressure amplitude at the walls (e.g., Fig. 2.41c). The corresponding fluctuation amplitude for the vertical velocity is also shown in this figure. The frequency f of this oscillation will be $f = \bar{a}/2H$, where H is the height of the channel and $\bar{a}$ an average speed of sound in the gas.

In this flow, the strong transverse wave interacts with the shear layers on either side of the flameholder to produce a regular train of vortices that are shed alternately from each lip of the flameholder in the manner suggested in Fig. 2.41a. Vortices are shed in an antisymmetric mode, in keeping with the antisymmetric shape of the velocity and pressure perturbations, and they grow rapidly as they convect downstream.

The coupling mechanism in this example appears to be the triggering of these vortices, which in turn are responsible for the unsteady heat addition that drives the oscillations. Rogers and Marble[37] suggested that the vortices sweep unburned gas deep within the recirculation zone and that this combustible material will burn rapidly after a short time delay. In this situation, the time delay associated with the entrainment of combustible fluid is short and the delay is primarily fixed by the chemical reaction. The entrained material will start to react τ_d s after the vortex is shed and the heat addition rate will receive a sharp pulse at that time. When τ_d is close to an integral multiple of the period of the resonant frequency for the acoustic disturbance, $2H/\bar{a} = 1/f$, the oscillation will be driven. In the present case, driving at the frequency of the first mode was strong enough to produce a large-amplitude oscillation. However, oscillations at higher harmonics were not observed.

In this example, the instability was not present at low values of the fuel-air ratio and started at a well-defined value when the fuel-air ratio was increased. This threshold value ϕ_s was found to vary with mixture temperature and fuel type. In both instances, the authors were able to predict the observed variations in ϕ_s by using the model suggested above. In making these predictions, values of the delay time τ_d were found by using the assumption that τ_d scaled with τ_c, the characteristic time described in the flame stabilization discussion in Sec. 2.5. Some of the experimental values of τ_c used in this process are shown in Fig. 2.15. The sharp boundary for ϕ_s is a result of the very rapid change of τ_d with the fuel-air ratio, which is inferred from the variation of τ_c shown by the data of this figure. In their

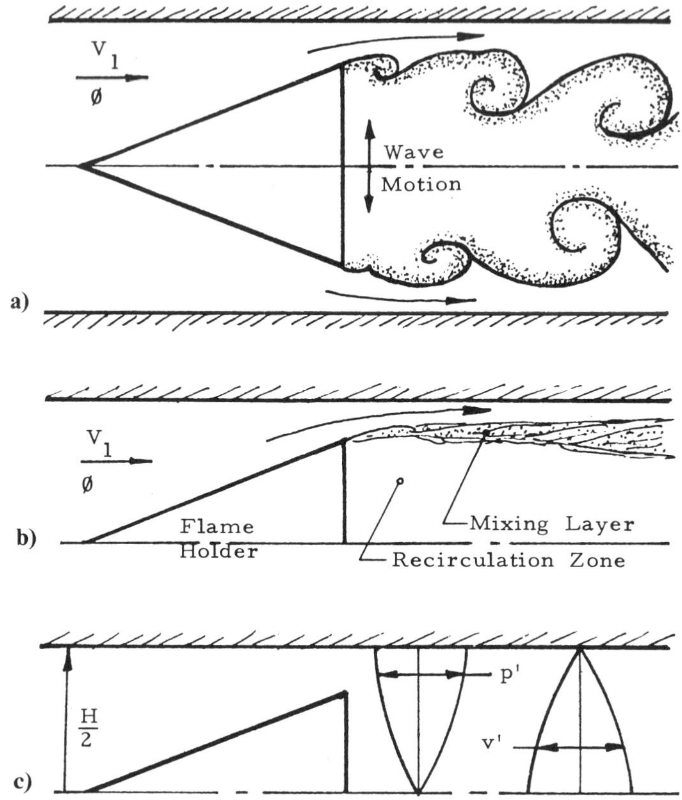

Fig. 2.41 Flame front geometry during unstable (a) and stable (b) combustion; mode shapes for perturbations in pressure p' and transverse velocity v'.

example, the resonant frequency was about 4000 Hz, which suggests that τ_d should be close to 0.25 ms (if the heat release rate rises after the first cycle of the oscillation). Thus, values of τ_d and of τ_c are almost equal (see Fig. 2.15).

The successful application of perforated liners for the suppression of high-frequency combustion instabilities has very often involved the use of liner perforations in the immediate neighborhood of the flame stabilization region. This suggests that the flame stabilization region is very important in coupling high-frequency disturbances to the heat addition process and that one important mechanism is the one described above.

Because of the large-scale inhomogeneities present in the flame spreading region, one would expect that coupling between the heat release and acoustic disturbances would be possible in this region too. However, no experimental evidence for such processes has yet been developed.

Finally, any nonsteady heat release occurring in the combustor will react with the nozzle to produce longitudinal disturbances. Note that the amplitude of these disturbances can be very large. For example, for wavelengths that are long compared to the nozzle inlet length and an inlet Mach number of 0.8, a total temperature fluctuation with an amplitude of 10% will produce an upstream propagating wave with an amplitude of about 5% of the mean pressure.

In summary, some of the aspects of the combustion instability problem and a few of the many possible driving, damping, and coupling mechanisms have been described. Given the many resonant acoustic modes, the many possibilities for coupling, the high rate of heat release per unit volume, and the small amount of energy required to drive a large-amplitude oscillation, it is surprising that any afterburner system operates without acoustic disturbances.

Supplementary Reading

The purpose of this section is to call attention to a few reference works that will give the reader a different perspective on some of the material covered in this chapter and, in several cases, provide a source of more detailed information.

A number of general reviews of jet propulsion systems appeared in the late 1950s and early 1960s. The material presented in these volumes is dated, but in many cases is still of interest today and still covers many topics as adequately as they can be covered now. Two volumes from the *High Speed Aerodynamics and Jet Propulsion* series are of interest. Volume XII, edited by Lancaster contains information on augmentation systems and also ramjet systems which are pertinent. Volume XI, edited by Hawthorne and Olson, contains a series of articles covering all the engine components and includes a discussion of flame stabilization in heterogeneous flows. A brief introduction to the Soviet work on combustion problems in gas turbines is given in the translation of the book by Zuyev and Skubachevskii. Material relevant to afterburners is discussed in the chapters on ramjet engines and reheat combustion chambers.

A good report on current diffuser research work is that of Dolan and Runstadler. They present a large body of data on conical diffusers covering the effects of inlet Mach number for the whole subsonic range, a wide range of inlet boundary-layer thicknesses, and diffuser geometry including area ratio, length, and divergence angle.

The specific problems concerning ignition in afterburner systems are not treated in detail in any combustion texts. A detailed treatment of a number of experiments concerned with ignition are presented by Lewis and von Elbe, who also discuss other material concerning flame stabilization and flame spreading.

The generation of fuel sprays and their vaporization and combustion is still a topic of vigorous research. Analytic treatments are available in the text by Williams and the *Fourteenth and Fifteenth Symposia (International) on Combustion* contain interesting collections of papers given at sessions on heterogeneous combustion. Beer and Chigier also discuss relevant material, although their chief interest is in industrial furnaces rather than aircraft engines.

Alternate views of the flame stabilization process are presented by Lewis and von Elbe and by Beer and Chigier.

Finally, five journals are particularly useful sources of information: the *AIAA Journal* of the American Institute of Aeronautics and Astronautics; *Combustion and Flame* (published by Elsevier, New York); and *Combustion Science and Technology* (published by Gordon and Breach, New York); and translations of two Soviet journals, *Combustion, Explosion and Shock Waves* (translated and published by Consultants Bureau, New York) and *Soviet Aeronautics* (translated and published by Elsevier, New York).

Beer, J. M. and Chigier, N. A., *Combustion Aerodynamics*, Halsted Press Div., John Wiley & Sons, New York, 1972.

Dolan, F. X. and Runstadler, P. W. Jr., "Pressure Recovery Performance of Conical Diffusers at High Subsonic Mach Numbers," NASA CR-2299, Aug. 1973.

Fifteenth Symposium (International) on Combustion, The Combustion Institute, Pittsburgh, Pa., 1974.

Fourteenth Symposium (International) on Combustion, The Combustion Institute, Pittsburgh, Pa., 1973.

Hawthorne, W. R. and Olson, W. T. (eds.), "Design and Performance of Gas Turbine Power Plants," *High Speed Aerodynamics and Jet Propulsion*, Vol. XI, Princeton University Press, Princeton, N. J., 1960.

Lancaster, O. E. (ed.), "Jet Propulsion Engines," *High Speed Aerodynamics and Jet Propulsion*, Vol. XII, Princeton University Press, Princeton, N. J., 1959.

Lewis, B. and von Elbe, G., *Combustion Flames and Explosions of Gases*, 2nd ed., Academic Press, New York, 1961.

Williams, F. A., *Combustion Theory*, Addison-Wesley Publishing Co. Reading, Mass., 1965.

Zuyev, V. S. and Skubachevskii, L. S., *Combustion Chambers for Jet Propulsion Engines* (translation), The MacMillan Co., New York, 1964.

References

[1]Johnson, R. L. and Cullom, R. R., "Altitude Test of Several Afterburner Configurations on a Turbofan Engine With a Hydrogen Heater to Simulate an Elevated Turbine Discharge Temperature," NASA TP 1068, 1977.

[2]Schetz, J. and Padhye, A., "Penetration and Breakup of Liquids in Subsonic Airstreams," *AIAA Journal*, Vol. 15, Oct. 1977, pp. 1385–1390.

[3]Ingebo, R. D. and Foster, H. H., "National Advisory Committee for Aeronautics," NACA TN 4087, Oct. 1957.

[4]Haddock, G. H., "Flame-Blowoff Studies of Cylindrical Flame Holders in Channeled Flow," Jet Propulsion Laboratory, Pasadena, Calif., Progress Rept. 3-24, May 14, 1951.

[5]Zukoski, E. E. and Marble, F. E., "The Role of Wake Transition in the Process of Flame Stabilization on Bluff Bodies," *Combustion Researches and Reviews 1955*, Butterworths Scientific Publications, London, 1955.

[6]Zukoski, E. E. and Marble, F. E., "Experiments Concerning the Mechanism of Flame Blowoff from Bluff Bodies," *Proceedings of the Gas Dynamics Symposium on Aerothermochemistry*, Northwestern University Press, Evanston, Ill., 1955.

[7]Wright, F. H., "Bluff Body Flame Stabilization: Blockage Effects," *Combustion and Flame*, Vol. 3, 1959, p. 319.

[8]Broman, G. E. and Zukoski, E. E., "Experimental Investigation of Flame Stabilization in a Deflected Jet," *Eighth Symposium (International) on Combustion*, The Williams and Wilkins Co., Baltimore, 1960, p. 944.

[9]Cornell, W. G., "The Flow in a Vee-Gutter Cascade," *Transactions of ASME*, Vol. 78, 1956, p. 573.

[10]Wright, F. H., "Multiple Flame Holder Arrays: Flame Interactions," *ARS Journal*, Feb. 1959.

[11]Potter, A. E. Jr. and Wong, E. L., "Effect of Pressure and Duct Geometry on Bluff Body Flame Stabilization," NASA TN 4381, Sept. 1958.

[12]Mullins, B. P., "A Spontaneous Ignition Theory of Combustion Intensity and Combustion Stability Behind a Baffle," *Combustion Researches and Reviews 1955*, Butterworths Scientific Publications, London, 1955.

[13]Mironenko, V. A., "The Use of Thermal Ignition Theory to Determine General Conditions for Flame Stabilization by Means of a Bluff Body," *Izvestiya VUZ, Aviatsionnaya Tekhnika*, No. 1, 1966, pp. 140–145 (translation).

[14]Solokhin, E. L. and Mironenko, V. A., "Stable Combustion Limits in GTE Reheat Combustion Chambers," *Izvestiya VUZ, Aviatsionnaya Tekhnika*, Vol. 15, No. 1, 1972, pp. 135–141 (translation).

[15]Hottel, H. C., Williams, G. C., Jensen, W. P., Tobey, A. C., and Burrage, P. M. R., "Modeling Studies of Baffle-Type Combustors," *Ninth Symposium (International) on Combustion*, Academic Press, New York, p. 923.

[16]Kosterin, V. A., Dudin, L. A., Motylinskii, I. P., Khismatullin, A. Ya., and Gilyazov, M. Sh., "Correlation of Experimental Data on the Limits of Flame Stabilization," *Izvestiya VUZ, Aviatsionnaya Tekhnika*, Vol. 11, No. 3, 1968, pp. 59–66 (translation).

[17]Marshall, R. L., Canuel, G. E., and Sullivan, D. J., "Augmentation Systems for Turbofan Engines," *Cranfield International Symposium Series*, Vol. 10, Pergamon Press, Ltd., London, 1968, p. 129.

[18]Brown, G. L. and Roshko, A., "On Density Effects and Large Structure in Turbulent Mixing Layers," *Journal of Fluid Mechanics*, Vol. 64, Pt. 4, 1974, pp. 775–816.

[19]Williams, G. C., Hottel, H. C., and Scurlock, A. C., "Flame Stabilization and Propagation in High Velocity Gas Streams," *Third Symposium on Combustion and Flame and Explosion Phenomena*, The Williams and Wilkins Co., Baltimore, 1949, p. 21.

[20]Wright, F. H. and Zukoski, E. E., "Flame Spreading from Bluff Body Flame Holders," *Eighth Symposium (International) on Combustion*, The Williams and Wilkins Co., Baltimore, 1960, p. 933.

[21]Solntsev, V. P., "Stabilization of a Flame and Development of the Process of Combustion in a Turbulent Stream," edited by G. M. Gorbunov, Oborongiz, Moscow, 1961, p. 75.

[22]Lefebvre, A. H. and Reid, R., "The Influence of Turbulence on the Structure

and Propagation of Enclosed Flames," *Combustion and Flame*, Vol. 10, 1966, p. 355.

[23] Wohl, K., Shore, L., Von Rosenberg, H., and Weil, C. W., "The Burning Velocity of Turbulent Flames," *Fourth Symposium (International) on Combustion*, The Williams and Wilkins Co., Baltimore, 1953, p. 620.

[24] Thurston, D. W., "An Experimental Investigation of Flame Spreading from Bluff Body Flameholders," Thesis, California Institute of Technology, Pasadena, 1958.

[25] Tsien, H. S., "Influence of Flame Front on the Flow Field," *Journal of Applied Mechanics*, June 1955, p. 188.

[26] Fabri, J., Siestrunck, R., and Foure, C., "On the Aerodynamic Field of Stabilized Flames," *Fourth Symposium (International) on Combustion*, The Williams and Wilkins Co., Baltimore, 1953, p. 443.

[27] Brown, G. L., "The Entrainment and Large Structure in Turbulent Mixing Layers," *Fifth Australasian Conference on Hydraulics and Fluid Mechanics*, University of Canterbury, Christchurch, New Zealand, 1974, p. 352.

[28] Spalding, D. B., "The Spread of Turbulent Flames Confined in Ducts," *Eleventh Symposium (International) on Combustion*, The Combustion Institute, Pittsburgh, Pa., 1967, p. 807.

[29] McAulay, J. E. and Abdelwahab, M., "Experimental Evaluation of a TF30-P-3 Turbofan Engine in an Altitude Facility: Afterburner Performance and Engine-Afterburner Operating Limits," NASA TN D-6839, July 1972.

[30] Useller, J. W., "Effect of Combustor Length on Afterburner Combustion," *Combustion and Flame*, Vol. 3, 1959, p. 339.

[31] Robinson, K., "Afterburning Regulation Concepts," *Power Plant Controls for Aero-Gas Turbine Engines*, AGARD CP 151, Feb. 1975, p. 1074.

[32] Carrierre, P., "Exhaust Nozzles," *Supersonic Turbo-Jet Propulsion Systems and Components*, AGARDograph 120, 1969, pp. 287–379.

[33] Beheim, M. A. et al., "VIII: Supersonic Exhaust Nozzles," *Aircraft Propulsion*, NASA SP-259, Nov. 1970, p. 233.

[34] Devriese, J. and Young, P. H., "Olympus in Concorde," *Aeronautical Journal*, Vol. 76, No. 774, Dec. 1972, pp. 683–694.

[35] Harrje, D. T. (ed.), *Liquid Propellant Rocket Combustion Instability*, NASA SP-194, 1972.

[36] Marble, F. E. and Candel, S. M., "Acoustic Disturbance from Gas Non-Uniformities Convected Through a Nozzle," *Journal of Sound and Vibration*, Vol. 55, No. 2, 1977, pp. 225–243.

[37] Rogers, D. E. and Marble, F. E., "A Mechanism for High-Frequency Oscillation in Ramjet Combustors and Afterburners," *Jet Propulsion*, Vol. 26, June 1956, pp. 456–462.

[38] Barker, C. L. R., "Experiments Concerning the Occurrence and Mechanism of High Frequency Combustion Instability," Thesis, California Institute of Technology, Pasadena, 1958.

CHAPTER 3. AXIAL FLOW COMPRESSOR AERODYNAMICS

George K. Serovy
Iowa State University, Ames, Iowa

3. AXIAL FLOW COMPRESSOR AERODYNAMICS

Nomenclature

A	= area
AVDR	= axial velocity-density ratio, $\rho_2 V_{x,2}/\rho_1 V_{x,1}$ for blade sections in cascade or blade row
a	= distance measured along chord line from leading end of section camber line to maximum camber point, Fig. 3.8
a	= acoustic velocity
b	= coordinate of maximum camber measured along perpendicular to chord line, Fig. 3.8
c	= chord length, length of straight line connecting leading and trailing end points of camber line, Fig. 3.8
D	= diffusion parameter, Eqs. (3.12–3.15)
D	= diameter
d	= distance of maximum thickness location from leading end of section camber line, Fig. 3.8
E	= energy transfer
F	= body force
g	= local gravitational acceleration
H	= total enthalpy
h	= static enthalpy
I	= rothalpy, Eq. (3.5)
i	= incidence angle, angle between inlet flow direction and line tangent to blade section camber line at leading edge
L	= length
LER	= leading-edge radius of blade section
M	= Mach number
m	= mass flow rate
N	= rotational speed of shaft
P	= total pressure
p	= static pressure
q	= heat transferred
R	= radius
R^*	= passage radial location, $(r - r_h)/(r_t - r_h)$
r	= radial coordinate, measured from compressor axis
r_c	= radius of curvature of stream surface trace in meridional plane, Fig. 3.24
s	= entropy
s	= blade spacing, Fig. 3.8

T = temperature
T_B, T_S = moments of forces, Eq. (3.10)
Tu = turbulence parameter
TER = trailing-edge radius of blade section
t = time
t = maximum thickness of blade section, Fig. 3.8
U = blade section velocity
V = velocity in fixed coordinate system
W = velocity relative to coordinate system attached to rotor
x = axial coordinate
β = angle between velocity vector and axial direction
γ = stagger angle, angle between chord line and axial direction, Fig. 3.8
γ = fluid specific heat ratio
δ = deviation angle, angle between cascade exit fluid velocity and line tangent to section camber line at trailing edge, Fig. 3.8
δ = ratio of total pressure at compressor inlet to standard sea-level pressure
ε = angle in meridional plane between line tangent to stream surface and axial direction
η = compressor or stage efficiency
θ = circumferential coordinate
θ = ratio of total temperature at compressor inlet to standard sea-level temperature
θ^* = total boundary-layer momentum thickness at section trailing edge
κ = angle between line tangent to section camber line and axial direction, Fig. 3.8
μ = fluid dynamic viscosity
ρ = fluid density
σ = cascade or blade row solidity, chord length/blade spacing
ϕ = flow coefficient, $(V_x/U)_{\text{mean}}$
φ = blade section camber angle, $\kappa_1 - \kappa_2$, Fig. 3.8
ψ = pressure coefficient, $\Delta H_{\text{isentropic}}/U^2_{\text{mean}}$
$\bar{\omega}$ = average total pressure loss coefficient, $\bar{P}_1 - \bar{P}_2/\bar{P}_1 - p_1$, Fig. 3.8

Subscripts

A, B, C, D = location of computing station
a, b = measurement location
act = actual
ad = adiabatic
an = annulus
CS = control surface
ex = exit
F = frontal cross-section
h = hub

i	= inlet
in	= inlet
is	= isentropic
m	= meridional component
R	= rotor
r	= radial component
rel	= relative
t	= tip
u	= blade velocity
x	= axial component
θ	= circumferential or tangential component
1	= inlet to blade row, Fig. 3.8
2	= exit of blade row, Fig. 3.8
1–20	= measurement or computing station location

Superscripts

$\rightharpoonup$	= vector quantity (on illustrations)
$\overline{(\ )}$	= average

3.1 Introduction

This chapter begins with qualitative descriptions of axial flow compressor geometry and of some of the features of the flow in this form of turbomachine. The purpose of these sections is to develop a working understanding of the specialized terminology associated with compressor aerodynamics and aeromechanics. Subsequent sections discuss the elements of axial flow compressor design, including design objectives, the requirements for a design system, and the characteristics of typical current and developing design methods. Finally, a brief outline of the quantitative status of axial flow compressor technology is given. This is done, not to guarantee the prompt obsolescence of the chapter, but to provide baseline data and illustrate trends.

Like most of the other subjects covered in the specialized chapters of these volumes, the axial flow compressor has a tremendous literary background. This chapter should be a book, but since it is not, readers should consult the Bibliography. These references cover a broad range of topics and points of view and each one makes a contribution to real axial flow compressor technology, which is the only kind worth writing about.

3.2 Axial Flow Compressor Nomenclature and Terminology

All gas turbine propulsion systems must have a compressor component that develops some or all of the pressure rise specified by the system cycle. Shaft work for the compression process is supplied by the turbine component of the system. In some system cycles, the compressor unit is divided into sections driven by separate shafts and identified by descriptive names such as fan, low-pressure compressor, intermediate-pressure compressor, and high-pressure compressor.

Meridional Plane Flow Passage Terminology

Three axial flow compressor units from current aircraft gas turbine engines are shown in the sectional views of Figs. 3.1–3.3. In each case, the section shown is one intersected by a plane including the rotational axis of the engine. This section is known as a meridional plane section. The compressor units of Figs. 3.1–3.3 are described as axial flow compressors because the throughflow direction of the working fluid is approximately parallel to the rotational axis.

Figure 3.1 shows a single-shaft compressor for a turbojet engine. This compressor has eight stages with each stage consisting of one rotating (rotor) row and one stationary (stator) row of blades. Each blade (Fig. 3.4) is made up of individual airfoil sections "stacked" on an axis so as to produce a prescribed flowfield. The first stage in Fig. 3.1 has a row of inlet guide vanes. The function of this type of row is to set up a design distribution of velocity vectors at the first rotor entrance. In some compressor units, one or more rows of stationary blades (exit guide vanes) may be used at the outlet of the last stator row to establish a desired flow pattern at the exit of the compressor.

Figure 3.2 shows a two-shaft (two-spool) compressor in a turbofan engine. In this case, a three-stage fan and six-stage low-pressure compressor section are driven by the low-pressure section of the turbine, while the seven-stage high-pressure compressor is driven by the high-pressure turbine section through a separate shaft. The fan has inlet guide vanes. Note that there is a splitter in the flow passage just downstream from the rotor of the third fan stage. This splitter separates the core flow into the low- and high-pressure compressor sections from the bypass flow in the outer duct.

A two-spool compressor for a large turbofan engine is shown in Fig. 3.3. The fan, low-pressure compressor, and high-pressure compressor have one, four, and eleven stages, respectively. In this case, the fan has no inlet guide vanes.

In order to clarify additional compressor terminology, Fig. 3.5 shows a meridional section of a three-stage fan at a larger scale. The rotor and stator rows are installed in an annular flow passage. The outer or tip casing wall holds the stator blade rows. The inner or hub ends of the stator blades are joined by circumferential shrouds that make up a part of the annulus boundary at the hub. The rotor rows are in this case installed in rotor disks so that the hub profile is made up of alternate rotating and stationary wall segments. In some compressors, the rotor blades are installed in a built-up series of disks and rings making up a continuous rotating hub wall for the flow path. In the latter case, there is a clearance between the ends of the stator blades and the hub, just as there is normally a tip clearance at the rotor outer radius. Occasionally, a rotating shroud is used at the rotor tips, but more frequently part-span shrouds or dampers are used at a location along the rotor span between the hub and tip (Figs. 3.2 and 3.3). If shrouds or dampers are used, the principal function is aeromechanical—to limit the amplitude of blade vibration. They are not likely to be an asset in an

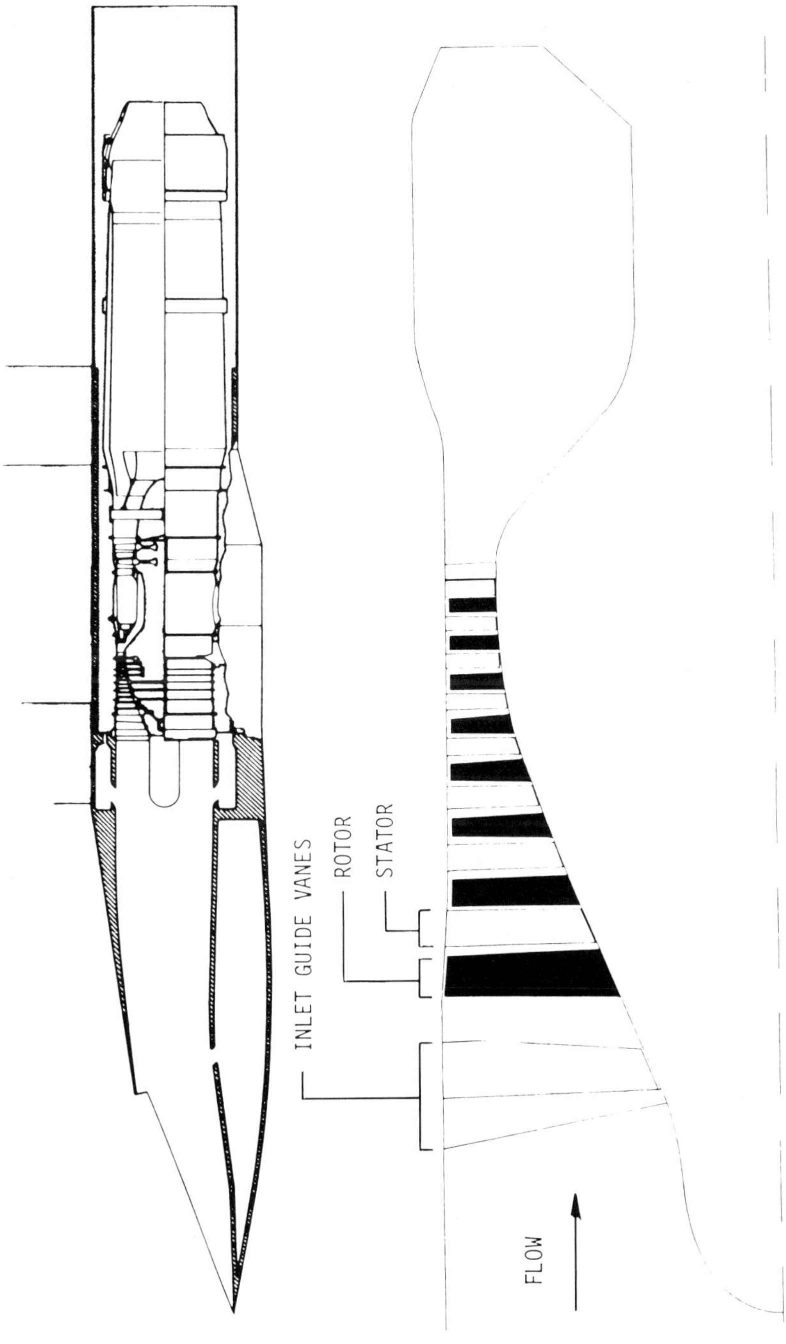

Fig. 3.1 Meridional plane sections of turbojet engine and compressor (from Refs. 1 and 2).

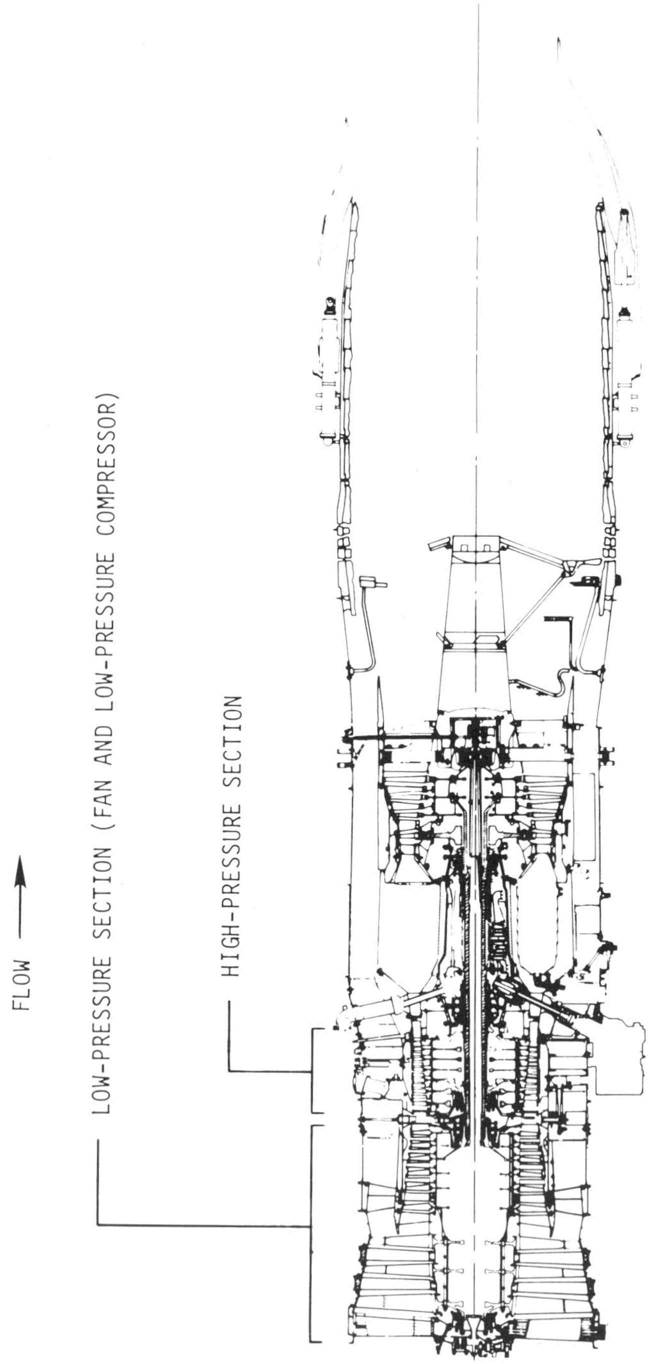

Fig. 3.2 Meridional plane section of a two-spool turbofan engine (from Ref. 3).

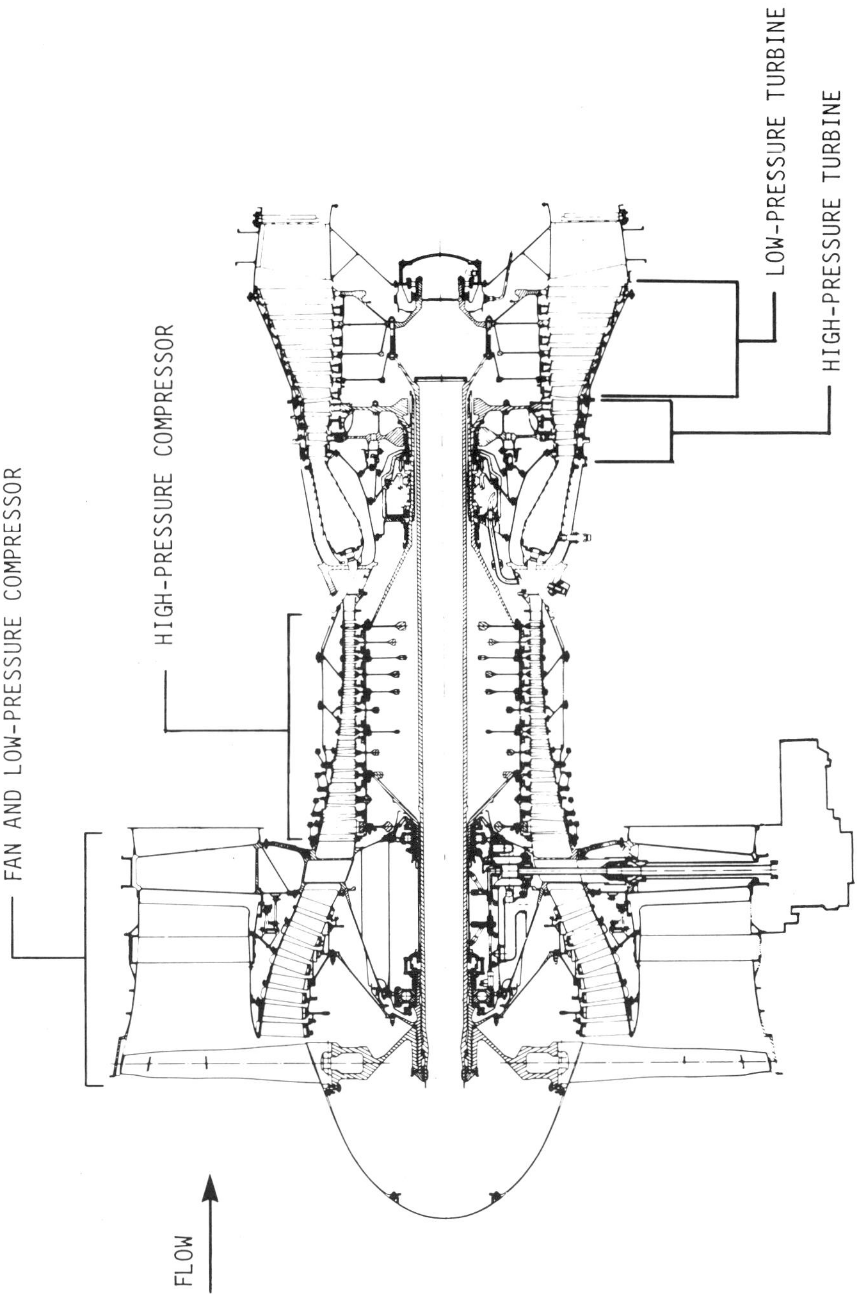

Fig. 3.3 Meridional plane section of high-bypass-ratio turbofan engine (from Ref. 4).

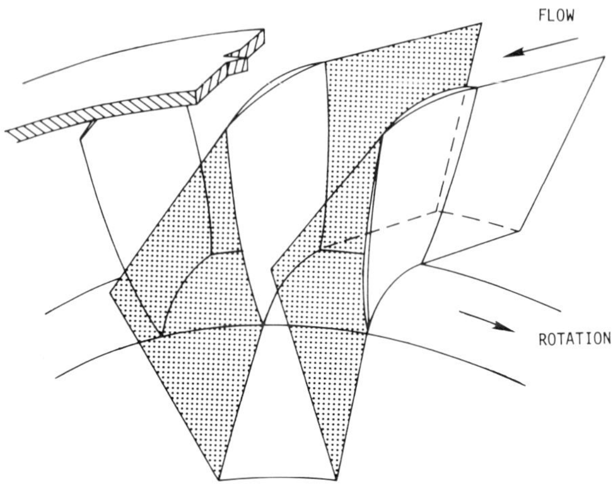

Fig. 3.4 Schematic view of axial flow compressor rotor blade row.

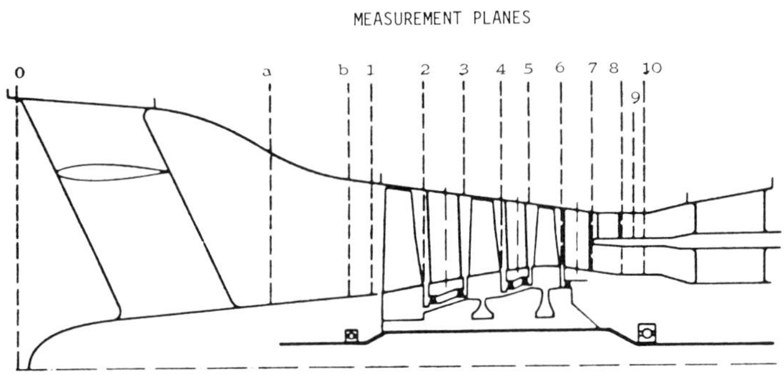

Fig. 3.5 Meridional plane section of three-stage axial flow fan (from Ref. 5).

aerodynamic sense. The number of blades in each rotor and stator row is specified in the aerodynamic and aeromechanical design of the unit. Each blade, just as an isolated airfoil, has a chord length and span. The chord (axial projection only is seen in the meridional plane) may vary between the hub and tip and the span normally varies between the entrance and exit of a given blade row because of design variations in fluid density and velocity levels. The hub-tip ratio for a specific stage is usually defined as the numerical ratio of the hub diameter to the tip casing diameter taken at the leading edge of the rotor row for the stage. The term aspect ratio is frequently used in describing compressor blade rows, but the definition may vary from one case to another. Typically, the aspect ratio might be defined as the ratio of an average blade span to an axial projection of the blade row length or chord at the hub.

Blade-to-Blade Flow and Cascade Nomenclature and Terminology

The flow through an axial flow compressor is an internal flow through an annular flow path, where the flow is influenced by both rotor and stator blade rows. The blades are generally nonuniform in the spanwise direction and consist of stacked airfoil sections as shown in Fig. 3.4. It might be expected that these individual airfoil sections would exhibit some of the same behavioral characteristics as isolated airfoils—and they do. They are subject to lift and drag forces, they stall, and they generate boundary layers, wakes, and under some circumstances shock waves. The blades frequently have dihedral and act as swept airfoils. However, they operate under conditions unlike typical isolated airfoils. They are located near other airfoils, both in the circumferential and axial directions, and they are influenced by these adjacent bodies. They are bounded at both ends by walls, which may be stationary or moving with the blades. In all rows except the first, the airfoils operate in an unsteady flow, where both the velocity magnitude and direction fluctuate. Most important, the airfoil rows in the compressor produce a significant flow deflection and change in the thermodynamic state of the fluid. Rows are generally designed so as to behave as diffusers, with an increase in static pressure from the upstream to the downstream flow region.

To aid in the discussion of blade row characteristics, consider a cylindrical surface at constant radius r from the rotational axis of the compressor stage of Fig. 3.6. If this surface is developed on a plane, the intersection traces of the blades produce two lattices or cascades of airfoil sections. There is no reason to believe that the flow in a compressor will follow the cylindrical surface of Fig. 3.6 and, in fact, the real paths followed by fluid particles are much more complex. Nevertheless, the concept of an airfoil cascade is not only convenient for the purpose of introducing compressor nomenclature, but it also is the basis for historically and currently significant models of the flow in compressors and turbines.

In both computation and experimentation, it has been common practice to track the flow through axial flow compressors as if it followed axisymmetric paths. In the earliest axial flow configurations, these paths were assumed to be cylindrical, but more recently have become more complex surfaces of revolution that change radius in an attempt to follow the flow more accurately. The validity of these paths will be discussed in more detail later, but in any case they define a cascade array of airfoil sections by their intersections with each blade row.

For many years the flows through compressor blade rows have been simulated experimentally using both annular and plane cascade test facilities. In an annular cascade facility, blades are installed in an annulus, just as in a real compressor, so that the infinite character of the blade row in a real machine is correctly modeled. In a plane or linear cascade, a finite number of blades are installed in a flow passage with a rectangular cross section as shown in Fig. 3.7. In both annular and plane cascades, measurements are made upstream and downstream from the blade row with the row arranged

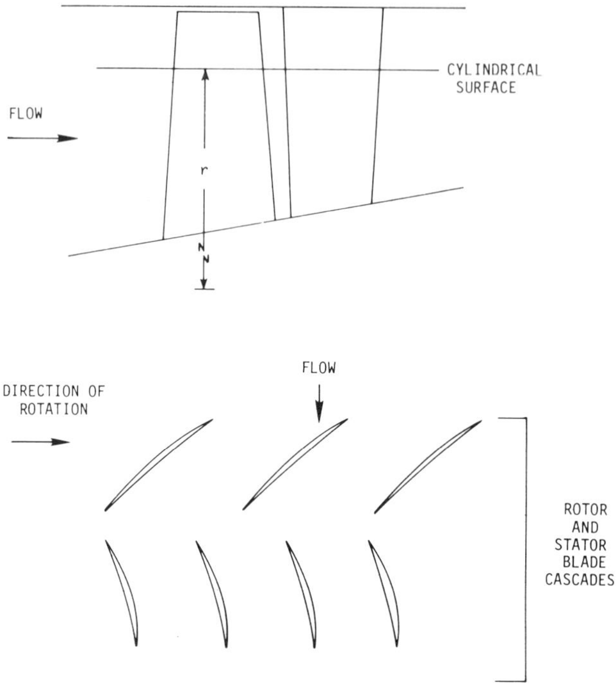

Fig. 3.6 Compressor stage with cascade intersection trace in developed cylindrical surface.

geometrically so that it is exposed to an entrance flow similar to that found in a compressor blade row. When reference is made to cascade data in the literature, the data may have been obtained in either an annular or plane cascade, but it is a fact that the vast majority of cascade data has originated in plane cascade facilities.

Figure 3.8 shows a plane cascade of airfoils with some important geometric and flow variables. Blades of a uniform blade section geometry are aligned at the blade setting angle or stagger angle γ with respect to the cascade axis. The cascade axis corresponds to the axial direction in the compressor. The blades are equally spaced at a distance s, called the gap or blade spacing. The cascade has a solidity σ, defined as the ratio of blade chord to blade spacing c/s. Although the three variables—blade section geometry, blade setting angle, and solidity—completely define the plane cascade geometry, there are, of course, a very large number of subsidiary variables required to describe the blade section geometry. Many of these have important effects in determining the flowfield in a cascade.

The plane cascade test facility is really a special form of wind tunnel. In cascade experiments, compressor blade row flowfields are simulated by varying or controlling the relative inlet flow angle β_1, the upstream Mach number M_1, the Reynolds number, and/or the upstream turbulence proper-

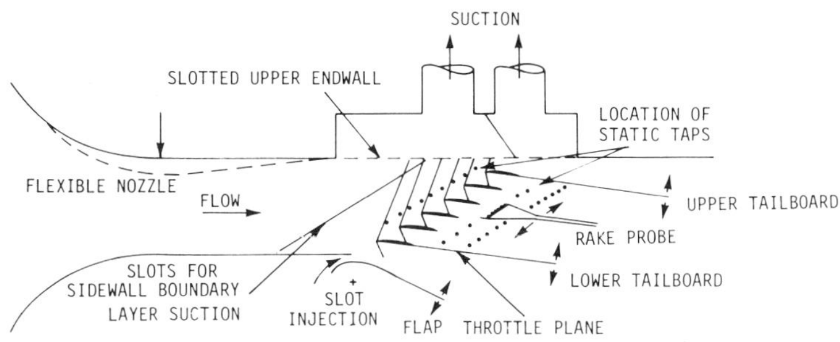

Fig. 3.7 Schematic layout of a plane cascade test section for transonic flow conditions.

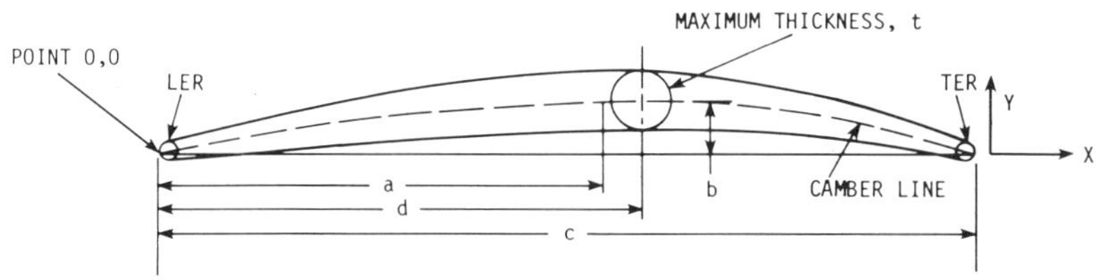

a = LOCATION OF MAXIMUM CAMBER
b = COORDINATE OF MAXIMUM CAMBER
c = CHORD LENGTH
d = LOCATION OF MAXIMUM THICKNESS
l = LENGTH OF CAMBER LINE MEASURED ALONG CAMBER LINE ARC
LER = LEADING EDGE RADIUS
TER = TRAILING EDGE RADIUS

Fig. 3.8a Blade section geometry and cascade terminology.

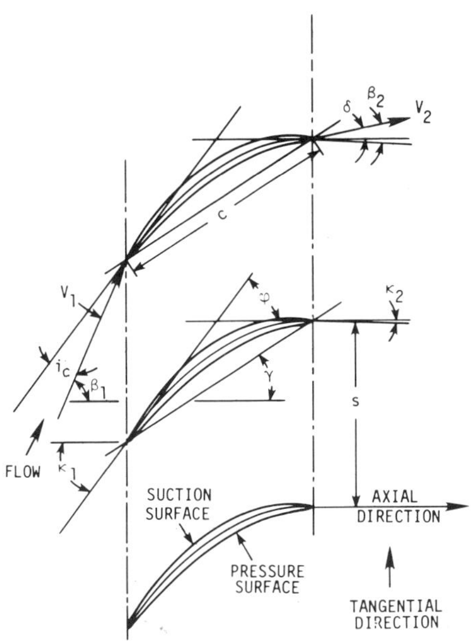

Fig. 3.8b Blade row geometry and cascade terminology.

ties. Changing the angle β_1 for a constant blade setting angle changes the angle of incidence i (or angle of attack) by an equal amount.

Test conditions may be varied to control the axial velocity × density ratio (AVDR) and the static pressure ratio across the cascade, because these parameters affect the cascade blade surface pressure distributions and the two cascade performance parameters that are of most interest in design. These dependent variables are the downstream fluid flow angle β_2 and a measure of the loss in total pressure across the cascade. The parameters most commonly used to quantify these measurements are the gap-averaged deviation angle δ and the gap-averaged total pressure loss coefficient $\bar{\omega}$. The averaging methods used have an effect on measured cascade performance, as does the location of the measuring station with respect to the cascade trailing-edge line.

There is disagreement about the value of cascade data in design, and there is some factual basis for a negative view of cascade experiments. However, cascade results represent an important part of the data base in most design systems and new cascade data continue to appear in the literature on a regular basis.

Figure 3.9 is a typical plot representing the results of a series of subsonic plane cascade experiments. The fact that the loss coefficient rises at both

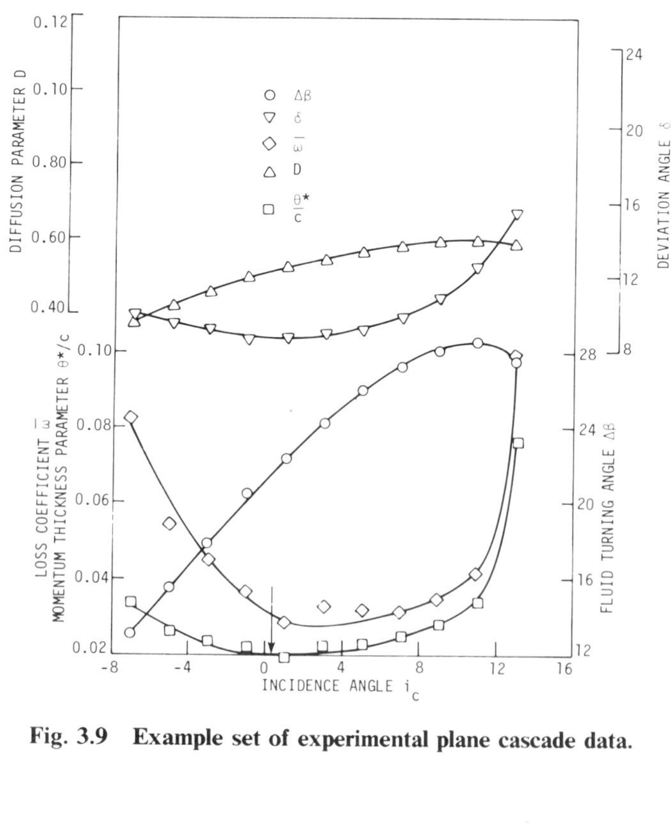

Fig. 3.9 Example set of experimental plane cascade data.

ends of the incidence angle range clearly shows that flow separation and stall are present in cascade flowfields. These cascade experiments were run under two-dimensional flow conditions (AVDR = 1.0) and with a constant entrance Mach number M_1. Some additional comments on cascade results will be made later in this chapter, especially in connection with the effects of Mach number and blade row geometry.

3.3 Characteristics of the Flow in Axial Flow Compressor Configurations

Velocity Diagrams, Energy Transfer, Path Lines, and Streamlines

It is essential to clearly understand velocity component diagrams and their use in describing the annulus flowfield. This is a good time to consider these diagrams and their relation to the complete flowfield.

At each point and at each instant of time in the compressor flowfield, a velocity vector $\mathbf{V}$ exists. A cylindrical coordinate system is generally used, with the coordinates r measured from the rotational axis, θ measured from an arbitrary meridional plane (with the positive θ direction in the direction of rotor rotation), and x measured parallel to the rotational axis from an arbitrary r, θ plane. In this coordinate system, $\mathbf{V}$ = function (x, r, θ, t) and for any point and time the velocity components V_x, V_θ, and V_r can be determined. Also, a meridional plane velocity component can be defined as the resultant of V_x and V_r. This is called the meridional plane velocity V_m.

With rotating blades, it is necessary to consider both absolute velocities and velocity components measured relative to points on the rotating blades. For example, to study the fluid velocity relative to the leading-edge point A on the rotor of Fig. 3.10a, subtract the local rotor velocity $\mathbf{U}$ from the absolute fluid velocity $\mathbf{V}$ to get the relative velocity $\mathbf{W}_\text{A}$,

$$\mathbf{W}_\text{A} = \mathbf{V}_\text{A} - \mathbf{U}_\text{A}$$

The relative velocity at any point has components W_x, W_θ, and W_r,

$$W_x = V_x, \qquad W_r = V_r, \qquad W_\theta = V_\theta - U$$

A typical velocity diagram for a rotor leading-edge point such as A is shown in Fig. 3.10a as it would appear in a plane tangent to the meridional velocity component V_m at A. If a particle entering the rotor at A followed a path such that it left the rotor at B and continued through the stator row along the path C-D (for convenience all points are rotated from their actual circumferential locations into the meridional plane shown), velocity diagrams could also be constructed at B, C, and D as in Fig. 3.10a. The four diagrams are combined in a single composite stage velocity diagram in Fig. 3.10b, showing the effect of the passage and blades on fluid particles following a path line A-B-C-D.

For a fluid particle of mass δm passing through a rotor row from A to B (see Fig. 3.10a), the energy transferred as work done on the fluid by the

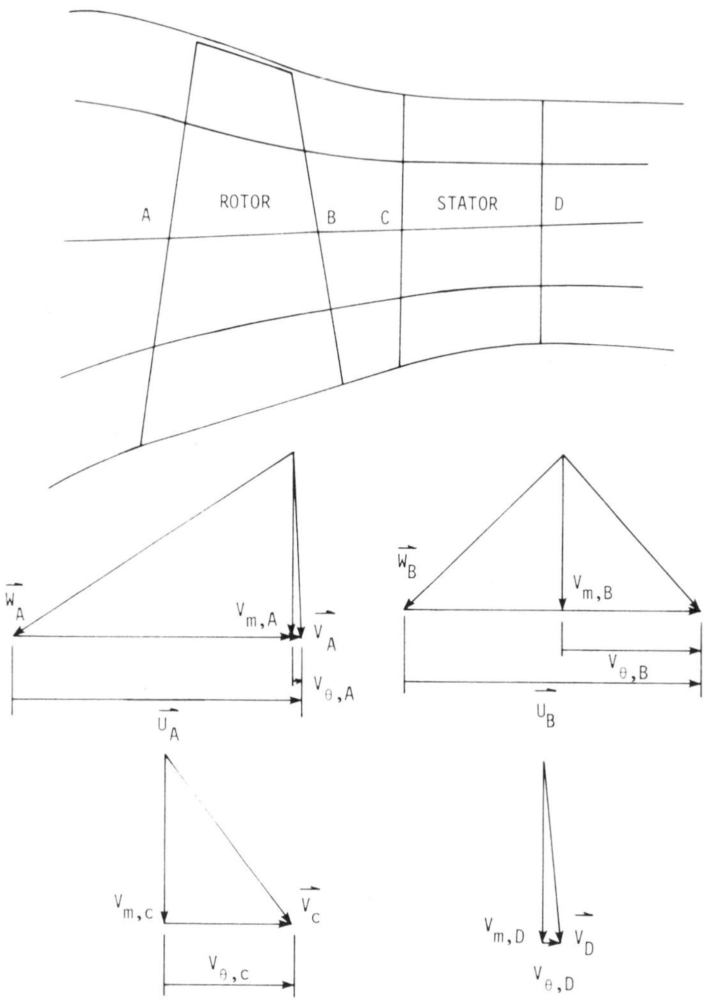

Fig. 3.10a Velocity diagrams for flow along path line A-B-C-D in a compressor stage.

blades is given by Euler's turbine equation

$$E = \delta m(\omega)[r_B V_{\theta,B} - r_A V_{\theta,A}] = \delta m[U_B V_{\theta,B} - U_A V_{\theta,A}] \qquad (3.1)$$

Equation (3.1) is a consequence of the law of conservation of angular momentum (see Sec. 3.6). In order to use this equation to determine the energy transfer or work done on δm by the rotating blades, both the exit point B and entrance point A for δm must be *known*. Determination of these points by computation or by experiment is no trivial matter and is a part of the flow modeling problem.

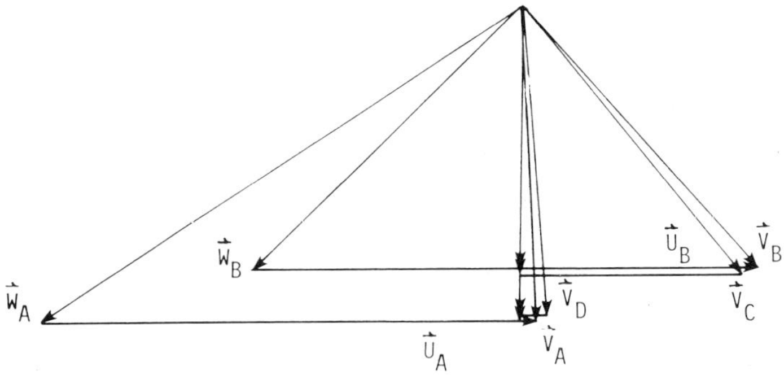

Fig. 3.10b Combined velocity diagram for flow along path line A-B-C-D in a compressor stage.

In the total flow passing through a rotor, each fluid particle will experience an energy transfer governed by Eq. (3.1) for the points corresponding to A and B at which it enters and leaves the rotor. Normally the work done on fluid passing through the rotor will vary as a function of entrance location. In fact, there is no reason why subsequent particles entering the rotor at A should experience the same energy transfer rate as the original δm, because they may leave at points other than B or at different values of V_θ due to an unsteady or nonaxisymmetric real flow.

In spite of the doubts about the real flow, to gain a simplified qualitative understanding of the performance of a compressor blade row or stage, assume that the flow A-B-C-D in the stage of Fig. 3.10 is steady and representative of the entire flow across that stage. The airfoils encountered by the flow are represented by cascades of blades similar to those already shown in Fig. 3.6. Consider the rotor to have a constant angular velocity so that the rotor blade speed will be a constant at a given radius. Vary the meridional velocity component, simulating a variation in the stage flow rate. The sequence of stage operating conditions is represented in Fig. 3.11 by velocity diagrams and by a plot of the total pressure rise against the flow rate for three flows, 1–3. For each of the three flows, V_m is assumed to remain constant along A-B-C-D. Some features to note are as follows:

(1) As the flow varies, the angle of incidence on the rotor blade airfoils will vary, with the highest angle of incidence at the lowest flow. For excessive angles of incidence, cascade airfoils will stall (unacceptable flow separation occurs). Compressor blade rows will also stall at extreme angles of attack. Just as cascades do, they operate most effectively at moderate values of incidence and less effectively at high and low values. This was shown in Fig. 3.9.

(2) The energy transfer rate as measured by the change in V_θ across the rotor (U is nearly constant) increases with decreasing flow rate. This assumes, as Fig. 3.4 showed, that the deviation angle for the rotor exit does not change a great deal until the losses begin to increase rapidly. The stage performance shows this effect as a higher stage total pressure rise at lower

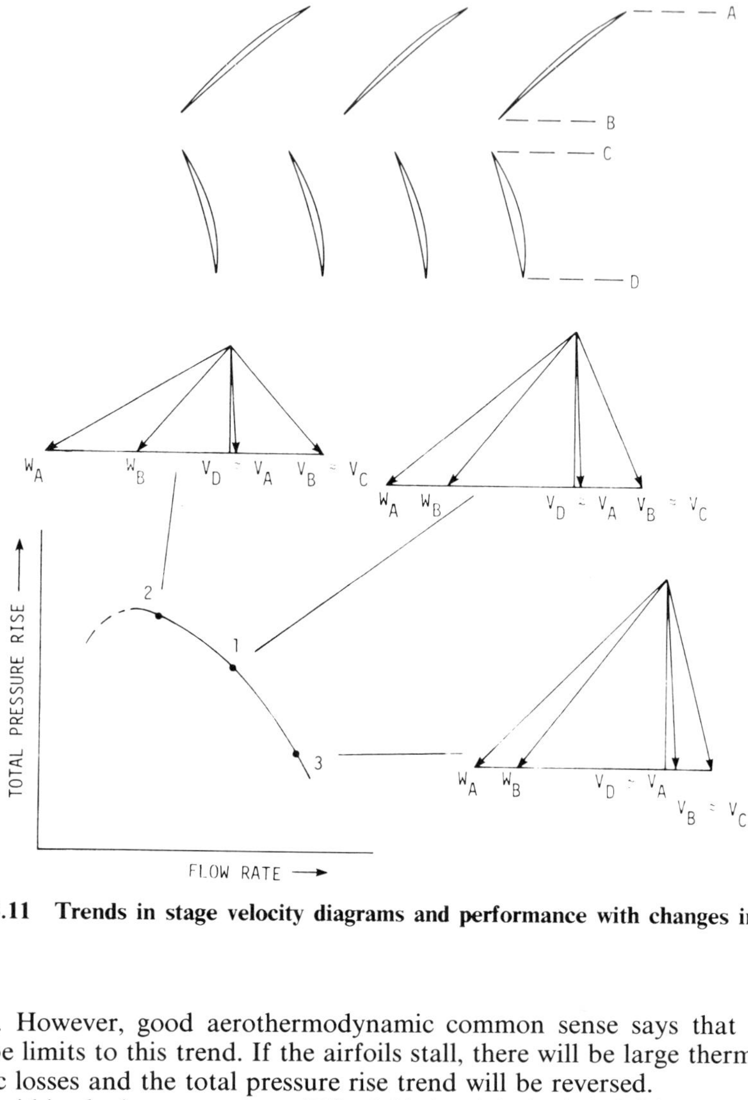

Fig. 3.11 Trends in stage velocity diagrams and performance with changes in flow rate.

flows. However, good aerothermodynamic common sense says that there will be limits to this trend. If the airfoils stall, there will be large thermodynamic losses and the total pressure rise trend will be reversed.

To aid in the interpretation of Fig. 3.11, it might be helpful for readers to write some simple steady, one-dimensional energy, continuity, and momentum equations for the flow A-B-C-D, considering the effect of work done by the blades on fluid total enthalpy and how this might, in turn, affect the total pressure changes from A to B and from C to D if the flow were isentropic or if it were nonisentropic. Draw an enthalpy-entropy diagram to represent the processes.

In the axial flow compressor, work is done on the fluid only through the action of the rotating blade rows and this transfer of energy results in an increased level of fluid stagnation or total enthalpy. An increase of total pressure also results in each rotor row, with this increase dependent on the losses or irreversibility associated with the flow process as well as on the energy transfer. In the stator rows, no energy is transferred as work and there can only be reduced values of total pressure because of irreversibility. Static pressure levels may increase or decrease in either rotor or stator row flows, but the predominant design intent is, naturally, to develop significant static pressure level increases in each stage. This aspect of blade row design is important and will receive substantial attention in later sections. It is important to observe that in constructing combined velocity diagrams, such as those of Fig. 3.10b and 3.11 for a stage, the path lines (such as A-B-C-D) are really streamlines and the layout of the blade rows as cascades is in terms of axisymmetric stream surfaces. When this extension of terminology is made, the requirement for steady flow should be recalled and understood. Subsequent sections in this chapter and many important referenced publications use the terms "streamline," "stream tube," and "stream surface." The assumption of a steady relative and/or absolute flow as a part of the development of a flow model is generally involved when these terms are used and it should be remembered in evaluating any analysis.

Performance of Axial Flow Compressors

Design point flowfield characteristics. This section is a qualitative discussion of the real flows in axial flow compressors. At the same time, further terminology and additional concepts useful in flowfield analysis and in experimental research and development will be defined.

A compressor geometry is usually specified initially for operation at a design point. This means that it is configured so that it will develop a design value of the exit-to-entrance pressure ratio with a predicted value of thermodynamic efficiency at a design value of the shaft rotational speed, while a design value of the mass flow rate exists in each stage. The flow rate may vary from one stage to the next because of bypass arrangements or extraction of flow. All of the design requirements are associated with a particular working fluid and with the design values of total pressure and total temperature distribution at the compressor inlet. At the design point, the flow path dimensions are fixed and the blade rows are arranged so as to produce an ordered distribution of fluid velocities and properties through the compressor. In other words, the design point velocity diagrams are specified. However, the real design point flowfield must be recognized as extremely complex. Every row following the first sees an entering flow that is unsteady. The flow is viscous and, in almost all modern compressors, it is compressible with regions of both supersonic and subsonic local flow. Boundary layers develop on the blade surfaces and on the hub and tip annulus walls. The annulus boundary layers are called end wall boundary layers. Local regions of laminar, transitional, and turbulent boundary-layer flow may exist, accompanied by separation of flow from the blades or end walls. Both the blade surface and the end wall boundary layers are generally three-dimensional in character. The flow throughout the annulus is, in fact, normally three-dimensional in nature. Secondary flows are generated, a fact that has been widely discussed, but is difficult to predict for design purposes. The real flow is by no means axisymmetric.

It must be recalled that all of the above complexities are associated with the design operating point. Away from the design point, the configuration must operate in a flowfield matching neither the annulus shape nor the angular settings of the stacked airfoils forming the blades.

Equilibrium performance maps. To demonstrate that axial compressor flows are not beyond redemption in a real situation, Fig. 3.12 shows what is called an equilibrium performance map for an eight-stage fixed geometry compressor. The performance parameters used are discussed in Chaps. 6 and 8 of *Aerothermodynamics of Gas Turbine and Rocket Propulsion*. They are more or less traditional in nature. The experimental performance of the compressor was measured at several constant values of equivalent rotational speed (corrected to standard sea-level compressor inlet conditions).

At each rotational speed, the mass flow rate was varied between two limits. The upper limit was close to the maximum obtainable at that speed. The lower limit was fixed by the existence of an aerodynamic instability common to all compressor types known as surge. The character of surge is associated with the nature of the compressor geometry and with aerodynamic and geometric boundary conditions imposed upon it by the installation arrangement and by the flowfield (e.g., inlet geometry, downstream struts or flow control devices, inlet flow nonuniformity). Surge is manifested by large-scale flow instability, often with pulsating reversal of flow involving the entire unit. In an aircraft engine application, compressor surge may produce highly disturbing acoustic evidence (violent bangs). Individual surge points at the various rotational speeds define a surge line that is universally considered as a boundary for acceptable compressor operation in an aircraft engine.

Surge occurs only when one or more blade rows in the compressor are operating so that flow separation or stall occurs in the flowfield. However, one or more blade rows *may* be stalled without the occurrence of surge. Stall in a blade row is unique in a number of respects, especially in the fact that zones of flow separation propagate from blade to blade in the annulus, resulting in a phenomenon known as rotating stall.

Contours of constant adiabatic efficiency* are plotted in Fig. 3.12. These contours show that the thermodynamic quality of compressor performance

---

*This efficiency is defined as

$$\eta_{ad} = \frac{(H_{20} - H_0)_{is}}{(H_{20} - H_0)_{\text{act}}}$$

where subscripts 20 and 0 refer to the compressor exit and inlet measuring stations, respectively.

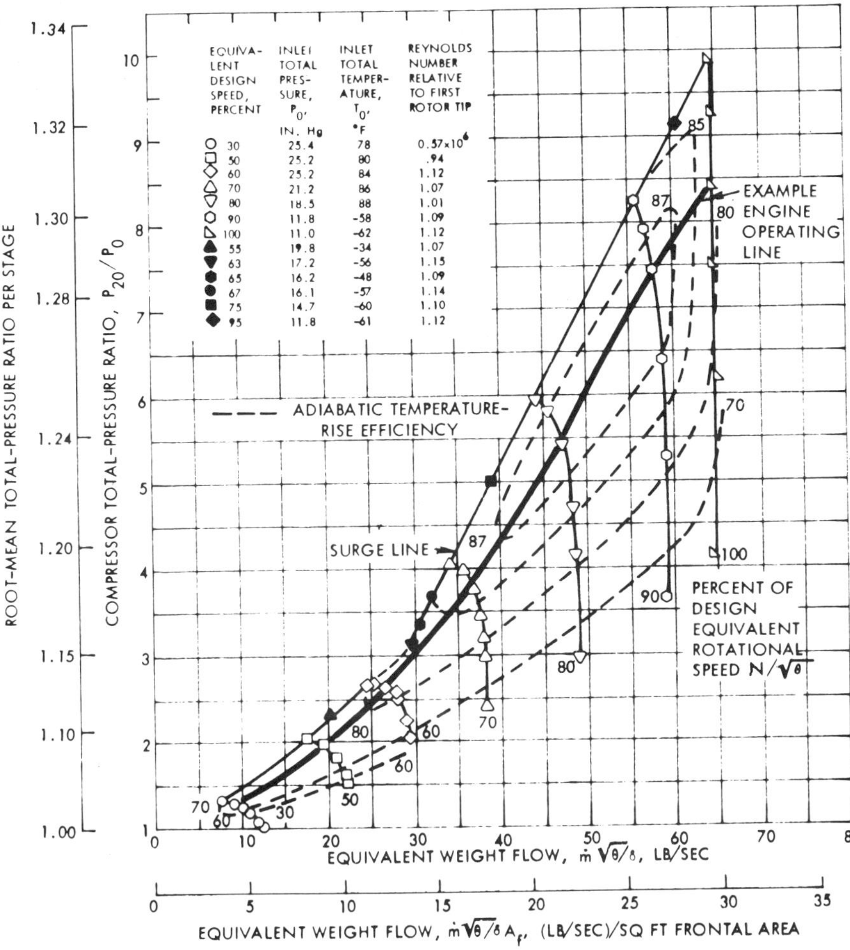

Fig. 3.12 Equilibrium performance map for an eight-stage axial flow compressor (from Ref. 6).

is reduced significantly when the compressor operates at flows and rotational speeds other than the design values.

The equilibrium performance map for a fixed geometry compressor is a classical example of the use of dimensional analysis and the principles of similitude. However, as plotted, it presents several possibilities for misinterpretation:

(1) The plotting parameters $m\sqrt{\theta}/\delta$ and $N/\sqrt{\theta}$ have been intentionally simplified to report the performance of a single compressor geometry at entrance conditions corresponding to standard sea-level temperature and pressure. These modified parameters are not dimensionless. They are, respectively, proportional to Mach numbers associated with the throughflow

components of velocity and with the velocities of points on the rotating blades.

(2) The map does not define the effect on performance of changes in the Reynolds number or the entrance turbulence parameter. These parameters are dynamic similarity variables and can have substantial effects on the map.

(3) The map is based on a single working fluid (air) and would be modified for other fluids with a different specific heat ratio γ or molecular weight.

(4) The map is based on uniform values of the fluid properties and velocities at the compressor entrance. Distorted or unsteady entrance conditions change the surge lines and other mapped performance values.

It should not be forgotten that, when relatively minor geometric changes in a compressor take place (for example, when tip clearances change or when blade surfaces are abraded or damaged), the alterations in the compressor performance may not be minor in terms of the effect on the engine specific fuel consumption or thrust.

Variations on the Basic Performance Map

Figure 3.13 is a simplified performance map showing two aspects of the equilibrium compressor performance not included in Fig. 3.12. The map shows the general effect on the compressor surge line caused by nonuniform

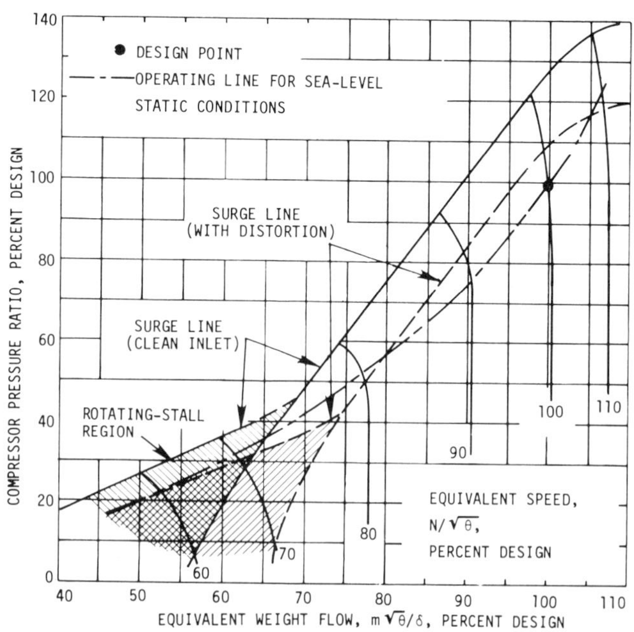

Fig. 3.13 Example of axial flow compressor map showing effect of inlet flow distortion and rotating stall region (from Ref. 7).

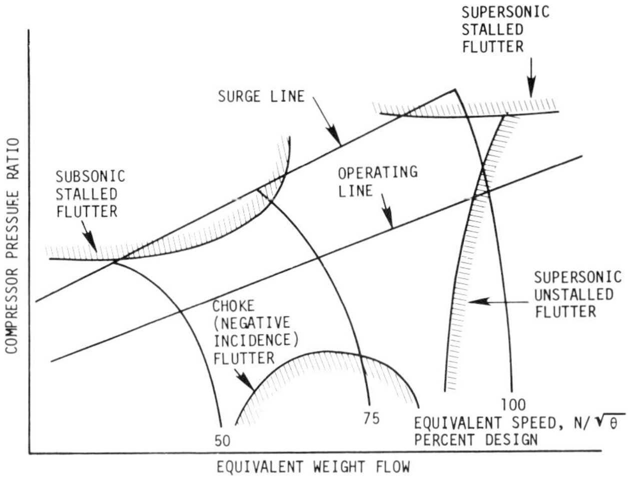

Fig. 3.14 Schematic compressor map showing boundaries for four types of blade flutter (from Ref. 8).

inlet flow (inlet distortion). The significance of this effect can be seen in terms of the estimated engine operating line superimposed on the map. Not shown are the possible adverse effects of inlet distortion on the constant $N/\sqrt{\theta}$ lines and compressor efficiency.

Figure 3.13 also defines regions at low rotational speeds in which rotating stall occurs in the entrance stages of the compressor. Although the background of rotating stall has not yet been discussed, it is necessary to determine why any forms of stall occur in the entrance stages at low speeds and why rotating stall appears to be tolerated by aerodynamicists. It can be inferred that measurements are made and that the engines operate in the presence of rotating stall. This indicates a toleration level greater than that for surge.

Figures 3.12 and 3.13 do not show another class of performance phenomena frequently creating problems in compressor development and operation —the aeromechanical or aeroelastic phenomena known as flutter. Figure 3.14 shows in schematic form some general regions on a compressor map in which specific varieties of flutter are encountered. Blade vibration amplitudes in flutter have frequently caused catastrophic compressor failures. It is necessary to understand flutter so that mechanisms for its elimination can be applied where possible or so that its occurrence can be limited to zones on the map where the compressor will not operate when installed in an engine.

Individual Stage Characteristics

Recalling Fig. 3.12 as a performance parameter plot showing some of the features of off-design operation for an eight-stage axial flow compressor, what the individual stages are doing in a multistage compressor as it runs at various locations on its overall map will be examined here. Naturally, it is expected or hoped that each stage will be running under favorable aerodynamic conditions when a multistage compressor operates at its design point flow rate and rotational speed. Ideally, this means that the velocity diagrams and properties of the fluid entering and leaving each blade row are those that the designer specified. Extending this line of reasoning, it is obvious that each stage must have a design point flow rate, pressure ratio, and efficiency that, in combination with the other stages, will give the overall compressor design point performance. To describe this situation, it is said that the stages are matched to function at certain operating points when the complete machine is at its design point. Studies related to changes in the operation of the individual stages in a multistage compressor are frequently called stage-matching studies.

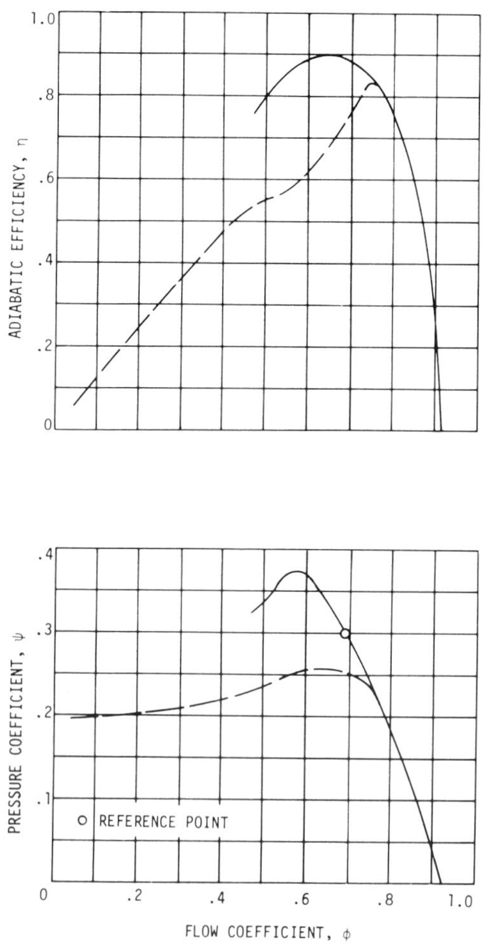

Fig. 3.15 Example of assumed stage performance characteristics used in stage-matching study (from Ref. 9).

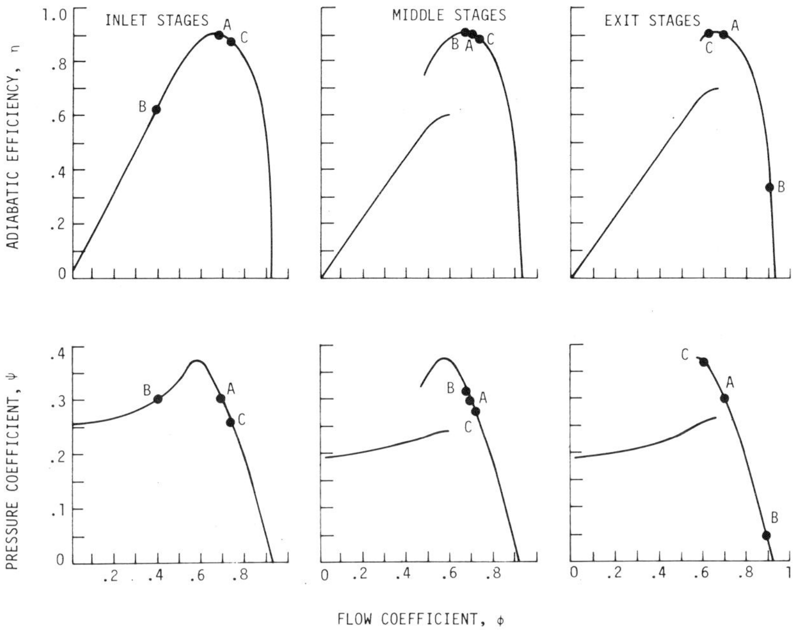

Fig. 3.16 Trends in stage performance for an axial flow compressor operating off-design conditions (from Ref. 10).

Although the individual stage performance could be plotted using the same parameter forms used for multistage maps, these stage plots for matching purposes are usually plotted as shown in Fig. 3.15. The parameters ϕ, ψ, and η are related to those historically used for fans and blowers in which the flow is approximately at a constant density (incompressible flow). At this point, note that a plot with single $\psi - \phi$ and $\eta - \phi$ curves (such as in Fig. 3.15) does not adequately represent the Mach or Reynolds numbers effects on the stage performance.

For a fixed geometry multistage compressor, a good qualitative understanding of off-design performance can be gained by considering the three stage plots of Fig. 3.16. Remember, it is assumed that when confronted with a given value of ϕ, the entrance, "middle," and exit stages will each follow the curves shown, regardless of what other stages are doing. In other words, some elements of the interaction of stages are neglected. The three stages are assumed to be matched so as to operate at the points marked A in Fig. 3.16 when the multistage compressor is at its design point. Under these conditions, in the fixed geometry compressor, each stage will develop exit levels of pressure, density, and velocity such that the downstream stage will be at its point A.

Now suppose that the compressor operates at a point B that is at a rotational speed substantially lower than the design value. It is evident that, at *any* value of ϕ, the pressure increase across any stage will be reduced because U^2 is reduced in the ψ parameter. In turn, this means that no stage can develop its design density rise. However, the annulus area variation through the compressor is based on the design point rate of the fluid density rise. The result will be to force the level of throughflow velocity up in the exit stages so that ϕ will be higher than the design values, with a tendency toward a limiting maximum at low rotational speeds. In turn, the entrance stages will be forced to operate at lower than design ϕ, with a tendency toward unstable (stalled) operation at low speeds. This is why the performance map of Fig. 3.13 showed a rotating stall zone at low $N/\sqrt{\theta}$ values. The trends will be reversed at rotational speeds greater than design (point C, Fig. 3.16), with the exit stages moving toward stall and the entrance stages moving toward a maximum limiting flow coefficient ϕ. This is one reason why Fig. 3.14 shows a region of supersonic stalled flutter at high rotational speeds.

3.4 Aerodynamic Design Objectives for Axial Flow Compressor Units

As an introduction to a discussion of design systems for axial flow compressors, some criteria for a *good* design should be stated. To do this in a logical way, look again at the equilibrium performance characteristics shown in Fig. 3.12. The first goal of a design system should be to generate a compressor geometry that will produce the design point total pressure ratio when the design point mass flow rate exists in each stage of the compressor. These design point values are developed in the process of selecting the best propulsion system cycle parameters for the aircraft application. Because each compressor section is driven by a turbine component, the rotational speed at the compressor design point must be consistent with the aerodynamic and aeromechanical design requirements of the turbine. Design point pressure ratio, flow rate, and rotational speed must be associated with an acceptable, if not superior, level of compressor thermodynamic efficiency and with adequate surge margin. Surge margin is a measure of the location of any operating point with respect to the surge line at a given rotational speed and should be sufficient so that no engine transient will force the compressor into surge.

However, while design point performance is a primary consideration, Fig. 3.12 and earlier chapters point out that the design operating point for the compressor is reached only after starting and accelerating the complete engine and that propulsion system operation may call for extended periods of performance at locations on the equilibrium performance map substantially different from the design point. Furthermore, transient system operation may create additional deviation from normal compressor operating points. In terms of Fig. 3.12, engine operation may be expected to require the compressor to run within a band of flows and pressure ratios at any rotational speed. In fact, the performance map itself will be altered by changes in parameters such as tip clearance and compressor entrance flow distribution. The compressor design process must therefore develop a compressor geometry (and possible variable compressor geometry) that demonstrates the following characteristics:

(1) Stable aerodynamic and aeromechanical operation, including sufficient surge margin and no excessive blade or disk vibrations for all potential operating conditions in the system installation.

(2) Acceptable thermodynamic efficiency for all expected operating conditions.

(3) Acceptable noise generation characteristics.

Subject to these constraints, the compressor configuration should be designed for

(4) The maximum possible mass flow rate per unit frontal area.

(5) The highest possible average pressure ratio per stage.

(6) Minimum values of axial stage length and number of blades specified for each row.†

These characteristics, of course, lead to the minimization of compressor unit weight and cost. Axial flow compressor configurations for aircraft propulsion systems have demonstrated good and improving characteristics measured against the six objectives listed, but they have remained very expensive, primarily because of the complex, sometimes variable geometry required to realize the best possible performance.

3.5 Elements of a Compressor Design System —Technical Requirements

Consideration of the required elements of a compressor design system leads directly to the definition of more detailed requirements for flowfield models and for aeromechanical models of the blading and other flow path components. Based on the objectives listed in Sec. 3.4, at a minimum the design system must include the following:

(1) A procedure for determining potentially satisfactory flow path geometries capable of stable and efficient operation at the design flow rate and at a rotational speed acceptable to the turbine and mechanical system designers, with the procedure sufficiently quantitative to permit.

(a) Specification of annulus dimensions throughout the compressor.

(b) Selection of blade row geometric parameters such that design point velocity diagrams and fluid property distributions are realized at acceptable efficiencies (low blade row thermodynamic losses).

(2) Procedures for predicting the flowfield throughout the compressor at the design point, with emphasis on details of the flow within the blade rows, so that forces acting on the blades may be computed and so that potential aerodynamic problem areas may be located (e.g., blade sections with excessive adverse pressure gradients).

---

†Objectives 4–6 may not apply in a few design cases. If the compressor is very small in size, manufacturing limits on geometry (e.g., blade dimensions) may force the design of a compressor larger than the minimum that aerodynamic criteria alone would allow.

(3) Methods for predicting compressor performance at off-design flows and rotational speeds and with nondesign entrance fluid properties and flow distributions.

(4) Stress analysis techniques to verify the stress-strain characteristics of blade row geometries, disks, and shrouds.

(5) Aeroelastic analysis methods to insure that blades, disks, and shrouds will not be subject to unexpected flutter or other vibrational effects at potential compressor operating points.

In subsequent sections these parts of the complete design system will be referred to by the following terminology:

(1) Preliminary design: hub-to-tip passage design and blade row geometry selection.

(2) Design analysis.

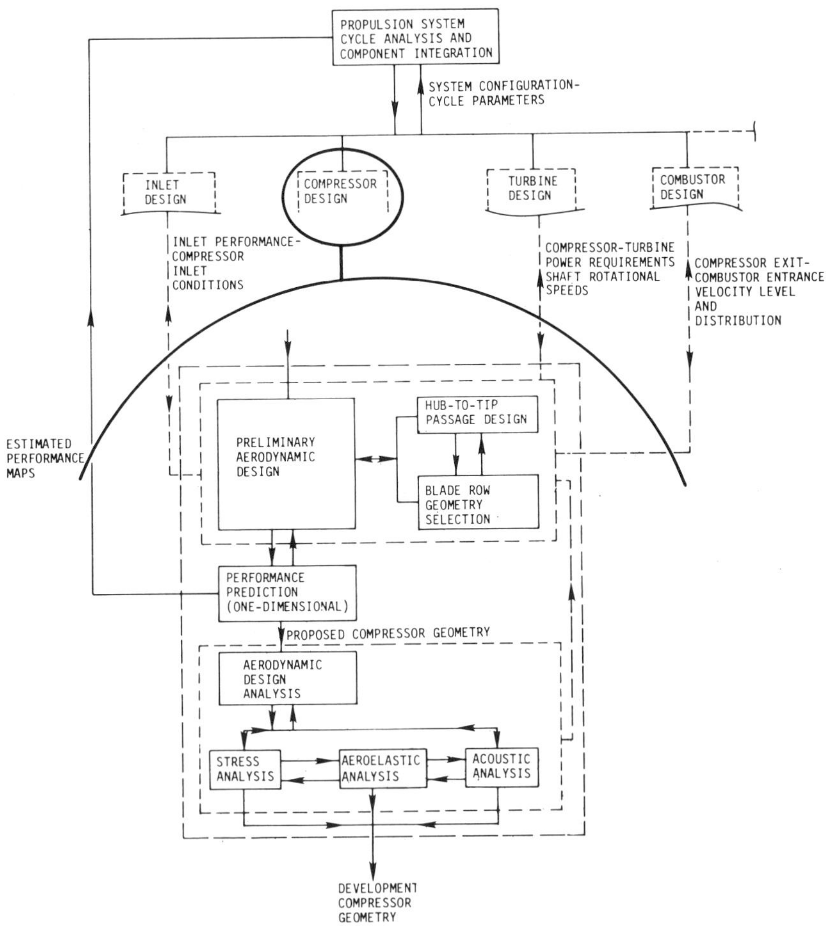

Fig. 3.17 Compressor design system flow chart.

(3) Performance prediction.
(4) Stress analysis.
(5) Aeroelastic analysis.

Figure 3.17 is a possible design system flow diagram that indicates an order of procedure and most probable iterations. However, in some ways, compressor design is likely to remain an artistic endeavor for many years to come and, as a result, it will be highly "specialist dependent" with a "local philosophy" in each competing organization. In terms of the flow diagram, specialist activity enters the picture primarily at the preliminary design level, where past experience in compressor development—including memories of performance maps and of mechanical, fabrication, and aeroelastic problems—has a large influence on the choices that produce candidate geometries for further study.

3.6 Content of Current and Developing Design Systems —The Technology Base

The previous sections reviewed the objectives and component parts of a design system. The next step is to outline in more detail the strategy followed in setting up a series of computational models for the flow as required by the various phases of design. In this process, the significant aerodynamic and aeromechanical problems and limits encountered in axial flow compressor design will be described.

The design and development process is not just a matter of deriving a set of equations and a technique for solution. It requires access to a significant amount and variety of experimental data in correlated form, which will aid in the estimation of (1) the flow turning and loss characteristics of the multiple cascades of airfoils encountered in the various blade rows, (2) the growth of boundary layers on the flow passage walls, and (3) the proximity of operating conditions to possible regions of aerodynamic and aeromechanical trouble. In most cases, it demands constant access to experimental installations such as plane or annular cascade tunnels, single-stage and stage group test units, and model or full-scale multistage compressor experimental facilities. While the experimental support for a design task was intentionally omitted from the discussion of the requirements for a design system, the nature of the necessary experimental base will be referred to frequently in this section.

An overwhelming array of background literature exists on the development and evaluation of flow models and associated computational methods for turbomachinery. There is no possibility of condensing the content of the literature in the present chapter. There is a possibility that the important technical features of the most-used models and methods can be briefly described and compared.

General Comments on Flow Models

Flowfield models—steady and adiabatic flow, continuous fluid medium. In axial flow compressor design and analysis the working fluid is almost invariably and without discussion assumed to behave as a continuous medium, allowing use of the standard texts on fluid mechanics as sources for the equations. Existing flowfield models for design and design analysis cases are almost always based on a line of reasoning that views the flow relative to each blade row as steady and adiabatic. While this assumption contradicts obvious physical facts, it has been essential in order to reach quantitative solutions and, fortunately, many of these solutions have been quantitatively acceptable. Only a few design analysis studies have approached the unsteady flow model problem in a realistic way and even fewer have seriously attacked the computational aspect of these unsteady flow models. In the following subsection, the design system components will be dealt with using adiabatic, steady or time-average relative flowfield assumptions.

Streamlines, stream surfaces, and stream tubes. The process of the development of mathematical models for the flowfield in axial flow turbomachines is often based on the concepts of streamlines, stream surfaces, and/or stream tubes in the flow path. These lines, surfaces, or tubes have been generally handled aerodynamically and mathematically, as if they were the paths of fluid particles or conduits through which specific fluid particles could be followed. In order for streamlines and path lines to coincide in a flow, the flow must be steady. Although additional restrictive assumptions are often made about streamlines and surfaces, the steady flow assumption is the most common.

There has been no lack of recognition in the past that the streamlines and surfaces in an axial flow compressor or turbine would be geometrically complex even if the steady flow assumption is made. In relatively recent times, the most widely referenced recognition of this complexity is found in the analyses of C.-H. Wu. His work brought the idea of S_1 and S_2 stream surfaces into the working vocabulary of the compressor specialist. Wu formulated equations for relative flows on families of hub-to-tip (S_2) surfaces between and within the blade rows and for the relative flows on families of periodic blade-to-blade (S_1) surfaces both between and inside the blade rows. These intersecting families of surfaces, shown schematically in Fig. 3.18, were proposed as a means for computation of the turbomachine flowfield by an iterative process.

S_1 and S_2 surfaces have been discussed in many publications on flow modeling and in many reports and papers in which quantitative results of flowfield computations have been presented. Readers, however, should be cautious in their study of any of these analyses because the vast majority do not determine or compute on true S_1 or S_2 surfaces, but only on approximations to these surfaces.

In the case of blade-to-blade (S_1) flows, frequent use is made of the idea of assumed axisymmetric stream surfaces or stream tubes. These lead to simplified, but not simple, equation formulation and add credibility to the thinking that justifies the use of "circumferentially averaged" velocities and fluid properties in hub-to-tip flow model equations.

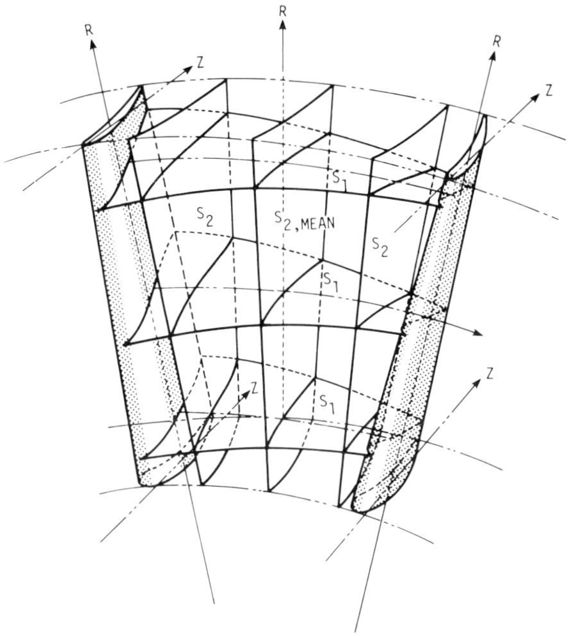

Fig. 3.18 S_1 and S_2 surfaces in blade-to-blade flow passage (from Ref. 11).

For hub-to-tip computations, the flow (when computed inside blade rows) is often handled as if it took place on a "mean stream surface" between the blades. When used, this mean surface is supposed to be representative of the family of S_2 surfaces inside the channel formed by adjacent blades.

Based on the preceding discussion, one should examine every proposed compressor flow analysis to reveal whether it is based on axisymmetric or true S_1 equation formulation and whether it works with a true S_2 surface, a mean surface, or in a meridional plane. In the end, remember that none of the above is more than a simulation of the real flow because of the equation forms assumed. In almost every case where the hub-to-tip and blade-to-blade solution surface idea is utilized (whether or not the surfaces are S_2 and S_1 stream surfaces), the result is what should be called a quasi-three-dimensional or pseudo-three-dimensional flowfield method.

Effects of viscosity and compressibility. The effects produced by fluid viscosity in real compressor flowfields are included in the flow models used for preliminary design and design analysis in a variety of ways. Because these real fluid effects are the ultimate source of irreversibility in the compression process, they must be simulated or modeled in some fashion in the flowfield computation process. It is essential in the preliminary design to set quantitative limits on the aerodynamic parameters guaranteeing that viscous and compressibility losses will not be excessive.

Also, it is necessary in design analysis to be able to estimate these effects as a function of aerodynamic parameters associated with the flowfield.

The sources of losses or thermodynamic irreversibility effects observed in compressor configurations are not different from those found in any other viscous, compressible internal flow. As already noted, three-dimensional boundary layers form on blade surfaces and on the hub and tip end walls. Complex shear flows occur in the corners (where the blades meet the end walls) and at the tip radial clearances (between rotors and stators and the end wall). Wakes develop downstream of the blade row trailing edges and vorticity is generated because of circulation gradients along the blade span and unsteady flow at the entrance to most blade rows.

All of these effects of viscosity lead to continuous and cumulative increases in entropy along the paths followed by fluid particles. If the existence of streamlines and stream surfaces is assumed, entropy increases should be allowed for in the flow direction in a consistent way.

In *most* contemporary compressors, absolute or relative velocities are sufficiently high so that regions of relative sonic flow occur on the surfaces of blades. In *many* contemporary compressors, the velocities relative to rotor and/or stator blade rows are supersonic along at least a portion of the blade span. These configurations are called transonic compressors. In *some* compressors, velocities relative to the rotor and/or stator blade rows have been supersonic along the entire span of the blade at the entrance and/or exit. These configurations are called supersonic compressors. Shock wave patterns developed in all of these cases are complex in shape and are sources of losses, as are all shock waves. At the same time, the shock wave field interacts with boundary layers and influences their development. Once again, the situation calls for consistent accounting for entropy increase in the direction of flow.

When methods for modeling the flowfield in a compressor flow path are studied, a strategy for formulation is seen that is to some extent controversial. The level of controversy is reduced somewhat by the fact that the strategy seems to work. In all real flows, there are local shear stresses acting on all individual fluid particles. These shear stresses are recognized and modeled, for example, in the Navier-Stokes equations of motion. As an individual particle moves along its path from a blade row entrance to the exit, the shear stresses acting on it at any particular point may be quite small compared to the other influences on its motion. However, if the cumulative thermodynamic effect of the shear stresses along a path line between the entrance and exit of a blade row is considered, a significant irreversibility can be measured. Readers will see evidence of a double-standard approach when viscous terms are dropped from the equation of motion, but the terms are effectively reintroduced by using entropy gradients in flows as a means of accounting for accumulated irreversibility along path lines.

Specific effects of blade row and passage geometry on the flowfield. Another frequently misunderstood or misrepresented influence on design and design analysis flow models for turbomachines are the force terms in the equations of motion. In a typical textbook set of component equations of motion for cylindrical coordinates, x, r, and θ, one might find

$$F_r - \frac{1}{\rho}\frac{\partial p}{\partial r} = \frac{\partial V_r}{\partial t} + V_r\frac{\partial V_r}{\partial r} + \frac{V_\theta}{r}\frac{\partial V_r}{\partial \theta} + V_x\frac{\partial V_r}{\partial x} - \frac{V_\theta^2}{r} \tag{3.2}$$

$$F_\theta - \frac{1}{\rho r}\frac{\partial p}{\partial \theta} = \frac{\partial V_\theta}{\partial t} + V_r\frac{\partial V_\theta}{\partial r} + \frac{V_\theta}{r}\frac{\partial V_\theta}{\partial \theta} + V_x\frac{\partial V_\theta}{\partial x} + \frac{V_r V_\theta}{r} \tag{3.3}$$

$$F_x - \frac{1}{\rho}\frac{\partial p}{\partial x} = \frac{\partial V_x}{\partial t} + V_r\frac{\partial V_x}{\partial r} + \frac{V_\theta}{r}\frac{\partial V_x}{\partial \theta} + V_x\frac{\partial V_x}{\partial x} \tag{3.4}$$

These equations are usually considered in fluid mechanics to model the flow of an inviscid, compressible, and continuous fluid medium. The $F_{x,r,\theta}$ terms are body force components; for example, those due to the influence of a gravitational force field. In turbomachine flow models, however, an F-component term may be used to (1) simulate the viscous or "dissipation" effects in the flowfield and (2) represent a distributed force field in the flow, which accounts for the effects on the flow due to blade forces (because blades have camber and lean or dihedral, there can be components in the x, r, and θ directions).

For hub-to-tip surface solutions inside blade rows the computation model should account for physical blockage of the flow path by the blades as well as for effective blockage by boundary layers on the blade surfaces. At the blade ends and on the passage walls between rows, effective blockage due to end wall and corner flows should be recognized. The mechanism for accounting for effective blockage will vary with the nature of the flow model.

Equations

With the initial assumption of a continuous fluid medium, recognize the existence of a number of mathematical formulations of physical laws that may be the basis for models of the real flow in turbomachines, and specifically in axial flow compressors. The physical laws are (1) conservation of mass, (2) conservation of linear and angular momentum, (3) first law of thermodynamics, and (4) second law of thermodynamics. These are supplemented for any working fluid by (5) equations of state, (6) relationships between thermodynamic properties of a pure substance, and (7) relationships between the flowfield velocities and velocity gradients and the shear stresses in the fluid.

Table 3.1 gives some equations based on these laws and other relationships. In the case of some of the differential equations, component equations are written for a cylindrical coordinate system. Table 3.1 will serve as a reference for subsequent sections.

Table 3.1 Reference Equations

From the law of conservation of momentum for steady flow in an inertial, cylindrical coordinate system, local shear stresses neglected: at any point in a flowfield,

$$F_r - \frac{1}{\rho}\frac{\partial p}{\partial r} = V_r \frac{\partial V_r}{\partial r} + \frac{V_\theta}{r}\frac{\partial V_r}{\partial \theta} + V_x \frac{\partial V_r}{\partial x} - \frac{V_\theta^2}{r} \tag{3.2}$$

$$F_\theta - \frac{1}{\rho r}\frac{\partial p}{\partial \theta} = V_r \frac{\partial V_\theta}{\partial r} + \frac{V_\theta}{r}\frac{\partial V_\theta}{\partial \theta} + V_x \frac{\partial V_\theta}{\partial x} + \frac{V_r V_\theta}{r} \tag{3.3}$$

$$F_x - \frac{1}{\rho}\frac{\partial p}{\partial x} = V_r \frac{\partial V_x}{\partial r} + \frac{V_\theta}{r}\frac{\partial V_x}{\partial \theta} + V_x \frac{\partial V_x}{\partial x} \tag{3.4}$$

From the law of conservation of momentum for steady flow measured relative to a cylindrical coordinate system rotating at angular velocity ω about x axis: at any point in a flowfield,

$$T\frac{\partial s}{\partial r} - \frac{\partial I}{\partial r} = -\frac{W_\theta}{r}\left[\frac{\partial}{\partial r}(rV_\theta) - \frac{\partial W_r}{\partial \theta}\right] + W_x\left[\frac{\partial W_r}{\partial x} - \frac{\partial W_x}{\partial r}\right]$$

$$\frac{T}{r}\frac{\partial s}{\partial \theta} - \frac{1}{r}\frac{\partial I}{\partial \theta} = \frac{W_r}{r}\left[\frac{\partial}{\partial r}(rV_\theta) - \frac{\partial W_r}{\partial \theta}\right] - W_x\left[\frac{1}{r}\frac{\partial W_x}{\partial \theta} - \frac{\partial W_\theta}{\partial x}\right]$$

$$T\frac{\partial s}{\partial x} - \frac{\partial I}{\partial x} = -W_r\left[\frac{\partial W_r}{\partial x} - \frac{\partial W_x}{\partial r}\right] + W_\theta\left[\frac{1}{r}\frac{\partial W_x}{\partial \theta} - \frac{\partial W_\theta}{\partial x}\right]$$

$$I = H - \omega(rV_\theta) = h + (\mathbf{V}^2/2) - \omega(rV_\theta) \tag{3.5}$$

From the law of conservation of mass for steady flow in a cylindrical coordinate system: at any point in a flowfield,

$$\frac{1}{r}\frac{\partial}{\partial r}(r\rho V_r) + \frac{1}{r}\frac{\partial}{\partial \theta}(\rho V_\theta) + \frac{\partial}{\partial x}(\rho V_x) = 0 \tag{3.6}$$

or

$$\frac{1}{r}\frac{\partial}{\partial r}(r\rho W_r) + \frac{1}{r}\frac{\partial}{\partial \theta}(\rho W_\theta) + \frac{\partial}{\partial x}(\rho W_x) = 0 \tag{3.7}$$

(Table continued on next page.)

Table 3.1 (cont.) Reference Equations

From the law of conservation of mass for a fixed control volume and steady flow:

$$\oiint_{CS}(\rho \mathbf{V} \cdot \mathrm{d}\mathbf{A}) = 0 \tag{3.8}$$

for region A_1 of control surface

$$m = \oiint_{A_1}(\rho \mathbf{V} \cdot \mathrm{d}\mathbf{A})$$

From the first law of thermodynamics for a fixed control volume and steady flow:

$$\frac{\delta q}{\mathrm{d}t} - \frac{\delta W_{\mathrm{shaft}}}{\mathrm{d}t} - \frac{\delta W_{\mathrm{shear}}}{\mathrm{d}t} = \oiint_{CS}\left(h + gz + \frac{\mathbf{V}^2}{2}\right)(\rho \mathbf{V} \cdot \mathrm{d}\mathbf{A}) \tag{3.9}$$

From the law of conservation of angular momentum for a fixed control volume and steady flow, component equation for x axis:

$$T_S + T_B = \oiint_{CS}(rV_\theta)(\rho \mathbf{V} \cdot \mathrm{d}\mathbf{A}) \tag{3.10}$$

where T_S is the sum of moments about x axis of all surface forces acting on contents of control volume and T_B the sum of moments about x axis of all body forces acting on contents of control volume.

For a pure substance:

$$\mathrm{d}h = \mathrm{d}p/\rho + T\mathrm{d}s \tag{3.11}$$

The second law of thermodynamics, of course, enters into the parameters that measure irreversibility or losses in the flowfield and, since in classical thermodynamics the property entropy is derived as a consequence of the second law, the law enters into the formulation of the Gibbs relation [Eq. (3.11)].

Equations of state for fluid media are available in mathematical or tabular forms. The relationships between velocities, velocity gradients, and fluid shear stresses cannot be covered here, but are discussed in Chap. 6 of this volume and Chaps. 2 and 10 of *Aerothermodynamics of Gas Turbine and Rocket Propulsion*.

Performance Limiting Flow Phenomena— Design Limit Parameters

Design limit parameters provide a quantitative basis for predicting the conditions under which unacceptable blade row or overall performance will occur. They are needed in preliminary design as arbitrary stop signs in setting up a geometry-flowfield combination and they are used in design analysis to estimate the degree of risk associated with a particular flow pattern. In terms of design system objectives, the following should be established:

(1) Whether any form of aerodynamic or aeromechanical instability will occur anywhere in the performance range of the compressor as the result of a design decision, with parameters that will permit estimation of the margin of safety between any operating point and the nearest unstable condition.

(2) Values of parameters influencing mass flow per unit frontal area that may be utilized in design to realize high specific mass flow rates without adversely affecting other elements of compressor performance.

(3) Quantitative values of parameters that will determine the maximum stage pressure ratios which can be realized at specified efficiency levels and that will allow estimation of the thermodynamic irreversibility associated with specified combinations of blade row geometry and flowfield conditions.

These are not three unrelated problems. For example, a decision made in design that leads to a high pressure ratio at an acceptable efficiency for a given stage may set up flow conditions which cause that stage to have an inadequate stall (instability) margin.

Historically, limiting aerodynamic parameters in axial flow compressors have been of four fundamental types:

(1) Limits related to aerodynamic "loading," especially to the generation of unacceptable losses or flow instability by possible blade row geometries and to the prediction of proximity of an operating condition to regions of high loss or instability.

(2) Limits related to the generation of high losses or flow instability as the result of the effects of compressibility.

(3) Parameters related to the limiting of flow rate through a flow path element as the result of compressibility.

(4) Limits related to the aeroelastic aspects of compressor blade rows, including prediction of the occurrence of the various flutter regimes.

The first three of these will be briefly dealt with in the following subsections. The aeroelastic limits are considered in Chap. 7 of *Aircraft Propulsion System Technology and Design*.

Aerodynamic design limits related to excessive velocity diffusion rates. The velocity diagrams discussed earlier in this chapter were used to aid in understanding the functions of blade rows in typical compressor configurations and to point out that some familiar parameters and characteristics of the flow in isolated airfoil technology are encountered again in the technology of airfoil cascades. It was noted that there were some fundamental differences in these technologies. One of these differences is that, in the cascade arrangements utilized in compressors and turbines, there is generally a change in static pressure level between the flowfield upstream and the flowfield downstream of the cascade. In compressors one expects to find rows in which static pressure rises. This is evident in typical velocity diagrams for axial flow compressors (e.g., Fig. 3.10) where the blade rows can be seen to act as diffusers of the relative fluid velocity. Diffusers are flow passages in which the relative velocity decreases from the entrance to exit while the static pressure level increases. Boundary layers in excessive adverse (positive) pressure gradients separate. This is the cause of stalling of isolated airfoils and of stalling and flow instability in airfoil cascades (as well as other diffusers). It might be well to recall here that in fluid mechanics there are both subsonic and supersonic diffusers. Both have the same function, but the geometries are significantly different. Both subsonic and supersonic diffusing situations are encountered in compressor blade rows.

A second look at the velocity diagram of Fig. 3.10 shows that some of the diffusion of relative flow takes place in these cascades because the flow is deflected (turned). Velocity diffusion is increased if the flow is turned more (while the axial velocity remains constant) or if the axial velocity across a blade row is reduced (with a constant turning).

There should be no question about the fact that higher stage pressure ratios will be related to increased diffusion rates across the blade rows and that there are quantitative limits on diffusion rates which in turn limit stage pressure ratios. What are these quantitative limits? A lot of suggestions have been made. Some are simple and direct. One used frequently is $p_2 - p_1/P_1 - p_1$, the static pressure rise coefficient. Another is the exit/entrance relative velocity ratio $(W_2/W_1)_{\text{rotor}}$ or $(V_2/V_1)_{\text{stator}}$. A slightly less direct parameter that has been widely used is D. This parameter was originally developed by Lieblein as a measure of the diffusion of velocity on the suction surfaces of cascades of blade sections in two-dimensional flow. The equation derived was

$$D = 1 - \frac{\mathbf{V}_2}{\mathbf{V}_1} + \frac{V_{\theta 1} - V_{\theta 2}}{2\sigma \mathbf{V}_1} \tag{3.12}$$

Initial application to rotating blade rows used two defining equations,

$$D = 1 - \frac{\mathbf{W}_2}{\mathbf{W}_1} + \frac{V_{\theta 2} - V_{\theta 1}}{2\sigma \mathbf{W}_1} \tag{3.13}$$

or

$$D = 1 - \frac{\mathbf{W}_2}{\mathbf{W}_1} + \frac{W_{\theta 1} - W_{\theta 2}}{2\sigma \mathbf{W}_1} \tag{3.14}$$

which are not equivalent unless $U_1 = U_2$.

Subsequently, for cases in which compressor blade rows were involved, a version allowing for flow on stream surfaces that change radius between row entrance and exit was developed,

$$D = 1 - \frac{\mathbf{W}_2}{\mathbf{W}_1} \pm \frac{r_1 V_{\theta 1} - r_2 V_{\theta 2}}{(r_1 + r_2)\sigma \mathbf{W}_1} \qquad \begin{matrix} + = \text{stators} \\ - = \text{rotors} \end{matrix} \tag{3.15}$$

In all of these definitions, the final term is a measure of blade row circulation.

The parameter D was intended to provide a quantitative limit for the diffusion in cascades operating near the minimum-loss angle of incidence. It could be calculated using flow velocities readily available from design point velocity diagrams. Through the years, its application has evolved by extrapolative thinking to include situations in which it is not correct. Alternative parameters have been suggested, but the final answer is not in hand and the ultimate diffusion limit remains a subject for research, just as there are not genuinely definitive design limits available for most other diffuser cases.

All of the diffusion or blade loading limit parameters mentioned are useless without numerical values that define regimes of acceptable and unacceptable blade row performance. Such values are established by correlation of experimental blade row performance data from plane cascades or from rotor and stator blade rows in compressors. Such correlations have been published for cascades and blade rows by the National Aeronautics and Space Administration (NASA), such as Figs. 3.19 and 3.20 and by others for specific compressor test data sets. Generally, the correlations allow estimation of an expected profile total pressure loss coefficient for blade rows at incidence angles in the low-loss region. The loss parameters used are in forms such as $\bar{\omega}\cos\beta_2/2\sigma$ or θ^*/c. Although the parameter D has probably been the most widely accepted blade loading parameter for preliminary design, others have been proposed to replace or supplement it

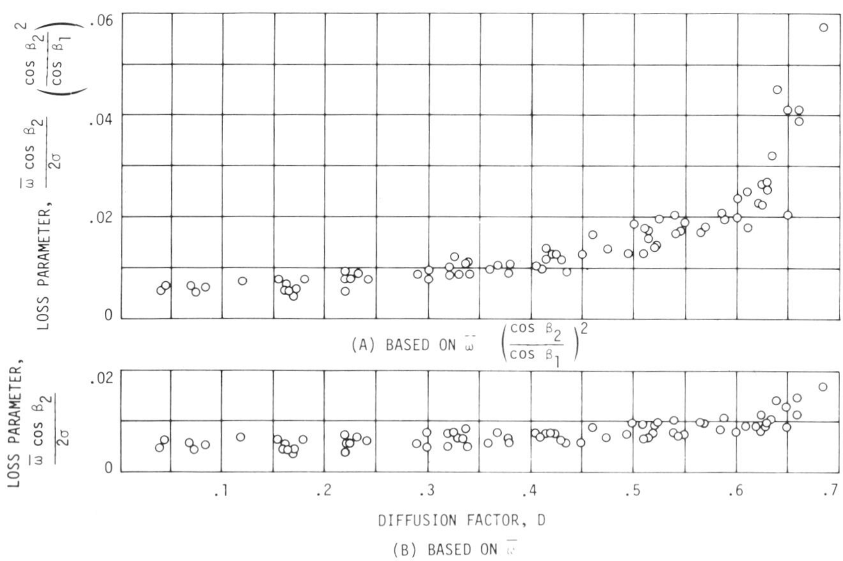

Fig. 3.19 Variation of loss parameters with diffusion factor at reference minimum loss incidence angle computed from low-speed cascade tests of NACA 65-(A_{10}) 10-cascade blades (from Ref. 12).

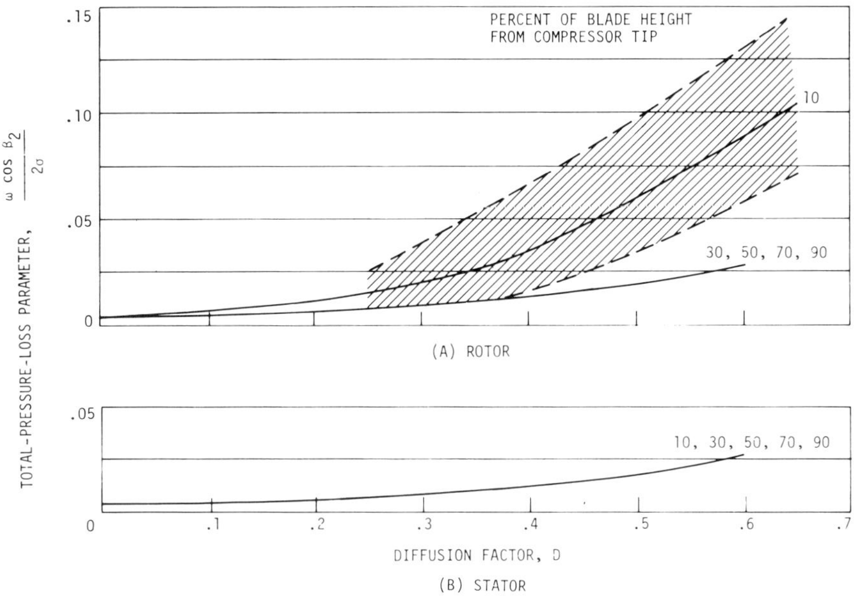

Fig. 3.20 Variation of loss parameters with diffusion factor for blade sections in compressors operating near minimum loss incidence angle (from Ref. 13).

in design analysis. These include a second Lieblein parameter, known as D_{eq}, which has been utilized in both design analysis and off-design point performance prediction.

In the regions near the hub and tip end walls, some design groups prefer to use the static pressure rise coefficient $(p_2 - p_1)/(P_1 - p_1)$ as a limit. This parameter is considered to be a useful criterion in the prediction of the onset of "wall stall."

Influence of Mach number levels on axial flow compressor design. For many years there have been continuous trends toward higher levels of Mach number in axial flow compressor blade rows. Figure 3.21 shows one reason why this trend exists. The figure demonstrates how the flow rate per unit frontal area (specific flow rate) for a compressor stage varies with average axial Mach number and hub-tip ratio. High flow rate per unit frontal area has been designated as an important design objective because of the obvious relationship between the required frontal area and the component weight. The gains to be made by increased axial Mach number and by decreased hub-tip ratios are evident in Fig. 3.21.

A second reason for the trend toward higher Mach number levels in compressor stages develops from the design objective calling for increased stage pressure ratios. As observed previously, the energy transfer in a compressor stage takes place in the rotor and is governed by the Euler

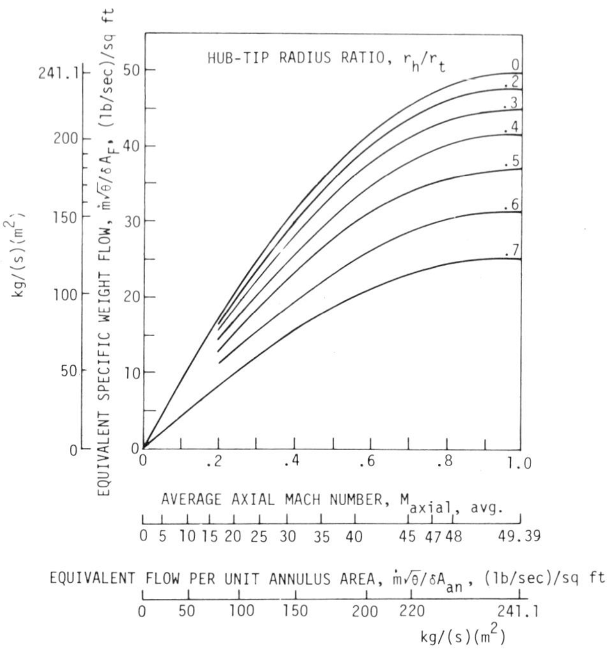

Fig. 3.21 Compressor specific flow rate as a function of axial Mach number and hub-tip ratio.

turbine equation, which represents an application of the law of conservation of angular momentum [Eq. (3.10)] to an axisymmetric stream tube,

$$E/\delta m = U_2 V_{\theta,2} - U_1 V_{\theta,1} \tag{3.16}$$

Either increased U or increased change in *absolute* tangential velocity is effective in increasing energy transfer. However, the loading limit D discussed above includes a term showing the direct contribution of $(V_{\theta,2} - V_{\theta,1})$ to diffusion rates. Estimation of the other diffusion limits shows that the change in $(V_{\theta,2} - V_{\theta,1})$ enters each limit in a similar way. Figure 3.22 shows how increased blade speed and increased rotor loading, as measured by Lieblein's diffusion factor, affect the pressure ratio available from a stage. The effect of increased blade speed is impressive. The data points are measured values for single stages.

The problem associated with high blade speed values has several facets. First, such values lead to high centrifugal force loadings on rotor blades. Second, there are direct relationships between the blade speed and noise generation. The third and most significant aerodynamic feature of the problem is that high blade speeds are likely to result in high flow Mach

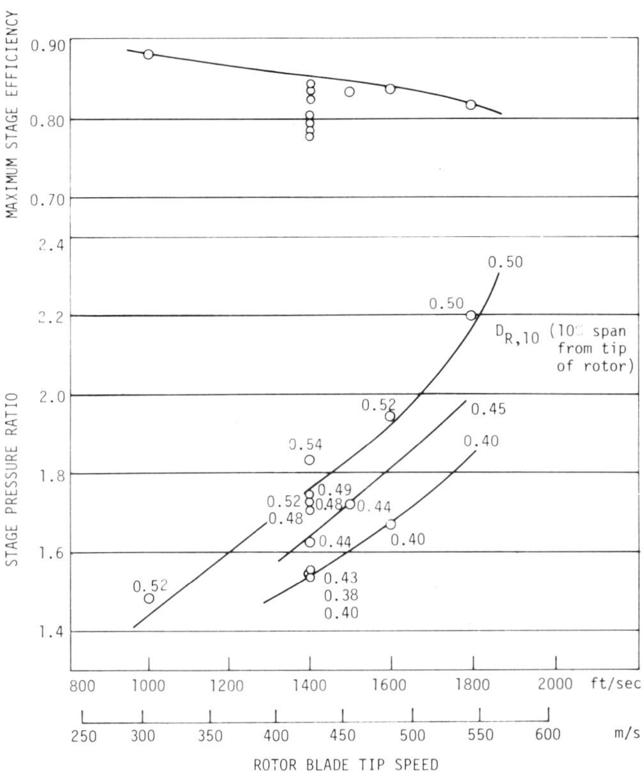

Fig. 3.22 Trends in low hub-tip ratio stage performance with rotor tip speed and loading level.

numbers measured relative to the rotating blades. Consider a rotor row with no tangential fluid velocity component at the leading edge. The relative velocity would be

$$W_1 = \sqrt{V_m^2 + U^2} \tag{3.17}$$

$$M_{\text{rel},1} = \sqrt{M_m^2 + M_u^2} \tag{3.18}$$

$$M_u = U/a_1 \tag{3.19}$$

Therefore, the objectives of high mass flow per unit frontal area and high pressure ratio per stage both force designers to move in the direction of high rotor relative Mach numbers. These same objectives also lead to high values of absolute Mach number at the entrance of stator blade rows.

The potential for advanced compressor performance levels associated with high average axial Mach number levels and high rotor blade speeds has been understood since the early 1940s; since that time numerous stage configurations utilizing supersonic Mach number levels have been suggested, designed, and tested. Until about 1950, the actual performance levels achieved were discouraging. Most of the rotors and stages involved were true supersonic configurations, with relatively high hub-tip ratios and supersonic relative Mach numbers along the entire blade span in rotors and/or stators at the design point. The poor performance included low, sometimes disastrous, efficiencies and vertical pressure ratio flow lines at design rotational speed. On the other hand, experimental subsonic rotors were designed to operate with high blade speeds, but with inlet guide vanes to reduce the level of the inlet Mach number at the tip to "advanced" subsonic values. Unfortunately, these stages generally used blade section geometries and design velocity diagrams that combined to produce excessive blade diffusion loading and poor stage performance.

Then, around 1950, the observation was made simultaneously in several countries that some supersonic stages were performing very well at rotational speeds below the design value. These stages were of the shock-in-rotor type, in which the design relative Mach number at the rotor inlet was supersonic and both the relative Mach number at rotor exit and the absolute Mach number approaching the stator row were subsonic. Encouraging performance was observed under flow conditions involving supersonic relative inlet Mach numbers along only part of the rotor span. Subsequently, stages were designed to operate in this transonic mode and were extremely successful.[‡] The key features of these stages were the selection of an aerodynamically desirable blade geometry and a velocity diagram that eliminated inlet guide vanes while limiting blade diffusion loading.

The transonic compressor design concept was a dramatic breakthrough, which can be seen in proper perspective by noting that, over about a two-year period, limiting relative Mach number levels for experimental rotor blade rows increased from about 0.75 to about 1.2 and a typical research inlet stage pressure ratio level increased from about 1.25 to 1.5. The transonic stage concept led directly to lower inlet stage hub-tip ratios, higher specific flow rates, and higher stage pressure ratios—all at acceptable efficiencies and with acceptable stability characteristics.

While the story of the transonic compressor has a certain romantic flavor, it is a long and difficult one beyond the first "breakthrough" years. Mach number levels approaching or exceeding 1.0 at the inlet and within or at the exit of a blade row tend to be associated with increased thermodynamic losses, particularly at incidence angles away from the minimum-loss values. Also associated with high subsonic and all supersonic Mach number levels is

[‡]In the United States, the early transonic compressor studies were carried out by the National Advisory Committee for Aeronautics (NACA). In the Soviet Union, the basic studies were apparently completed in the Central Institute for Aerohydrodynamics (TsAGI) and the Central Institute for Aviation Engines (TsIAM).

the problem of blade-channel "choking." As a result, it was necessary to develop models for the losses due to shock waves and for the incorporation of both shock and profile losses into the design computation method. Furthermore, the higher stage pressure ratio levels caused more rapid passage area changes and wall curvature, again calling for changes in design procedures.

Mach number levels in axial flow compressor designs have increased steadily during the past 25 years, so that average *axial Mach number* levels of 0.60–0.70 at the first rotor entrance are found in a number of aircraft engine applications, accompanied by compressor exit Mach numbers as high as 0.45. Rotor blade tip relative Mach numbers have been increased to levels from 1.4 to as high as 2.0 in experimental stages. This has required extensive use of special blade sections and detailed consideration of flowfield control in the region of the blade row. Stator blade Mach number levels have also increased, but not as rapidly, with most design values in the high subsonic regime.

The quantitative effects of compressibility on blade row performance have been studied extensively in plane cascades and compressor rotor and stage facilities. As a result, the computation of losses due to shock patterns has been enhanced considerably and the quantitative understanding of the effects of incidence angle, blade row geometry, and downstream conditions on both losses and passage flow rate limits has been improved. Currently, absolute limits are not set on Mach number levels in compressor design, but the control of the flowfield and the magnitude of expected losses at all required levels is considered carefully.

Preliminary Design

In preliminary design, an immediate confrontation occurs between elegance in flow modeling and the requirement to complete the design task. Several approaches are possible. Two will be mentioned here.

Suppose that you were given the task of designing and developing an axial flow compressor for an aircraft gas turbine propulsion system. The given design point information might include only the following: the thermodynamic state of the working fluid in a duct upstream from the compressor; the total pressure increase required from this location to the compressor exit; and mass flow rates at the compressor entrance, in any bypass ducts, at extraction points and at the exit. A table of thermodynamic properties of the working fluid should be available and there will probably be an estimated overall efficiency used by the cycle analysts in preliminary system studies. As suggested by Fig. 3.17, an early step should be to insure frequent exchanges of information with the designers and developers of other components. The turbine shaft rotational speed is important to compressor designers and the fluid velocity levels and patterns at the compressor exit will be a major concern in combustor design.

What is required of the compressor design group is a geometry meeting objectives outlined. The first question is how and where to start.

Input information—first output. The given information suggested above was the following:

m = compressor entrance flow rate
m_a, m_b = all flow rates bypassed or extracted from the configuration with the total pressure levels involved
P_{ex}/P_{in} = overall total pressure ratio for configuration
T_{in} = total temperature at compressor entrance
P_{in} = total pressure at compressor entrance
η = compressor component efficiency or efficiencies used in cycle analysis

The first geometric output information needed includes general configuration dimensions and a proposed shaft rotational speed, as follows:

$D_{\text{tip, max}}$ = maximum tip diameter
$(D_{\text{hub}}/D_{\text{tip}})_{\text{stage 1}}$ = hub-to-tip diameter ratio, first stage
N = shaft rotational speed
N_{st} = estimated number of stages required
L_{comp} = estimated total length of bladed compressor flow path

Component scaling as an approach to preliminary design. At this point, no compressor designer should overlook the possibility or underestimate the advantages of scaling an existing compressor geometry of known performance to meet his current design goals. Recalling the similarity parameter used in comparing performance for a given working fluid

$$\left.\begin{matrix} \dfrac{P_{ex}}{P_{in}} \\ \eta \end{matrix}\right\} = \text{functions of}\left(\text{geometry}, \frac{m\sqrt{T_{in}}}{P_{in}D^2}, \frac{ND}{\sqrt{T_{in}}}, \frac{P_{in}D}{\mu\sqrt{T_{in}}}, Tu, \gamma\right)$$

it is easy to set up a procedure for predicting the geometric scale ratio required to meet any flow requirement if a specific operating point to be scaled is picked from an existing performance map. It will probably not be easy to satisfy all of the equalities needed for exact similitude, nor will it always be possible to find an existing geometry that produces the right pressure ratio at a good operating point. However, judicious scaling can often be supplemented by the addition or deletion of stages at the front (e.g., adding "zero" stages) and rear of existing multistage geometries and by other geometric modifications.

There are dangers in scaling. Aerodynamic, aeroelastic, and stress considerations are all involved and numerous examples of expensive problems in compressor development testify to the need for careful analysis of all these areas of similarity.

Preliminary design by computational methods using a simple flow model. Geometric scaling and related techniques for performance modification do not add significantly to the state-of-the-art and, aside from the satisfaction of creating a genuinely superior compressor configuration, there are other reasons for the decision to design "from scratch." Although the flow models for preliminary design described here are not very sophisticated, it should be understood that real skill is necessary in selecting parameters for a preliminary design study.

From the input and first geometric output listed earlier in this section, as well as the design objective list of Sec. 3.4, it can be seen that a number of questions will be answered when target levels are set for mass flow per unit frontal area and pressure ratio per stage. There is no way to set these target levels except on the basis of past experience and trends.

In the case of mass flow per unit frontal area, the target level is generally set for and by the first stage. Figure 3.21 shows the values of equivalent specific flow rate attainable for various levels of average axial Mach number and hub-tip radius ratio for a first-stage rotor. One significant point demonstrated by Fig. 3.21 is that there are only small gains available in going below hub-tip ratios of about 0.3 and above axial Mach number values of about 0.7. While values of hub-tip ratio of 0.4 and below and $M_{\text{axial, av}}$ greater than 0.6 are found in current compressor designs, it will become evident that there are problems created by pushing too far in either of the directions leading to higher specific flow rates.

Setting target levels for pressure ratio per stage requires the designer to make additional decisions. The attainable total pressure ratio in a given stage is a function of the energy transfer in the rotor of that stage as determined by Eq. (3.1) for each fluid particle and by the thermodynamic losses in the stage determining how much of the energy transfer is converted to an increase in total pressure. However, the total pressure ratio is not in itself a completely valid goal. The real goal should be to realize adequate static pressure rises in each stage so that excessive velocity increases across stages do not result. In gas turbine applications, the compressor exit Mach number will be limited by combustor design requirements, and both entrance $M_{\text{axial, av}}$ and $M_{\text{axial, av}}$ changes from stage to stage must be influenced by this limit.

An estimate of the pressure ratio levels that might be achieved must be made and experience-based plots such as Fig. 3.22 can be of help in this regard. This figure might, for example, be used to select several possible combinations of first rotor tip speed, blade loading parameter, and average stage efficiency to be studied. These selections will be only trial values, subject to change by iteration as the design process continues.

Based on the trial values of $(M_{\text{axial}})_{\text{av}}$, $(D_{\text{hub}}/D_{\text{tip}})_{\text{stage 1}}$, U_{tip}, and $D_{R,10}$, along with the given design point conditions, the size of the compressor and its rotational speed can be estimated by setting rough values for blade row aspect ratios and axial row spacing. At this level in the design, obvious misfit configurations can be eliminated by consultation with the turbine and engine configuration integration groups. If all trial configurations are misfits, some accommodation may be possible by adding inlet guide vanes or by other strategies.

Successful first output configuration combinations may be used as partial input to a less empirical second phase of the preliminary design in which the first real use of flowfield model ideas takes place.

The overall design of an axial flow compressor requires the solution of an inverse problem in aerodynamics. The problem is to determine the compressor geometry for a proposed flowfield. In real compressor design, the only time when the problem is seriously worked on in this fashion is at the preliminary design level. This is done in a restricted way by specifying a few features of the flowfield at selected calculation locations in the compressor, usually located in the spaces between the blade rows.

The computation process ordinarily is based on a very simple form of the radial component of the equation of motion [Eq. (3.2)], on the principal of conservation of mass [Eq. (3.8)], and on the use of Euler's turbine equation [Eq. (3.10)] with the first law of thermodynamics [Eq. (3.9)]. The flow is assumed steady, adiabatic, axisymmetric, and free from radial gradients in entropy at any of the between-row calculation planes. An efficiency level is assumed for each rotor and stage. Experience-based estimates are made for the tip casing radius variation through the compressor, blade row loadings, solidities, and stage-to-stage velocity variations. Approximate velocity diagram shapes are set. The effect of the stream surface shape and curvature is omitted. Under these much simplified conditions, the differential equation governing the radial variation in axial velocity in each calculation plane is

$$V_x \frac{\partial V_x}{\partial r} = \frac{\partial H}{\partial r} - \frac{V_\theta^2}{r} - V_\theta \frac{\partial V_\theta}{\partial r} \tag{3.20}$$

This equation (known as the isentropic simple radial equilibrium equation), along with conservation of mass, must be satisfied by iteration in each calculation plane throughout the compressor. This is a hub-to-tip or meridional plane solution that provides sufficient information about the compressor internal flow pattern to allow the first trial blade row geometries to be determined.

At this level in design, correlated cascade and stage blade row data are generally used for blade selection. Because the solidities and chord lengths were estimated for the meridional plane solution, some degrees of freedom of choice have been eliminated. However, a great many have not. These include blade section geometry (profile shape, thicknesses, and camber line shape) and incidence angle.

Of course, computer programs exist for both meridional plane flowfield equations and for blade selection. Blade sections are, in preliminary design, usually specified on conical stream surfaces as shown in Fig. 3.23.

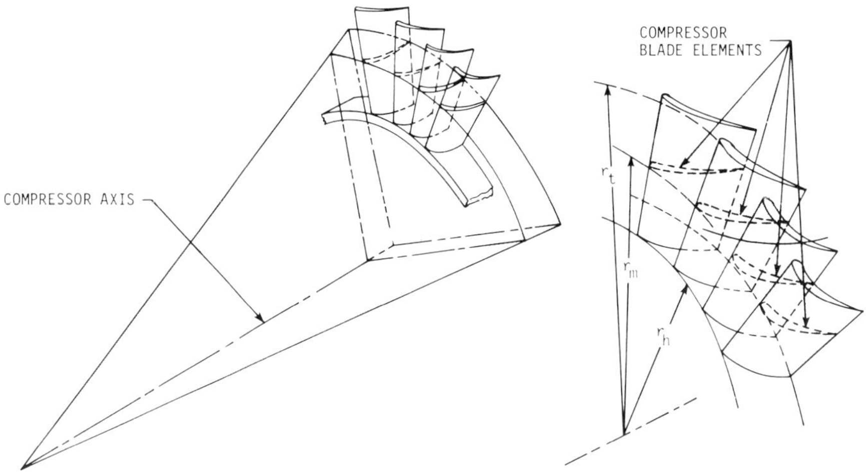

Fig. 3.23 Layout of compressor blade sections on conical surface through stream surface intersection radii at blade row leading and trailing edge.

The result of the immense array of choices made and a significant amount of computation is a rough configuration geometry or series of configuration geometries that can be subjected to the initial scrutiny and criticism of stress analysts, aeroelasticians, and engine layout designers. Any approval at this stage can be only tentative, because the flowfield solution is almost totally incomplete within the blade rows. The next step is design analysis.

Design Analysis

Input and output information. The inputs to the design analysis phase of overall compressor design are the configuration geometries generated in the preliminary design. This represents a complex and extensive collection of information, supplemented by the same design point flow, rotational speed, working fluid, and compressor entrance condition data used in the preliminary design. Now, however, the result for each geometry will be an estimated value of overall pressure ratio and efficiency, with detailed velocity and fluid property distributions throughout the configuration, including those within the blade rows. This information is used to:

(1) Determine whether the configuration geometry, when operating at the design point, will be capable of producing the correct input flow to the next cycle component.

(2) Locate any aerodynamically dangerous flowfield conditions, such as regions of excessive diffusion rate or passage choking.

(3) Provide adequate input for stress, aeroelastic, and acoustic analysis.

Current design analysis methods. The nature of design analysis output requirements calls for a much more inclusive flow model. These

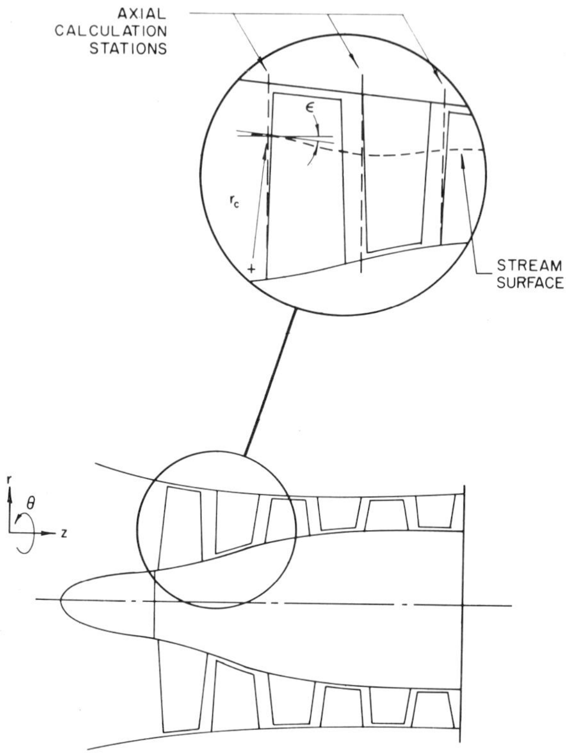

Fig. 3.24 Meridional plane section of axial flow compressor showing axisymmetric stream surface notation.

models are so detailed in content that the numerical solutions can be dealt with only by digital computers.

Most design analysis methods are based on a loosely coupled combination of hub-to-tip flowfield equations and empirically based blade-to-blade surface row performance correlations. The hub-to-tip solution model is generally based on adiabatic, steady relative flow on axisymmetric approximations to stream surfaces. Computations are often based on equations written for quasi-S_2 surfaces. Stream surface geometry (curvature, lean, slope) are computed and subjected to iteration. Values computed are usually projected into the meridional plane. The equations of motion for fixed coordinates [Eqs. (13.2–13.4)] or relative coordinates [Eq. (3.5)] control the distribution of velocity subject to the conservation of mass.

Calculation control surfaces may be radial as in Figs. 3.24 and 3.25 or nonradial to conform to a leading- or trailing-edge shape. As a result, the component equations of motion become complex and difficult to interpret and compare. "Body" force field terms, as discussed earlier, are encountered inside blade rows.

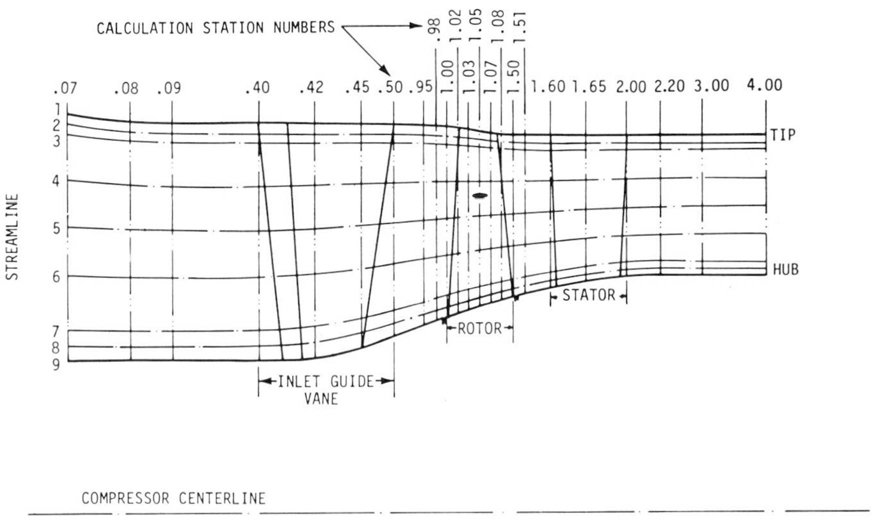

Fig. 3.25 Calculation control surfaces for an axial flow compressor stage with computed stream surface shapes (from Ref. 14).

Equilibrium equations of motion for a radial calculation surface such as those of Fig. 3.24 may be as simple as

$$\frac{1}{\rho}\frac{\partial p}{\partial r} = F_r + \frac{V_\theta^2}{r} \pm \frac{V_m^2 \cos \varepsilon}{r_c} \tag{3.21}$$

but are much more difficult when calculation surfaces are nonradial. In any case, the solution of the equations of motion, conservation of mass, and energy transfer and of the equations resulting from conditions imposed by the second law of thermodynamics is iterative, numerical, and subject to mathematical problems in convergence and stability. The iterations related to the stream surface shape and curvature often must be subject to stability control by damping adjustments.

In design analysis, both analytical and empirical procedures for estimating cascade performance are potentially useful. Whatever procedures are used must predict shock and profile (diffusion rate) losses and flow angle distributions through and downstream of each stream surface/cascade intersection. This is also an iterative process because of the dependence of both losses and downstream flow angle on the real downstream conditions (see Fig. 3.26). A final problem in design analysis method development and application is in the treatment of end wall flows. In this region, the typical main channel flow model assumptions are not adequate, especially with regard to viscous loss accumulation. Several auxiliary end wall models have

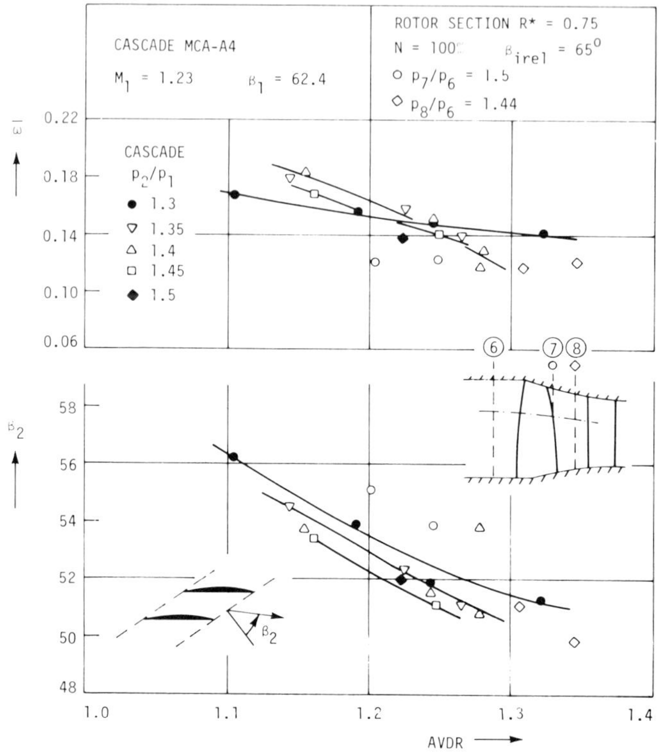

Fig. 3.26 **Comparison of cascade experiments and rotor blade section data (from Ref. 33).**

been suggested, varying from simple empirical blockage allowances through three-dimensional boundary-layer computation methods.

Development of advanced methods for internal flowfield computation. Recently, serious attempts have been made to develop three-dimensional internal flowfield computation methods, all based on digital computers of large capacity and high speed. Most of these methods involve the modeling of viscous effects, including turbulent shear stresses. Results have been obtained for certain geometries and flow situations that, unfortunately, do not correspond to those occurring in axial flow compressors. Over a time span of many years, it seems probable that meaningful progress will be made in realistic problem situations. However, in the meantime, these methods will assist in the design analysis process, and in correlation of data.

Performance Prediction

Design analysis methods as defined in the previous section attempt to predict the compressor configuration flowfield for the design operating point only. This is by no means easy, because the flow model and numerical methods must adapt to some degree of nonoptimum blade section operation. However, if in design analysis a significant region of separated flow was indicated, the calculation would stop until geometric adjustments were made.

However, in performance prediction, the primary objective is to estimate the overall compressor unit performance map (including the surge line) and this requires a capability for accommodating the flow phenomena, which are not easily modeled, in the computation system. Therefore, it is not easy to use a design analysis flow model and computation system for general performance prediction.

Current performance map prediction methods are normally based on one of two simplified approaches. The first is "map generalization," in which families of performance maps are correlated on a nondimensional basis so that similarity methods can be used for proposed compressor geometries. The second is a stage characteristic "stacking" method using approximate stage or stage group performance (see Fig. 3.16) as a starting point. Both the map generalization and stage-stacking procedures are relatively easy to apply, but are inadequate for most advanced configuration applications because they are based on past experience.

Over a period of several years, some of the current design analysis methods will be developed into true performance prediction systems. This will require some adjustments to the flow model and an especially improved capability in estimating cascade/stream surface flow turning and loss for nonoptimum entrance and exit conditions.

Stress, Aeroelastic, and Acoustic Analysis

Flowfield property and velocity distributions from the design analysis phase are utilized as partial input to stress and aeroelastic evaluation methods. During recent years the interaction between "pure" aerodynamic analysis and "pure" aeroelastic predictions has become better understood and as a result genuine progress has been made. Reference should be made to Chap. 7 of *Aircraft Propulsion System Technology and Design* for more detailed comments. Compressor acoustic analysis is covered in Chap. 7 of this volume.

3.7 Component and Configuration Experimental Development

A high-quality design system is under continuous development and one of the primary sources of change lies in the base of experimental information supporting all elements of the system. Cascade experimental results and data correlation have been briefly discussed. There is some evidence that rotor and stage test results supply important adjustments to cascade correlations by introducing the effects of rotation (end wall relative motion, adjacent blade section influence, and other secondary flow features). There is, in fact, an awesome accumulation of experimental design system support.

As advanced and exploratory axial flow compressor configurations are designed and evaluated, there has been a continual requirement to upgrade

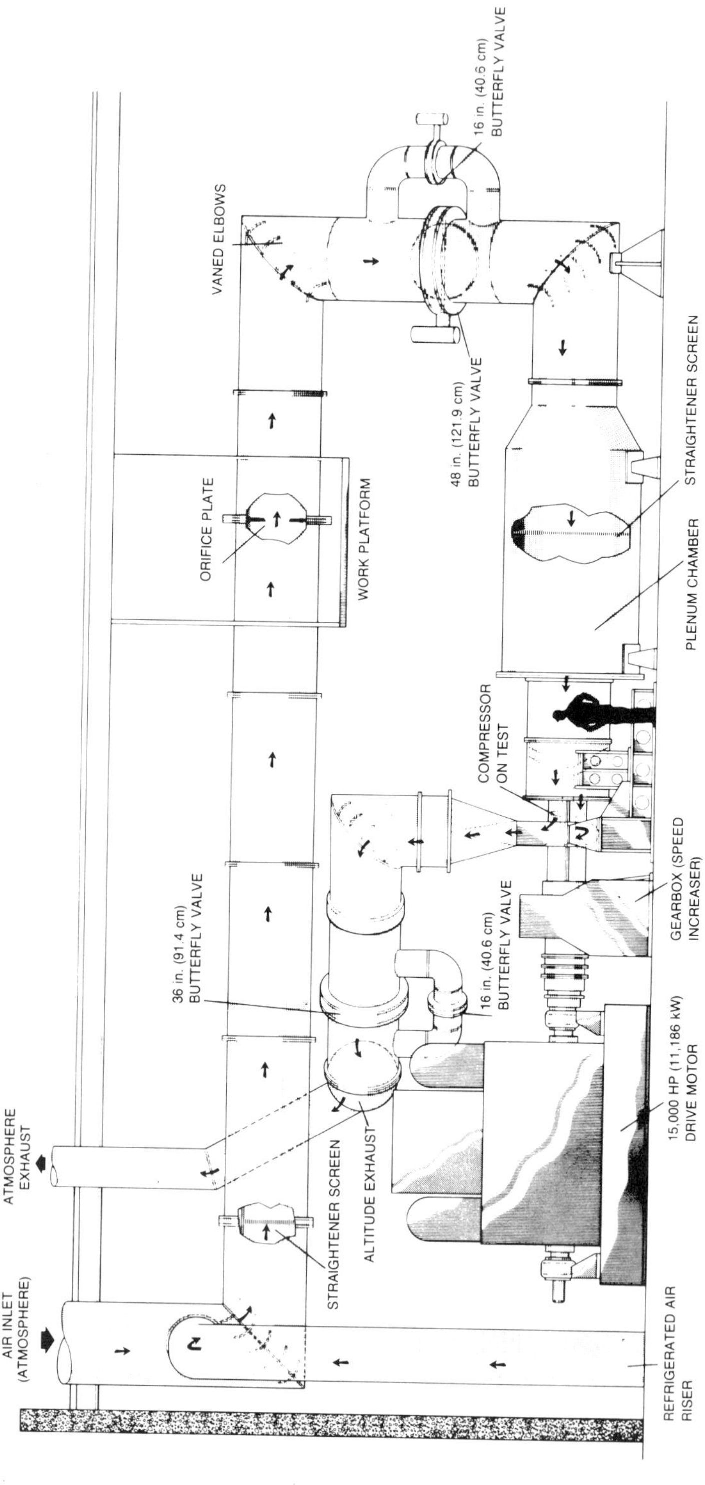

Fig. 3.27 Compressor test facility.

test facility and measurement quality. Plane cascade test section design has been improved radically by adding new flowfield control techniques as well as performance measurement strategy for both subsonic and supersonic flow regimes. Open- and closed-loop rotor, stage, and multistage compressor facilities have been built (see Fig. 3.27) and equipped with extensive instrumentation to reveal transient and unsteady flow phenomena.

However, the greatest direct element of support for compressor design systems comes from steady-state or time-averaged flow passage measurements, both between blade rows and, where possible, inside rows. Review of any preliminary design or design analysis method will quickly demonstrate the dependence of the method on correlated experimental information.

Because a compressor unit installed in an aircraft engine can operate over only a limited part of its performance map and because configuration adjustments are much more difficult in an engine installation, test facilities such as those in Fig. 3.27 are used directly in component development programs. In these programs, only limited instrumentation may be possible, but if some characteristics of stage behavior (frequently only casing static

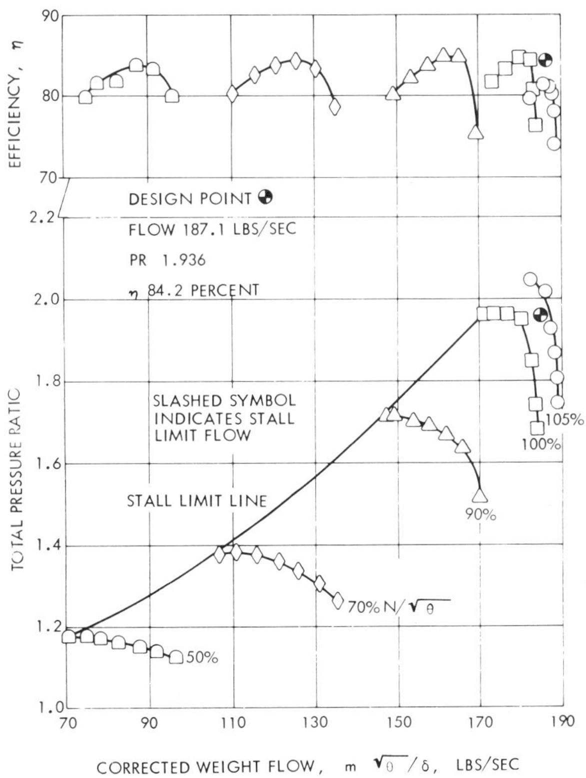

Fig. 3.28 Overall performance of single-stage axial flow compressor designed to operate at a tip speed of 1600 ft/s (488 m/s) (from Ref. 15).

pressure patterns) can be measured along with overall performance, significant improvements in performance maps can often be made by rematching individual stages or by alterations to specific blade row geometries.

3.8 Axial Flow Compressor Performance Trends

Figures 3.28–3.30 have been included to show in a very limited way the performance results representative of recent compressor research and development activities. They also give a qualitative idea of the capabilities and limitations of compressor design systems. Reference should be made to Figs. 3.21 and 3.22 as the performance is reviewed to see where the individual units fall in relation to preliminary design choices.

Figure 3.28 is a plot of measured overall performance for a single-stage transonic compressor design. This stage had a rotor entrance tip diameter of

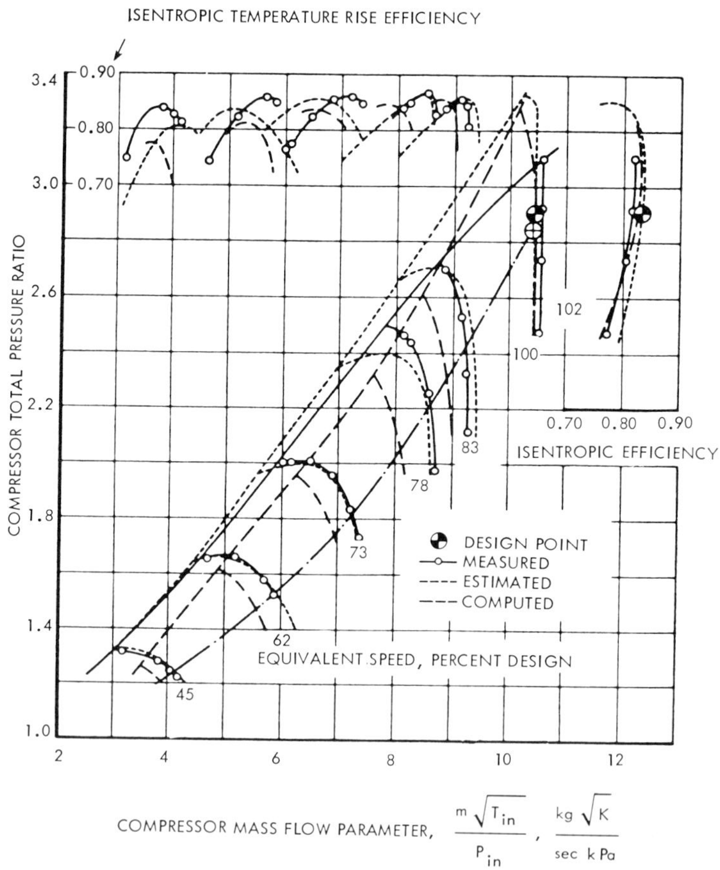

Fig. 3.29 Predicted and measured performance characteristics of three-stage axial flow compressor (from Ref. 5).

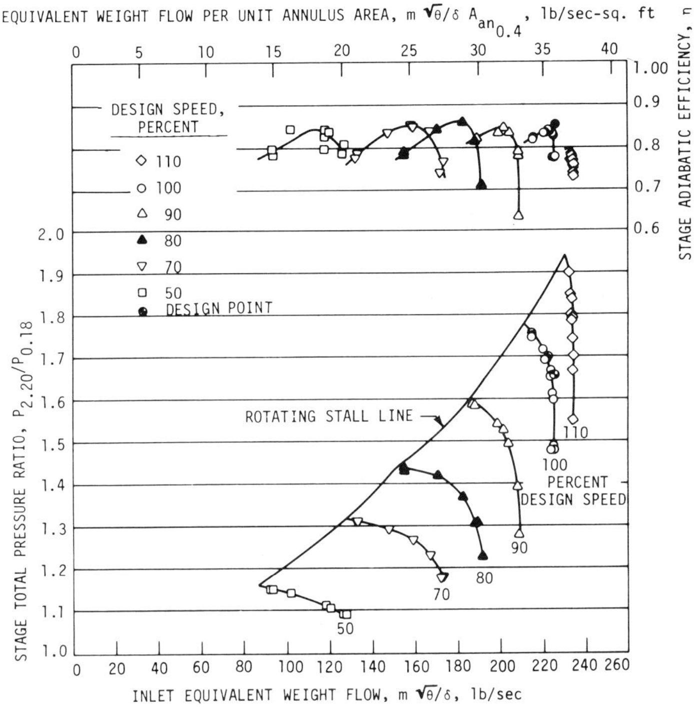

Fig. 3.30a Stage performance map for IGV/stator schedule 0/0 deg (see Fig. 3.25 and Ref. 14).

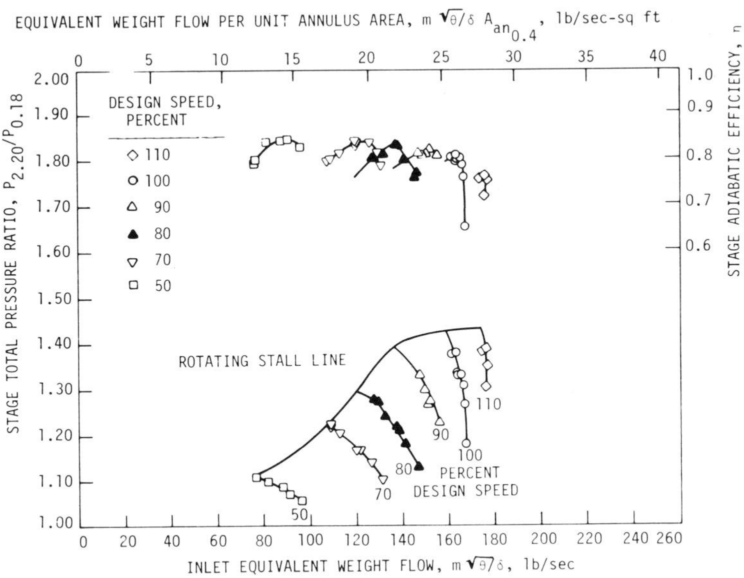

Fig. 3.30b Stage performance map for IGV/stator schedule 40/8 deg (see Fig. 3.25 and Ref. 14).

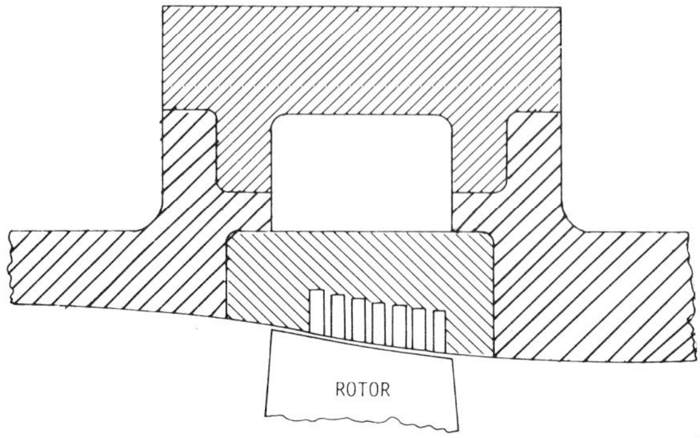

Fig. 3.31 Schematic of one form of compressor rotor tip casing treatment.

33.0 in. and a hub-tip ratio at the rotor entrance of 0.5. No guide vanes were used.

Figure 3.29 shows the predicted and measured performance of the three-stage axial flow compressor (fan) previously shown in the meridional plane section of Fig. 3.5.

Figure 3.30 shows performance plots for a compressor stage with variable inlet guide vane geometry and variable stator blading (meridional section in Fig. 3.25). The two maps show that significant performance changes are possible with variable blade geometry. This stage had a 0.5 hub-tip ratio and a tip diameter of 36.5 in.

In multistage compressors, the trend has been and is toward higher flow rates per unit annulus area and higher stage pressure ratios. This has been true partly because the higher tip speeds associated with transonic entrance stages are also present in subsequent stages. This leads to higher stage pressure ratios for a given blade row loading level. Although the general Mach number levels decrease as the temperature levels rise through a compressor, transonic stage operation frequently exists in several stages at the design point.

Continual emphasis has been placed on the development of broad-range stable engine core compressors. Variable geometry blading has entered many configurations and other geometric techniques such as casing treatment (see Fig. 3.31) have been introduced into both core and fan stages. The technology of high hub-tip ratio, high-pressure stages is a complex and difficult research area.

Throughout the relatively short technological history of the axial flow compressor, the most significant overall trends have been toward the reduction of size and cost by increasing the average Mach number levels and stage pressure ratios. All of the configuration maps shown in this chapter have been the direct result of attempts to advance these trends. Transonic and supersonic compressor development have demonstrated that there are certain design features common to most successful stages. These are summarized in Fig. 3.32. A good way to review the content of this chapter would be to consider these features and to justify them in terms of the flow parameters and phenomena exposed herein.

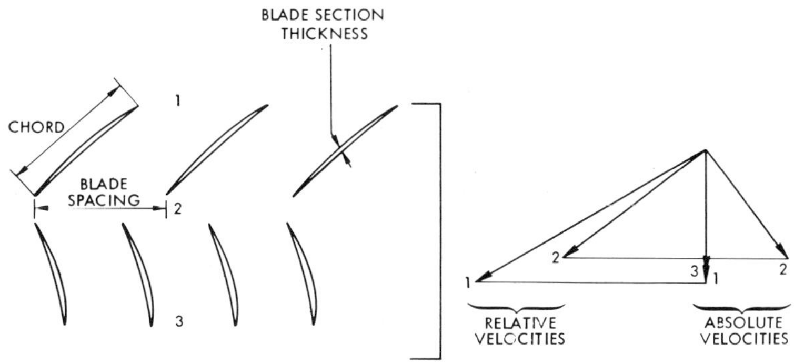

TRANSONIC STAGE DESIGN CHARACTERISTICS

VELOCITY DIAGRAMS

- AXIAL ENTRANCE FLOW
- NEAR AXIAL EXIT FLOW
- MAXIMUM RELATIVE M AND MINIMUM FLOW TURNING IN ROTOR TIP REGION
- MAXIMUM FLOW TURNING IN ROTOR HUB REGION
- MAXIMUM STATOR M IN HUB REGION

BLADE ROW GEOMETRY

- LOW RATIOS OF MAXIMUM THICKNESS/CHORD
- HIGHER LEVELS OF SOLIDITY IN MAXIMUM M REGIONS THAN WOULD BE USED FOR SUBSONIC BLADING
- BLADE CAMBER (TURNING) MOVED TOWARD REAR OF AIRFOIL IN REGIONS OF HIGH M
- REDUCED ASPECT RATIO

Fig. 3.32 Design characteristics of advanced transonic compressor stages.

Acknowledgments

Some good friends among axial flow compressor scholars have provided information, advice, and data to support the preparation of this chapter. Included in this list are A. W. Stubner of Pratt & Whitney Aircraft Div., United Technologies Corporation; Robert O. Bullock; Arthur J. Wennerstrom of the Air Force Aero Propulsion Laboratory; and Melvin J. Hartmann and Donald M. Sandercock of NASA Lewis Research Center. The data shown in Fig. 3.22 were assembled by D. M. Sandercock from the results of NASA research.

Bibliographic Notes and Guide to Reference Materials

Many references to technical papers and books were considered necessary to give a balanced review of axial flow compressor aerodynamic technology. It was, therefore, decided that the main text should be presented without interruption and that all bibliographic notes should be assembled and references classified by section at the end of the chapter.

References 1–16 are sources of figures as well as information. In some cases, the figures were redrawn or modified to illustrate a specific point.

[1]Mitchell, G. A., "Effect of Inlet Ingestion of a Wing Tip Vortex on Compressor Face Flow and Turbojet Stall Margin," NASA TM X-3246, 1975.

[2]Tesch, W. A. and Steenken, W. G., "*J85* Clean Inlet Flow and Parallel Compressor Models," *Blade Row Dynamic Digital Compressor Program*, Vol. I, NASA CR-134978, March 1976.

[3]"Fact Sheet, TF30-P-412A Turbofan Engine," Pratt & Whitney Div., United Technologies Corp., 1973.

[4]"Fact Sheet, JT9D-70/-70A Turbofan Engine," Pratt & Whitney Div., United Technologies Corp., 1974.

[5]Volkmann, H., Föttner, L., and Scholz, N., "Aerodynamische Entwicklung eines dreistufigen Transsonik-Frontgebläses," *Zeitschrift für Flugwissenschaften*, Vol. 22, 1974, pp. 135–144.

[6]Geye, R. P., Budinger, R. E., and Voit, C. H., "Preliminary Analysis of Over-All Performance," *Investigation of High-Pressure-Ratio Eight-Stage Axial-Flow Research Compressor with Two Transonic Inlet Stages*, Vol. II, NACA RM E53J06, 1953.

[7]Bullock, R. O. and Prasse, E. I., "Compressor Design Requirements," *Aerodynamic Design of Axial-Flow Compressors*, NASA SP-36, 1965, Chap. II.

[8]Snyder, L. E. and Commerford, G. L., "Supersonic Unstalled Flutter in Fan Rotors; Analytical and Experimental Results," *Journal of Engineering for Power, Transactions of ASME*, Ser. A, Vol. 96, 1974, pp. 379–386.

[9]Benser, W. A., "Analysis of Part-Speed Operation for High-Pressure-Ratio Multistage Axial-Flow Compressors," NACA RM E53I15, 1953.

[10]Huppert, M. C. and Benser, W. A., "Some Stall and Surge Phenomena in Axial Flow Compressors," *Journal of the Aeronautical Sciences*, Vol. 20, Dec. 1953, pp. 835–845.

[11]Herzig, H. Z. and Hansen, A. G., "Three-Dimensional Compressor Flow Theory and Real Flow Effects," *Aerodynamic Design of Axial-Flow Compressors*, NASA SP-36, 1965, Chap. XIV.

[12]Lieblein, S., "Experimental Flow in Two-Dimensional Cascades," *Aerodynamic Design of Axial-Flow Compressors*, NASA SP-36, 1965, Chap. VI.

[13]Robbins, W. H., Jackson, R. J., and Lieblein, S., "Blade-Element Flow in Annular Cascades," *Aerodynamic Design of Axial-Flow Compressors*, NASA SP-36, 1965, Chap. VII.

[14]Doyle, V. L. and Koch, C. C., "Evaluation of Range and Distortion Tolerance for High Mach Number Transonic Fan Stages, Design Report," NASA CR-72720, 1970.

[15]Sulam, D. H., Keenan, M. J., and Flynn, J. T., "Data and Performance Multiple-Circular-Arc Rotor," *Single-Stage Evaluation of Highly-Loaded High-Mach-Number Compressor Stages*, Vol. II, NASA CR-72694, 1970.

[16]Bilwakesh, K. R., "Evaluation of Range and Distortion Tolerance for High Mach Number Transonic Fan Stages," *Task II, Stage Data and Performance Report for Undistorted Inlet Flow Testing* (Vols. I and II), NASA CR-72787, Jan. 1971.

3.1 Introduction

3.2 Axial Flow Compressor Nomenclature and Terminology

References 17–23 are textbooks and summary reports having significant content on axial flow compressors. They are historically valuable and represent several independent points of view on compressor design.

[17]Horlock, J. H., *Axial Flow Compressors*, Butterworths, London, 1958. (Reprint with supplemental material, Krieger, Huntington, N.Y., 1973.)

[18]Johnsen, I. A. and Bullock, R. O. (eds.), *Aerodynamic Design of Axial-Flow Compressors*, NASA SP-36, 1965.

[19]Kholshchevnikov, K. V., "Theory and Design of Aircraft Turbomachines," Translation of *Teoriya i Raschet Aviatsionnykh Lopatochnykh Mashin*, 1970, Foreign Technology Div., Wright-Patterson AFB, Ohio, FTD-HC-23-754-71, 1972.

[20]Klein, A., "Aerodynamics of Cascades," translation and revision of *Aerodynamik der Schaufelgitter* by N. Scholz, AGARDograph 220, 1977.

[21]Stechkin, B. S., Kazandzhan, P. I., Alekseev, L. P., Govorov, A. N., Nechaev, Yu. N., and Fedorov, R. M., *Theory of Reaction Engines (Bladed Machines)*, Oborongiz, Moscow, 1956.

[22]Traupel, W., *Thermische Turbomaschinen*, Vol. I, 2nd ed. Springer-Verlag, Berlin/Heidelberg/New York, 1966.

[23]Vavra, M. H., *Aero-Thermodynamics and Flow in Turbomachines*, John Wiley & Sons, New York, 1960.

Cascade test facilities and experimental techniques are described in Refs. 24–29 and some examples of systematic cascade test programs are contained and referenced in Refs. 30–34. References 35–37 are thorough explanations of supersonic flows in cascades.

[24]Schilhansl, M. J., "Survey of Information on Two-Dimensional Cascades." Wright Air Development Center, Air Research and Development Command, Tech. Rept. 54-322, March 1955 [Aero. Res. Lab. Contract AF 33(616)-16, Proj. 3066].

[25]Erwin, J. R. and Emery, J. C., "Effect of Tunnel Configuration and Testing Technique on Cascade Performance," NACA Rept. 1016, 1951.

[26]Stubner, A. W., Case, L. F., and Blake, T. R., "Supersonic Cascade Tunnel Used to Evaluate Compressor Blade Performance," *Journal of Engineering for Power, Transactions of ASME*, Ser. A, Vol. 88, 1966, pp. 153–156.

[27]Breugelmans, F. A. E., "High Speed Cascade Testing and Its Application to Axial Flow Supersonic Compressors," ASME Paper 68-GT-10, March 1968.

[28]Dzung, L. S., "Averaging Procedure in the Theory of Blade Cascades," *Brown Boveri Review* Vol. 54, No. 1, Jan. 1967, pp. 36–43.

[29]Sieverding, C., Starken, H., Lichtfuss, H.-J., and Schimming, P., "Aerodynamic Measurements in Cascades," *Modern Methods of Testing Rotating Components for Turbomachines (Instrumentation)*, AGARD-AG-207, 1975, pp. 1–76.

[30]Emery, J. C., Herrig, L. J., Erwin, J. R., and Richard, F. A., "Systematic Two-Dimensional Cascade Tests of NACA 65-series Compressor Blades at Low Speeds," NACA Rept. 1368, 1958.

[31]Savage, M., Boxer, E., and Erwin, J. R., "Resume of Compressor Research at the NACA Langley Laboratory," *Journal of Engineering for Power, Transactions of ASME*, Ser. A, Vol. 83, 1961, pp. 269–285.

[32]Schlichting, H. and Das, A., "Recent Research on Cascade Flow Problems," *Journal of Basic Engineering, Transactions of ASME*, Ser. D, Vol. 88, 1966, pp. 221–228.

[33]Schreiber, H. A., "Comparison between Flows in Cascades and Rotors in the Transonic Range," *Investigation of Two Transonic Compressor Cascades and Comparison with Rotor Data*, Vol. II, von Kármán Institute Lecture Series 84, Feb. 1976.

[34]Starken, H., Breugelmans, F. A. E., and Schimming, P., "Investigation of the Axial Velocity Density Ratio in a High Turning Cascade," ASME Paper 75-GT-25, 1975.

[35]Starken, H. and Lichtfuss, H.-J. "Supersonic Cascade Performance," *Advanced Compressors*, AGARD-LS-39-70, 1970, Paper 8.

[36]Starken, H. and Lichtfuss, H.-J., "Some Experimental Results of Two-Dimensional Compressor Cascades at Supersonic Inlet Velocities," *Journal of Engineering for Power, Transactions of ASME*, Ser. A, Vol. 92, 1970, pp. 267–274.

[37]Lichtfuss, H.-J. and Starken, H., "Supersonic Cascade Flow," *Progress in Aerospace Science*, Vol. 15, Pergamon Press Ltd., New York, 1974, pp. 37–149.

3.3 Characteristics of the Flow in Axial Flow Compressor Configurations

References 17 and 38 develop the similarity parameters used in plotting compressor performance. Performance map features and additional influences on map characteristics are discussed as follows: Reynolds number and scaling effects in Refs. 39–42; aerodynamic instability rotating stall, surge, and inlet distortion in Refs. 1 and 43–52; and aeroelastic phenomena in Refs. 8 and 53.

Stage performance and stage-matching methods are covered by Refs. 9, 10, and 54–57.

[38]Hodge, J., *Cycles and Performance Estimation*, Academic Press, New York, 1955.

[39]Bullock, R. O., "Analysis of Reynolds Number and Scale Effects on Performance of Turbomachinery," *Journal of Engineering for Power, Transactions of ASME*, Ser. A, Vol. 86, 1964, pp. 247–256.

[40]Wassell, A. B., "Reynolds Number Effects in Axial Compressors," *Journal of Engineering for Power, Transactions of ASME*, Ser. A, Vol. 90, 1968, pp. 149–156.

[41]Pampreen, R. C., "Small Turbomachinery Compressor and Fan Aerodynamics," *Journal of Engineering for Power, Transactions of ASME*, Ser. A, Vol. 95, July 1973, pp. 251–256.

[42]Schaffler, A., "Experimental and Analytical Investigation of the Effects of Reynolds Number and Blade Surface Roughness on Multistage Axial Flow Compressors," *Journal of Engineering for Power, Transactions of ASME*, Ser. A, Vol. 102, 1980, pp. 5–13.

[43]Emmons, H. W., Kronauer, R. E., and Rockett, J. A., A Survey of Stall Propagation-Experiment and Theory, *Journal of Basic Engineering, Transactions of ASME*, Ser. D, Vol. 81, 1959, pp. 409–416.

[44]Fabri, J., "Rotating Stall in Axial Flow Compressors," *Internal Aerodynamics (Turbomachinery)*, Institution of Mechanical Engineers, London, 1970, pp. 96–110.

[45]Yershov, V. N., "Unstable Conditions of Turbodynamics: Rotating Stall," FTD-MT-24-04-71, Aug. 12, 1971.

[46]Greitzer, E. M., "Surge and Rotating Stall in Axial Flow Compressors, Pt. I: Theoretical Compressor System Model," *Journal of Engineering for Power, Transactions of ASME*, Ser. A, Vol. 98, 1976, pp. 190–198.

[47]Greitzer, E. M., "Surge and Rotating Stall in Axial Flow Compressors, Pt. II: Experimental Results and Comparison With Theory," *Journal of Engineering for Power, Transactions of ASME*, Ser. A, Vol. 98, 1976, pp. 199–217.

[48] Langston, C. E., "Distortion Tolerance—by Design Instead of by Accident," ASME Paper 69-GT-115, 1969.

[49] Mikolajczak, A. A. and Pfeffer, A. M., "Methods to Increase Engine Stability and Tolerance to Distortion," *Distortion Induced Engine Instability*, AGARD LS-72, Oct. 1974, Paper 7.

[50] Mikolajczak, A. A., "The Practical Importance of Unsteady Flow," *Unsteady Phenomena in Turbomachinery*, AGARD-CP-177, 1976, Paper 1.

[51] Hetherington, R. and Moritz, R. R., "Influence of Unsteady Flow Phenomena on the Design and Operation of Aero Engines," *Unsteady Phenomena in Turbomachinery*, AGARD-CP-177, 1976, Paper 2.

[52] Greitzer, E. M., "The Stability of Pumping Systems—The 1980 Freeman Scholar Lecture," *Journal of Fluids Engineering, Transactions of ASME*, Vol. 103, 1981, pp. 193–242.

[53] Mikolajczak, A. A., Arnoldi, R. A., Snyder, L. E., and Stargadter, H., "Advances in Fan and Compressor Blade Flutter Analysis and Predictions," *Journal of Aircraft*, Vol. 12, 1975, pp. 325–332.

[54] Finger, H. B. and Dugan, J. F. Jr., "Analysis of Stage-Matching and Off-Design Performance of Multi-Stage Axial-Flow Compressors," NACA RM E52D07, 1952.

[55] Benser, W. A., "Compressor Operation with One or More Blade Rows Stalled," *Aerodynamic Design of Axial-Flow Compressors*, NASA SP-36, 1965, Chap. XIII.

[56] Doyle, M. D. C. and Dixon, S. L., "The Stacking of Compressor Stage Characteristics to Give an Overall Compressor Performance Map," *Aeronautical Quarterly*, Vol. 13, 1962, pp. 349–367.

[57] Gray, S., "Stage Matching, Stall, and Surge in Multi-Stage Axial-Flow Compressors," *Thermodynamics and Fluid Mechanics Convention 1970*, Vol. II: *Axial and Radial Turbomachinery*, *Proceedings*, *Institution of Mechanical Engineers*, Vol. 184, Pt. 3G(II), 1970, pp. 49–56.

3.4 Aerodynamic Design Objectives for Axial Flow Compressor Units

3.5 Elements of a Compressor Design System—Technical Requirements

References 7 and 58 outline these subjects.

[58] Bullock, R. O. and Johnsen, I. A., "Compressor Design System," *Aerodynamic Design of Axial-Flow Compressors*, NASA SP-36, 1965, Chap. III.

3.6 Content of Current and Developing Design Systems

The development of design systems can be best studied with an understanding of how present thinking evolved. References 59–63 are some important British base points and Refs. 58 and 64–66 give some background on compressor design systems in the United States.

[59] Howell, A. R., "Fluid Dynamics of Axial Compressors," *Proceedings*, *Institution of Mechanical Engineers*, Vol. 153, 1945, pp. 441–452.

[60] Howell, A. R., Design of Axial Compressors. *Proceedings*, *Institution of Mechanical Engineers*, Vol. 153, 1945, pp. 452–462.

[61] Carter, A. D. S., Three-Dimensional-Flow Theories for Axial Compressor and Turbines. *Proceedings*, *Institution of Mechanical Engineers*, Vol. 159, 1948, pp. 255–268.

[62] Carter, A. D. S., Turner, R. C., Sparkes, D. W., and Burrows, R. A., "The Design and Testing of an Axial-Flow Compressor Having Different Blade Profiles in Each Stage," British Aeronautical Research Council R&M 3183, 1957.

[63] Carter, A. D. S., "Blade Profiles for Axial-Flow Fans, Pumps, Compressors, etc.," *Proceedings, Institution of Mechanical Engineers*, Vol. 175, 1961, pp. 775–806.

[64] Johnsen, I. A., "Investigation of a 10-Stage Subsonic Axial-Flow Research Compressor, I: Aerodynamic Design," NACA RM E52B18, 1952.

[65] Giamati, C. C. Jr. and Finger, H. B., "Design Velocity Distribution in Meridional Plane," *Aerodynamic Design of Axial-Flow Compressors*, NASA SP-36, 1965, Chap. VIII.

[66] Serovy, G. K., "Recent Progress in Aerodynamic Design of Axial-Flow Compressors in the United States," *Journal of Engineering for Power, Transactions of ASME*, Ser. A, Vol. 88, 1966, pp. 251–261.

The subject of flowfield models for the hub-to-tip problem is brought forward from the period beginning about 1950 until very recent years in Refs. 67–89 and for the blade-to-blade problem in Refs. 20, 24, and 90–100. Important reference material on these subjects may also be found in the listed references for design analysis, beginning with Ref. 160.

[67] Wu, C.-H. and Wolfenstein, L., "Application of Radial-Equilibrium Condition to Axial-Flow Compressor and Turbine Design," NACA Rept. 955, 1950.

[68] Hatch, J. E., Giamati, C. C., and Jackson, R. J. "Application of Radial-Equilibrium Condition to Axial-Flow Turbomachine Design Including Consideration of Change of Entropy With Radius Downstream of Blade Row," NACA RM L54A20, 1954.

[69] Holmquist, C. O. and Rannie, W. D., "An Approximate Method of Calculating Three-Dimensional Compressible Flow in Axial Turbomachines," *Journal of the Aeronautical Sciences*, Vol. 23, 1956, pp. 543–556, 582.

[70] Wu, C.-H., "General Through-Flow Theory of Fluid Flow with Subsonic or Supersonic Velocity in Turbomachines of Arbitrary Hub and Casing Shapes," NACA TN 2302, 1951.

[71] Wu, C.-H., "A General Theory of Three-Dimensional Flow in Subsonic and Supersonic Turbomachines of Axial-, Radial-, and Mixed-Flow Types," NACA TN 2604, Jan. 1952.

[72] Hamrick, J. T., Ginsburg, A., and Osborn, W. M., "Method of Analysis for Compressible Flow Through Mixed-Flow Centrifugal Impellers of Arbitrary Design," NACA Rept. 1082, 1952.

[73] Smith, L. H. Jr., "The Radial-Equilibrium Equation of Turbomachinery," *Journal of Engineering for Power, Transactions of ASME*, Ser. A, Vol. 88, 1966, pp. 1–12 (discussion in Vol. 88, 1966, pp. 282–283).

[74] Novak, R. A., "Streamline Curvature Computing Procedures for Fluid-Flow Problems," *Journal of Engineering for Power, Transactions of ASME*, Ser. A, Vol. 89, 1967, pp. 478–490.

[75] Sirotkin, Ya. A., "Aerodynamic Calculation of Blades in Axial-Flow Turbomachines," FTD-MT-24-1809-72, May 31, 1973.

[76] Silvester, M. E. and Hetherington, R., "Three Dimensional Compressible Flow Through Axial Flow Turbomachines," *Numerical Analysis: An Introduction*, edited by J. Walsh, Academic Press, New York, 1966.

[77]Horlock, J. H., "On Entropy Production in Adiabatic Flow in Turbomachines," *Journal of Basic Engineering, Transactions of ASME*, Ser. D, Vol. 93, 1971, pp. 587–593.

[78]Marsh, H., "A Digital Computer Program for the Through-Flow Fluid Mechanics in an Arbitrary Turbomachine Using a Matrix Method," NGTE Rept. R.282, 1966.

[79]Marsh, H., "The Through-Flow Analysis of Axial Flow Compressors," *Advanced Compressors*, AGARD-LS-39-70, 1970, Paper 2.

[80]Horlock, J. H. and Marsh, H., "Flow Models for Turbomachines," *Journal of Mechanical Engineering Science*, Vol. 13, 1971, pp. 358–368.

[81]Marsh, H., "The Uniqueness of Turbomachinery Flow Calculations Using the Streamline Curvature and Matrix Through-Flow Methods," *Journal of Mechanical Engineering Science*, Vol. 13, 1971, pp. 376–379.

[82]Katsanis, T., "Use of Arbitrary Quasi-Orthogonals for Calculating Flow Distribution in the Meridional Plane of a Turbomachine," NASA TN D-2546, 1964.

[83]Katsanis, T., "Use of Arbitrary Quasi-Orthogonals for Calculating Flow Distribution in a Turbomachine," *Journal of Engineering for Power, Transactions of ASME*, Ser. A, Vol. 88, 1966, pp. 197–202.

[84]Smith, D. J. L. and Barnes, J. F., "Calculation of Fluid Motion in Axial Flow Turbomachines," ASME Paper 68-GT-12, 1968.

[85]Davis, W. R. and Millar, D. A. J., "A Discussion of the Marsh Matrix Technique Applied to Fluid Flow Problems," *CASI Transactions*, Vol. 5, No. 2, Sept. 1972, pp. 64–70.

[86]Smith, D. J. L., "Computer Solutions of Wu's Equations for Compressible Flow Through Turbomachines," *Fluid Mechanics, Acoustics, and Design of Turbomachinery*, NASA SP-304, Pt. I, 1974, pp. 43–74.

[87]Wennerstrom, A. J., "On the Treatment of Body Forces in the Radial Equilibrium Equation of Turbomachinery," *Traupel-Festschrift*, edited by P. Suter and G. Gyarmathy, Juris-Verlag, Zurich, 1974.

[88]Wu, C.-H., "Three Dimensional Turbomachine Flow Equations Expressed with Respect to Non-Orthogonal Curvilinear Coordinates and Methods of Solution," Paper presented at Third International Symposium on Air Breathing Engines, Munich, March 1976.

[89]Serovy, G. K., "Axial-Flow Turbomachine Through-Flow Calculation Methods," *Through Flow Calculations in Axial Turbomachines*, AGARD-AR-175, 1981.

[90]Wu, C.-H. and Brown, C. A., "Method of Analysis for Compressible Flow Past Arbitrary Turbomachine Blades on General Surface of Revolution," NACA TN 2407, 1951.

[91]Roudebush, W. H., "Potential Flow in Two-Dimensional Cascades," *Aerodynamic Design of Axial-Flow Compressors*, NASA SP-36, 1965, Chap. IV.

[92]Roudebush, W. H. and Lieblein, S., "Viscous Flow in Two-Dimensional Cascades, *Aerodynamic Design of Axial-Flow Compressors*, NASA SP-36, 1965, Chap. V.

[93]Wu. C.-H. and Brown, C. A., "A Theory of Direct and Inverse Problems of Compressible Flow Past Cascade of Arbitrary Airfoils," *Journal of the Aeronautical Sciences*, Vol. 19, 1952, pp. 183–196.

[94]Scholz, N., "A Survey of the Advances in the Treatment of the Flow in Cascades," *Internal Aerodynamics (Turbomachinery)*, The Institution of Mechanical

Engineers, London, 1970, pp. 20–31, discussion by K. Barsun and D. J. L. Smith, 1970, pp. 41–56.

[95]Katsanis, T., "Use of Arbitrary Quasi-Orthogonals for Calculating Flow Distribution on a Blade-to-Blade Surface in a Turbomachine," NASA TN D-2809, 1965.

[96]Gostelow, J. P., "Review of Compressible Flow Theories for Airfoil Cascades," *Journal of Engineering for Power Transactions of ASME*, Ser. A, Vol. 95, 1973, pp. 281–292 (discussion by D. C. Prince Jr., Vol. 95, 1974, pp. 77–79).

[97]Gostelow, J. P., "Trailing Edge Flows Over Turbomachine Blades and the Kutta-Joukowsky Condition," ASME Paper 75-GT-94, 1975.

[98]Wilkinson, D. H., "Stability, Convergence, and Accuracy of Two-Dimensional Streamline Curvature Methods Using Quasi-Orthogonals," *Proceedings of the Institution of Mechanical Engineers*, Vol. 184, Pt. 3G (1), 1969–1970, pp. 108–119.

[99]Hansen, E. C., Serovy, G. K., and Sockol, P. M., "Axial-Flow Compressor Turning Angle and Loss by Inviscid-Viscous Interaction Blade-to-Blade Computation," *Journal of Engineering for Power, Transactions of ASME*, Vol. 102, 1980, pp. 28–34.

[100]Calvert, W. J., "Application of an Inviscid-Viscous Interaction Method to Transonic Compressor Cascades," *Viscous Effects in Turbomachines*, AGARD-CP-351, 1983.

Design limit parameters are developed and discussed in Refs. 101–109.

[101]Voit, C. L. and Thomson, A. R., "An Analytical Investigation Using Aerodynamic Limitations of Several Designs of High Stage Pressure Ratio Multistage Compressors," NACA TN 2589, 1951.

[102]Lieblein, S., Schwenk, F. C., and Broderick, R. L., "Diffusion Factor for Estimating Losses and Limiting Blade Loadings in Axial-Flow-Compressor Blade Elements," NACA RM E53D01, 1953.

[103]Lieblein, S. and Roudebush, W. H., "Theoretical Loss Relations for Low-Speed Two-Dimensional-Cascade Flow," NACA TN 3662, 1956.

[104]Lieblein, S. and Roudebush, W. H., "Low-Speed Wake Characteristics of Two-Dimensional Cascade and Isolated Airfoil Sections," NACA TN 3771, 1956.

[105]Lieblein, S., "Analysis of Experimental Low-Speed Loss and Stall Characteristics of Two-Dimensional Compressor Blade Cascades," NACA RM E57A28, 1957.

[106]Savage, M., "Analysis of Aerodynamic Blade-Loading-Limit Parameters for NACA 65-($C_{10}A_{10}$)10 Compressor-Blade Sections at Low Speeds," NACA RM L54L02a, 1955.

[107]Lieblein, S., "Review of High-Performance Axial-Flow-Compressor Blade-Element Theory," NACA RM E53L22, 1954.

[108]Klapproth, J. F., "General Considerations of Mach Number Effects on Compressor Blade Design," NACA RM E53L23a, 1954.

[109]NATO Advisory Group for Aerospace Research and Development, Propulsion and Energetics Panel Working Group 12, *Through Flow Calculations in Turbomachines*, AGARD-AR-175, 1981.

Supersonic and transonic compressor progress beginning in about 1945 can be followed in Refs. 110–129.

[110]Kantrowitz, A., "The Supersonic Axial Flow Compressor," NACA TR 974, 1950.

[111]Erwin, J. R., "Early Work on Supersonic Compressor," *Aerodynamics of Turbines and Compressors*, Vol. X of *High Speed Aerodynamic and Jet Propulsion*, edited by W. R. Hawthorne, Princeton University Press, Princeton, N.J., 1964, Sec. G, Chap. 1.

[112]Ferri, A., "Aerodynamic Properties of Supersonic Compressors," *Aerodynamics of Turbines and Compressors*, Vol. X of *High Speed Aerodynamics and Jet Propulsion*, edited by W. R. Hawthorne, Princeton University Press, Princeton, N.J., 1964, Sec. G, Chap. 2.

[113]Wilcox, W. W., Tysl, E. R., and Hartmann, M. J., "Resume of the Supersonic-Compressor Research at NACA Lewis Laboratory," *Journal of Basic Engineering, Transactions of ASME*, Ser. D, Vol. 81, 1959, pp. 559–568.

[114]Klapproth, J. F., "A Review of Supersonic Compressor Development," *Journal of Engineering for Power, Transactions of ASME*, Ser. A, Vol. 83, 1961, pp. 248–268.

[115]Conrad, O., "Statistische Untersuchung von Überschallverdichtern," *Jahrbuch 1965 der WGLR*, 1965, pp. 374–383.

[116]Conrad, O., Pleiss, J., and Weber, T., "Study of Supersonic Compressors," FTD-HT-23-1436-68, Feb. 18, 1969.

[117]Chauvin, J., Breugelmans, F., and Janigro, A., "Supersonic Compressors," *Supersonic Turbo-Jet Propulsion Systems and Components*, edited by J. Chauvin, AGARDograph 120, 1969.

[118]Lieblein, S. and Johnsen, I. A., "Resume of Transonic-Compressor Research at NACA Lewis Laboratory," *Journal of Engineering for Power, Transactions of ASME*, Ser. A, Vol. 83, 1961, pp. 219–234.

[119]Bullock, R. O., "Critical High Lights in the Development of Transonic Compressor," *Journal of Engineering for Power, Transactions of ASME*, Ser. A, Vol. 83, 1961, pp. 243–257.

[120]Calmon, J., "Le Developpement du Compresseur Transsonique a la SNECMA," *Technique et Science Aeronautiques et Spatiales*, No. 5, 1965, pp. 431–445.

[121]Breugelmans, F. A. E., "The Supersonic Compressor Research at the von Karman Institute for Fluid Dynamics," Paper presented at 1st International Symposium on Air Breathing Engines, Marseilles, France, June 1972.

[122]Thiaville, J.-M., "Tassement des Compresseurs Axiaux," *L'Aeronautique et l'Astronautique*, Vol. 44, 1974, pp. 21–31.

[123]Benser, W. A., "Transonic Compressor Technology Advancements," *Fluid Mechanics, Acoustics, and Design of Turbomachinery*, edited by B. Lakshminarayana et al., NASA SP-304, Pt. II, 1974, pp. 581–602.

[124]Simon, H., "A Contribution to the Theoretical and Experimental Examination of the Flow Through Plane Supersonic Deceleration Cascades and Supersonic Compressor Rotors," *Journal of Engineering for Power, Transactions of ASME*, Ser. A, Vol. 95, 1973, pp. 233–243.

[125]Simon, H. and Bohn, D., "Experimental Investigation of a Recently Developed Supersonic Compressor Stage (Rotor and Stator Performance)," ASME Paper 74-GT-116, 1974.

[126]Wennerstrom, A. J. and Hearsey, R. M., "The Design of an Axial Compressor Stage for a Total Pressure Ratio of 3 to 1," ARL 71-0061, 1971.

[127]Wennerstrom, A. J., Frost, G. R., and DeRose, R. D., "Test of an Axial Compressor Stage Designed for a Total Pressure Ratio of Three to One," ARL TR 74-0001, 1974.

[128]Wennerstrom, A. J. and Frost, G. R., "Design of a Rotor Incorporating Splitter Vanes for a High Pressure Ratio Supersonic Axial Compressor Stage," ARL 74-0110, 1974.

[129]Wennerstrom, A. J., Buzzell, W. A., and DeRose, R. D., "Test of a Supersonic Axial Compressor Stage Incorporating Splitter Vanes in the Rotor," ARL TR 75-0165, 1975.

Reference 130 describes a typical computer program that may be used for preliminary design of the hub-to-tip flow passage.

[130] Bryans, A. C. and Miller, M. L., "Computer Program for Design of Multistage Axial-Flow Compressors," NASA CR-54530, 1967.

For blade row geometry selection, Refs. 131–134 summarize some methods originally developed in the United Kingdom based on cascade data correlation.

[131]Howell, A. R., "The Present Basis of Axial Flow Compressor Design, Pt. 1: Cascade Theory and Performance," British Aeronautical Research Council, R&M 2095, 1942.

[132]Andrews, S. J., "Tests Related to Profile Shape and Camber Line on Compressor Cascade Performance," British Aeronautical Research Council, R&M 2743, 1949.

[133]Carter, A. D. S., "The Low Speed Performance of Related Aerofoils in Cascades," British Aeronautical Research Council, CP 29, 1950.

[134]Howell, A. R., "Flow in Cascades," *High Speed Aerodynamics and Jet Propulsion*, Vol. X of *Aerodynamics of Turbines and Compressors*, Princeton University Press, Princeton, N.J., 1964.

References 12, 13, and 135–139 cover methods and considerations involved in cascade data correlation in the United States.

[135]Lieblein, S., "Incidence and Deviation-Angle Correlations for Compressor Cascades," *Journal of Basic Engineering, Transactions of ASME*, Ser. D, Vol. 82, 1960, pp. 575–587.

[136]Lieblein, S., "Loss and Stall Analysis of Compressor Cascades," *Journal of Basic Engineering, Transactions of ASME*, Ser. D, Vol. 81, 1959, pp. 387–400.

[137]Wright, L. C., "Blade Selection for a Modern Axial-Flow Compressor," *Fluid Mechanics, Acoustics, and Design of Turbomachinery*, edited by B. Lakshminarayana et al., NASA SP-304, Pt. II, 1974, pp. 603–626.

[138]Serovy, G. K., "Utilization of Cascade Data in Axial-Flow Compressor Design and Analysis—A Critical Review," ASME Paper 70-GT-108, 1970.

[139]Serovy, G. K., "Deviation Angle/Turning Angle Prediction for Advanced Axial-Flow Compressor Blade Row Geometries," AFAPL-TR-77-81, 1978.

Cascade data procedures developed in the USSR are reviewed in Refs. 140–142.

[140]Bunimovich, A. I. and Svyatogorov, A. A., "Generalization of Results of Investigation of Foil Compressor Cascades at Subsonic Speed," FTD-MT-24-69-68, 1968.

[141]Bunimovich, A. I. and Svyatogorov, A. A., "Aerodynamic Characteristics of Foil Compressor Cascades at High Subsonic Speed," FTD-MT-24-69-68, 1968.

[142] Komarov, A. P., "Investigation of Flat Compressor Grids," FTD-MT-24-69-68, 1968.

Losses produced by transonic and supersonic operation are studied in Refs. 143–147.

[143] Miller, G. R., Lewis, G. W. Jr., and Hartmann, M. J., "Shock Losses in Transonic Blade Rows," *Journal of Engineering for Power, Transactions of ASME*, Ser. A, Vol. 83, 1961, pp. 235–242.

[144] Mikolajczak, A. A., Morris, A. L., and Johnson, B. V., "Comparison of Performance of Supersonic Blading in Cascade and in Compressor Rotors," *Journal of Engineering for Power, Transactions of ASME*, Ser. A, Vol. 93, 1971, pp. 42–48.

[145] Huffman, G. D. and Tramm, P. C., "Airfoil Design for High Tip Speed Compressors," *Journal of Aircraft*, Vol. 11, 1974, pp. 682–689.

[146] Prince, D. C. Jr., "Three-Dimensional Shock Structures for Transonic/Supersonic Compressor Rotors," *Journal of Aircraft*, Vol. 17, 1980, pp. 28–37.

[147] Wennerstrom, A. J. and Puterbaugh, S. L., "A Three-Dimensional Model for the Prediction of Shock Losses in Compressor Blade Rows," ASME Paper 83-GT-216, 1983.

Some additional papers on subjects related to blade selection are Refs. 109 and 148–154.

[148] Smith, L. H. Jr., "Secondary Flow in Axial-Flow Turbomachinery," *Transactions of ASME*, Vol. 77, 1955, pp. 1065–1976.

[149] Dzung, L. S., "Consistent Mean Values for Compressible Media in the Theory of Turbomachines," *Brown Boveri Review*, Vol. 58, No. 10, Oct. 1971, pp. 3–10.

[150] Brown, L. E., "Axial Flow Compressor and Turbine Loss Coefficients: A Comparison of Several Parameters," *Journal of Engineering for Power, Transactions ASME*, Ser. A, Vol. 94, 1972, pp. 193–201.

[151] Esgar, G. M. and Sandercock, D. M., "Some Observed Effects of Part-Span Dampers on Rotating Blade Row Performance Near Design Point," NASA TM X-2696, 1973.

[152] Smith, L. H. Jr. and Yeh, H., "Sweep and Dihedral Effects in Axial-Flow Turbomachinery," *Journal of Basic Engineering, Transactions of ASME*, Ser. D, Vol. 85, 1963, pp. 401–416.

[153] Smith, L. H. Jr., "Some Aerodynamic Design Considerations for High Bypass Ratio Fans," General Electric Co., Advanced Fan and Compressor Aerodynamic Operation, Tech. Memo 74-131, Feb. 1, 1974.

[154] Koch, C. C. and Smith, L. H. Jr., "Loss Sources and Magnitudes in Axial-Flow Compressors," *Journal of Engineering for Power, Transactions of ASME*, Ser. A, Vol. 98, 1976, pp. 411–424.

Methods for blade section and blade row computation and stacking are described in Refs. 155–158.

[155] Frost, G. R., Hearsey, R. M., and Wennerstrom, A. J., "A Computer Program for the Specification of Axial Compressor Airfoils," ARL 72-0171, 1972.

[156] Frost, G. R. and Wennerstrom, A. J., "The Design of Axial Compressor Airfoils Using Arbitrary Camber Lines," ARL 73-0107, 1973.

[157] Frost, G. R., "Modifications to ARL Computer Programs Used for Design of Axial Compressor Airfoils," ARL 74-0060, 1974.

[158]Crouse, J. E., "Computer Program for Definition of Transonic Axial-Flow Compressor Blade Rows," NASA TN D-7345, 1974.

A useful system for estimating compressor size and weight is given in Ref. 159.

[159]Sagerser, D. A., Lieblein, S., and Krebs, R. P., "Empirical Expressions for Estimating Length and Weight of Axial-Flow Components of VTOL Powerplants," NASA TM X-2406, 1971.

Programs for design analysis are the subject of Refs. 160–180.

[160]Hetherington, R., "Computer Calculations of the Flow in Axial Compressors," *Internal Aerodynamics (Turbomachinery)*, Institution of Mechanical Engineers, London, 1970, pp. 57–63.

[161]Marsh, H., "A Digital Computer Program for the Through-Flow Fluid Mechanics in an Arbitrary Turbomachine Using a Matrix Method," British Aeronautical Research Council, R&M 3509, 1968.

[162]Davis, W. R., "A Computer Program for the Analysis and Design of Turbomachinery—Revision," Div. of Aerothermodynamics, Carleton University, Ottawa, Canada, Rept. ME/A 71-5, Sept. 1971.

[163]Davis, W. R., "A Computer Program for the Analysis and Design of the Flow in Turbomachinery, Part B—Loss and Deviation Correlations," Div. of Aerothermodynamics, Carleton University, Ottawa, Canada, Rept. ME/A 70-1, July 1970.

[164]Davis, W. R. and Millar, D. A. J., "A Comparison of the Matrix and Streamline Curvature Methods of Axial Flow Turbomachinery Analysis, from a User's Point of View," *Journal of Engineering for Power, Transactions of ASME*, Ser. A, Vol. 97, 1975, pp. 549–560.

[165]Smith, D. J. L. and Frost, D. H., "Calculation of the Flow Past Turbomachine Blades," *Axial and Radial Turbomachinery, Thermodynamics and Fluid Mechanics Convention 1970*, Vol. II, *Proceedings of Institution of Mechanical Engineers*, Vol. 184, Pt. 3G(II), 1970, pp. 72–85.

[166]Glenny, D. E., "An Application of Streamline Curvature Methods to the Calculation of Flow in a Multistage Axial Compressor," Aeronautical Research Laboratories (Australia), Note ARL/M.E. 346, 1974.

[167]Glenny, D. E., "Performance Evaluation of a Multistage Axial Compressor Using a Streamline Curvature Model," ASME Paper 75-WA/GT-10, 1975.

[168]Hearsey, R. M., "A Revised Computer Program for Axial Compressor Design, Vol. I: Theory, Descriptions and User's Instructions," ARL TR 75-0001, Vol. I, 1975; "Vol. II: Program Listing and Program Use Example," ARL TR 75-0001, Vol. II, 1975.

[169]Novak, R. A. and Hearsey, R. M., "A Nearly Three-Dimensional Intrablade Computing System for Turbomachinery," *Journal of Fluids Engineering, Transactions of ASME*, Ser. I, Vol. 99, 1977, pp. 154–166.

[170]Novak, R. A., "Through-Flow Calculations in Axial Turbomachinery," AGARD-CP-195, 1976.

[171]Biniaris, S., "The Calculation of the Quasi-Three-Dimensional Flow in an Axial Gas Turbine," *Journal of Engineering for Power, Transactions of ASME*, Ser. A, Vol. 97, 1975, pp. 283–294.

[172]Hirsch, C. and Warzee, G., "An Integrated Quasi-3D Finite Element Calcu-

lation Program for Turbomachinery Flows," *Journal of Engineering for Power, Transactions of ASME*, Ser. A, Vol. 101, 1979, pp. 141–148.

[173]Katsanis, T. and McNally, W. D., "Fortran Program for Calculating Velocities and Streamlines on the Hub-Shroud Mid-Channel Flow Surface on an Axial- or Mixed-Flow Turbomachine, I: User's Manual," NASA TN D-7343, 1973; "II: Programmer's Manual," NASA TN D-7344, 1974.

[174]Wilkinson, D. H., "Calculation of Blade-to-Blade Flow in a Turbomachine by Streamline Curvature," British Aeronautical Research Council, R&M 3704, 1972.

[175]Katsanis, T., "Fortran Program for Calculating Transonic Velocities on a Blade-to-Blade Stream Surface of a Turbomachine," NASA TN D-5427, 1969.

[176]Katsanis, T. and McNally, W. D., "Fortran Program for Calculating Velocities in a Magnified Region on a Blade-to-Blade Stream Surface of a Turbomachine," NASA TN D-5091, 1969.

[177]McDonald, P. W., "The Computation of Transonic Flow Through Two-Dimensional Gas Turbine Cascades," ASME Paper 71-GT-89, 1971.

[178]Dodge, P. R., "The Use of a Finite Difference Technique to Predict Cascade, Stator, and Rotor Deviation Angles and Optimum Angles of Attack," *Journal of Engineering for Power, Transactions of ASME*, Ser. A, Vol. 95, 1973, pp. 185–190.

[179]Dodge, P. R., "A Non-Orthogonal Numerical Method for Solving Transonic Cascade Flows," ASME Paper 76-GT-63, 1976.

[180]Hantman, R. G., Mikolajczak, A. A., and Camarata, F. J., "Application of a Two-Dimensional Supersonic Passage Analysis to the Design of Compressor Rotors," ASME Paper 74-GT-97, 1974.

Reference 181 describes a computer program that functions in the design mode, combining an axisymmetric solution for the component of the equation of motion (nonisentropic, streamline curvature) at hub-to-tip computation stations fitted to blade row leading- and trailing-edge shapes with a blade row definition system based on Ref. 158. This program requires more detailed input than that of Ref. 130.

[181]Crouse, J. E. and Gorrell, W. T., "Computer Program for Aerodynamic and Blading Design of Multistage Axial-Flow Compressors," NASA Tech. Paper 1946, AVRARCOM TR 80-C-121, 1981.

References 182–187 outline methods for end wall boundary-layer prediction.

[182]Hanley, W. T., "A Correlation of End Wall Losses in Plane Compressor Cascades," *Journal of Engineering for Power, Transactions of ASME*, Ser. A, Vol. 90, 1968, pp. 251–257.

[183]Stratford, B. S., "The Use of Boundary Layer Techniques to Calculate the Blockage from the Annulus Boundary Layers in a Compressor," ASME Paper 67-WA/GT-7, 1967.

[184]Jansen, W., "The Application of End-Wall Boundary Layer Effects in the Performance Analysis of Axial Compressors," ASME Paper 67-WA/GT-11, 1967.

[185]Horlock, J. H. and Hoadley, D., "Calculation of the Annulus Wall Boundary Layers in Axial Flow Turbomachines," British Aeronautical Research Council, CP 1196, 1972.

[186]Bradshaw, P., "Prediction of Turbulent Shear Layers in Turbomachines," *Fluid Mechanics, Acoustics, and Design of Turbomachinery*, NASA SP-304, Pt. I, 1974, pp. 251–277.

[187]Smith, L. H. Jr., "Casing Boundary Layers in Multistage Axial-Flow Compressors," *Flow Research on Blading*, Elsevier Publishing Co., Amsterdam, 1969.

Some potentially attractive three-dimensional flowfield computation methods are described in Refs. 188–191.

[188]Stuart, A. R. and Hetherington, R., "The Solution of Three-Variable Duct-Flow Equations," *Fluid Mechanics, Acoustics, and Design of Turbomachinery*, NASA SP-304, Pt. I, 1974, pp 135–153.

[189]Erdos, J., Alzner, E., Kalben, P., McNally, W., and Slutsky, S., "Time-Dependent Transonic Flow Solutions for Axial Turbomachinery," *Aerodynamic Analyses Requiring Advanced Computers*, NASA SP-347, Pt. I, 1975, pp. 587–621.

[190]Briley, W. R. and McDonald, H., "Computation of Three-Dimensional Turbulent Subsonic Flow in Curved Passages," United Aircraft Research Laboratories, Rept. R75-911596-8, March 1975.

[191]Eiseman, P. R. and McDonald, H., "A Method for Computing Three-Dimensional Flow in Ducts," *Prediction Methods for Jet V/STOL Propulsion Aerodynamics*, Naval Air Systems Command Research and Technology Group, Vol. 1, July 1975, pp. 83–113.

Performance prediction by simplified methods is discussed in Refs. 192–196.

[192]Howell, A. R. and Bonham, R. P., "Overall and Stage Characteristics of Axial-Flow Compressors," *Proceedings, Institution of Mechanical Engineers*, Vol. 163, 1950, pp. 235–248.

[193]Robbins, W. H. and Dugan, J. F. Jr., "Prediction of Off-Design Performance of Multistage Compressors," *Aerodynamic Design of Axial-Flow Compressors*, NASA SP-36, 1965, Chap. X.

[194]Southwick, R. D., "A Stage-Stacking Simulation of Axial Flow Compressors with Variable Geometry," Aeronautical Systems Division, U.S. Air Force Systems Command, ASD-TR-74-38, 1974.

[195]Serovy, G. K., "Compressor and Turbine Performance Prediction System Development—Lessons from Thirty Years of History," *Modern Prediction Methods for Turbomachine Performance*, AGARD-LS-83, 1976.

[196]Howell, A. R. and Calvert, W. J., "A New Stage Stacking Technique for Axial-Flow Compressor Performance Prediction," ASME Paper 78-GT-139, 1978.

Performance prediction by extension of design analysis methods is discussed in Refs. 197–203. For cases where inlet distortion is present, Refs. 204–206 are suggested.

[197]Serovy, G. K., "A Method for the Prediction of the Off-Design Performance of Axial-Flow Compressors," Ph.D. Dissertation, Iowa State University, Ames, 1958.

[198]Serovy, G. K. and Anderson, E. W., "Method for Predicting Off-Design Performance of Axial-Flow Compressor Blade Rows," NASA TN D-110, 1959.

[199]Swan, W. C., "A Practical Engineering Solution of the Three-Dimensional Flow in Transonic Type Axial Flow Compressors," WADC Tech. Rept. 58-57, Feb. 1958.

[200] Swan, W. C., "A Practical Method of Predicting Transonic-Compressor Performance," *Journal of Engineering for Power, Transactions of ASME*, Ser. A, Vol. 83, 1961, pp. 322–330.

[201] Jansen, W. and Moffatt, W. C., "The Off-Design Analysis of Axial-Flow Compressors," *Journal of Engineering for Power, Transactions of ASME*, Ser. A, Vol. 89, 1967, pp. 453–462.

[202] Grahl, K., "Teillastberchnung für Axialverdichterstufen," *Zeitschrift für Flugwissenschaften*, Vol. 20, 1972, pp. 42–51.

[203] Novak, R. A., "Flow Field and Performance Map Computation for Axial-Flow Compressors and Turbines," *Modern Prediction Methods for Turbomachine Performance*, AGARD-LS-83, 1976.

[204] Novak, R. A. and Hearsey, R. M., "The Performance Prediction of Multistage Axial Compressor of Arbitrary Geometry Operating with Combined Radial and Circumferential Distributions," *Proceedings of the Air Force Airborne-Propulsion Compatibility Symposium*, June 1969, AFAPL-TR-69-103, 1970, pp. 627–672.

[205] Novak, R. A. and Hearsey, R. M., "Model for Predicting Compressor Performance with Combined Circumferential and Radial Distortion Computer Program," APL-TR-68-142, Pt. VIII, Dec. 1968.

[206] Mazzawy, R. S., "Multiple Segment Parallel Compressor Model for Circumferential Flow Distortion," Paper presented at AGARD Conference on Unsteady Phenomena in Turbomachinery, Monterey, Calif., Sept. 1975, pp. 22–26.

Aeroelastic analysis methods should be understood by the aerodynamicist. References 53 and 207–210 will help.

[207] Jeffers, J. D. II and Meece, C. E. Jr., "F100 Fan Stall Flutter Problem Review and Solution," *Journal of Aircraft*, Vol. 12, 1975, pp. 350–357.

[208] Danforth, C. E., "Distortion-Induced Vibration in Fan and Compressor Blading," *Journal of Aircraft*, Vol. 12, 1975, pp. 216–225.

[209] Danforth, C. E., "Blade Vibration: Same Key Elements in Design Verification," *Journal of Aircraft*, Vol. 12, 1975, pp. 333–342.

[210] Fleeter, S., McClure, R. B., Sinnet, G. T., and Holtman, R. L., "Supersonic Inlet-Torsional Cascade Flutter," *Journal of Aircraft*, Vol. 12, Aug. 1975, pp. 664–669.

3.7 Component and Configuration Experimental Development

3.8 Axial Flow Compressor and Performance Trends

Some specific programs of continuing and current interest can be located and are summarized by configuration.

NACA eight-stage compressor: Refs. 211–219
NACA five-stage compressor: Refs. 220–225
J-83 turbojet engine compressor: Refs. 226 and 227
CJ805-23 fan component: Ref. 228
NASA-GE rotor research program: Refs. 229–231
NASA-PW 1600 ft/s tip speed stage: Refs. 15 and 232
NASA-GE 1500 ft/s tip speed stage: Refs. 14–16
NASA-PW 1800 ft/s tip speed stage: Refs. 233 and 234

NASA-AiResearch 1500 ft/s fan stage: Refs. 235–237
NASA-PW two-stage fan: Refs. 238–240
NGTE four-stage compressor: Ref. 241
AFAPL 1500 ft/s tip speed stage: Refs. 242 and 243

[211]Voit, C. H., "Investigation of a High-Pressure-Ratio Eight-Stage Axial-Flow Research Compressor with Two Transonic Inlet Stages, I—Aerodynamic Design," NACA RM E53124, 1953.

[212]Geye, R. P. Budinger, R. E., and Voit, C. H., "Investigation of High-Pressure-Ratio Eight-Stage Axial-Flow Research Compressor with Two Transonic Inlet Stages, II—Preliminary Analysis of Over-All Performance," NACA RM E53J06, 1953.

[213]Voit, C. H. and Geye, R. P., "Investigation of a High-Pressure-Ratio Eight-Stage Axial-Flow Research Compressor with Two Transonic Inlet Stages, III—Individual Stage Performance Characteristics," NACA RM E54H17, 1954.

[214]Geye, R. P. and Voit, C. H., "Investigation of a High-Pressure-Ratio Eight-Stage Axial-Flow Research Compressor with Two Transonic Inlet Stages, IV—Modification of Aerodynamic Design and Prediction of Performance," NACA RM E55B28, 1955.

[215]Standahar, R. M. and Geye, R. P., "Investigation of High-Pressure-Ratio Eight-Stage Axial-Flow Research Compressor with Two Transonic Inlet Stages, V—Preliminary Analysis of Over-All Performance of Modified Compressor," NACA RM E55A03, 1955.

[216]Standahar, R. M., Hanson, M. P., and Geye, R. P., "Investigation of a High-Pressure-Ratio Eight-Stage Axial-Flow Research Compressor with Two Transonic Inlet Stages, VI—Over-All Performance, Rotating Stall, and Blade Vibration at Low and Intermediate Compressor Speeds," NACA RM E55I13, 1955.

[217]Geye, R. P. and Lucas, J. G., "Investigation of Effects of Reynolds Number on Over-All Performance of an Eight-Stage Axial-Flow Research Compressor with Two Transonic Inlet Stages," NACA RM E56L11a, 1957.

[218]Sievers, G. K., Geye, R. P., and Lucas, J. G., "Preliminary Analysis of Over-All Performance of an Eight-Stage Axial-Flow Research Compressor with Two Long-Chord Transonic Inlet Stages," NACA RM E57H14, 1958.

[219]Sievers, G. K., Geye, R. P., and Lucas, J. G., "Investigation of Reynolds Number Effect on Performance of an Eight-Stage Axial-Flow Research Compressor with Long- and Medium-Chord Lengths in the Two Transonic Inlet Stages," NACA RM E57J30, 1958.

[220]Sandercock, D. M., Kovach, K. and Lieblein, S., "Experimental Investigation of a Five-Stage Axial-Flow Research Compressor with Transonic Rotors in All Stages, I—Compressor Design," NACA RM E54F24, 1954.

[221]Kovach, K. and Sandercock, D. M., "Experimental Investigation of a Five-Stage Axial-Flow Research Compressor with Transonic Rotors in All Stages, II—Compressor Over-All Performance," NACA RM E54G01, 1954.

[222]Sandercock, D. M. and Kovach, K., "Experimental Investigation of a Five-Stage Axial-Flow Research Compressor with Transonic Rotors in All Stages, III—Interstage Data and Individual Stage Performance Characteristics," NACA RM E56G24, 1956.

[223]Sandercock, D. M., "Experimental Investigation of a Five-Stage Axial-Flow

Research Compressor with Transonic Rotors in All Stages, IV—Blade-Element Performance," NACA RM E57B12, 1957.

[224] Robbins, W. H. and Glaser, F. W., "Experimental Investigation of the Effect of Circumferential Inlet Flow Distortion on the Performance of a Five-Stage Axial-Flow Research Compressor with Transonic Rotors in All Stages," NACA RM E57J17, 1958.

[225] Kovach, K. and Sandercock, D. M., "Aerodynamic Design and Performance of Five-Stage Transonic Axial-Flow Compressor," *Journal of Engineering for Power, Transactions of ASME*, Ser. A, Vol. 83, 1961, pp. 303–321.

[226] King, J. A., "The J-83 Seven-Stage Transonic Compressor," *Journal of Engineering for Power, Transactions of ASME*, Ser. A., Vol. 83, 1961, pp. 291–302.

[227] Swan, W. C., "An Experiment with Aspect Ratio as a Means of Extending the Useful Range of a Transonic Inlet Stage of an Axial-Flow Compressor," *Journal of Engineering for Power, Transactions of ASME*, Ser. A, Vol. 86, 1964, pp. 243–246.

[228] Wright, L. C. and Novak, R. A., "Aerodynamic Design and Development of the General Electric CJ805-23 Aft Fan Component," ASME Paper 60-WA-270, 1960.

[229] Seyler, D. R. and Smith, L. H. Jr., "Single Stage Experimental Evaluation of High Mach Number Compressor Rotor Blading, Pt. 1—Design of Rotor Blading," NASA CR-54581, 1967.

[230] Gostelow, J. P., Krabacher, K. W., and Smith, L. H. Jr., "Performance Comparisons of High Mach Number Compressor Rotor Blading," NASA CR-1256, 1968.

[231] Gostelow, J. P., "Design and Performance Evaluation of Four Transonic Compressor Rotors, *Journal of Engineering for Power, Transactions of ASME*, Ser. A, Vol. 93, 1971, pp. 33–41.

[232] Monsarrat, N. T., Keenan, M. J., and Tramm, P. C., "Design Report. Single-Stage Evaluation of Highly-Loaded High-Mach-Number Compressor Stages," NASA CR-72562, 1969.

[233] Morris, A. L., Halle, J. E., and Kennedy, E., "High-Loading, 1800 Ft/Sec Tip Speed Transonic Compressor Fan Stage, I: Aerodynamic and Mechanical Design," NASA CR-120907, 1972.

[234] Morris, A. L. and Sulam, D. H., "High-Loading, 1800 Ft/Sec Tip Speed Transonic Compressor Fan Stage, II: Final Report," NASA CR-120991, 1972.

[235] Wright, L. C., Vitale, N. G., Ware, T. C., and Erwin, J. R., "High-Tip-Speed, Low-Loading Transonic Fan Stage, Pt. 1—Aerodynamic and Mechanical Design," NASA CR-121095, 1973.

[236] Ware, T. C., Kobayashi, R. J., and Jackson, R. J., "High-Tip-Speed, Low-Loading Transonic Fan Stage, Pt. 2—Data Compilation," NASA CR-121262, 1974.

[237] Ware, T. C., Kobayashi, R. J., and Jackson, R. J., "High-Tip-Speed, Low-Loading Transonic Fan Stage, Pt. 3—Final Report," NASA CR-121263, 1974.

[238] Messenger, H. E. and Kennedy, E. E., "Two-Stage Fan, I: Aerodynamic and Mechanical Design," NASA CR-120859, 1972.

[239] Messenger, H. E. and Keenan, M. J., "Two-Stage Fan, II: Data and Performance with Redesigned Second Stage Rotor Uniform and Distorted Inlet Flows," NASA CR-134710, 1974.

[240] Ruggeri, R. S. and Benser, W. A., "Performance of a Highly-Loaded Two-Stage Axial Flow Fan," NASA TM X-3076, 1974.

[241] Dransfield, D. C. and Calvert, W. J., "Detailed Flow Measurements in a Four Stage Axial Compressor," ASME Paper 76-GT-46, 1976.

[242] Wennerstrom, A. J. and Frost, G. R., "Design of a 1500 Ft/Sec, Transonic, High-Through-Flow, Single-Stage Axial-Flow Compressor with Low Hub/Tip Ratio," AFAPL-TR-76-59, 1976.

[243] Wennerstrom, A. J., "Experimental Study of a High-Through-Flow Transonic Axial Compressor Stage," *Proceedings of Sixth International Symposium on Air Breathing Engines*, 1983, pp. 447–457.

Experimental development of compressors and special methods for performance improvement are described in Refs. 244–250.

[244] Stone, A., "Effects of Stage Characteristics and Matching on Axial-Flow-Compressor Performance," *Transactions of ASME*, Vol. 80, 1958, pp. 1273–1293.

[245] Watkins, J., "Stage Performance and Radial Matching of Axial Compressor Blade Rows," *Journal of Engineering for Power*, *Transactions of ASME*, Ser. A, Vol. 81, 1959, pp. 13–23.

[246] Brown, L. E. and Groh, F. G., "Use of Experimental Interstage Performance Data to Obtain Optimum Performance of Multistage Axial Compressors," *Journal of Engineering for Power*, *Transactions of ASME*, Ser. A, Vol. 84, 1962, pp. 187–194.

[247] Brown, L. E., "The Use of Experimental Interstate Data for Matching the Performance of Axial Compressor Stages Having Variable-Setting-Angle Blade Rows," ASME Paper 76-GT-42, 1976.

[248] Kovach, K. and Griffiths, P. R.,"Engine Compression System Surge Line Evaluation Techniques," Paper presented at Third International Symposium on Air Breathing Engines, Munich, 1976.

[249] Prince, D. C. Jr., Wisler, D. C., and Hilvers, D. E., "A Study of Casing Treatment Stall Margin Improvement Phenomena," ASME Paper 75-GT-60, 1975.

[250] Holman, F. F. and Kidwell, J. R., "Effects of Casing Treatment on a Small, Transonic Axial-Flow Compressor, ASME Paper 75-WA/GT-6, 1975.

CHAPTER 4. TURBINE AERODYNAMICS

Robert P. Dring
United Technologies Research Center, East Hartford, Connecticut
and
William H. Heiser
General Electric Aircraft Engine Business Group, Evendale, Ohio

4. TURBINE AERODYNAMICS

Nomenclature

c = true airfoil chord
c_p = specific heat at constant pressure
f = function
m = mass flow
n = exponent
P = total pressure
q = dynamic head
S = airfoil surface distance
s = specific entropy
T = total temperature
U = freestream velocity
γ = ratio of specific heats
Γ = boundary layer dimensionless parameter
δ = boundary layer thickness
δ_1 = boundary layer displacement thickness
δ_2 = boundary layer momentum thickness
Δ = unit process change (exit-inlet)
η = thermodynamic efficiency
ν = kinematic viscosity
Σ = summation

Subscripts

e = exit
i = inlet
j = stream number
sep = separation

4.1 Introduction

Few devices have changed as drastically during the past decade as the aircraft gas turbine. The changes are the natural result of improving jet engine performance by means of new cycle parameters or configurations. Such advances are usually tied to increasingly demanding turbine operating conditions, including higher temperature, pressure, and output power levels. In order to meet these requirements with acceptable cost, weight, and durability, turbine cooling has been reduced to practice and the designer has sought to extract as much work as possible from each stage and, indeed, from each airfoil. Because these cycle changes spring from a fundamental thermodynamic foundation, there is every reason to expect the progression to continue into the next century.

A curious fact is that no comprehensive review of turbine aerodynamic technology has appeared since Horlock published *Axial Flow Turbines* in 1966.[1] That book still provides an excellent framework for the study of aircraft gas turbines, and many of the concepts developed there will be employed here. The main purpose of this chapter is to describe what has happened to turbine aerodynamics during the past decade. The remarks are largely concerned with two types of knowledge: (1) improved understanding of familiar but complex phenomena (e.g., transonic flows and transitional boundary layers) and (2) information regarding areas that have received essentially new attention (e.g., cooling flows and end wall flows).

We will be seen to have a bias in favor of sophisticated analysis supported by realistic testing. Experience has shown that the pressures of the marketplace have made these tools not only desirable, but essential. On one hand, cooling has driven the number of design permutations to the point where correlations based on the existing data cannot provide the required design information. On the other hand, building an experimental turbine has become so costly and time consuming that a modern engine development program is under a tremendous handicap if the initial turbine design falls short of the mark. Both arguments lead to the conclusion that confident analysis and accurate model and component testing are highly desirable. As described subsequently, many important advances have already been made and others now seem likely. Much of the chance has been removed from turbine design, and we are hopeful that this trend will continue.

General Trends

The leading trends in turbine technology can be illustrated by means of the information contained in Table 4.1, which compares the Pratt & Whitney JT3D and JT9D engines operating under sea-level, hot-day takeoff conditions. Both engines were designed for commercial usage and employ similar cycles, but were developed almost a decade apart.

The most remarkable feature of the data is that the high-pressure turbine (HPT) inlet temperature rose beyond the working temperatures of available materials during the period in question. This required the extensive use of cooling air and led to new airfoil shapes capable of providing suitable coolant passages. Meanwhile, the average stage loading coefficient (i.e., stage specific work/wheel speed squared), a direct measure of the associated aerodynamic difficulties, remained essentially constant. One is led to the conclusion that the HPT designers were well occupied with the novel problems of cooling.

The low-pressure turbine (LPT) presents a completely different picture. In order to satisfy the greatly increased demand for power by the fan without either increasing the wheel speed or adding a large number of stages, the average stage loading had to be increased. In fact, if the JT9D LPT was

Table 4.1 Comparison of Pratt & Whitney Engines

Parameter	JT3D	JT9D
Year of introduction	1961	1970
Engine bypass ratio	1.45	4.86
Engine overall pressure ratio	13.6	24.5
Core engine flow, lb/s	187.7	272.0
High-pressure turbine		
Inlet temperature, °F	1745	2500
Power output, hp	24,100	71,700
Number of stages	1	2
Average stage loading coefficient	1.72	1.76
Coolant plus leakage flow, %	2.5	16.1
Low-pressure turbine		
Inlet temperature, °F	1410	1600
Power output, hp	31,800	61,050
Number of stages	3	4
Average stage loading coefficient	1.44	2.47
Coolant plus leakage flow, %	0.7	1.4

based upon JT3D aerodynamic technology, six or seven stages would be required to replace the existing four. In order to avoid the excessive corresponding cost and weight penalties, LPT designers concentrated on applying advanced uncooled turbine aerodynamics.

The foregoing examples lead to a key point, namely that natural forces have combined to drive turbine aerodynamics into unexplored and uncertain territory. This has increased the risks being taken and increased the demand for methods to help assure success. The remainder of this chapter is largely devoted to an elaboration of the current understanding of the art.

Related Disciplines

Although this chapter deals principally with aerodynamics, there are other aspects of turbine design of equal importance. Since turbines are expected to run for thousands of hours without major overhaul, it follows that they cannot be based upon aerodynamic considerations alone. A successful machine results only from a highly iterative series of thoughtful aerodynamic, heat transfer, materials, and structural evaluations. The best solution to each design problem effectively couples respect for the important factors together in the correct proportions.

The literary history of turbines is rather out of balance in this regard, the overwhelming emphasis having been upon aerodynamics until recently. It now appears that this situation is slowly changing for the better, largely because persistent development difficulties of modern machines have shown that turbine durability can make or break an engine program. References 2–4 are compilations of papers reviewing heat transfer, high-temperature materials, and structural integrity, respectively. They provide a good starting point for those wishing to pursue these topics further.

Table 4.2 Airfoil Parameters

Parameter	JT3D	JT9D
High-pressure turbine		
Exit Mach number	0.7–1.0	0.6–0.9
Reynolds number (millions)	0.8–1.5	0.9–2.4
Aspect ratio (span/chord)	1.7–2.6	1.0–2.2
Turning angle, deg	68–110	70–111
Maximum thickness/axial chord	0.24	0.27
Low-pressure turbine		
Exit Mach number	0.6–0.9	0.5–0.8
Reynolds number (millions)	0.34–0.87	0.29–0.53
Aspect ratio (span/chord)	2.0–4.8	2.7–6.0
Turning angle, deg	36–97	67–107
Maximum thickness/axial chord	0.15	0.15

A deep appreciation of these sister technologies is essential to anyone seriously interested in gas turbine design. In fact, several specific examples of the interaction between structures and aerodynamics will be found in Sec. 4.3.

4.2 Turbine Airfoil Characteristics

This section takes a somewhat closer look at the prevailing turbine aerodynamic situation. This will be done by means of well-known airfoil dimensionless parameters, some of the most revealing of which are gathered together in Table 4.2. The JT3D and JT9D again serve as examples because of the fair comparison they provide.

This information should immediately convey the message that turbine airfoils have unique geometrical characteristics. In conventional terms, they would be called thick, highly cambered, and stubby. This means that they will exhibit a wide variety of fluid phenomena and are not especially susceptible to many classical analytical approaches.

This impression will be confirmed by the following remarks regarding the aerodynamic parameters.

Mach Number

When high subsonic or low supersonic exit Mach numbers are found in combination with airfoils of large thickness and/or deflection, local regions of supersonic flow (sometimes called bubbles) will often be found on the suction (or convex) surface. This is evidently the case for several HPT and LPT airfoils of Table 4.2 and one can safely presume that the trend will be toward higher exit Mach numbers and greater turning angles in order to increase stage work. There are, in fact, several "transonic" (i.e., subsonic inlet, supersonic exit) turbine airfoils already in service, a good example being the inlet guide vane of the JT15D, and the maximum JT9D turning angles have already been exceeded.

Predicting the behavior of "mixed" (i.e., partly subsonic, partly supersonic) flows around cascaded airfoils is not any easier than it is for isolated airfoils and the consequences of mistakes can be unacceptably severe.

Reynolds Number

A convenient reference point here is that natural transition occurs on a flat plate at a Reynolds number of about 2×10^6. This might lead to the conclusion that the turbine airfoils of Table 4.2 are dominated by laminar flow, but that would be wrong. One of the main reasons for this is that they are immersed in an extraordinarily turbulent environment, the turbulence level being in the range of 3–20%. The disturbances are initially generated in the combustion process and any tendency to decay is resisted by the boundary layers and wakes produced by the successive rows of airfoils. Large pressure gradients, particularly adverse ones appearing near the trailing edge on the suction surface, also exert a strong effect on the boundary layer transition.

An intense competition therefore often takes place along the suction surface between the Reynolds number, turbulence level, and pressure gradient over the state of the boundary layer. The result is difficult to predict in advance, but crucial to both the aerodynamic and thermal performance of the airfoil. It is even frequently found that there are long stretches which do not resemble either purely laminar or fully turbulent behavior and which are best thought of as extended transition regions.

Ample evidence of these effects exists in the open literature. For example, Hebbel[5] and Kiock[6] have carried out classical experiments demonstrating the sensitivity of turbine airfoils to suction surface boundary layer behavior. Much of what is now known began with their fine research.

The data of Table 4.2 show that the Reynolds number has increased in the HPT and decreased in the LPT with time. Since these were due, respectively, to increasing the engine overall pressure ratio and increasing the bypass ratio, the future will probably offer more of the same.

End Wall and Leakage Flows

The above remarks apply to that part of the flow that can be described in two-dimensional terms. A very important part of turbine aerodynamic behavior, however, is governed by three-dimensional effects. These latter flows generally lead to lost performance or inefficiency and therefore require special mention.

Passage end wall flows are one of the most common forms of three-dimensional flows found in turbines. They are caused by the blade-to-blade pressure difference that the mainstream flow impresses upon the end wall boundary layer and they result in an unrecoverable skewing of the exit plane velocity field. As the airfoil aspect ratio is reduced, end wall flow effects occupy an increasing portion of the span and must therefore become more significant. Table 4.2 reveals that higher cycle temperatures and pressures have acted to reduce the aspect ratio of HPT airfoils. The effect of increasing inlet temperature has been to require a longer airfoil chord in order to provide a sufficient coolant flow passage area. The effect of increasing the inlet pressure has been to reduce the necessary throughflow area and, correspondingly, the span. When the aspect ratio nears unity, as it has in the JT9D HPT, the end wall flow assumes equal importance with the mainstream flow and the boundary layer approach breaks down.

Similar remarks can be made for leakage or clearance effects. Since clearances have minima that are independently determined by manufacturing tolerances and are harder to maintain in the more hostile environments, they have a relatively greater impact on the shorter airfoils of high-pressure-ratio machines. Leakage losses have become, in fact, a determining factor in turbine engine development.

Conclusion

One clear message emerges from the foregoing facts: turbine design simultaneously involves most of those things that aerodynamicists have learned to dread. This includes at least transonic, transitional flow over thick, highly cambered cascaded airfoils in the presence of strong turbulence and three-dimensional effects. Contrast this with the recent comments of the authoritative Smith[7] regarding high-lift isolated airfoils, "That is not to say that nearly all problems have been solved, but to a certain extent the remaining problems amount to just irritating details." In this context modern machines seem intent upon overwhelming us with irritating details.

4.3 Design Considerations

At this point, it will be profitable to stand back and view the actual design approach and procedure so that the various advances can be appreciated in their proper context. In this regard, the work by Horlock[1] is an excellent starting point. He has provided a concise derivation and description of the relationships between the stage loading coefficient, flow coefficient (axial velocity/wheel speed), reaction, turning, and velocity triangles in general. In addition, he has included a discussion of structural considerations such as the centrifugal tensile stress at blade roots (blade pull stress) and the hoop stress due to centrifugal loading on the rotor disks. These derivations will not be repeated here. The reader who is unfamiliar with these terms would profit greatly by examining that work.

The designer has before him the task of satisfying certain requirements and remaining within certain constraints, while meeting or exceeding his performance goal. The requirements are generally specified by the cycle and they include, for example, producing a given power at given values of inlet pressure, temperature, and mass flow. The constraints stem from many different sources, including structural, mechanical, and aerodynamic considerations. They could include limits on rotor blade pull stress, disk rim speed, airfoil maximum thickness and trailing edge thickness, flow turning, and Mach number. The performance goal that must be met is efficiency. Thus, while satisfying the requirements, the designer must consider a multitude of configurations in order to establish the trends that lead to maximum efficiency. When the final design emerges, it is invariably true that the performance is limited as a direct result of one or more of the constraints. With this in mind, several typical and commonly occurring design constraints will be considered here. In particular, attention will be given to the way in which structural constraints can bring about limitations on the blade tip and disk rim speeds and hence on the stage loading coefficient. Following this, a brief description of the design system itself shall be presented. Among other things, the importance of the aerodynamic loss system shall be discussed.

Design Constraints

The material limitations of maximum allowable blade pull stress and disk stress are not difficult to appreciate. What shall become clear is that these stresses can pose severe restrictions on the aerodynamicist in his selection of a flow path and consequently they can have a direct impact on the ability to reach the goal efficiency. As shown by Horlock,[1] the blade pull stress is proportional to the product of the flow path annulus area and shaft speed squared. The disk stress is proportional to the disk rim velocity squared. Thus, three variables that the designer has at his disposal in configuring a stage consistent with these two constraints are the annulus inner and outer radii and the shaft speed. He would look at various combinations of these variables (and others, such as reaction) that lead in the direction of increasing efficiency. In the design of a high-pressure turbine, it is frequently true that the search for higher efficiency places the design on the limit of allowable blade pull stress. The reason for this is the desire to minimize the stage loading factors (by raising the wheel speed) and the Mach number level (by increasing the annulus area). In addition, if the shaft speed is specified by the compressor requirements, the designer's freedom is considerably restricted. In the ultimate case, it can occur that with a specified shaft speed the highest achievable efficiency occurs when both the blade pull and disk stresses are at their limiting values. In this case, there is only one flow path that can satisfy all of the constraints simultaneously and the highest efficiency the turbine can reach has been limited by the structural and mechanical constraints.

Although low-pressure turbine design is seldom hampered by these stress considerations, the shaft speed is frequently limited to a very low value because of restrictions on the fan tip speed. Further, the maximum diameter of the turbine may be limited by restrictions on engine size and weight. The evolution of low-pressure turbines with high stage loading coefficients (see Table 4.1) is a direct result of these constraints. The consequence is a trend toward airfoils with increased turning, decreased flow acceleration, and in general a more severe aerodynamic environment.

The discussion up to this point has addressed the impact of constraints on the flow path and the velocity triangles, but there are also constraints on the airfoil design. The leading and trailing edge diameters must be large enough to allow the incorporation of cooling devices as necessary. The airfoil thickness distribution is often dictated by the necessity of allowing room for internal cooling devices and the entire spanwise layout of an airfoil may be dictated by the need to insert and withdraw these devices. It is only a slight exaggeration to say that many cooled airfoils are a result of the best aerodynamic design that could be wrapped around their internal cooling devices.

Although only a very few of the constraints on a turbine design have been mentioned here, it should be clear that they can have a profound impact on the final configuration. The not uncommon misconception that turbines present few difficult aerodynamic challenges stems in part from a lack of appreciation of the constraints placed upon them. It is the constraints that have brought about the need for advanced analytical design tools.

The Design System

Before one can appreciate the importance of an improved analytical capability, one must have some appreciation of the design system in which it is incorporated. The system is a sequence of different types of calculations with inputs and constraints from various sources. It is a sequence in that each analysis provides input to the next. It is also an iterative sequence in that the results of each analysis alter inputs used earlier. What follows here is a brief summary of this sequence. A far more in-depth discussion of a modern computer interactive turbine design system has been presented by Thomas and Piendel.[8]

The initial input is in the form of cycle performance requirements. These would typically include parameters such as inlet and exit pressures and temperatures, primary and secondary (coolant) mass flows, and shaft speeds. The constraints at this point might include, for example, blade pull stress, airfoil chord, and trailing edge thickness. These serve as input to the meanline analysis.

The meanline analysis provides midspan velocity triangles. Preliminary root and tip velocity triangles may be determined with an assumption such as free-vortex flow. The variables include the flow path definition, blade speed, stage work, and reaction. Input must include sufficient information about the airfoil geometry such that the various loss systems can be employed. This analysis may be carried out for a single turbine or for a complete system including the high- and low-pressure turbines and the exit guide vane. As mentioned above, the objective of the analysis is to point out the highest performance configuration (flow path and velocity triangles) that will satisfy the cycle requirements while remaining within all the constraints.

The performance is determined by the loss system. This system is a series of calculations and correlations that determine the magnitude of the various losses (profile, end wall, parasitic, and cooling loss) based on the aerodynamic and geometric input provided. The problem is, for example, to make an accurate estimate of the profile loss without having any detailed idea of the airfoil contour, but rather having only the velocity triangles, pitch, chord, and trailing edge thickness. The loss system is the most crucial feature of a meanline optimization study. For example, the number of stages in a turbine will be dictated to a large degree by the way the predicted efficiency responds to variations in the stage loading coefficient. These efficiency variations are dictated by the loss system. Thus, a loss system with an excessive penalty for higher turning will cause the final configuration to have lower stage loading coefficients and hence more stages. For this reason, improvements in the loss systems can lead to significant changes in the final turbine configuration. Great care must be exercised to insure that the best aerodynamics are reflected in the loss system. Otherwise, they need never appear in the final configuration. Results of the meanline analysis provide updated input to the cycle analysis. They also provide input to the streamline analysis.

The streamline analysis is a two-dimensional axisymmetric meridional plane solution by, for example, streamline curvature or matrix inversion techniques. It brings the spanwise variations into the design. The streamline analysis may simply serve as a refinement of the results of the meanline analysis or it may be employed in place of the meanline analysis as a basic optimization tool that includes spanwise variations. An example of this approach can be seen in the work by Schlegel et al.[9] Regardless of the way it is employed, the basic virtue of the streamline analysis is that it permits one to eliminate many of the adverse effects accompanying a simple free-vortex design. The large radial variations of reaction that occur in a free-vortex design, for example, can be eliminated by "controlled-vortex" design.[10] The result is that the difficulties of very low reaction (impulse) blade roots and the leakage penalties of very high reaction shrouded blade tips can be minimized. The results of the streamline analysis serve as input to further iterations with the meanline analysis and also as input to the airfoil contour design calculation.

Turbine airfoil contours are generally uniquely designed for each application. This is a major departure from the approach in many compressor designs where the airfoils are selected from standard families of contours. With the velocity triangles dictated at each radial section by the streamline analysis and a first estimate of the gap-chord ratio from lift coefficient considerations, the objective is to devise a contour that will have a pressure distribution with desirable boundary layer characteristics. One aspect of turbine airfoil design that simplifies matters is that the flow deviation, and hence the mass flow, can be determined with sufficient accuracy from the geometry of the airfoil. Thus, turbine airfoil design is a matter of defining the contour that will most efficiently provide the desired lift.

Two approaches are currently being followed in airfoil design. In the "direct" approach, the contour is defined and the pressure distribution is determined. In the "indirect" approach, a pressure distribution is specified and the contour is determined. The important point to be appreciated is that accurate and fast cascade flow calculations are becoming increasingly available (e.g., see Ives and Liutermoza[11]).

The results of the airfoil design serve as input to the meanline analysis in terms of updated profile loss estimates and to the heat transfer, structural, and mechanical designers for all of their various analyses.

Summary

The turbine design sequence is a highly analytical procedure with the objective of simultaneously satisfying a wide variety of requirements and constraints while optimizing on high performance, low weight, and low cost. It is an unusual turbine that is not pressed hard against one or more of the imposed constraints. In addition, the basic configuration of the turbine (for example, the number of stages) will be largely defined by the nature of the aerodynamic loss system used to predict efficiency. These are but a few examples of the numerous tradeoffs that occur during a turbine design. The tradeoff between increased stress and higher efficiency is only one example. Others include: (1) reductions in airfoil solidity that cause reductions in weight and the number of airfoils (cost), but bring with them the risk of flow separation and reduced efficiency and (2) increases in rotor blade cooling that can permit increased operating stress levels and higher aerodynamic performance, but also bring about increased cooling losses.

The advances in technology that will be discussed in following sections are generally associated with either moving a constraint or reducing the penalty associated with a constraint. Advances in turbine aerodynamics have, for example, pushed to much higher values the historical constraints on maximum turning and exit Mach number. The result has been much higher performance in machines with high stage loading than had ever been anticipated.

4.4 Performance

In this section, the turbine aerodynamic phenomena will be examined in detail. The attendant complexities can best be unraveled by organizing our thinking about the factors determining performance or efficiency. This approach leads directly to a comprehensive framework for the remainder of this chapter.

Uncooled Turbine Efficiency

When the engine cycle calls for turbine inlet temperatures below about 1800°F, airfoils can usually be safely manufactured from available materials requiring no protective cooling. In such cases, the weight flow does not change throughout the turbine. If the working fluids are taken to behave as perfect gases with constant properties, some useful results can be obtained in a convenient algebraic form. Translation to real gases is, of course, required for any actual turbine design, but this is easily accomplished in practice. The effects of changing streamline radial location can also be rigorously included in these formulations. Since they are relatively minor for axial flow machines and unnecessarily complicate the resulting expressions, they are omitted here.

The uncooled turbine is best thought of as an open-cycle machine working between two reservoirs of different total pressure levels. The total pressure and total temperature levels are not only the most thermodynamically meaningful, but are also far easier to measure than static properties.

Application of the first and second laws of thermodynamics leads directly to the relation

$$\eta = \frac{\text{actual power}}{\text{ideal power}} = \frac{1 - \left(\frac{P_e}{P_i}\right)^{(\gamma-1)/\gamma} \exp\left(\frac{s_e - s_i}{c_p}\right)}{1 - \left(\frac{P_e}{P_i}\right)^{(\gamma-1)/\gamma}} \leq 1 \tag{4.1}$$

Even though the turbine appears to have both steady and unsteady flow processes taking place when viewed from any single frame of reference, the path traversed by each particle of fluid in passing through the machine can be thought of as consisting of a series of steady flow processes viewed from appropriate frames of reference. Each of these sequential steady flow processes, viewed in its appropriate frame of reference (alternately stationary or rotating), will be referred to here as a unit process. The special advantage of this approach is that the total entropy change for any particle of fluid is equal to the sum of the entropy changes for all the unit processes, for entropy is a state property and does not depend upon the frame of reference. Since the entropy only rises in each unit process, then

$$s_e - s_i = \Sigma \Delta s \geq 0 \tag{4.2}$$

This reasoning and Eq. (4.1) lead to the usual conclusion that unit process entropy increases must be held to a minimum in order to maximize turbine efficiency. Finally, since in each unit process the total temperature remains constant, then

$$\frac{\Delta s}{c_p} = -\left(\frac{\gamma - 1}{\gamma}\right) \ell n \left(1 + \frac{\Delta P}{P_i}\right) \tag{4.3}$$

so that entropy increases and total pressure losses go hand-in-hand.

Cooled Turbine Efficiency

The same reasoning that led to a useful definition of thermodynamic efficiency for uncooled turbines can be applied to each of the individual streams entering a cooled turbine for the same purpose. Care must be taken to provide a proper control volume completely surrounding the turbine and all inlet and outlet flows must be accounted for. The resulting expression, equivalent to that of Eq. (4.1), is

$$\eta = \frac{\sum_j \text{actual power}}{\sum_j \text{ideal power}}$$

$$= \frac{\sum_j T_{ji}\left[1 - \left(\frac{P_{je}}{P_{ji}}\right)^{(\gamma-1)/\gamma} \exp\left(\frac{s_{je} - s_{ji}}{c_p}\right)\right] m_j}{\sum_j T_{ji}\left[1 - \left(\frac{P_{je}}{P_{ji}}\right)^{(\gamma-1)/\gamma}\right] m_j} \leq 1 \tag{4.4}$$

where j is the minimum number of separate streams necessary to describe the operation of the turbine. Equation (4.4) shows that the thermodynamic efficiency of the machine in question is entirely dependent upon the total entropy changes experienced by the individual streams while traversing the turbine. This is a satisfying and valuable result. Among other things, it shows that the unit process reasoning still applies and that each stream contributes to inefficiency in proportion to its mass flux and entropy rise. However, as was previously mentioned and as will be elaborated upon later, the entropy change of a unit process now also includes irreversibility due to thermal mixing.

Specifically, unit process entropy changes for each of the j separate streams are given by

$$\frac{\Delta s}{c_p} = \ell n\left(1 + \frac{\Delta T}{T_i}\right) - \left(\frac{\gamma - 1}{\gamma}\right) \ell n\left(1 + \frac{\Delta P}{P_i}\right) \tag{4.5}$$

rather than by Eq. (4.3).

In some cases, Eq. (4.4) can be further simplified. These cases require that the inlet and exit total pressures are each uniform over the j streams. They occur, for example, when the exit flow is taken to be completely mixed and the inlet flow either comes from a single source or the turbine is included as part of a full accounting of the losses which take place after the compressor discharge station. The corresponding expression for efficiency becomes

$$\eta = \frac{\sum_j T_{ji}\left[1 - \left(\frac{P_e}{P_i}\right)^{(\gamma-1)/\gamma} \exp\left(\frac{s_{je} - s_{ji}}{c_p}\right)\right] m_j}{\left[1 - \left(\frac{P_e}{P_i}\right)^{(\gamma-1)/\gamma}\right] \sum_j m_j T_{ji}} \leq 1 \tag{4.6}$$

which contains some useful guidance. First, the denominator reveals that the temperature of thermodynamic significance is the mass average value, regardless of the distribution between streams. Second, it emphasizes the fact that all of the streams passing through the turbine are capable of producing work. This leads, in turn, to the conclusion that all losses experienced by all streams coming from the total pressure reservoir to the turbine must be understood and considered if maximum efficiency is to be obtained. These losses include, for example, those generated by the metering holes and slots or within the cooled airfoils. Although this point should be fairly obvious, it is not always honored in practice.

Total Pressure Loss Breakdown

By now it should be clear that increases in entropy are the root cause of inefficiency and that they occur as a result of total pressure loss and thermal mixing [see Eq. (4.5)]. The entropy change due to thermal mixing is both inevitable and easily calculated. The contribution of total pressure loss, on the other hand, is neither completely under control nor easy to predict. The designer will seek to minimize the total pressure losses. His most fruitful avenue of approach is through an understanding of the phenomena that generate them.

The sources of total pressure or aerodynamic losses are the same as those encountered throughout fluid mechanics. They include: skin-friction drag, pressure or form drag, shock losses, leakage, and mixing. Of these only the first four matter in uncooled turbines. The last must be accounted for in cooled turbines, where gases of considerably different properties are irreversibly mixed together [see Eq. (4.5)].

It is convenient and has become customary to partition the losses into four other categories, namely: profile, end wall, parasitic, and cooling. This method associates losses with the location of their production rather than with their phenomenological origins. A rough but useful idea of the relative importance of each category can be found in Table 4.3.

Although profile losses are evidently important, it is clear that the others cannot be disregarded. The surrounding world seems to have overlooked these simple facts, for the bulk of turbine aerodynamic research has been aimed at profile design.

The above information has been used to shape the remainder of this chapter. Careful attention is given to each category of loss, cooling losses also being dealt with in Chap. 5.

Table 4.3 Loss Categories

Loss Category	Percent Cooled HPT	Percent Uncooled LPT
Profile	35	60
End wall + parasitic	35	40
Cooling	30	0
Total	100	100

4.5 Profile Aerodynamics

Well-designed airfoil contours are an essential feature of any gas turbine that is to reach its performance goals, not to mention its goals in weight and cost. The vast amount of effort that has been expended in this area in recent years is an indication of the recognition of this fact. To say that profile design is simplified by the generally accelerating nature of the flow is highly misleading. Turbine airfoils do have strong recompressions and boundary-layer separation is a real and present problem that must be addressed in every design. Furthermore, the obvious advantages of reduced weight and cost that can be achieved with highly loaded (low solidity) designs brings pressure to have even stronger local recompressions. Keeping this in mind and remembering also, as mentioned earlier, that turbine airfoils can be thick, highly cambered, and transonic and can have transitional boundary layers leads one to the conclusion that the design of good turbine airfoils is a challenging and difficult task.

Design Approach

At the outset of an airfoil design, the velocity triangles are available as a result of the streamline analysis. Included in them are the effects of the anticipated total pressure loss across the cascade and also the stream tube thickness variations (due to meridional plane streamline curvature and/or annulus area divergence). In addition, constraints such as on leading and trailing edge diameters, the value and location of the maximum or minimum thickness, and others may have already been specified. If, for example, the airfoil is a highly stressed (centrifugally) high-pressure turbine rotor blade, it will be necessary to taper the airfoil. This will require an airfoil with a very thin, low solidity tip and a relatively thick, high solidity root.

The first step in the airfoil design is to set the cascade throat dimension and/or the trailing edge mean camber line angle in order to provide the proper mass flow through the cascade. In subsonic flow, there is no unique relationship between the throat area and mass flow and one is totally dependent on the correlations. The work by Ainley and Mathieson[12] is a good starting point in this area. In transonic flow, once the cascade is choked, the simple one-dimensional relationship between flow area, throat area, and Mach number can be applied with confidence. Care must be taken in both cases to properly account for things that can alter the continuity relationship between the throat and downstream. These include, for example, total pressure loss, mass addition, and thermal dilution. With an estimate on the solidity based on something like the Zweifel[13] coefficient, a first guess on the profile contour is made.

The potential flow calculation is then executed, keeping in mind various aerodynamic requirements. There will be a need to hold down the maximum suction (convex) surface Mach number in order to reduce the adverse pressure gradient at the trailing edge. In transonic flow, pains must be taken to minimize the effects of shocks, both from considerations of shock loss and the impact on the suction surface boundary layer. One final require-

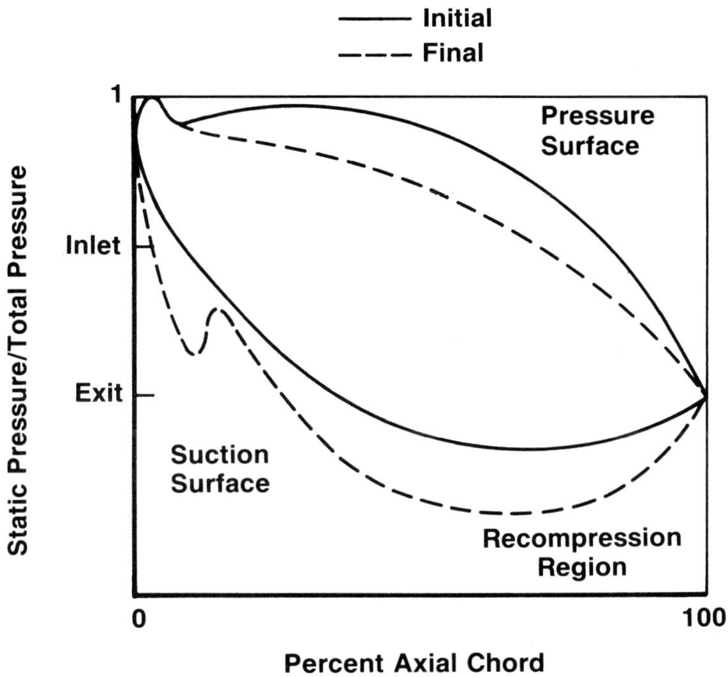

Fig. 4.1 Cascade airfoil pressure distributions.

ment might be related to the leading edge overspeeds and how they affect performance at incidence.

With an acceptable potential flow, a boundary layer calculation may be carried out. The aerodynamic requirements here include the avoidance of gross separation and possibly that the region of laminar flow be as long as possible. The boundary layer results at the trailing edge are coupled with other information (e.g., trailing edge blockage) and a wake mixing calculation is performed, as described by Stewart,[14] to determine the cascade loss.

An extremely useful aspect of cascade aerodynamics is that there are several rules to which the pressure distribution around any airfoil designed for a specific purpose must conform. Portrayed in Fig. 4.1 are the pressure distributions for two differently shaped cascade airfoils intended to accomplish the same purpose. The two curves will have at least the following things in common:

(1) The static pressure equals the stagnation pressure on the leading edge stagnation point.

(2) The areas circumscribed by the curves are equal (to the circumferential force on the airfoil).

(3) The trailing edge pressure approximates the ultimate downstream static pressure (which is available from prior streamline calculations).

(4) There is a minimum pressure point along the suction surface from which recompression is required to reach the trailing edge (this can be shown to be true by means of streamline curvature arguments and is due to the generally convex shape of that surface).

The last point is crucial, for it guarantees that there are regions within every turbine where the vulnerable boundary layer must "run uphill." When the adverse pressure gradient in the recompression region is sufficiently large, the boundary layer separates. The profile losses are then controlled by pressure or form drag rather than skin friction drag and are usually unexpectedly and unacceptably large.

By this time, it should be obvious that the designer must sometimes reduce the trailing edge pressure rise. This is accomplished by altering the geometric shape of the airfoil and checking the results. The two pressure distributions of Fig. 4.1 are intended to illustrate the beginning and end of a successful iterative search process. The reader can also use Fig. 4.1 to understand why increasing the number of airfoils (which reduces the circumscribed area) or increasing the airfoil reaction (by lowering the exit static pressure) are other tricks that can help reduce the strength of the recompression.

This entire process is iterative in nature and may require a large number of passes before a suitable contour is defined satisfying all (or most) of the aerodynamic requirements. It should be noted here that this iteration could be altered significantly by a "synthesis" or "indirect" analysis. In this case, the pressure distribution is specified and the contour is computed from it. The discussion in the following section will be limited to the "direct" approach since it has been the main focus of attention in the literature.

Potential Flow

The progress during the past decade in the area of potential flow in cascades has been remarkable. For flows that are compressible but subsonic, a large number of fast, versatile, and accurate calculations are now available. The works by Katsanis and McNally[15] and Van den Braembussche[16] give ample evidence of this progress. The state-of-the-art today is such that one is even able to perform detailed analysis on the leading edge flow[17] to determine the impact of incidence on the leading edge overspeed.

In the area of transonic cascade flow, progress has been equally remarkable. The complicated nature of these flows is illustrated in Fig. 4.2. The pressure distribution and the locations of the sonic lines and shocks are shown schematically for several Mach numbers. At an exit Mach number of 0.7, the flow is everywhere subsonic. As the back pressure is lowered (holding the inlet flow angle and total pressure fixed), the area within the pressure distribution (i.e., the work-producing tangential force) increases until at an exit Mach number of 0.9 there is a thin supersonic region with a sonic line extending from 35 to 75% axial chord. This type of flow is typical of a large number of existing gas turbines. As the back pressure is further lowered, the exit Mach number passes through 1.0. Near this condition, several things happen. The inlet Mach number, which has been increasing

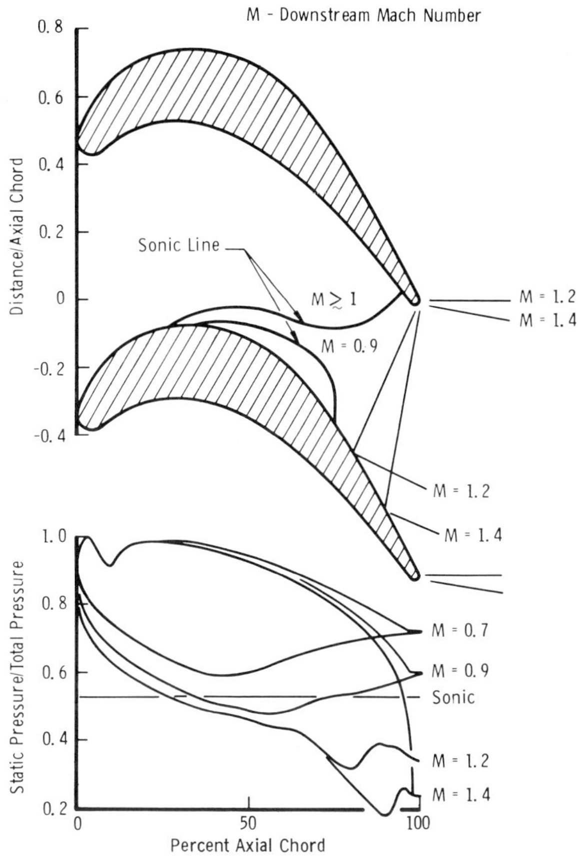

Fig. 4.2 Transonic cascade flow.

up to this point, reaches an asymptotic upper limit and will not respond to further reductions in back pressure. In addition, the sonic line, which has expanded across the throat and moved forward on the suction surface, also reaches an asymptotic position. At this condition, the sonic line is highly curved. This is a result of the relatively high loading on this airfoil. With lower loading on the airfoil (higher solidity), the area within the pressure distribution would be reduced. The suction surface would have gone through sonic velocity further aft and the sonic line would have come nearly straight across the throat. As the back pressure is further reduced, the pressure distribution on the pressure surface and most of the suction surface remains unchanged. What is occurring is that a shock has formed emanating from near the trailing edge and reflecting off the suction surface. As the Mach number is increased, this shock moves further aft, leaving the flow upstream of it unchanged. The condition when the shock reaches the trailing edge is referred to as "limit loading" since the force on the airfoil has reached a

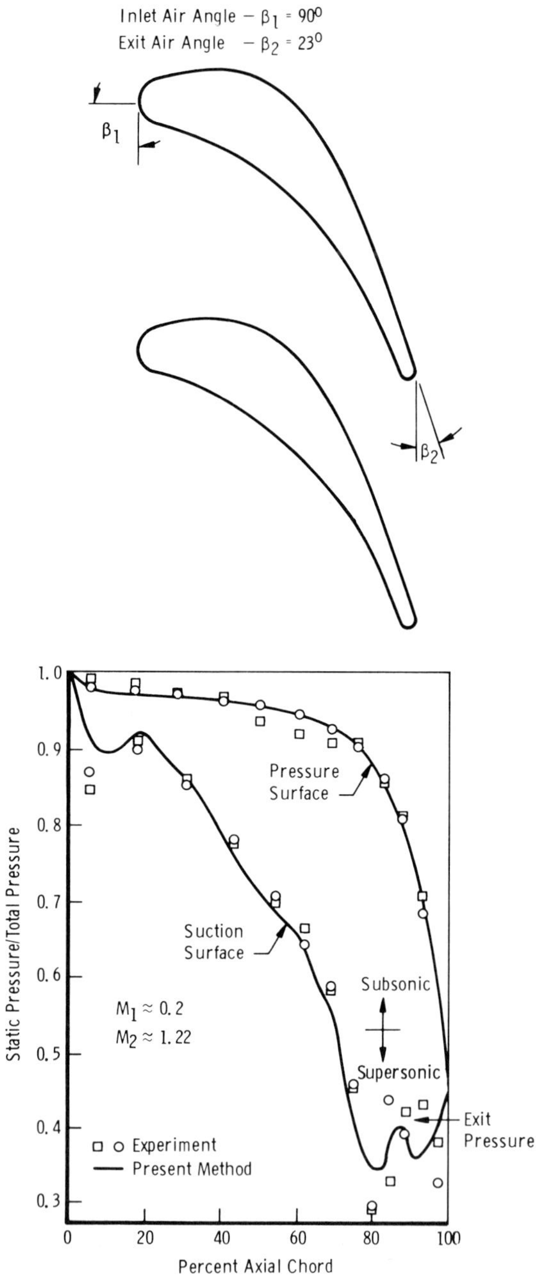

Fig. 4.3 Predicted and measured stator pressure distribution.

maximum. This condition corresponds to an exit axial Mach number of 1.0 and hence the back pressure can be reduced no further.

As one would expect, the prediction of transonic cascade flows has lagged behind that of subsonic flows. Presently, however, there are a number of analyses available. A time-marching approach has been followed by McDonald[18] and Delaney and Kavanagh.[19] The results often seem too good to be true, as the portions of McDonald's work contained in Figs. 4.3 and 4.4 will attest. As a matter of fact, these flows can be so complex that the

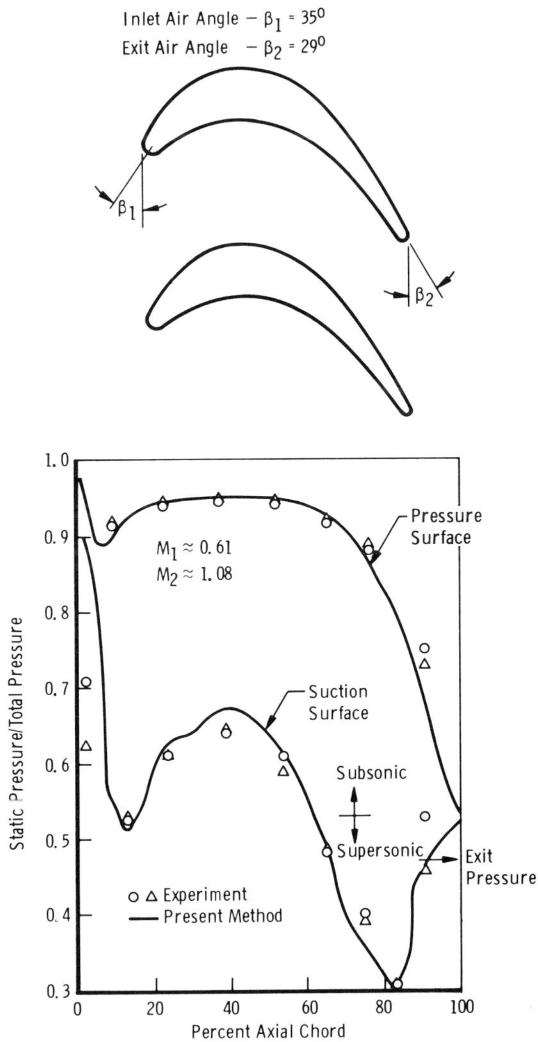

Fig. 4.4 Predicted and measured rotor pressure distribution.

experimental data shown there were considered suspect until later verified and clarified by the analysis. In recent work, such as Refs. 11 and 20, conformal mapping and relaxation techniques have been applied to achieve equally remarkable results.

The present status of turbine cascade potential flow analysis is that the problems that existed 10 years ago have been for the most part solved. Although there are still a number of areas being worked in transonic cascade flow, they are in the detail, or quality, of the results and not in the basic computational capability. Unfortunately, many of the missing details are intimately tied to the near-wake flow and can therefore only be predicted by a viscous analysis.

Boundary Layer

Another area of aerodynamics that has seen great advances during the past 10 years has been boundary layer theory. The gains in the understanding of the various phenomena and the analytical modeling of them has drawn together into a rational system what had earlier been a large quantity of unexplained and confusing data. It is becoming increasingly clear that the wide variations in cascade performance, which had been previously attributed to the basic nature of the velocity triangles and overall cascade parameters (e.g., solidity), can now be explained in detail by boundary layer analysis. The very large losses conventionally associated with high turning can to a great extent be avoided by proper airfoil design and boundary layer analysis. In order to understand this more clearly, consider the phenomena occurring in a cascade profile boundary layer.

The pressure distribution in Fig. 4.5 is not necessarily a good one from a performance point of view, but it will serve as an example. Consider the

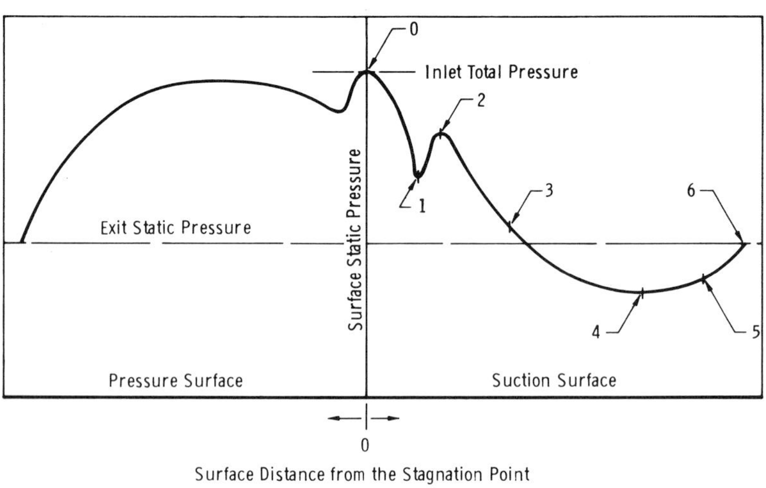

Fig. 4.5 Cascade airfoil pressure distribution.

suction surface boundary layer. It starts at the stagnation point 0 as laminar and begins to accelerate strongly around the leading edge circle. Leaving the circle the flow goes into a strong recompression as a result of the leading edge overspeed. This recompression may cause the boundary layer to separate at point 1 and to form a bubble that may undergo transition and reattach fully turbulent. The very large penalties associated with large incidences occur when this overspeed is sufficiently strong such that the bubble cannot reattach. Assuming that the bubble has reattached (fully turbulent), the boundary layer accelerates strongly at point 2 and may undergo relaminarization, or reverse transition, by point 3. The boundary layer continues to accelerate down to the minimum pressure point 4 and there the final recompression begins. At this point several things may occur. The boundary layer may undergo transition. It may separate to form a bubble that reattaches fully turbulent or it may not reattach at all. If the boundary layer does reattach, it will proceed all the way to the trailing edge at point 6 or will undergo turbulent separation (without reattachment) at point 5 before it reaches the trailing edge. A similar story can easily be imagined for the pressure surface boundary layer, but without the final recompression. For a given cascade of airfoils, this sequence of events can be altered greatly by variations in the Reynolds number, Mach number, freestream turbulence, and incidence. Also, for a given set of flow conditions, the sequence of events can be altered by relatively small variations in the airfoil design. It is this latter point that makes it difficult to construct a truly universal loss system. In a meanline analysis, the detailed contour does not exist yet and hence some assumptions must be made in the construction of the loss system. An analytical loss system might, for example, be based

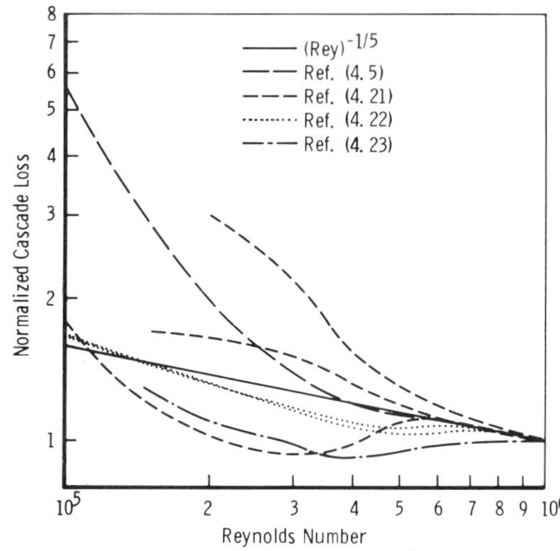

Fig. 4.6 Effect of Reynolds number on cascade loss.

on the assumption of boundary layers that are fully turbulent at all points. An experimentally based loss system can be misleading unless the tests are carefully controlled and the results properly analyzed and generalized. The significance of this latter point cannot be overemphasized. In Fig. 4.6 the variations with Reynolds number of the normalized performance of a number of cascades from various sources (Refs. 5 and 21-23) are illustrated. As would be expected from the complicated sequence of events that can occur in a cascade boundary layer as described above, no single scaling rule appears to apply. It is upon application of the analyses that are now available that the situation begins to become clear.

The cascade designer has at his disposal two basically different approaches to the boundary layer calculation. The first is based on a simplified set of equations, correlations, and criteria. The second is based on a more fundamental approach that, although more complicated and time consuming, will provide a more definitive result. As an example of the first type of approach, the closed form integral boundary layer analysis by Dring[24] is typical. Although it is presented in the context of a turbulent boundary layer, it can also be used for laminar flows with an appropriate choice of constants. Typical of the many conditions used in the literature to correlate the occurrence of boundary layer separation is

$$\Gamma = \delta_2\left(\frac{U\delta_2}{\nu}\right)^{1/n}\left(\frac{1}{U}\frac{\mathrm{d}U}{\mathrm{d}S}\right) \tag{4.7}$$

For laminar flow $n = 1$ and $\Gamma_{sep} = -0.1567$ correspond to the von Kármán-Pohlhausen solution and for turbulent flow $n = 4$ and $\Gamma_{sep} = -0.06$ correspond to the Buri solution.[25] Similar conditions for the occurrence of transition and relaminarization have been published by Dunham[26] and by Launder,[27] respectively. Finally, the work of Horton[28] and Roberts[29] provides empirical criteria for determining the onset and size of separation bubbles as well as the nature of the boundary layer upon reattachment, if it occurs. With an analytical arsenal such as this, even the complicated boundary layer situation described above can be attacked.

The second, more fundamental, approach to the analysis of turbine airfoil boundary layers is composed of various sophisticated solutions of the boundary layer and Navier-Stokes equations. An exhaustive discussion of this entire area of computational fluid dynamics is presented in Chap. 6. Only brief mention of some typical work that relates to the prediction of airfoil boundary layers shall be related here. The prediction procedure of McDonald and Fish[30] and Kreskovsky et al.[31] is an excellent example of the high-quality techniques that have been perfected in recent years. This single computational tool provides accurate predictions of laminar, transitional, and fully turbulent boundary layer flow as well as relaminarization. The effects of compressibility, Reynolds number, pressure gradients, and freestream turbulence are included in the analysis. The calculation is a solution of a finite difference model of the boundary layer equations with an integral form of the turbulence kinetic energy equation. An equally rigorous

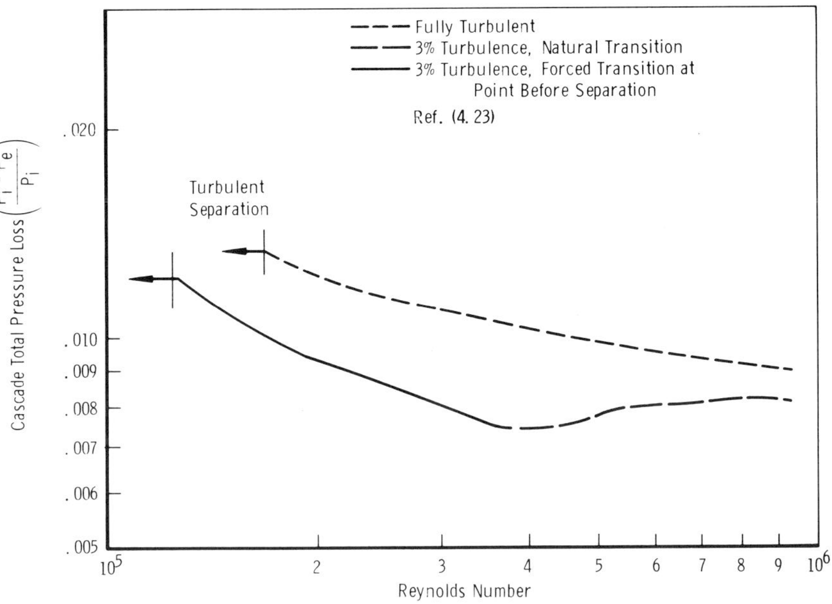

Fig. 4.7 Cascade profile loss vs Reynolds number.

treatment of separation bubbles is presented by Briley and McDonald.[32] This consists of a time-dependent solution of the Navier-Stokes equations employing the McDonald-Fish turbulence model. The intention here is not to summarize all of the work in the computational area, but simply to point out that the analytical power does exist for the problems at hand.

As an example of the type of analytical results that have been produced, consider Fig. 4.7 from Ref. 23. The variation of loss with Reynolds number is shown for a cascade and the effect of the nature of the boundary layer is demonstrated. The transitional boundary layer case has a loss level over much of the range 30% below the fully turbulent level. The incentive is clear for designing airfoils with pressure distributions that encourage as much laminar and transitional flow as possible.

The conclusion regarding the present level of understanding and predictability of the boundary layers on gas turbine airfoils is that the tools are now available to make reasonably accurate predictions. However, there are still a number of areas in need of further work. One of the examples pointed out by Brown and Martin[33] in their review of heat-transfer predictions is that of the relatively subtle and competing influences of freestream turbulence and favorable pressure gradient on the onset of transition.

Wake Mixing

The final step in computing cascade performance is the wake mixing analysis. This is a conceptually straightforward control volume calculation that establishes the uniform, far downstream conditions once the conditions have been specified at the cascade trailing edge plane. Having specified the gapwise variation of flow properties (static pressure, speed, and direction) in the potential flow region between the airfoils, the boundary layer displacement and momentum thicknesses at the trailing edge, and finally the trailing edge blockage and base pressure, one can compute exactly the far downstream Mach number, flow direction, and ultimate total pressure loss.[14] This type of analysis can be generalized to include the effects of trailing edge discharge of cooling air.[34] The feature of the wake flow presently not at all well understood is the base pressure. This influence may be thought of as a form drag. For most cases, the trailing edge base pressure is very nearly equal to the downstream static pressure. However, Prust and Helon[35] demonstrated that trailing edge shape can have a pronounced effect on the airfoil performance. The fundamental reason for the effect was that the base pressure was varying with the trailing edge shape. The conclusion from their work is that a sharply squared-off trailing edge can degrade performance significantly by lowering the base pressure. The advantages of rounded trailing edges become apparent. Another, far more serious, situation that can cause a lower base pressure (increased loss) is a high exit Mach number. MacMartin and Norbury[36] report base pressures 40–50% below the downstream static pressure for exit Mach numbers between 1.0 and 1.3. Base pressure levels as low as this have a major impact on the transonic performance of turbine cascades. Until a basic analysis is devised or sufficient experimental data are amassed, the uncertainty of base pressure is going to plague the design of transonic turbine airfoils. Evidence of such difficulty can even now be seen in the literature. In the work of Waterman,[37] a surprisingly large increase in base pressure contributed to an equally significant improvement in cascade performance. The source of the change was a relatively subtle change in the airfoil contour.

The conclusion is that the major area of wake mixing requiring study is the base pressure excursions that occur in transonic flow. The problem is further complicated by the fact that most transonic airfoils are in high-pressure turbines and hence are likely to have a trailing edge discharge of cooling air. The present situation is simply that there is no design control.

Cascade Testing

During the years prior to the development of the analytical capabilities discussed in the preceding sections, the most powerful research and development tool available to the turbine aerodynamicist was the cascade test. The types of information that could be produced in such testing include the airfoil pressure distribution, loss, surface flow visualization (to indicate separation and the nature of the boundary layer), and schlieren photographs (to indicate shock locations). A large amount of parametric testing was carried out with the goal of improving airfoil performance and developing general profile loss systems. Cascade testing is still used today as a basic profile aerodynamic research tool. The value of the results, however, has been greatly enhanced by the analysis that now accompanies them. The analysis has helped to point out the importance of running a truly controlled test. The misleading conclusions that can be drawn from data taken without the proper simulation of such variables as Reynolds number, freestream turbulence, and spanwise stream tube divergence can be made clear by an analysis in which there is a far greater degree of control over the individual variables.

Today cascade testing focuses on those areas on the fringes of aerodynamics technology where analyses are either being developed or are nonexistent. Transonic aerodynamics is receiving much attention as a result of the need for efficient high-work HPT stages. The specific areas include the trailing edge shock system and the impact of the shock on the suction surface pressure distribution and boundary layer. Trailing edge base pressure is also receiving attention in that it has direct effects on both the performance and the shock system. Interest in the area of low Reynolds number LPT performance has been on the rise as a result of trends in this direction caused by higher bypass ratio engines. In this regime, the rather subtle interactions of Reynolds number, freestream turbulence, and pressure gradient are under study. As a final example, there is the area of spanwise stream tube divergence. Virtually all turbines have diverging flow paths. The effect of this divergence on a high-turning, low-reaction airfoil can be very large. There are implications on both the potential flow and the boundary layer. It is not difficult to imagine that accurate simulation of these effects is extremely difficult to achieve.

The successful execution of a cascade test program has a number of key ingredients. The phenomenon under study and the important parameters must be defined as well as possible. The actual cascade test situation should be analyzed to make sure it provides an adequate simulation. Before the test is carried out, analytical predictions of the expected measured results should be available for immediate comparison with test results as they become available. In terms of the test facility, its characteristics should be well understood ahead of time. For example, inlet distortions not only in the total pressure but also in the flow direction must be documented before one can begin to evaluate the airfoil performance. Simple flow visualization at an early stage of the testing can provide some early and frequently startling insight. Finally, the importance of a high-accuracy, dependable data system with rapid reduction of the results to an intelligible form cannot be overemphasized. The best check on data is an immediate comparison with the best prediction available.

The conclusion is that cascade testing continues to be a valuable research tool. An intimate marriage between testing and analytical prediction serves to mutually enhance both.

Cooling Effects

As turbine inlet temperatures have risen, the need for increasingly effective airfoil cooling configurations has driven the turbine designer to what are referred to as film cooling and multihole cooling schemes. Both are discussed at length in Chap. 5. The only point to be made here is that these cooling schemes have a first-order effect on airfoil profile performance. In both schemes, cooling air is brought into the flowfield through arrays of discharge holes in the airfoil surface. When 5–10% of the flow passing through a cascade is being discharged through the airfoil surface, it is not surprising that performance is affected. Progress in understanding these effects would be facilitated if cascade loss data were reported in their entirety, rather than being reduced to a single elusive parameter.[34]

The impact of coolant discharge on pressure distributions under most circumstances is not very significant. As a result, there has been no major analytical effort in this area. It would not be difficult, however, to modify any one of the numerous existing potential flow calculations to include the effect of mass addition.

Far and away the major influence of coolant discharge on cascade performance is as a result of total pressure losses due to the mixing of the coolant and mainstream flows and alterations to the nature of the boundary layer. Both effects are described in more detail in Chap. 5. The point to remember is, however, that cooling losses can equal or exceed all other contributions to the profile loss of highly cooled airfoils.

Conclusions

During the past 10 years, turbine profile aerodynamics has passed from a state of being almost entirely dependent on a relatively small body of experimental data to its present condition of being based almost totally on a fundamental analytical sequence of design procedures. Although there are a number of areas still badly in need of attention, the majority of the former problem areas are now subject to a high degree of analytical design control.

4.6 End Wall Aerodynamics

In the context used here the term end wall aerodynamics will refer to all the complex three-dimensional flow phenomena occurring in the hub and tip regions of a turbine flow path. The end wall loss relating to leakage and other such flows that leave and/or enter the flow path is referred to separately as parasitic loss and shall be discussed in Sec. 4.7. The concern of the present section is the large concentration of loss and flow distortion that occurs near the end walls of both shrouded and unshrouded turbine airfoils.

The state-of-the-art of understanding and predicting end wall aerodynamics is extremely primitive relative to that of profile aerodynamics. Strongly three-dimensional end wall flows occur as a result of the interacting effects of inviscid and viscous flow mechanisms. In addition, whereas profile loss is purely a function of the row of airfoils in question, the end wall loss will depend on the nature of the end wall flow entering the row and hence it depends on the end wall region performance of all of the upstream rows. Thus, the end wall performance of each row cannot be uniquely related to the geometry of that row. For this reason, one is in the position of having to consider net (that generated within the airfoil row) vs gross (the total appearing at the exit plane) loss. With this rather challenging situation in mind, some of the developments that have occurred in the area of end wall aerodynamics during the past 10 years will be considered here.

Experimental Evidence

The evolution of the manner in which investigators have studied end wall aerodynamics reflects an increasing degree of respect for the complexity of the problem. At the outset, end wall loss was measured downstream of cascades in much the same manner as profile loss, that is, with little or no regard for the incoming end wall boundary layer. Concern then began to expand toward what was entering the cascade on the end wall as well as what was exiting from it. An extremely fine example of this can be seen in the work by Armstrong,[38] where the exit plane loss was shown to depend strongly on the nature of the inlet boundary layer. A review of cascade data, especially that including the effect of the inlet boundary layer, has been given by Dunham.[39] He presented a composite gross loss correlation of the available data with an inlet boundary layer thickness dependence in the form

$$f(\delta_1/c) = 0.0055 + 0.078\sqrt{\delta_1/c} \tag{4.8}$$

The scatter in the data, however, is rather large. Subsequent work by Came[40] and by Morris and Hoare[41] lead them to a function in the form

$$f(\delta_1/c) = 0.011 + 0.294\,(\delta_1/c) \tag{4.9}$$

Both forms lead to roughly similar conclusions over the range of the bulk of the data ($0 \leq \delta_1/c \leq 0.07$). Two other important points are mentioned in these papers. The first is that the loss continues to rise for a distance downstream of the cascade. This rise cannot be explained by end wall skin friction alone. The second point is that far more success was had in correlating the cascade gross loss (including the inlet end wall loss) than in correlating the net loss (inlet loss subtracted out). Both observations severely aggravate the problem of trying to develop a reliable end wall loss correlation for design purposes. As an example of this, when Dunham and Came[42] examined the performance of a number of turbine rigs, they were forced to correlate the data with a single constant instead of a function of inlet boundary layer thickness. Such boundary layer data are rarely available in a turbine environment. The constant, $f = 0.0334$, corresponds to a correlation for net loss since it is based on turbine performance. Hence it is not comparable, even in magnitude, to either of the above correlations for gross loss.

The trend toward more highly detailed testing is proceeding to its logical next step in that now careful studies are being conducted within the cascade passage. Examples of this approach are the fine works by Langston et al.,[43] Carrick,[44] and Sjolander.[45] With these studies, a far more complete appreciation of the truly complex nature of the flow has been achieved. As an example, consider the flow visualization of the end wall limiting streamlines

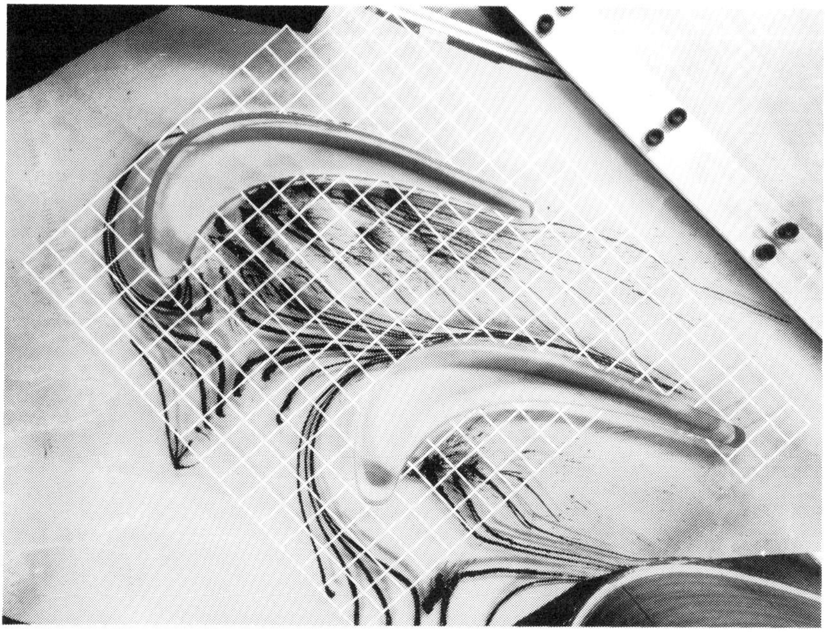

Fig. 4.8 Cascade end wall limiting streamlines.

taken by Langston et al.[43] and presented in Fig. 4.8. As the boundary layer ($\delta/c \approx 0.1$) approaches the cascade, a horseshoe vortex is formed on the end wall. It is similar to that which the boundary layer forms in front of a circular cylinder, but is displaced away from the extension of the mean camber line toward the pressure (concave) surface. This displacement is a result of the potential flow around the airfoil. This vortex is a feature peculiar to cascades. It would not be found, for example, in a simple turning duct. One leg of the vortex proceeds down the channel, while the other is wrapped around the leading edge and is carried down the adjacent channel. The result is that in some areas the limiting streamlines at the wall are flowing in the opposite direction of the mainstream potential flow above them. This is particularly true in the region near the pressure surface just behind the leading edge circle. Most of the fluid in the inlet boundary layer is carried to the suction surface in the vortex. It reaches the suction surface approximately halfway (axially) through the cascade. The end wall region in the channel between the airfoils downstream of this vortex is virtually void of fluid coming from the inlet boundary layer. The boundary layer in this region is extremely thin ($\delta/c \approx 0.002$) and is skewed toward the suction surface. The fluid swept to the suction surface forms an intense vortex that has a first-order effect on the pressure distributions on both the end wall and the airfoil suction surface. The unloading that occurred in work by Langston et al.[43] is shown in Fig. 4.9. The flow into the suction surface/end

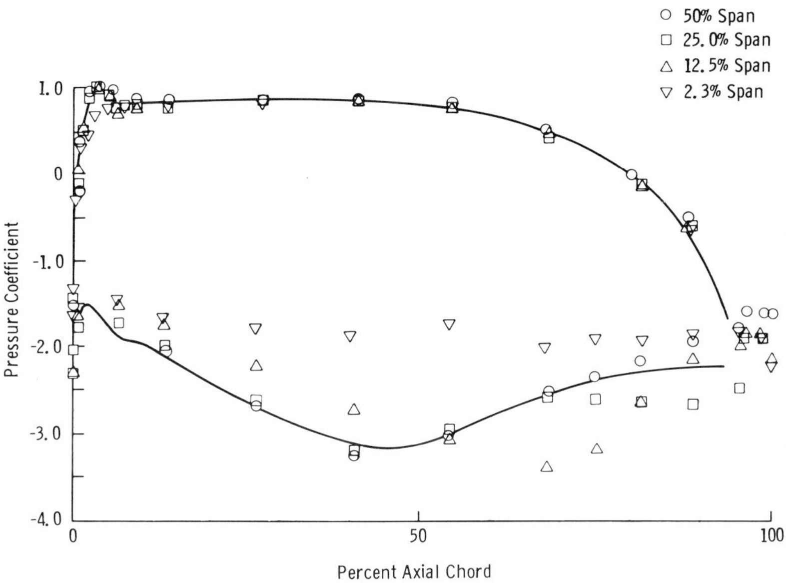

Fig. 4.9 Cascade static pressure distribution showing effect of end wall flow.

wall corner is so strong that the occurrence of corner stall (back flow) in turbine cascades is highly unlikely. The entire process can be seen clearly in Fig. 4.10. At a station just downstream of the leading edge, the distortion of the inlet boundary layer due to the legs of the horseshoe vortex is visible (Fig. 4.10a). By midchord, the bulk of the inlet boundary layer has been swept over to the suction surface, leaving an extremely thin end wall boundary layer behind it (Fig. 4.10b). Near the trailing edge plane, the flow has moved up the suction surface and a vortical motion is apparent (Fig. 4.10c). As the vortex proceeds downstream of the cascade, it decays slowly (relative to a two-dimensional wake) and in the process produces significant additional loss.

This is a description of what has been observed in a plane cascade with a collateral (unskewed) inlet boundary layer. In a turbine environment, however, the inlet boundary layer could be highly skewed. As pointed out by Moore and Richardson,[46] inlet skew can have a favorable effect in compressors, where it tends to drive the low-momentum boundary layer fluid away from the suction surface and toward the pressure surface (opposing the end wall flow). In a turbine just the opposite occurs. The inlet skewing drives the inlet boundary layer fluid toward the suction surface and as a result enhances the end wall flow. The data taken by Carrick[44] demonstrate that inlet skewing can cause a significant increase in cascade end wall loss.

Another troublesome feature of end wall flows is the spanwise mixing they produce. The streamwise vortices produced near the end walls cause strong mixing of the incident radial profiles of total pressure and total

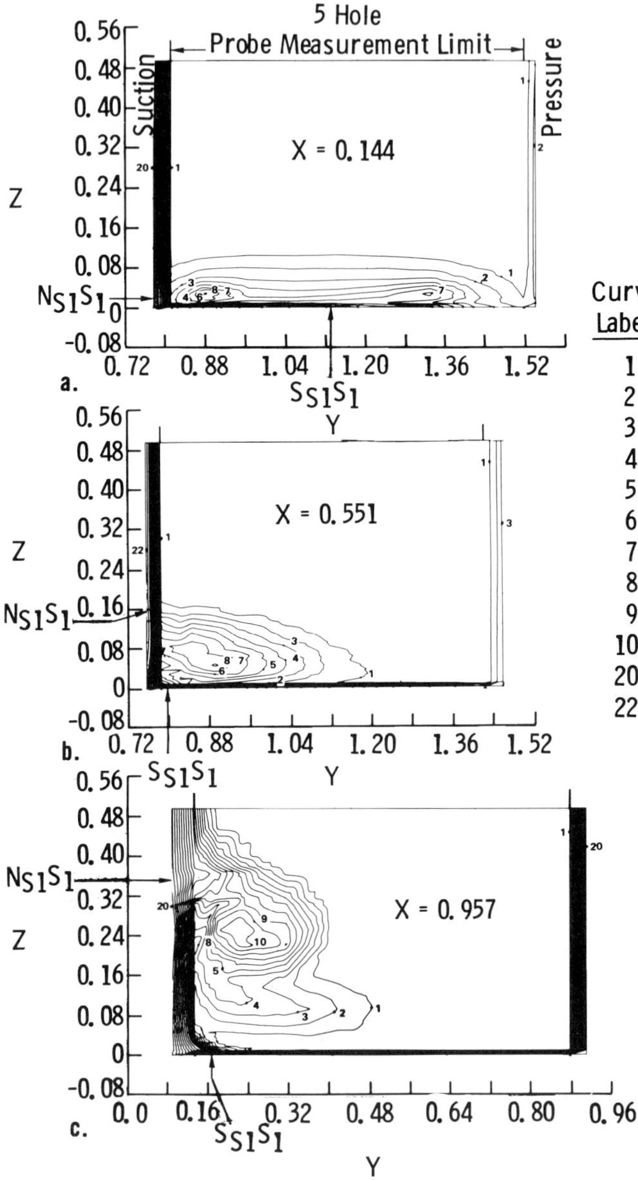

Curve Label	$\frac{P_i - P}{q_i}$
1	0.1
2	0.2
3	0.3
4	0.4
5	0.5
6	0.6
7	0.7
8	0.8
9	0.9
10	1.0
20	3.0
22	4.0

Fig. 4.10 Isobar plot of total pressure loss coefficients at three axial planes.

temperature. This causes problems both in the axisymmetric meridional plane streamline analysis, where the conventional assumption is that there is no transport between the stream tubes, and in designing rotor blades where it is desirable to have a particular spanwise temperature distribution so that the highly stressed sections are not in the highest-temperature flow region.

The spanwise mixing is very intense close to the end wall due to the vortices discussed above. Even well above the end wall, however, there can also be relatively strong streamwise vorticity. The source of this is the streamwise vorticity produced by putting relatively weak shear flows through high-deflection cascades. This phenomenon is discussed in the following section.

Classical Secondary Flow Analysis

The aspect of end wall aerodynamics to first receive analytical attention was the secondary flow. Inviscid rotational flows produce a streamwise component of vorticity when they are put through a deflection. The original work by Squire and Winter[47] and the classic by Hawthorne[48] provide ample clear evidence of the nature of this phenomenon. The theory of secondary flow has been generalized by Hawthorne[49] to include the effects of stratified, compressible fluids in a rotating frame of reference. This is an essential feature when considering the secondary flow effects in turbine rotor blade passages. The recent work by Came and Marsh[50] presents a simple derivation, not only of the distributed secondary vorticity but also of the trailing filament and shed vorticity. Taking the normal assumptions of secondary flow, most notably that it does not alter the primary flow, Belik[51] performed a very thorough parametric analysis of secondary flow losses. In this inviscid analysis, the kinetic energy of the relative secondary motion was taken to be lost. His analysis included variations in the inlet velocity profile shape, velocity triangles, aspect ratio, and solidity. The results showed remarkably good qualitative agreement with end wall loss data from a number of sources. In an effort to provide a more generally applicable analysis, Stuart and Hetherington[52] devised a method that permitted a strong coupling between the primary and secondary flows. The result was that severe distortions of the Bernoulli surfaces could be accounted for.

Classical secondary flow analysis has the potential of being helpful in understanding spanwise mixing in that viscosity may not necessarily have a strong role away from the end walls. In light of the complex viscous nature of the end wall flow described earlier, it would seem optimistic to expect this mechanism to make anything more than a relatively small contribution to the total end wall loss picture.

End Wall Boundary Layer Analysis

The application of boundary layer theory to the end wall problem in turbomachinery has followed two distinctly different paths. The first consists of tangentially averaging the annulus boundary layer equations and then, with additional approximations, following the growth of the boundary layer axially through the machine. The work by Mellor and Wood[53] is an example of this approach. This method has met with some success in compressor applications where the solidity is high and the turning is low relative to the turbines. Due to the very strong cross flows and the large gapwise variation in the end wall boundary layer in a turbine cascade, it would appear that the gap-averaged approach offers little hope.

The second approach consists of solving the three-dimensional boundary layer equations over the cascade end wall. Mager[54] presented a study of the momentum-integral form of the equations that was not limited to weak cross flows. This work was applied to the case of a turbine cascade end wall boundary layer by Dring.[55] The flow path between the cascade airfoils was treated as a duct. In this way, the complexity of the leading edge flow, most notably the horseshoe vortex, was avoided. The solution gave qualitatively reasonable predictions of the end wall flow in that for the case presented the incoming limiting streamlines (on the end wall surface) reached the suction surface at an axial location approximately halfway through the cascade. The predicted losses, however, were generally well below the measured values. Booth[56] extended this approach to include the effects of compressibility. His loss calculation included an additional effect due to the displaced inviscid flow. For several first-vane test cases, his predictions were also significantly below measured loss values. Carrick[44] applied the general integral method of Smith[57] to the cascade end wall problem. Steps were taken to include a degree of periodicity in the boundary layer ahead of the cascade leading edge, but not to the extent necessary to account for the complex nature of the flow observed by Langston et al.[43] Even more powerful three-dimensional boundary layer analyses are available (e.g., Nash and Patel[58]), but one must keep in mind that many aspects of the turbine end wall flow do not satisfy the basic boundary layer approximations. The horseshoe vortex, the back flow region behind it, and the alterations to the two-dimensional pressure distribution are features beyond the scope of boundary layer analysis.

Three-Dimensional Flow Analysis

The recent availability of larger and faster computational facilities has made possible the fully three-dimensional flow analysis. This area of computational fluid dynamics is described at length in Chap. 6. Attention here shall be restricted to an example of such a calculation as it has been applied to the turbine cascade end wall flow problem. The calculation by Waterman[37] and by Waterman and Tall[59] is a fully viscous solution of the three-dimensional, incompressible fluid flow equations. The calculation was executed for several geometrically complex first-vane annular cascades. The full equations were "split" into separate equations containing the viscous and pressure effects. The viscous equation was made parabolic by the assumption that the streamwise diffusion terms are negligible. This permitted a standard marching solution with each step containing a solution of the elliptic in-plane equation. The equation containing the pressure effects is fully elliptic and is solved by relaxation. It includes a displacement effect due to the viscous equation. By this means a coupling has been achieved between the viscous and inviscid phenomena. The advantages over the inviscid secondary flow and viscous boundary layer analyses become clear. Although there are many features of the analysis that one might argue about, the fact remains that this approach has the potential of accounting for more of the observed features of turbine end wall flows than any other approach currently available.

Unshrouded Rotor Tip Leakage

The clearance between the tip of a turbine blade and the outer air seal, or rubstrip, has long been recognized as a source of large turbine inefficiencies. Every effort is made to reduce this clearance mechanically. However, as a result of mechanical and thermal transients, eccentricities, and other reasons, the running clearance on most gas turbine rotor tips is 1–5% of the blade span. In turbines where for cycle reasons the flow area is to be varied, a similar situation occurs. A relatively simple means of varying the flow area is to rotate the vanes about a radial axis. The resulting mechanical configuration can have clearances between the airfoil and the flow path annulus at both the root and the tip. Thus, as many as three clearances can be present in a single turbine stage. The topic of clearance losses in shrouded rotors is discussed in Sec. 4.7. This mechanism is included under parasitic loss since it involves flow that leaves and re-enters the flow path.

The impact of these clearances on turbine performance has been under continuous study for years. This source of end wall loss has a unique feature that makes it relatively easy to study experimentally. That is, that the clearance in a rig or cascade test can usually be easily varied and the impact of clearance on performance can be isolated. The result is that there is a large body of cascade and rotating rig data available (e.g., see Refs. 60–62) that permits one to construct a relatively accurate correlation of the performance penalty as a function of relative clearance and other variables. A simple order-of-magnitude study will indicate that, except for very small clearances, the flow in a clearance gap is dominated by the pressure difference across an airfoil and not by the viscous effects in the gap. For this reason, the effects of relative wall motion, as in the case of a rotor tip, are usually weak. Hence, a correlation based on rotor tip data would be expected to be adequate for the prediction of variable-area vane clearance losses where there is no relative wall motion.

It can be shown from the references cited above that there is a reduction in turbine stage efficiency of approximately 1.5% for every percent of relative clearance. Thus it is an unusual unshrouded turbine stage that is not giving up at least 2% in performance to this mechanism. This powerful impact has led turbine designers to strive for minimal clearances, as well as to minimize the penalty occurring for a given clearance. This and other end wall loss reduction schemes are discussed in the following section.

Loss Reduction Devices

As a result of the powerful impact end wall loss mechanisms have on turbine performance, a number of attempts have been made to reduce these penalties. Some success has been seen in these attempts, but it is difficult to extrapolate a successful device from one turbine configuration to another. There is the distinct risk that a loss reduction scheme, if improperly executed, may result in a significant loss increase. With this in mind, consider a few of the successes that have been reported.

A method of contouring the annulus of a first vane to reduce end wall loss has been presented by Deich et al.[63] Incorporation of this scheme into turbine designs has produced conflicting results. Ewen et al.[64] applied the scheme to a small axial turbine and saw a remarkable improvement in performance. The results of Okapuu,[65] however, indicate no effect on performance at the design point, but some improvement at lower pressure ratios.

Another trial-and-error approach to the reduction of end wall loss has been followed by Prumper.[66] He employed boundary layer fences of varying size, shape, and location in an attempt to reduce the end wall loss in an annular cascade of first vanes. The result was that after testing over 400 configurations, an optimum was established that gave a significant improvement in performance. The underlying physics and its universality, however, remain to be determined.

The area of unshrouded rotor tips and rubstrips has also received considerable attention. Grooves and recesses have been cut into the airfoil tips in an attempt to make them behave as labyrinth seals. Rubstrips have been recessed radially in an attempt to keep them in a low-flow, separated, environment. Rubstrips have also been made out of honeycomb in an effort to reduce the running clearance, as well as to reduce the loss for a given clearance.

Turbine end wall loss reduction is at this point a black art and without any real design discipline. It is at best a trial-and-error approach with no guarantee that success in one application can be extrapolated to another application. The underlying problem is the basic lack of understanding of the composite of the numerous mechanisms that go into producing end wall loss.

Conclusions

At this point, the conclusions with regard to the state-of-the-art of end wall aerodynamics relative to that of profile aerodynamics must be obvious. Although the mechanisms by which one can lose total pressure can be counted on the fingers of one hand, these individual effects have not been tied together into a meaningful, unified picture.

The loss correlations that have been derived are at best inadequate. Scatter on the order of 100% is not uncommon. This is not surprising considering the questions of net and gross loss and the large increase in end wall loss downstream of a cascade. Accurate end wall loss correlations may eventually become available, but there is little evidence of it today. One notable exception is unshrouded rotor tip clearance loss, which appears to be well within the realm of correlation.

Analytical attacks on the problem stem from a number of different approaches. Those based on conventional secondary flow and three-dimensional boundary layer arguments are at present inadequate. These approaches have and probably will continue to contribute to the understanding of the situation, but the ultimate answer is probably to be found in the fully, or quasi-, three-dimensional numerical approach. Although still in its infancy, this approach has the potential to be ultimately successful.

The situation today is that design control does not exist and that the trends of end wall loss with only a few gross parameters can be predicted with any confidence.

4.7 Parasitic Loss

In the present context, the term "parasitic loss" is used to include all the numerous losses arising from flows that leave and/or enter the mainstream flow path as a direct result of gaps or clearances between adjacent pieces of hardware. One distinguishing feature of parasitic loss relative to end wall loss is the paucity of work on the subject. Relative to all other sources of aerodynamic loss, parasitic loss has received the least attention and yet, as shall be seen, it can have a powerful impact on turbine performance.

Sources and Mechanisms

Even a casual inspection of any typical turbine will reveal a surprisingly large number of gaps and clearances that can produce parasitic loss. The only type to receive any attention at all in the literature is the tip clearances that occur on shrouded rotors (e.g., Dunham and Came[42]). As illustrated in Fig. 4.11a, mainstream air can leave the flow path, flow through the tip (or labyrinth) clearance, and re-enter the flow path. The loss mechanisms include the throttling as the flow passes through the tip seal and the momentum mixing that occurs as the leakage flow re-enters the mainstream flow path. The driving potential for the leakage flow is the static pressure drop that occurs in the mainstream between the inlet and exit to the rotor tip. An aerodynamic approach to minimizing the leakage flow can be found in controlled-vortex meridional plane streamline design. The reduction in the tip reaction from the level produced by free-vortex design results in a reduced rotor tip static pressure drop. Thus, the driving potential for the leakage flow is reduced. Exotic sealing devices are also employed in practice to minimize the leakage flow.

An identical and equally important situation occurs at the roots of the stator vanes. The presence of the rotor shaft beneath them necessitates clearances and sealing devices such as those at the shrouded rotor tips (Fig. 4.11b). The losses produced by vane root leakage have the same impact on turbine performance as the rotor tip leakage losses. It is worthy of note that controlled vortexing can also reduce the static pressure drop across the vane root, which in turn reduces these leakage flows.

Another type of leakage flow occurs as a result of the gaps between the platforms of adjacent airfoils in a row (Fig. 4.11c). Mainstream flow can enter these gaps in the high static pressure region toward the front of an airfoil row and then re-enter the mainstream flow path in the low static pressure region at the aft end of the row. Attempts to minimize these flows

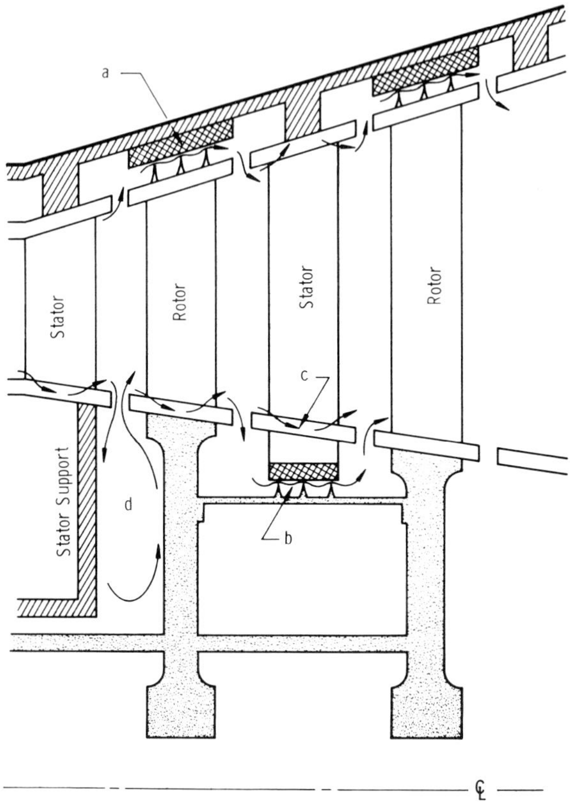

Fig. 4.11 Schematic diagram of turbine sealing arrangements.

have included bonding adjacent platforms, mechanical sealing devices, and conforming the mechanical design of the gap to the pressure gradients in the flow above it in such a way as to minimize leakage. The problem has upon occasion been completely eliminated by making the end wall platforms out of continuous hoops and then inserting the airfoil profiles.

The examples of parasitic loss discussed above all had in common the fact that they were driven by mainstream flow path pressure differences. This is not always the case. For example, the cavities surrounding the rotor disks (Fig. 4.11d) are linked to the mainstream by clearances on the flow path hub. These clearances are governed by the minimum gap necessary to prevent a rub between the rotating and stationary hardware during transient operation. The rotating disk acts like a pump on the air in the cavity as a result of surface shear forces. This in itself causes a not insignificant dissipation of power. In addition, however, this higher energy air is pumped out of the cavity into the flow path where momentum mixing losses occur. The effect is to thicken the end wall boundary layer entering the downstream row of airfoils and hence increase the end wall loss of that row. The situation is further complicated by the presence of cooling air deliberately leaked through the cavity in order to protect the disk.

A point that should be mentioned here is that part of the reason it is difficult to calibrate an end wall loss system on turbine rotating rig results is that it is difficult to isolate the true end wall loss from the frequently unavoidable parasitic loss. As an example of this, the turbine loss system calibration by Dunham and Came[42] considers only the effects of what is referred to here as end wall loss and the parasitic loss produced by shrouded blade tip clearance. One must conclude that the loss contributions of all the other types of parasitic loss have been lumped in with these two. A more thorough analysis of rotating rig turbine data would require careful documentation of the flow characteristics of all the various leakage paths and an assessment of the loss produced by each. Thus, not only must the aerodynamics of the turbine be well understood, but so also must be the mechanical details of the specific test hardware being analyzed. One can now begin to appreciate the difficulty of such an undertaking.

The prediction of the losses requires a prediction of the amount of flow through the gap and then of the loss per unit of leakage flow. The amount of flow may be predicted by various orifice flow calculations. In the case of disk pumping, useful analyses of the flowfield are available (e.g., Rott and Lewellen[67]). The various loss mechanisms may also be attacked analytically. For example, the momentum mixing losses occurring as the leakage or pumped flow re-enters the flow path may be estimated using the same type of analysis as is used to calculate airfoil cooling losses (see Chap. 5). However, as mentioned above, this problem is complicated by interaction of the thickened end wall boundary layer with the downstream row of airfoils.

Conclusions

The most significant advance in the area of parasitic loss is that the nature and importance of the problem is becoming appreciated. Once the mechanical designer has done his best to minimize clearances, it is up to the aerodynamicist to do the best he can with what remains. The potential for success in this regard is clearly evident in the work of Deich et al.[68] In this example, large improvements in turbine efficiency were gained by using end wall platform extensions. They had the effect of forcing the air entering the flow path at the hub to have an axial rather than a radial direction. The point is that, even with the clearances and gaps specified, there are often aerodynamic means by which one can reduce both the amount of leakage flow and the loss per unit leakage flow.

4.8 Structural Excitation

The useful life of a turbine airfoil is greatly affected by the flow in which it is immersed. Both the thermal stresses (see Chap. 5) and the steady-state gas loadings, for example, have aerodynamic origins. The flow is also responsible for airfoil vibratory stresses, the least publicized consumer of structural endurance.

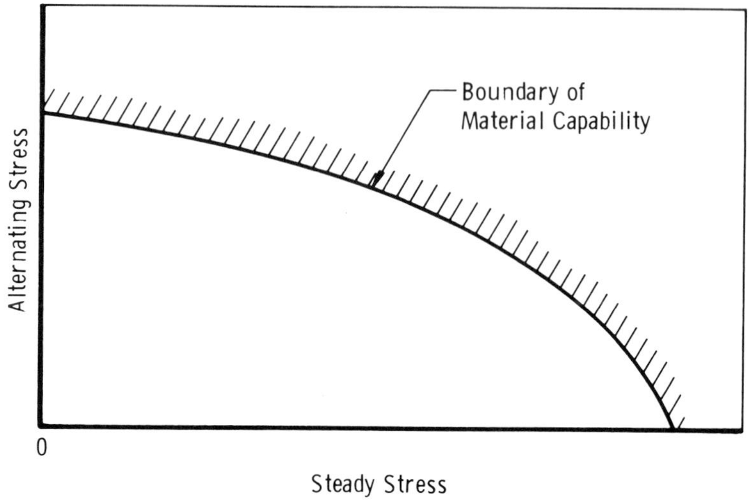

Fig. 4.12 Goodman diagram for combined static load and fatigue life.

Turbine airfoils ordinarily fail as a combined result of steady-state and cyclic stresses. Their interaction can be quickly appreciated by means of the Goodman diagram (Fig. 4.12), a commonly used structural design device. The design point must lie within the safe region in order to be found acceptable.

It is interesting to note that no methods now exist to predict the magnitude of the aerodynamically induced vibratory stresses. Most designers simply assume that they will not exceed the values they have previously experienced because there is no other choice. The fact that many engines have had to employ dampers, inserts, and other fixes in order to reduce cyclic stresses to tolerable levels shows that this approach is not entirely satisfactory. Premature turbine airfoil failure is an expensive and dangerous problem and deserves serious attention.

Another interesting aspect of aerodynamic vibrations is that they can be caused in two entirely different ways, as described below.

Buffeting

An airfoil will experience buffet stresses when exposed to any unsteadiness in the surrounding flow. The unsteadiness may be due to an axially propagating disturbance (e.g., combustor turbulence or afterburner transients), but the most important effects are usually obtained when the row of airfoils cuts through the spatial disturbances (e.g., wakes, vortices, or potential flowfield) of its neighboring rows. Because of the latter, designers carefully avoid tuning any natural airfoil vibratory mode to at least the first 10 orders of blade passing frequency at all operating conditions.

The principal problem here is the almost complete lack of relevant research. It would seem that the high stakes would justify a concerted attack.

Flutter

Under some conditions, a row of airfoils operating in a completely uniform flow can enter into a spontaneous (or self-excited) oscillation known as flutter. The motion is sustained by net energy extracted from the uniform flow during each vibratory cycle, the observed frequency generally corresponding to one of the lower blade or coupled blade/disk natural frequencies.

The outstanding feature of flutter is that the accompanying stresses are usually much higher than were allowed for in the design, with the result that the life is correspondingly reduced.

Very little has been said or written about turbine flutter, leaving the impression that it never occurs. Perhaps the thick, heavy turbine parts do not seem capable of responding to the wind in that manner. However, these impressions are incorrect—turbines can and do flutter. Our experience indicates, in fact, that turbines behave much like compressors in the sense that flutter should be anticipated whenever the reduced natural frequency (flow velocity/axial chord times blade natural frequency) approaches unity. If that is the case, it can be immediately concluded that the trend toward higher flow velocities may also increase the probability of flutter. This source of airfoil structural excitation is worthy of study.

4.9 Stage Performance

The time has come to face the crucial questions about turbine aerodynamic design. Have reliable means been found for increasing efficiency? If so, what approaches have been most fruitful and have they been fully exploited? What are the most promising future directions?

One solid measure of the control over turbine aerodynamics is stage performance. Since each stage is affected by many of the phenomena previously discussed, it challenges the totality of the knowledge of the designer.

Limits of Uncooled Stage Performance

It is fortunate that Smith[69] provided a 1965 benchmark against which uncooled turbine efficiency progress can be gauged. Smith used existing single-stage data to establish efficiency contours on a stage loading coefficient/flow coefficient chart, based on mean flow conditions and mean wheel speed. The chart reveals that there is a maximum efficiency that can be obtained at any given stage loading. When the locus of these points is plotted on an efficiency/stage loading coefficient chart (see the lower curve on Fig. 4.13), the upper bound for possible turbine performance is estab-

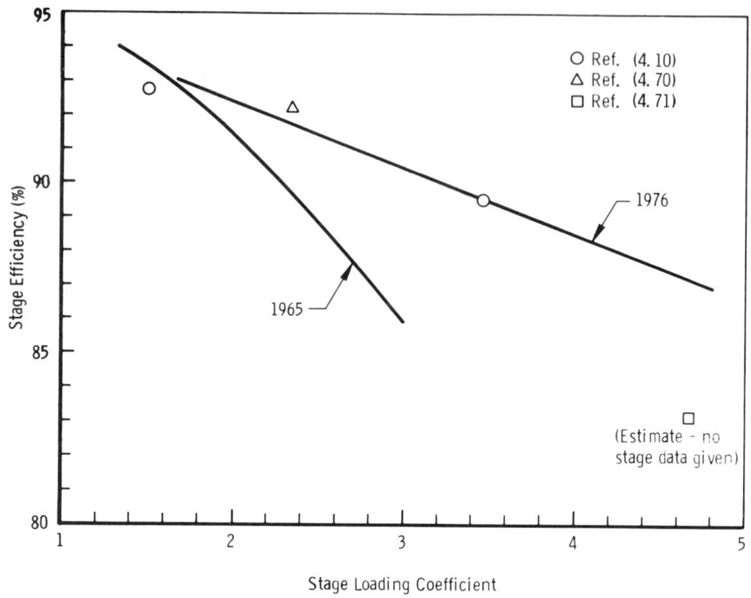

Fig. 4.13 Limits of turbine stage performance.

lished. Note that Smith's data were corrected to zero tip leakage and that most of his stages had axial entry, which gives them the somewhat artificial benefit of about 2% in efficiency.

A particularly interesting aspect of the Smith contour in Fig. 4.13 is the rapid decrease of the stage efficiency with the increasing stage loading coefficient. With data such as this in hand, it is small wonder that designers hesitated to venture much beyond a stage loading coefficient of two.

The upper curve in Fig. 4.13 represents our best estimate as to where the boundary stands today. It is based primarily upon data (Refs. 10, 70, and 71), but it is supported by considerable analysis and experience. Special note should be taken of the fact that the data are from stages with nonaxial inlet flows and realistic clearances, trailing edge thicknesses, platforms, and sealing arrangements. Comparison of these curves leads to two major conclusions: (1) earlier methods are perfectly adequate for designing lightly loaded stages (stage loading coefficient < 2) and (2) the penalty associated with highly loaded stages (stage loading coefficient > 2) is considerably less than previously thought. Special note should be taken of the apparent flatness of the presently recognized boundary. This has made it possible, in fan-drive turbines, for example, to trade several stages for only a few points of efficiency, as was done in the case of the JT9D. These improvements are neither accidental nor beyond the grasp of any serious designer and open the door to many possibilities once considered forbidden. Described below are some of the methods that have contributed to these advances.

Controlled-Vortex Aerodynamics

Turbine aerodynamic design systems originally used a free-vortex calculation procedure largely because of its mathematical simplicity. Unfortunately, free-vortex designs exhibit an undesirable radial distribution of rotor reaction. Rotor reaction decreases from the tip to the root in such a turbine and can even become negative at the root. Small root reactions can make it extremely difficult to contour a passage that remains attached. Large blade tip reactions lead to excessive leakage losses in shrouded machines.

During the mid-1960s, it became abundantly clear that a fundamental barrier to designing efficient, highly loaded stages was the flow separation in the low-reaction region of the rotor blade root. Not only did this cause large localized losses, but the unexpected blockage upset the intended flow pattern across much of the airfoil span. In this manner, additional losses appear away from the wall, as well as in subsequent rows of a multistage machine.

This led to a search for flow patterns that would alleviate the problem. Fortunately, streamline calculation procedures were already available to permit the design of turbines with carefully tailored radial distributions of reaction. Systematic analytical studies showed that the best solutions increased the root reaction by altering the spanwise work distribution. By reducing the work done at the root, less turning is required. The opposite situation holds true at the tip, where additional work can be done without difficulty. Such non-constant work machines have come to be known as controlled-vortex turbines.

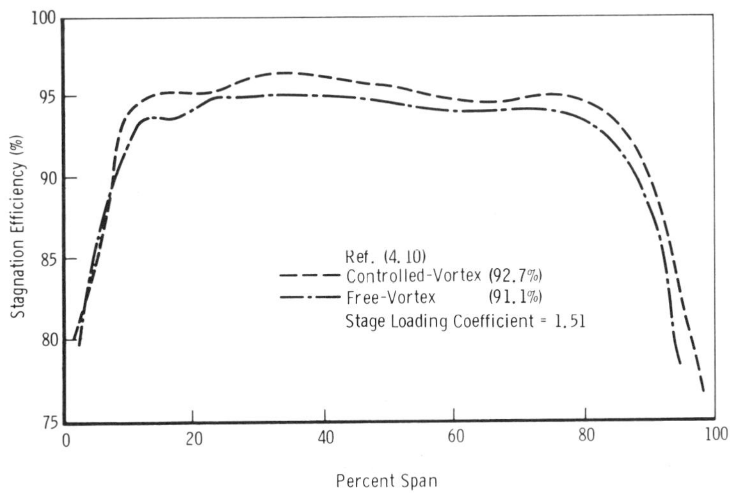

Fig. 4.14 Measured free-vortex and controlled-vortex spanwise traverse efficiency.

The success of this approach has been amply documented by Dorman et al.[10] In one program, they demonstrated a highly loaded, controlled-vortex turbine stage that considerably raised the efficiency standard to 89.5% at a stage loading coefficient of 3.46 (Fig. 4.13). In another program, they compared controlled-vortex and free-vortex stages under exactly the same conditions and realized a 1.6% gain in efficiency at a stage loading coefficient of 1.51. They repeatedly found there that controlled-vortex design led to better performance across the entire span (Fig. 4.14). There was no indication that an unusually distorted exit condition would be produced in a multistage controlled-vortex turbine due to the stacking of the nonuniform work profile effects. This is due to the relatively small change in each stage, relative to a free-vortex design, and the improved efficiency near the end walls that actually results in an improved temperature distribution. In addition, an analysis of the actual flow distribution has shown that the controlled-vortex stage has less tendency to force the flow away from the end walls; measurements have shown that the flow per unit area is more uniform than for the free-vortex stages.

Some Related Considerations

The above remarks are not intended to imply that the rest is now easy. On the contrary, many other techniques must be brought to bear in order to insure continuing progress. Some of them will be discussed here.

Correct airfoil contouring has already been mentioned frequently and certainly affects achievable stage performance. As loading increases, even the midspan airfoil section becomes more and more difficult to design. Once gross separation occurs, the aerodynamic losses become prohibitively high and nothing approaching the upper stage efficiency limit can be reached. Successful design of highly loaded stages is best accomplished with the combined help of controlled-vortex flow patterns and good airfoil contouring methods.

Small gas turbines have special problems of their own, one of the outstanding being the excessive end wall flow losses that accompany their low-aspect-ratio airfoils. In fact, as cycle temperatures and pressures increase, this situation is spreading to bigger machines. As has been seen so often, the large disturbances near the end walls upset the entire flowfield and strip the designer of his ability to control the outcome. One method for dealing with this situation might be to reduce the end wall loss and cross flows at the walls by reducing the turning and work done there. This can be thought of as controlled-vortex aerodynamics of a different sort. Midspan turning must, of course, be slightly increased in order to maintain constant-work extraction. In this way, performance would be improved both by doing more of the work in the more efficient region of the airfoil and by unloading the end walls. This notion was put to the test by Schlegel et al.[9] with excellent results. The design point performance of a research turbine stage revealed a 1.2% efficiency improvement over the baseline stage that it directly replaced. Furthermore, the flowfield behaved much more in accordance with the design intent in the research stage.

One may freely speculate that the very best turbines of the future will contain all of the ideas described above. This was distinctly the case of Welna and Dahlberg[70] (Fig. 4.13), in which these approaches were combined and applied in a straightforward way to their turbine. It should be particularly borne in mind that they exceeded the ambitious goal of their program on the first try. Better testimony on behalf of modern methods can hardly be expected.

Not all ideas are good ones, however. Many tricks have been proposed and tried through the years with no success. Among these are various schemes which tilt, lean, sweep, bow, serrate, place in tandem, or slit turbine airfoils, none of which has demonstrated superior performance.

It is difficult if not hopeless to attempt to define an efficiency boundary for cooled stages because the number of degrees of freedom is so large. Furthermore, the lack of completeness and consistency of the reported results has made it difficult to establish an adequate data base. Suffice it to say that the best aerodynamics are applied to maintain cooled stage performance, and yet a stage efficiency of even less than 85% might be adequate in some high-temperature applications.

4.10 Looking Ahead

It is already clear that the future holds many opportunities and challenges for gas turbine technology. Some of the outstanding expected developments will be outlined in this section, as well as their meaning to turbine aerodynamics.

The most exciting new concept in propulsion technology is the variable-cycle engine (VCE). Providing sufficient flexibility to flow path geometry will enable the VCE to continuously match its characteristics to the needs of the airframe while in flight. The large gains of system performance offered by the VCE make its development within this century almost certain. In fact, several propulsion companies have already modified existing engines in order to gain early experience with the VCE concept.

An essential ingredient of the VCE is the variable-area turbine (VAT). As presently conceived, the VAT will replace one or more rows of fixed airfoils with rows of airfoils that can be adjusted by rotation about a radial axis. In this manner, the turbine operating map can be continuously altered to suit the immediate needs of the engine. Although the early attempts to capitalize on this concept are likely to be confined to uncooled turbine stages, progression into the cooled stages seems inevitable.

The reader can well imagine what these new degrees of freedom mean to the turbine designer. Each VAT will have to meet its aerodynamic and heat-transfer goals over a wide range of flow conditions. Continuously changing cascade geometry, as well as marked increases in incidence angle and Reynolds number variation, will make great demands upon our design systems.

The progression to higher temperature and bypass ratio engines will continue throughout the century. During this period, the stoichiometric limit for hydrocarbon fuels (about 3600°F) will be reached and bypass ratios will double or triple. As far as the turbine is concerned, this continues the trends already discussed in this chapter. A thorough understanding of all facets of turbine aerodynamics will be essential to the orderly and successful development of such machines.

The turbine will ultimately become involved in the movement toward quieter engines. As the other components are made more and more silent, the noise floor of the engine will tend to become set by turbine noise, and acoustical tools tailored specifically for turbines will be required.

The turbine exit guide vane (EGV) diffuses and removes the swirl from the turbine exit flow. Since the EGV must often carry structural loads or contain fuel and oil lines, it can resemble a very-low-aspect-ratio compressor airfoil. Because exit swirl increases with stage loading, the EGV will become a limiting design factor and attract increased attention.

As fuel consumption becomes an increasingly critical item, performance will increase in importance relative to engine cost and weight. To the turbine designer this means pushing the limits of aerodynamics and reaching efficiencies near 95%.

The above remarks lead to a rather simple conclusion. Our appetite for reliable turbine aerodynamic prediction techniques will only grow as the years go by. There is every reason to concentrate on their successful development and to exploit them as they become available.

The gas turbine is one of the most demanding machines known. It embodies a combination of power, precision, and endurance which is at least awesome. But the modern devices are not the result of a single technological leap or revolutionary breakthrough. They are the product of simultaneous application of many small advances and a lot of hard work. Research has played an important role in this unfolding drama and has much to offer in the future.

Rather than being the docile, well-behaved devices of our mythology, turbines have become the bête noir of many recent development programs. The course is inextricably set so that this will continue to occur. But the struggle against former limits has brought success again and again and the future holds more of the same. For turbine designers, these are the good old days.

4.11 Looking Back—Developments Since 1977

During the six years that have passed since this chapter was originally written, much progress has taken place, in spite of the occasional feeling that the pace is excruciatingly slow. Important advances have been made in the areas of basic understanding and the development of reliable, sophisticated, computational design tools. The clearest evidence of this is a remarkable increase in the success rate for new designs, even when they have significant departures from past practice and ambitious goals. We are actually coming to believe that turbomachines will work on the first try. Furthermore, since the improvements are rooted in the fundamentals, there is reason to be optimistic about the future.

We have neither discovered nor been informed of any errors in the original text, either in philosophy or detail. Consequently, the purpose of this new section is only to bring the existing material up to date. This will be accomplished by the addition of Refs. 72–95 along with the brief introductions that follow. Please note that the focus remains, as in the original work, on the tools to make better design possible.

Significant advances have been made in the area of two-dimensional blade-to-blade subsonic and transonic flow calculations.[72–75] As a result of these and other studies, the accuracy of the results has been improved significantly and the computation time reduced. Perhaps more importantly, these analyses have provided a much firmer mathematical foundation. Three-dimensional blade-to-blade calculations have also entered the realm of design practice.[76,77] Their application has been focused not only in the obvious areas of strong flow path convergence and divergence, but also in modeling the three-dimensional effects due to distortions in the inflow.

There has been a continuing experimental and analytical effort underway in the area of boundary layer research.[78–83] These studies are becoming increasingly aimed at the physical mechanisms that tend to set turbomachinery airfoil boundary layers apart from other types of boundary layer flows. This is the reason for the emphasis on such effects as high levels of freestream turbulence, strong favorable and adverse pressure gradients, long transitional regions, and curvature. Turbine airfoil surface flows frequently also have small local regions of separated flow, i.e., closed separation bubbles. These may occur due to either the recompression following a leading edge overspeed or to suction surface shock/boundary layer interaction. The work in this area[84,85] has generally been based on the viscous-inviscid interaction approach.

Great progress has been experienced in the area of the calculation of the three-dimensional viscous flow on turbine end walls.[86–89] It is clear from these results that most of the physical mechanisms are being modeled appropriately and that the experimentally observed flows are being computed with reasonable accuracy. The next objective will be to use these analytical capabilities to deliberately design configurations with improved end wall performance (i.e., reduced end loss).

There has also been significant progress in several other areas of turbine aerodynamics technology. These include: (1) predicting airfoil trailing edge base pressure[90–92]; (2) understanding structural excitation, where the effect of rotor-stator interaction on airfoil pressure distributions has been documented[93]; and (3) utilizing cascade testing, an excellent overview being presented in Ref. 94.

Finally, turbomachinery design has become fully involved in the application of supercomputers to computational fluid dynamics (CFD). A representative summary of the situation can be found in Ref. 95, which is a 15 year projection of the impact of CFD on all aspects of aerodynamic design prepared by the National Research Council. An important conclusion of that report is that, even though internal (turbomachinery) aerodynamics is complex and forbidding, it will yield to CFD in the foreseeable future. This will make possible controlled advances that are beyond our reach today. We applaud these developments and look forward to their benefits. They are the logical extension of what has brought us so far so quickly.

Acknowledgment

The authors wish to acknowledge the generous and valuable support and cooperation of the Pratt & Whitney Aircraft, Engineering Division, United Technologies Corporation and the General Electric Aircraft Engine Business Group in the preparation of this chapter.

References

[1]Horlock, J. H., *Axial Flow Turbines*, Butterworths, London, 1966.

[2]*High Temperature Turbines*, AGARD CP 73, Jan. 1971.

[3]Sahn, P. R. and Speidel, M. O. (eds.), *High-Temperature Materials in Gas Turbines*, Elsevier, New York, 1974.

[4]"Special Issue: Propulsion System Structural Integration and Engine Integrity," *Journal of Aircraft*, Vol. 12, April 1975.

[5]Hebbel, H., "The Influence of the Mach Number and the Reynolds Number on the Aerodynamic Coefficients of Turbine Cascades at Various Turbulence Intensities of the Flow," *Forschung im Ingenieurwesen*, Vol. 30, 1964, pp. 65–77.

[6]Kiock, R., "Influence of the Degree of Turbulence on the Aerodynamic Coefficients of Cascades," AGARD-AG-164, Dec. 1972.

[7]Smith, A. M. O., "High-Lift Aerodynamics," *Journal of Aircraft*, Vol. 12, June 1975, pp. 501–530.

[8]Thomas, K. M. and Piendel, J. J., "An Automated Interactive Design System for Advanced Gas Turbines," ASME Paper 74-GT-82, 1974.

[9]Schlegel, J. C., Liu, H. C., and Waterman, W. F., "Reduction of End-Wall Effects in a Small Low-Aspect-Ratio Turbine by Radial Work Redistribution," *Journal of Engineering for Power, Transactions of ASME*, Ser. A, Vol. 98, Jan. 1976, pp. 130–136.

[10]Dorman, T. E., Welna, H., and Lindlauf, R. W., "The Application of Controlled-Vortex Aerodynamics to Advanced Axial Flow Turbines," *Journal of Engineering for Power, Transactions of ASME*, Ser. A, Vol. 90, July 1968, pp. 245–250.

[11]Ives, D. C. and Liutermoza, J. F., "Analysis of Transonic Cascade Flow Using Conformal Mapping and Relaxation Techniques," AIAA Paper 76–370, 1976.

[12]Ainley, D. G. and Mathieson, G. C. R., "A Method of Performance Estimation for Axial Flow Turbines," British Aeronautical Research Council, R&M 2974, 1957.

[13]Zweifel, O., "The Spacing of Turbomachine Blading, Especially with Large Angular Deflection," *Brown Boveri Review*, Vol. 32, Dec. 1945, pp. 436–444.

[14]Stewart, W. L., "Analysis of Two-Dimensional Compressible Flow Loss Characteristics Downstream of Turbomachine Blade Rows in Terms of Basic Boundary Layer Characteristics," NACA TN-3515, July 1959.

[15]Katsanis, T. and McNally, W. D., "Revised FORTRAN Program for Calculating Velocities and Streamlines on a Blade-to-Blade Stream Surface of a Turbomachine," NASA TM X-1764, 1969.

[16]Van den Braembussche, R. A., "Calculation of Compressible Subsonic Flow in Cascades with Varying Blade Height," ASME Paper 73-GT-59, 1973.

[17]Katsanis, T. and McNally, W. D., "FORTRAN Program for Calculating Velocities in a Magnified Region on a Blade-to-Blade Stream Surface of a Turbomachine," NASA TN D-5091, 1969.

[18]McDonald, P. W., "The Computation of Transonic Flow Through Two-Dimensional Gas Turbine Cascades," ASME Paper 71-GT-89, 1971.

[19]Delaney, R. A. and Kavanagh, P., "Transonic Flow Analysis in Axial-Flow Turbomachinery Cascades by a Time-Dependent Method of Characteristics," *Journal of Engineering for Power, Transactions of ASME*, Ser. A, Vol. 98, July 1976, pp. 356–364.

[20]Luu, T. S. and Coulmy, G., "Relaxation Solution for the Transonic Flow Through a Cascade," *Symposium Transsonicum II*, edited by K. Oswatitsch and D. Rues, Springer-Verlag, New York, 1976.

[21]Jaumotte, A. L. and Devienne, P., "Influence du Nombre de Reynolds sur les Pertes dans les Grilles D'Aubes," *Technique et Science Aeronautiques*, No. 5, 1956.

[22]Mukhtarov, M. K., "An Experimental Study of Losses in Turbine Cascades at Low Reynolds Numbers," *Teploenergetika*, Vol. 15, Sept. 1968, pp. 56–68.

[23]Webster, P., "Design of a 4½ Stage Turbine with a Stage Loading Factor of 4.66 and High Specific Work Output," NASA CR-2659, Aug. 1975.

[24]Dring, R. P., "A Closed Form Solution for Compressible Turbulent Boundary Layer Problems," *Zeitschrift fuer Flugwissenschaften*, Vol. 19, No. 5, 1971.

[25]Schlichting, H., *Boundary Layer Theory*, 6th ed., McGraw-Hill Book Co., New York, 1968.

[26]Dunham, J., "Predictions of Boundary Layer Transition on Turbomachine Blades," AGARD-AG-164, Dec. 1972.

[27]Launder, B. E., "Laminarization of Turbulent Boundary Layer by Acceleration," Gas Turbine Lab., Massachusetts Institute of Technology, Cambridge, Rept. 77, Nov. 1964.

[28]Horton, H. P., "A Semi-Empirical Theory for the Growth and Bursting of Laminar Separation Bubbles," British Aeronautical Research Council CP 1073, June 1967.

[29]Roberts, W. B., "The Effect of Reynolds Number and Laminar Separation on Axial Cascade Performance," *Journal of Engineering for Power, Transactions of ASME*, Ser. A, Vol. 97, April 1975, pp. 261–274.

[30]McDonald, H. and Fish, R. W., "Practical Calculations of Transitional Boundary Layers," *International Journal of Heat and Mass Transfer*, Vol. 16, No. 9, 1973, pp. 1729–1744.

[31]Kreskovsky, J. P., Shamroth, S. J., and McDonald, H., "Application of a General Boundary Layer Analysis to Turbulent Boundary Layers Subjected to Strong Favorable Pressure Gradients," *Journal of Fluids Engineering, Transactions of ASME*, Vol. 97, June 1975, pp. 217–224.

[32]Briley, R. W. and McDonald, H., "Numerical Prediction of Incompressible Separation Bubbles," *Journal of Fluid Mechanics*, Vol. 69, Pt. 4, 1975, pp. 631–656.

[33]Brown, A. and Martin, B. W., "A Review of the Bases of Predicting Heat Transfer to Gas Turbine Rotor Blades," ASME Paper 74-GT-27, 1974.

[34]Anderson, L. R. and Heiser, W. H., "Systematic Evaluation of Cooled Turbine

Efficiency," ASME Paper 69-GT-63, 1969.

[35]Prust, H. W. and Helon, R. M., "Effect of Trailing Edge Geometry and Thickness on the Performance of Certain Turbine Stator Blading," NASA TN D-6637, Jan. 1972.

[36]MacMartin, I. P. and Norbury, J. F., "The Aerodynamics of a Turbine Cascade with Supersonic Discharge and Trailing Edge Blowing," ASME Paper 74-GT-120, 1974.

[37]Waterman, W. F., "Low-Aspect-Ratio Turbine Technology Program," Garrett AiResearch, Rept. 75-211701 (1), Aug. 1975.

[38]Armstrong, W. D., "The Secondary Flow in a Cascade of Turbine Blades," British Aeronautical Research Council, R&M 2979, Mar. 1955.

[39]Dunham, J., "A Review of Cascade Data on Secondary Losses in Turbines," *Journal of Mechanical Engineering Science*, Vol. 12, No. 1, 1970, pp. 48–59.

[40]Came, P. M., "Secondary Loss Measurements in a Cascade of Turbine Blades," Institution of Mechanical Engineers, Conf. Pub. 3, 1973.

[41]Morris, A. W. H. and Hoare, R. G., "Secondary Loss Measurements in a Cascade of Turbine Blades with Meridional Wall Profile," ASME Paper 75-WA/GT-12, 1975.

[42]Dunham, J. and Came, P. M., "Improvements to the Ainley-Mathieson Method of Turbine Performance Prediction," *Journal of Engineering for Power, Transactions of ASME*, Ser. A, Vol. 92, July 1970, pp. 252–256.

[43]Langston, L. S., Nice, M. L., and Hooper, R. M., "Three-Dimensional Flow Within a Turbine Cascade Passage," *Journal of Engineering for Power, Transactions of ASME*, Ser. A, Vol. 99, Jan. 1977, pp. 21–28.

[44]Carrick, H. B., "Secondary Flow and Losses in Turbine Cascades with Inlet Skew," Ph.D. Thesis, Cambridge University, Cambridge, England, Oct. 1975.

[45]Sjolander, S. A., "The Endwall Boundary Layer in an Annular Cascade of Turbine Nozzle Guide Vanes," Dept. Mechanical/Aeronautical Engineering, Carleton University, Ottawa, Canada, Tech. Rept. ME/A 75-4, Dec. 1975.

[46]Moore, R. W. and Richardson, D. L., "Skewed Boundary Layer Flow Near the End Walls of a Compressor Cascade," *Transactions of ASME*, Vol. 79, Nov. 1957, pp. 1789–1800.

[47]Squire, H. B. and Winter, K. G., "The Secondary Flow in a Cascade of Airfoils in a Nonuniform Stream," *Journal of the Aeronautical Sciences.*, Vol. 18, 1951, pp. 271–277.

[48]Hawthorne, W. R., "Secondary Circulation in Fluid Flow," *Proceedings of the Royal Society of London*, Ser. A, Vol. 206, 1951, p. 1086.

[49]Hawthorne, W. R., "Secondary Vorticity in Stratified Compressible Flow in Rotating Systems," Cambridge University, Cambridge, England, Rept. CUED/A-Turbo/TR 63, 1974.

[50]Came, P. M. and Marsh, H., "Secondary Flow in Cascades: Two Simple Derivations for the Components of Vorticity," *Journal of Mechanical Engineering Science*, Vol. 16, No. 6, 1974, pp. 391–401.

[51]Belik, L., "An Approximate Solution for Kinetic Energy of Secondary Flow in Blade Cascades," *International Journal of Mechanical Sciences*, Vol. 10, 1968, pp. 765–782.

[52]Stuart, A. R. and Hetherington, R., "The Solution of Three Variable Duct Flow Equations," *Fluid Mechanics, Acoustics and Design of Turbomachinery*, NASA

SP-304, Pt. I, 1974, pp. 135–153.

[53]Mellor, G. L. and Wood, G. M., "An Axial Compressor Endwall Boundary Layer Theory," *Journal of Engineering for Power, Transactions of ASME*, Ser. A, Vol. 93, June 1971.

[54]Mager, A., "Generalization of Boundary-Layer Momentum-Integral Equations to Three-Dimensional Flows Including Those of Rotating System," NACA Rept. 1067, 1952.

[55]Dring, R. P., "A Momentum-Integral Analysis of the Three-Dimensional Turbine End-Wall Boundary Layer," *Journal of Engineering for Power, Transactions of ASME*, Ser. A, Vol. 93, Oct. 1971, pp. 386–396.

[56]Booth, T. C., "An Analysis of the Turbine Endwall Boundary Layer and Aerodynamic Losses," ASME Paper 75-GT-23, 1975.

[57]Smith, P. D., "An Integral Prediction Method for Three-Dimensional Compressible Turbulent Boundary Layers," British Aeronautical Research Council, R&M 3739, 1974.

[58]Nash, J. F. and Patel, V. C., *Three-Dimensional Turbulent Boundary Layers*, SBC Technical Books, Atlanta, Ga., 1972.

[59]Waterman, W. F. and Tall, W. A., "Measurement and Prediction of 3-D Viscous Flows in Low-Aspect-Ratio Turbine Nozzles," ASME Paper 76-GT-73, 1976.

[60]Szanca, E., "Research Turbine for High-Temperature Core Engine Application; Part II—Effect of Rotor Tip Clearance on Overall Performance," NASA TN D-7639, April 1974.

[61]Marshall, R. and Rogo, C., "Experimental Investigation of Low Aspect Ratio and Tip Clearance on Turbine Performance and Aerodynamic Design," Continental Aviation and Engineering Corp., Rept. 1043, May 1968.

[62]Kofskey, M. and Nussbaum, W., "Performance Evaluation of a Two-Stage Axial-Flow Turbine For Two Values of Tip Clearance," NASA TN D-4388, Jan. 1968.

[63]Deich, M. E., Zaryankin, A. E., Filippov, G. A., and Zatsepin, M. F., "Method of Increasing Efficiency of Turbine Stages With Short Blades," *Teploenergetika*, Vol. 7, No. 2, 1960, pp. 18–24.

[64]Ewen, J. S., Huber, F. W., and Mitchell, J. P., "Investigation of Aerodynamic Performance of Small Axial Turbines," ASME Paper 73-GT-3, 1973.

[65]Okapuu, U., "Some Results From Tests on a High Work Axial Gas Generator Turbine," ASME Paper 74-GT-81, 1974.

[66]Prumper, H., "Application of Boundary Layer Fences in Turbomachinery," AGARD-AG-164, Dec. 1972.

[67]Rott, N. and Lewellen, W. S., "Boundary Layers in Rotating Flows," Aerospace Corp., Gen. Res. Rept. ATN-64 (9227)-6, 1964.

[68]Deich, M. E., Frolov, V. V., Krugleukov, G. A., and Kustov, O. P., "Influence of the Pattern of Leakage of the Medium Through the Root Clearance on the Stage Efficiency," *Teploenergetika*, Vol. 19, No. 6, 1972, pp. 81–83.

[69]Smith, S. F., "A Simple Correlation of Turbine Efficiency," *Journal of the Royal Aeronautical Society*, Vol. 69, 1965, pp. 467–470.

[70]Welna, H. and Dahlberg, D. E., "Investigation of a Highly Loaded Two-Stage Fan-Drive Turbine," Vol. VI, Final Tech. Rept., AFAPL-TR-69-92, Nov. 1971.

[71]Whitney, W. J., Behning, F. P., Moffitt, T. P., and Holtz, G. M., "Investigation

of a $4\frac{1}{2}$ Stage Turbine with a Stage Loading Factor of 4.66 and High Specific Work Output, I—Overall Performance," NASA TMX-3498, 1977.

[72]Ives, D. C. and Liutermoza, J. F., "Second Order Accurate Calculation of Transonic Flow Over Turbomachinery Cascades," *AIAA Journal*, Vol. 17, Aug. 1979, pp. 870–876.

[73]Caspar, J. R., Hobbs, D. E., and Davis, R. L., "Calculation of Two-Dimensional Potential Cascade Flow Using Finite Area Methods," *AIAA Journal*, Vol. 18, Jan. 1980, pp. 103–109.

[74]Habashi, W. G. and Hafez, M. M., "Finite Element Methods for Transonic Cascade Flows," AIAA Paper 81-1472, July 1981.

[75]Caspar, J. R., "Unconditionally Stable Calculation of Transonic Potential Flow Through Cascades Using an Adaptive Mesh for Shock Capture," ASME Paper 82-GT-238, April 1982.

[76]Denton, J. E., "A Time Marching Method for Two and Three Dimensional Blade to Blade Flow," British Aeronautical Council, R&M 3775, Oct. 1974.

[77]Denton, J. D., "An Improved Time Marching Method for Turbomachinery Flow Calculations," ASME Paper 82-GT-239, April 1982.

[78]Abu Ghannam, B. J. and Shaw, R., "Natural Transition of Boundary Layers—The Effects of Turbulence, Pressure Gradient, and Flow History," *Journal of Mechanical Engineering Science*, Vol. 22, 1980, pp. 213–228.

[79]Gaugler, R. E., "Some Modifications To, and Operational Experiences With, the Two-Dimensional, Finite-Difference Boundary Layer Code, STAN5," ASME Paper 81-GT-89, March 1981.

[80]Blair, M. F., "Influence of Free-Stream Turbulence on Turbulent Boundary Layer Heat Transfer and Mean Profile Development, Part I—Experimental Data," *Journal of Heat Transfer, Transactions of ASME*, Vol. 105, Feb. 1983, pp. 33–40.

[81]Blair, M. F., "Influence of Free-Stream Turbulence on Turbulent Boundary Layer Heat Transfer and Mean Profile Development, Part II—Analysis of Results," *Journal of Heat Transfer, Transactions of ASME*, Vol. 105, Feb. 1983, pp. 41–47.

[82]Blair, M. F., "Influence of Free-Stream Turbulence on Boundary Layer Transition in Favorable Pressure Gradients," *Journal of Engineering for Power, Transactions of ASME*, Ser. A, Vol. 104, Oct. 1982, pp. 743–750.

[83]So, R. M. C., Edelfelt, I. H., Elovic, E., and Kercher, D. M., "A Two-Dimensional Boundary Layer Program for Turbine Airfoil Heat Transfer Calculation," ASME Paper 82-GT-93, April 1982.

[84]Carter, J. E. and Vatsa, V. N., "Analysis of Separated Boundary Layer Flows," Paper presented at Eighth International Conference on Numerical Methods in Fluid Dynamics, Aachen, FRG, June 28–July 2, 1982.

[85]Calvert, W. J. and Herbert, M. V., "An Inviscid-Viscous Interaction Method to Predict the Blade-to-Blade Performance of Axial Compressors," *Aeronautical Quarterly*, Vol. 31, Aug. 1980, pp. 173–196.

[86]Moore, J. and Moore, J. G., "A Calculation Procedure for Three-Dimensional, Viscous Compressible Duct Flow, Part II—Stagnation Pressure Losses in a Rectangular Elbow," ASME Paper 79-WA/FE-5, Dec. 1978.

[87]Briley, W. R. and McDonald, H., "Analysis and Computation of Viscous Subsonic Primary and Secondary Flows," AIAA Paper 79-1453, July 1979.

[88]Kreskovsky, J. P., Briley, W. R., and McDonald, H., "Prediction of Laminar and Turbulent Primary and Secondary Flows in Strongly Curved Ducts," NASA CR 3388, Feb. 1981.

[89]Hah, C., "Navier-Stokes Analysis of Three-Dimensional Turbulent Flows Inside Turbine Blade Rows at Design and Off-Design Conditions," ASME Paper 83-GT-40, March 1983.

[90]Sieverding, C. H., Stanislaw, M., and Snoek, J., "The Base Pressure Problem in Transonic Turbine Cascades," *Journal of Engineering for Power, Transactions of ASME*, Ser. A, Vol. 102, July 1980, p. 711.

[91]Sieverding, C. H., "The Influence of Trailing Edge Ejection on the Base Pressure in Transonic Turbine Cascades," ASME Paper 82-GT-50, April 1982.

[92]Motallebi, F., Edwards, S. J., and Norbury, J. F., "Base Pressure on a Blunt Base in Transonic Flow—Some Effects of Base Geometry and Bleed Air," ASME Paper 82-GT-317, April 1982.

[93]Dring, R. P., Joslyn, H. D., Hardin, L. W., and Wagner, J. H., "Turbine Rotor-Stator Interaction," *Journal of Engineering for Power, Transactions of ASME*, Vol. 104, Oct. 1982, pp. 729–742.

[94]Starken, H., "Airfoil Cascade Aerodynamics—Subsonic, Transonic and Supersonic Flow Regimes," ASME Turbomachinery Institute's *Fluid Mechanics of Turbomachinery* Ames, Iowa, July 1983.

[95]*The Influence of Computational Fluid Dynamics on Experimental Aerospace Facilities*, National Academy Press, Washington, D. C., 1983.

CHAPTER 5. TURBINE COOLING

Mikio Suo
General Electric Company, Cincinnati, Ohio
(Formerly with United Technologies Research Center, East Hartford, Connecticut)

5. TURBINE COOLING

Nomenclature

A_g	= gas side surface area
A_h	= coolant hole surface area
A_i	= inside surface area
a_1, a_2, a_3	= constants given in Table 5.1
b	= equivalent slot width, $(\pi/4)(d^2/p)$ or $(\pi/4)(D^2/p)$
b_1, b_2, b_3	= constants given in Table 5.1
c	= axial chord
c_{p_c}	= coolant specific heat
c_{p_e}	= approach gas specific heat
c_{p_g}	= mainstream gas specific heat
c_1, c_2	= constants given in Table 5.1
D	= film cooling hole diameter
d	= impingement coolant hole diameter or internal coolant hole diameter
e	= roughness height in Fig. 5.15
F	= coolant mass flux parameter of Eq. (5.22)
F_a	= empirical parameter of Eq. (5.9)
f	= half the heat-transfer region in Fig. 5.12
G_{cf}	= cross-flow mass velocity
G_h	= impingement hole mass velocity
H	= leading-edge dimension in Fig. 5.12
$h(x)$	= heat-transfer coefficient (independent of z)
$h(z, x)$	= local heat-transfer coefficient
h_g	= local gas side heat-transfer coefficient in Fig. 5.35
$\bar{h}_g$	= lateral average gas side heat-transfer coefficient in Eq. (5.21)
h_h	= coolant hole heat-transfer coefficient
h_i	= inside heat-transfer coefficient
I	= coolant-to-mainstream momentum flux ratio
L	= leading-edge impingement dimension in Figs. 5.10 and 5.12
l	= impingement hole to impingement point distance
M	= coolant-to-mainstream mass flux ratio
$\mathcal{M}$	= Mach number
m	= empirically determined parameter of Eq. (5.5)
Nu_{av}	= average Nusselt number based on characteristic dimension as used
Nu_{stag}	= stagnation point Nusselt number based on characteristic dimension as used

Nu_x = local Nusselt number based on characteristic dimension as used
n = wall roughness ring spacing in Fig. 5.15
P_{tg} = mainstream total pressure
ΔP_{tg} = mainstream total pressure loss
Pr = Prandtl number
p = center-to-center hole spacing
q = heat flux
$\bar{q}(x)$ = lateral average heat flux
Re = Reynolds number with characteristic dimension and velocity defined as used
r = leading-edge radius in Fig. 5.12
St = Stanton number defined as used
St_{av} = average Stanton number defined as used
s = slot height
T_c = coolant supply total temperature
T_{ce} = coolant total temperature at exit of coolant hole
T_e = mainstream gas total temperature
$T_f(z, x)$ = local film temperature
T_g = mainstream gas adiabatic wall temperature (also low-velocity gas static and total temperature)
T_m = temperature of metal defined as used
$T_m(x)$ = temperature of metal (independent of z)
T_{tg} = mainstream gas total temperature
$\bar{T}$ = film layer average total temperature
$\bar{T}_f(x)$ = lateral average film temperature
U_c = coolant velocity
U_e = cascade approach velocity
U_g = mainstream gas velocity
W_c = coolant flow rate
W_g = mainstream mass flow mixed with film cooling air for aerodynamic losses
W_{ge} = mainstream mass flow entrained in film cooling boundary layer
x = surface distance in streamwise direction
x' = surface distance in streamwise direction from effective starting point for boundary layer
z = surface distance in lateral or spanwise direction
β = angle of coolant flow from mainstream flow
γ = ratio of specific heats
δ^* = boundary-layer displacement thickness
ζ = defined by Eq. (5.13)
η_c = convective effectiveness defined by Eq. (5.19)
η_f = local film effectiveness
$\bar{\eta}_f$ = lateral average film effectiveness
θ = defined by Eq. (5.24)
μ_c = coolant viscosity
μ_e = approach gas viscosity
μ_g = mainstream viscosity

ν_g	= mainstream kinematic viscosity
ξ_1, ξ_2	= empirically determined parameter of Eq. (5.5)
ρ_c	= coolant density
ρ_e	= approach gas density
ρ_g	= mainstream gas density
Φ	= local cooling effectiveness defined by Eq. (5.18)
$\bar{\Phi}$	= average cooling effectiveness defined by Eq. (5.1)

5.1 Introduction

High turbine inlet temperatures have significant advantages for aircraft gas turbine engines. The reasons for these advantages have been discussed in Chap. 7 of *Aerothermodynamics of Gas Turbine and Rocket Propulsion*. Engine manufacturers have recognized this for some time and have been continuously increasing turbine inlet temperature, especially during the last 15 years. Figure 5.1 illustrates this trend.

A vital part of going to high turbine inlet temperatures is the technology of turbine cooling. This was recognized by some almost from the inception of the first turbojet engine. Cooling studies were first performed in the 1940s and many investigations were carried on in the 1950s. Around 1960, turbine cooling was first used in a commercial aircraft engine.[1] Since that time, there has been a very rapid rise in turbine inlet temperatures, which has placed an even greater emphasis on turbine cooling.

The turbine inlet temperature trend shown in Fig. 5.1 has been made possible not only by turbine cooling, but also by continuous improvements

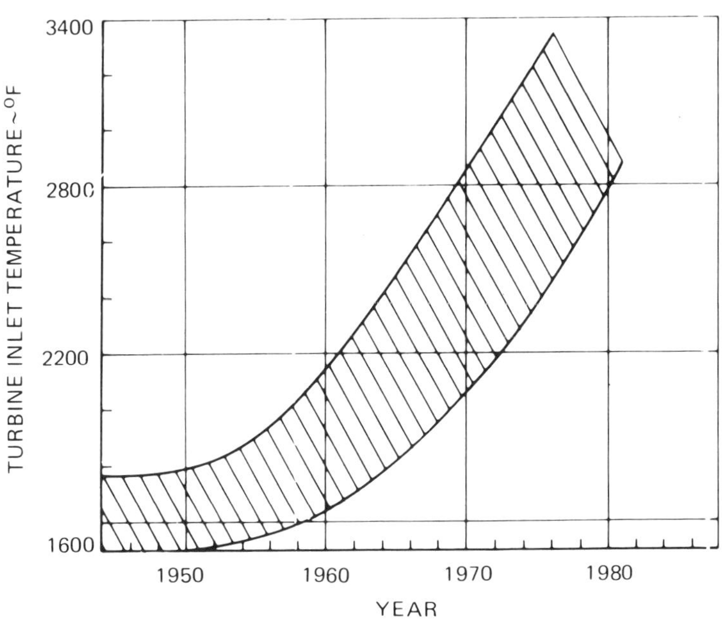

Fig. 5.1 Turbine inlet temperature history and projection.

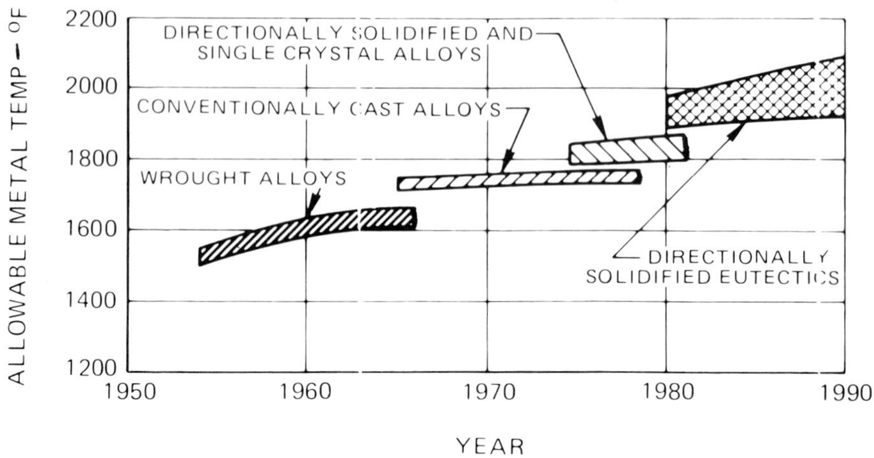

Fig. 5.2 Blade material temperature capability history and projection.

in high-temperature materials. Figure 5.2 shows the trend in improvements in rotor blade materials. As can be seen, materials improvements have played and will continue to play an important part in the increasing turbine inlet temperature trend. The materials shown in Fig. 5.2 are nickel-based alloys used for stators and rotors. A similar figure could be shown for the cobalt-based alloys that are used to a more limited extent in stator vanes and uncooled rotor blades. Many other types of materials have been proposed for use in turbines. Examples of these are refractory metal alloys and ceramics. Although much research has been performed on these materials, they have yet to receive widespread utilization. In this chapter, the nickel- or cobalt-based alloys are presumed to be used.

The cooling flows currently required on new aircraft engines are very large, on the order of 20% of the compressor discharge flow. This large quantity of flow has two effects that tend to reduce the advantages of high turbine inlet temperatures. The first is that the air used for cooling reduces the capability of the turbine to drive the compressor because of the lower temperature at which it enters the turbine. The second is that the cooling air mixes with the mainstream turbine air and causes aerodynamic losses. Both of these effects are strong and, therefore, vigorous steps are taken to minimize them by minimizing both the quantity of cooling air and the aerodynamic penalty associated with that quantity of cooling air used. The technical question addressed in this chapter is: how to maintain turbine durability while minimizing the quantity of cooling air and the aerodynamic losses associated with it.

In the above discussion, reference was made to turbine inlet temperature. There appears to be some disagreement on the definition of this term in the industry, but in this chapter the turbine inlet temperature will always refer to the average combustor exit temperature into the first stator vane.

5.2 Cooling Design Problem

Modes of Thermal Failure

There are several modes of turbine failure associated with turbine cooling. These are oxidation/corrosion/erosion, creep, and thermal fatigue. Oxidation/corrosion/erosion occurs because of chemical and particulate attack from the hot turbine gases. The materials used for the parts exposed to turbine gases, when used at a temperature commensurate with the other modes of failure, cannot survive this mode of attack for very long unless covered with special high-temperature coatings. In general, these coatings provide a layer of aluminum oxide on the surface to prevent chemical attack. Beyond this, the method of control is to limit the maximum surface temperature on the part.

Creep occurs as a result of prolonged exposure of materials to high stresses at high temperatures. This is a particularly acute problem on highly stressed rotating turbine blades. This mode of failure is, in general, controlled by limiting the average temperature of a turbine blade and thus the average creep. Creep also occurs as a local problem in airfoils, a mode of failure controlled by limiting the local temperature within an airfoil.

Thermal fatigue occurs as a consequence of repeated cycling through high thermal stresses. These stresses are large enough to cause local plastic deformation in each cycle. As a result, failure can result in a relatively few cycles (of the order of thousands rather than hundreds of thousands). The problem is the most severe in highly cooled airfoils, which must operate through rapid transients on the order of several seconds. The different parts of the airfoils respond at different rates to the transients and suffer large thermal stresses as a result. Thermal stresses are minimized by consideration of the overall cooling distribution, mass distribution within the airfoil (which sets thermal response), and the interior geometrical configuration.

In addition to creep and thermal stress as separate entities, there is a combined effect of the two acting together. Airfoils incurring high cyclic thermal stresses as well as prolonged exposure at high temperatures and high stresses fail before failure would occur if each mode were acting alone. This is called creep/thermal fatigue failure and is a mode of failure receiving increasing attention in turbine design.

Thermal Environment

The combustion gases enter the turbine in a nonuniform manner. A typical distribution of the temperature coming out of an annular burner is shown in Fig. 5.3. The distribution is random in a circumferential direction, but has some degree of order in the radial direction. The randomness of the temperature distribution causes the first stator vanes to be designed for the maximum temperature (hot spot) coming out of the burner. This maximum temperature is generally a specified burner design parameter. Typically, in a 2500°F turbine inlet temperature engine the hot spot may be 3000°F. The

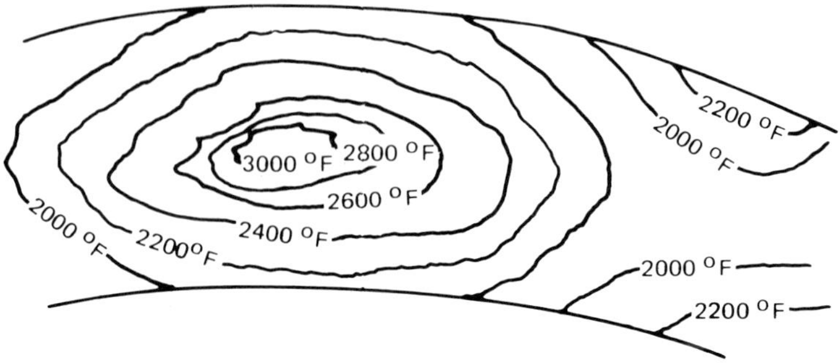

Fig. 5.3 Typical radial and circumferential temperature distribution for an annular burner at 2500°F.

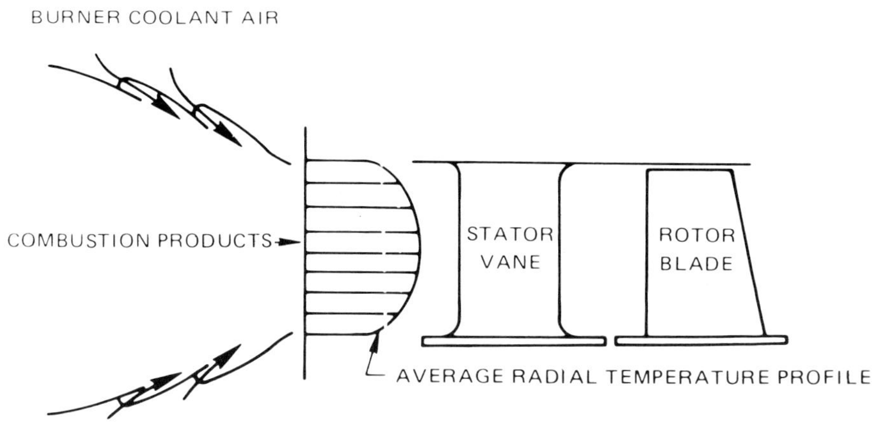

Fig. 5.4 Burner average radial temperature profile.

rotating structure may be designed for an average temperature in a circumferential direction. The radial distribution of the circumferentially averaged temperature is typically as shown in Fig. 5.4. This profile tends to occur naturally because the burner cooling air reduces temperatures at the end walls. In addition to this natural tendency, the burner is generally specifically designed to produce such a profile to ease the turbine cooling problem by reducing the blade root and end wall cooling requirements.

The first stator vanes are exposed to the highest turbine temperatures (including the hot spot from the burner). The first rotating blade row is exposed to somewhat lower temperatures because of circumferential averaging, dilution of turbine gases with first stator vane cooling air, and relative velocity effects. The second stator vane is exposed to a lower temperature because of cooling air dilution, work extraction from the turbine gases, and mixing that dilutes the hot spots. The turbine temperature decreases in a similar manner through each succeeding stage.

Cooling Air Supply

The air cooling the turbine is bled from the compressor. A typical cooling system is shown in Fig. 5.5. The first stator vane is fed by compressor discharge air that has bypassed the combustor. This is because the first stator vane requires a very high supply pressure. The first rotor blade is also fed by compressor discharge air. However, this air is accelerated through a row of nozzles pointed in the direction of rotation. The effect of this is to reduce the amount of work required to pump the cooling air and to reduce the cooling air temperature within the blade. In the case considered here, the second stator vane and second rotor blade are also cooled by compressor discharge air, but in general these airfoils can be fed from the compressor bleed air further up the compressor. This helps engine performance because of the reduced compressor work requirements and reduced cooling air supply requirements because of the lower cooling air temperature.

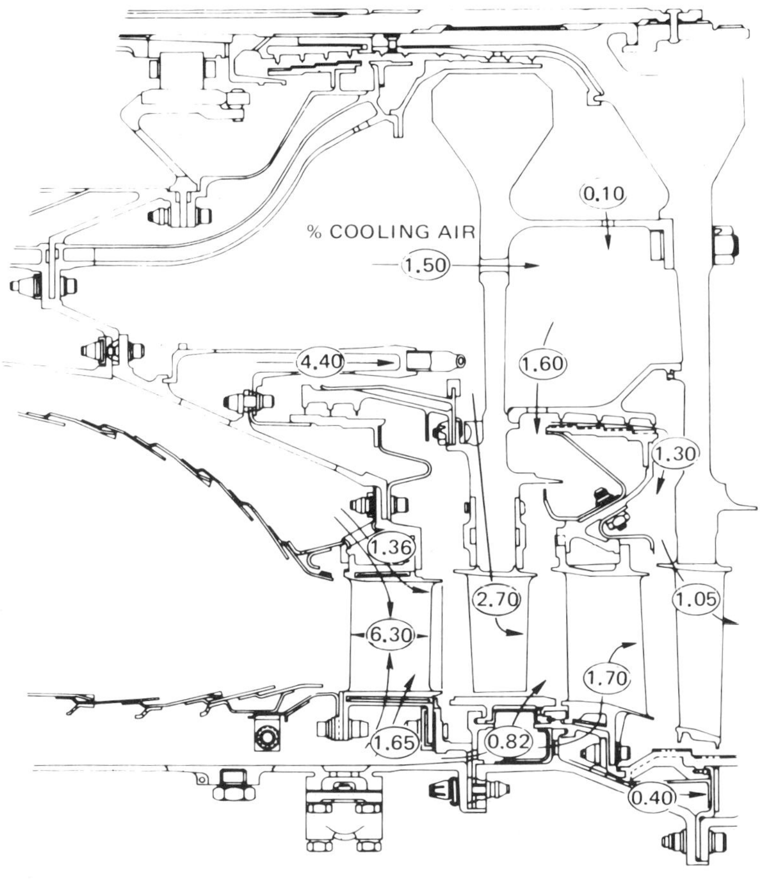

Fig. 5.5 Cooling air supply system.

Many variations on this delivery system are in use today. The basic principle is to deliver the required cooling air pressure at the minimum temperature possible. There is great potential in this area for improving engine performance by clever design. For example, it has been proposed in some applications to cool the cooling air with fan air through a heat exchanger. This would reduce the requirement for cooling flow, but would add complexity and weight.

The cooling air delivery system also performs another function. The cavities adjacent to the gas path are not directly cooled, but are kept at low temperatures because they are surrounded by cool air. This air is furnished by the cooling air delivery system through controlled leakages from the delivery system. An interesting thought is that the cooling air is often around 1000°F. One does not normally think of cooling with air that hot.

In this chapter, only air cooling of turbines will be considered. There are obviously many other ways to cool turbines, such as heat pipes, water, fuel, etc. These are of some interest for aircraft applications, but by far air cooling is of the greatest practical interest. Therefore, this chapter will not address other methods of cooling.

Cooling Design Methodology

By necessity, the cooling design process is an iterative one. The designer starts by either being given or assuming certain life requirements on oxidation/corrosion/erosion, creep, thermal fatigue, and creep/thermal fatigue. Simultaneously, he is given a mission under which these life requirements must be met. An initial aerodynamic design up to and including airfoil contours and cooling airflows and temperatures is also given. The last two quantities are generally closely estimated from past design practices. The designer can usually decide from past experience what type of cooling will be necessary (convective or film cooling) on each surface to be cooled, at which point the detailed design problem begins. The process is one of analysis and not one of synthesis, so the designer must first lay out a configuration. Having done this, he must then analyze it to determine the life and cooling penalties and, in general, he must do this several times, modifying the design each time.

To determine metal temperature and thus life, the designer must consider three aspects of the problem: (1) the heat load to the surface from the mainstream gas, which will be by convection through a viscous boundary layer or through a film cooling layer; (2) the convective cooling of the part by the coolant within the interior of the part; and (3) the thermal conduction and absorption within the part. The designer calculates the metal temperatures, taking into account each of these considerations and utilizing whatever tools he has available. In going through this, he generally finds that he must request relief from some of the initial constraints on the airfoil shape or cooling flows because of practical fabrication considerations or because he simply cannot cool the part adequately under the given conditions. During this time, a continuous re-evaluation of the penalties caused by the cooling air is also made. After several configurations are laid out and analyzed, the designer generally can "home in" on a design that meets most if not all of the original criteria.

In the following sections, the required information to calculate heat loads, internal convective cooling, and cooling penalties are discussed. They are discussed in a way that, hopefully, a designer new to the field might see what information is available in the literature and, perhaps more importantly, determine what additional information is required to execute a successful design.

5.3 Airfoil Cooling

Convective Cooling

The simplest way to cool airfoils is by convective cooling. In this process, the air is brought in at the root or tip of the airfoil and then discharged at the other end of the airfoil or through the trailing edge. This method of cooling is quite efficient and is always used where it is adequate to cool the airfoils. Cutaway views of two convectively cooled airfoils are shown in Fig. 5.6.

At this point, it is useful to introduce a definition. The average cooling effectiveness for an airfoil is defined by

$$\bar{\Phi} \equiv \frac{T_{tg} - T_m}{T_{tg} - T_c} \tag{5.1}$$

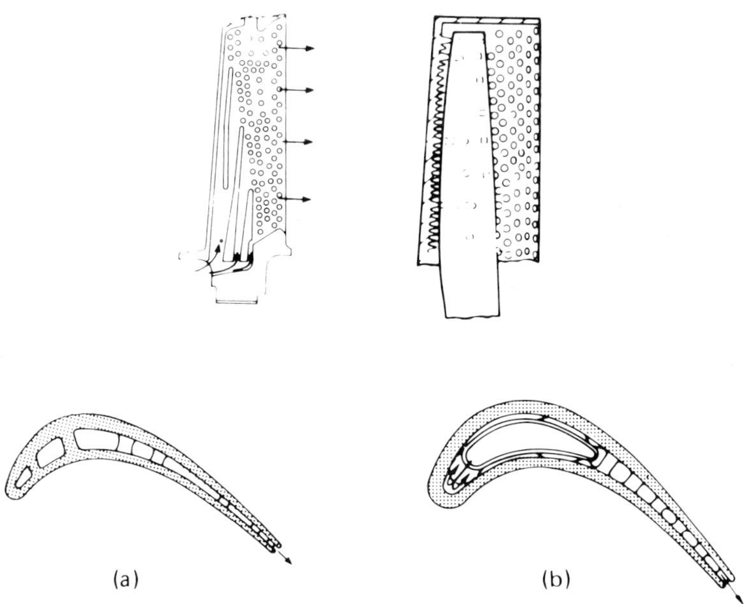

Fig. 5.6 Convectively cooled blades.

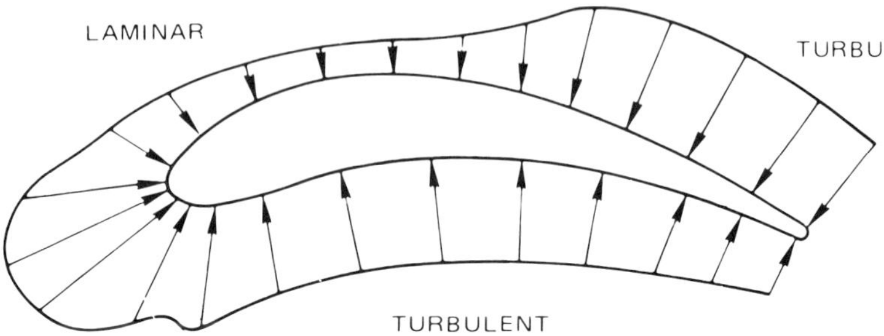

Fig. 5.7 Airfoil heat-transfer coefficient distribution.

where T_m is the average temperature of the airfoil. A value of cooling effectiveness of one is the maximum possible and a value near zero represents little or no cooling. The required average cooling effectiveness is the strongest parameter determining the methods of cooling and the quantities of cooling air.

Convective cooling is used where the average cooling effectiveness levels required are less than about 0.5. This limitation exists for two reasons: (1) the air supply pressure is limited and a much higher effectiveness would be possible only with higher supply pressures; and (2) with high effectiveness levels and convective cooling, the temperature gradients tend to become very large and thus aggravate thermal stress problems.

Airfoil heat-transfer coefficients. In gas turbines, the heat-transfer coefficients to the airfoils are very high and the conductivities of the materials are fairly low. This combination makes it imperative that the heat-transfer coefficient distribution over a whole airfoil be known in great detail. One cannot depend upon conductivity to smooth out any errors in distribution.

A typical heat-transfer coefficient distribution to an airfoil is shown in Fig. 5.7. At the stagnation point, this coefficient is very high. Its level can be correlated as that of the stagnation point of a cylinder in a cross flow that is affected by the freestream turbulence raising its level. On the concave or pressure side of the airfoil, the boundary layer almost always trips to turbulent flow and attains a heat-transfer coefficient level corresponding to a low Reynolds number turbulent flow. On the convex or suction side of the airfoil, the boundary layer is often at first laminar, but has heat transfer coefficients somewhat higher than those corresponding to laminar flow because of the effects of freestream turbulence. The boundary layer then goes through a gradual transition to turbulent flow. This gradual transition appears to occur because of the very high pressure gradients that tend to force the boundary layer to stay laminar. The curvature of the airfoil also tends to maintain laminar flow.

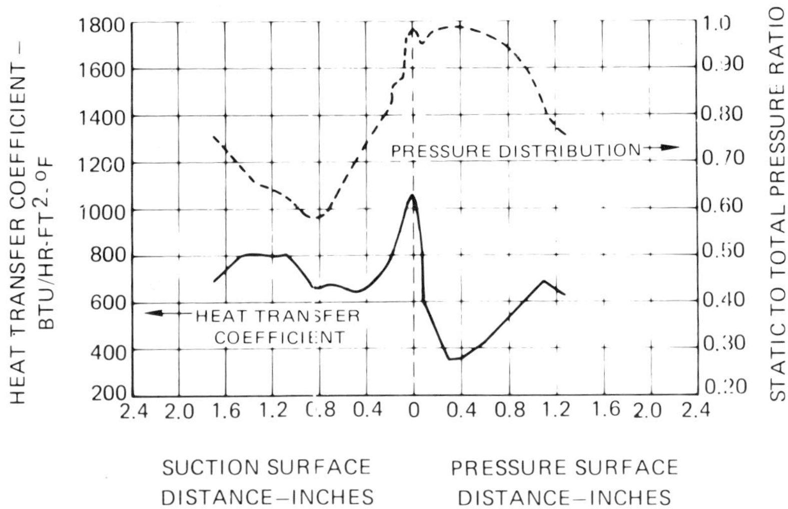

Fig. 5.8 Typical pressure and heat-transfer coefficient distributions.

The heat-transfer coefficient distribution to an airfoil is calculated utilizing two-dimensional boundary-layer theory. The pressure distribution within the cascade is first calculated using an inviscid, two-dimensional, compressible, numerical solution. Such solutions have become readily available within the past few years. The boundary layer is then calculated in a manner described in Chap. 6. The pressure distribution and heat-transfer coefficient distribution calculated from boundary-layer theory for a typical airfoil are shown in Fig. 5.8. The heat-transfer coefficient distribution thus calculated, although by no means exact, is reasonably valid except for the stagnation point and the laminar region on the suction side. There the heat-transfer coefficient distribution must be increased by a factor of 1.2–1.8 to account for the effects of the freestream turbulence mentioned above. The very high levels of the heat-transfer coefficient distribution in this example are typical of modern turbines.

The two-dimensional calculations have inherent drawbacks because of the three-dimensional effects that are present to varying degrees of importance. For example, the two-dimensional calculation assumes that the gas is at a constant total temperature. Because of secondary flows and the consequent radial mixing, the gas temperature at a given radial location tends to change as it goes through an airfoil row. This effect is fairly strong and is generally not calculable to any degree of accuracy. There are also radial flows in boundary layers caused by the radial pressure gradients. However, there is some theoretical evidence[2] that radial boundary-layer flows do not significantly affect the heat transfer. Near the tips of unshrouded airfoils, a significant amount of radial flow occurs because of leakage over the tip. This effect is strong and not well quantified.

In addition to three-dimensional flow effects, radiation plays a part in the heat transfer. This effect is generally limited to the leading edge of the first stator vanes. Beyond that, convection effects far outweigh radiation effects.

Impingement cooling at the leading edge. The airfoil external heat-transfer coefficient distribution discussed above must be counterbalanced with a comparable heat-transfer coefficient distribution on the inside of the airfoil. At the leading edge, the high heat-transfer coefficient is often counterbalanced with impingement cooling. Figure 5.9 shows how this is achieved in an airfoil. This method of cooling has been recognized for some time as being highly effective and, therefore, has received much attention from researchers. As a result, a large quantity of data and correlations are available in the literature. Of the rather large quantity of literature, only two works will be discussed here. Chupp et al.[3] measured and correlated the heat transfer in the geometry shown in Fig. 5.10. They correlated the heat-transfer coefficient over the leading-edge radius region. They also obtained information on the distribution of the heat-transfer coefficient. The

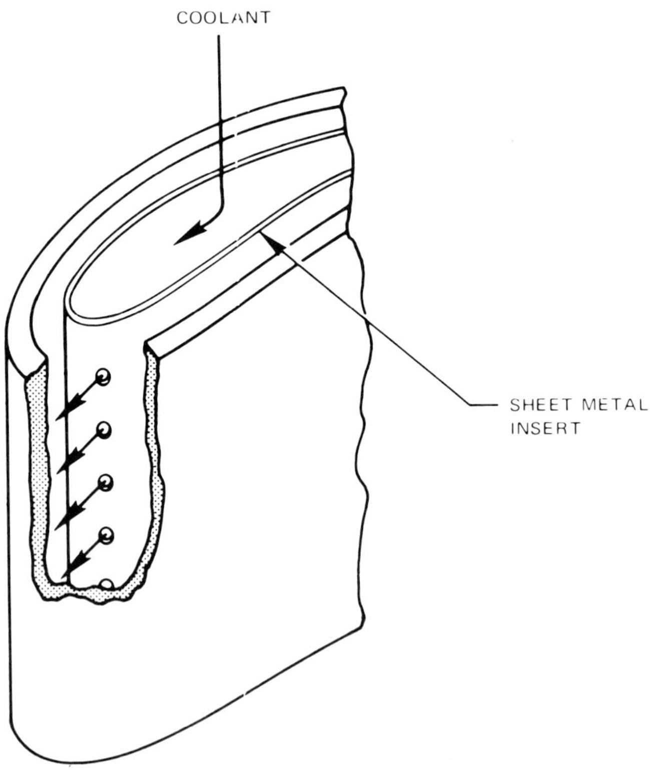

Fig. 5.9 Leading-edge impingement cooling.

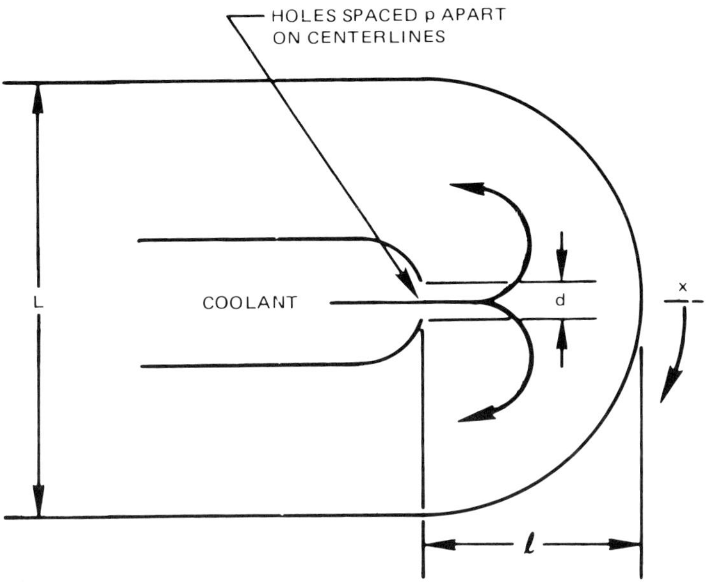

Fig. 5.10 Leading-edge impingement cooling model (from Ref. 3).

average and stagnation point Nusselt numbers were correlated by

$$Nu_{av} = 0.63Re^{0.7}(d/p)^{0.5}(d/L)^{0.6}$$

$$\exp\left[-1.27(l/d)(d/p)^{0.5}(d/L)^{1/2}\right] \qquad (5.2)$$

$$Nu_{stag} = 0.44Re^{0.7}(d/p)^{0.8}$$

$$\exp\left[-0.85(l/d)(d/p)(d/L)^{0.4}\right] \qquad (5.3)$$

where d is the characteristic dimension in the Nusselt and Reynolds numbers and the impingement hole velocity is the characteristic velocity in the Reynolds number. The thermal conductivity is evaluated at the average of wall and coolant temperatures and the viscosity is evaluated at the coolant temperature. The range of variables tested was

$$3000 < Re < 15{,}000$$
$$0.006 < d < 0.026 \text{ in.}$$
$$0.0312 < p < 0.125 \text{ in.}$$
$$0.040 < L < 0.094 \text{ in.}$$
$$0.006 < l < 0.160 \text{ in.}$$

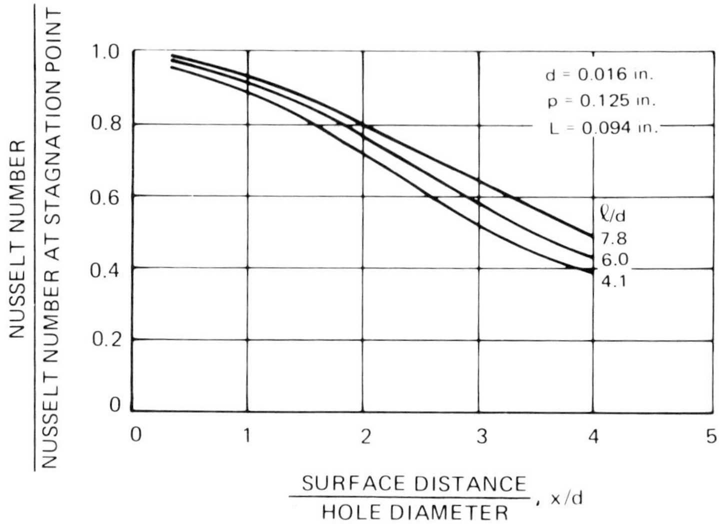

Fig. 5.11 Heat-transfer distribution from impingement cooling (from Ref. 3).

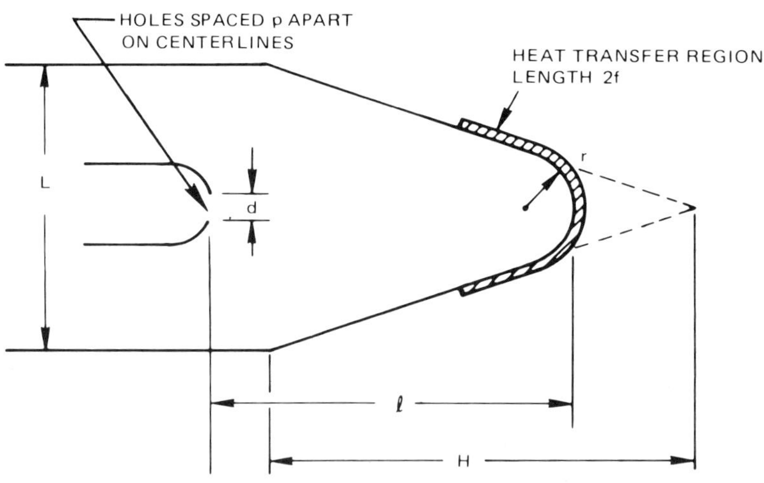

Fig. 5.12 Leading-edge impingement cooling model (from Ref. 4).

The distribution away from the stagnation point is shown in Fig. 5.11. This distribution was found to be relatively independent of (d/p) and *Re*.

Metzger et al.[4] performed a more complete investigation in that they correlated more geometric variables. Their geometry is shown in Fig. 5.12, which in a gross way approximates most of the geometric variables in an airfoil design. Of particular interest in Ref. 4 are the effects of a small leading-edge radius. Small internal leading-edge radii tend to occur in airfoils because of requirements to design small external leading-edge radii to give good aerodynamic performance and thick leading-edge walls to prevent overly rapid thermal response to transients. Unfortunately, their correlations are not easily put into a simple form. Their correlations have the functional form of

$$St_{av} = St_{av}\left(Re, \frac{p}{d}, \frac{r}{b}, \frac{l}{b}, \frac{2H}{L}, \frac{2r}{L} \right) \tag{5.4}$$

where St_{av} is based on the impingement hole velocity and Re is a Reynolds number based on twice an equivalent slot b and on the impingement hole velocity. The reader is referred to the original work for the correlations. This work measured only average heat-transfer coefficients over the region $2f$ rather than local heat-transfer coefficients, but fortunately these are generally adequate for airfoil design.

Impingement cooling at midchord. High levels of heat-transfer coefficients are desirable in the midchord region, as well as the leading edge. This is often accomplished with impingement cooling, as shown in Fig. 5.13. This

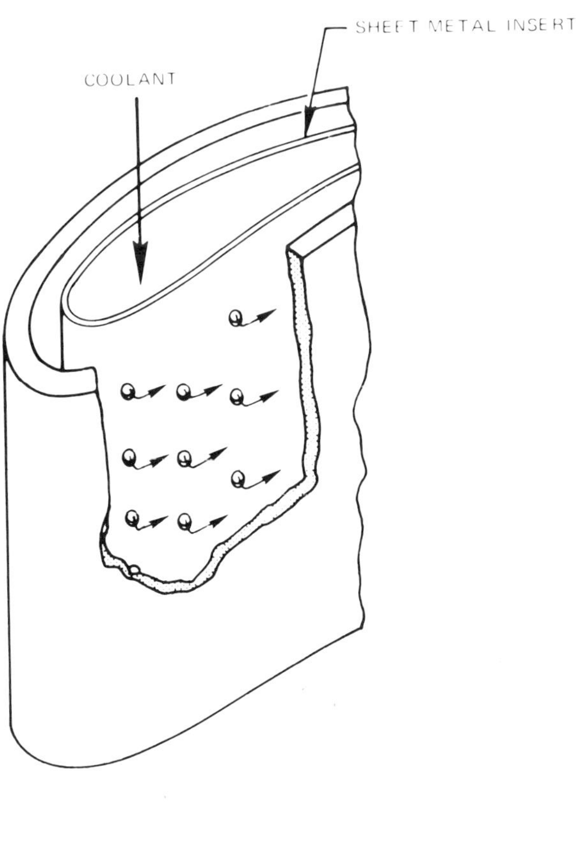

Fig. 5.13 Midchord impingement cooling.

is similar in some respects to impingement cooling at the leading edge, but is different in two major aspects: the radius of curvature of the impingement surface approaches infinity and the spent impingement air causes a strong cross flow.

Kercher and Tabakoff[5] performed experiments on impingement with cross flow on a flat plate. They correlated the data with the relationship

$$Nu_x = \xi_1 \xi_2 Re^m Pr^{\frac{1}{3}} (l/d)^{0.091} \tag{5.5}$$

where Nu_x is the local Nusselt number with the impingement hole diameter d as the characteristic length, Re the Reynolds number based on the mass velocity through the hole and on impingement hole diameter, Pr the Prandtl number, l the length from impingement hole to wall, p the hole pitch (square arrays), and ξ_1, ξ_2, and m are empirically determined parameters.

All properties are evaluated at the average temperature of the wall and coolant. The range of variables tested was

$$1 < l/d < 4.8$$

$$332 < Re < 3.7 \times 10^4$$

$$3.1 < p/d < 12.5$$

Colladay[6] has taken the correlations of Ref. 5 for ξ_1, ξ_2, and m and presented them as

$$m = a_1(p/d)^2 + b_1(p/d) + c_1 \tag{5.6}$$

$$\xi_1 = \exp\left[a_2(p/d)^2 + b_2(p/d) + c_2\right] \tag{5.7}$$

$$\xi_2 = \frac{1}{1 + a_3\left[(G_{cf}/G_h)(p/d)\right]^{b_3}} \tag{5.8}$$

where a_1, a_2, a_3, b_1, b_2, b_3, c_1, and c_2 are given in Table 5.1 and G_{cf} is the cross-flow mass velocity and G_h the impingement hole mass velocity.

Roughened walls. Another technique commonly used to obtain high heat-transfer coefficients is to place fins normal to the coolant flow path. Such a cooling scheme is shown in Fig. 5.14. Little data are available in the literature for the exact geometries used in airfoils, but data for circular pipes are available and may be applied utilizing the concept of a hydraulic diameter. For example, Webb et al.[7] investigated the geometry shown in

Table 5.1 Impingement Cooling Correlation (from Ref. 6)

Coefficient	Reynolds Number 300–3000	Reynolds Number 3000–30000
a_1	−0.0015	−0.0025
b_1	0.0428	0.0685
c_1	0.5165	0.5070
a_2	0.0126	0.0260
b_2	−0.5106	−0.8259
c_2	−0.2057	0.3985
a_3	0.4215	0.4696
b_3	0.580	0.965

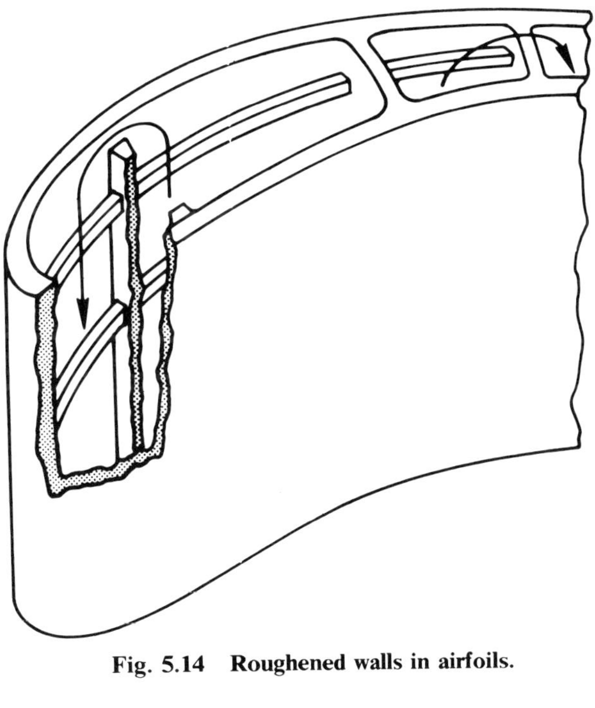

Fig. 5.14 Roughened walls in airfoils.

Fig. 5.15 with the range of parameters

$$0.01 < e/d < 0.04$$

$$10 < n/e < 40$$

and obtained heat-transfer and friction factor correlations. The reader is referred to the original work for the correlations.

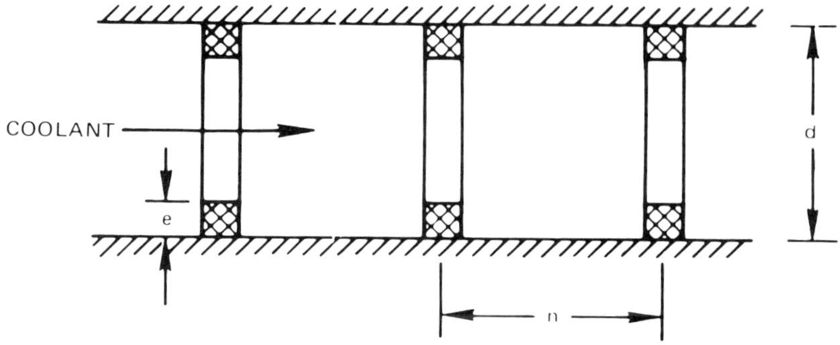

Fig. 5.15 Roughened wall model (from Ref. 7).

Trailing edges. The aerodynamic requirement for thin trailing edges makes it very difficult to cool the trailing edge convectively unless the coolant air ejects through the trailing edge. One way in which this is accomplished is shown in the airfoils of Fig. 5.6. In these airfoils, the two halves of the trailing edge are held together with pins (commonly called pedestals). These pins serve two purposes: they structurally hold the airfoil together and they augment the heat transfer. Halls[8] reports the use of the correlation from Grimison[9] for tube banks

$$Nu_{\text{av}} = F_a\, 0.284 Re^{0.61} \tag{5.9}$$

where $0.7 < F_a < 1.2$ and the reference velocity is in the minimum flow area and the reference dimension is the pedestal diameter. Coolant properties are evaluated at the average temperature of the wall and coolant. In Ref. 9 F_a is correlated as well as the friction factor for inline and staggered tube banks.

Another way to get the coolant out of the trailing edge is to use drilled holes. These are normally calculated using standard pipe heat-transfer correlations and an entry length correction term. This technique cannot achieve as high an effectiveness as pedestals can, but nevertheless is often used when a high degree of trailing-edge cooling is not required.

Flow distribution. In airfoil internal cooling design, one of the most important parameters is the flow distribution. The calculation of flow distribution is generally performed through the use of one-dimensional branched circuits, which account for friction, expansion-contraction losses, area change, heating, and rotation. The necessary correlations for each of the pressure loss mechanisms come from the data and correlations discussed above.

A quick look at airfoils such as shown in Fig. 5.6 will lead one to question the acceptability of such a simple model of the flow. In a complex flow situation (which seems to be almost every airfoil), it is useful to back up the one-dimensional flow calculations with model testing. Two types of models are often used: a large-scale (10 ×) transparent plastic model for flow visualization and a large-scale sheet metal model for heat transfer measurements. In the former, regions of low flow are determined by the use of water flow and dye injection. In the latter, heat transfer is measured by placing the model, instrumented with heat flux sensors, in a steam bath and passing cooling air through the model at low velocity. Neither of these two modeling techniques is completely satisfactory because they do not account for rotation and compressibility. However, they are found to be useful in that they often point out local problem areas that would undoubtedly fail had they not been detected and corrected. One-dimensional branched circuit analysis, when used with physical model testing, gives a reasonably good account of itself. Airfoils designed using these techniques tend to behave adequately close to design.

General. The general types of convection cooling have been covered in the above discussions. There are, of course, many variations on the detailed geometries and no attempt has been made to cover them all. The correlations presented are from the open literature. In most cases, manufacturers of gas turbines have further proprietary data from which airfoils are designed. These data are generally more directly focused on the range of geometries and flow variables required for airfoils.

Improvements in computational methods for viscous, turbulent, compressible, three-dimensional flows are being made at a rapid pace. It is conceivable that these will some day replace the one-dimensional branched circuit analyses currently used. However, because of the geometric complexity of the internal flows, the one-dimensional branched circuit analyses will probably be used for some time to come.

Film Cooling

With the levels of cooling supply pressures generally available in turbines, it becomes very difficult to convectively cool airfoils at average cooling effectiveness values greater than about 0.5. When the turbine gas temperature, coolant temperature, and allowable metal temperature require a higher effectiveness level, film cooling is utilized.

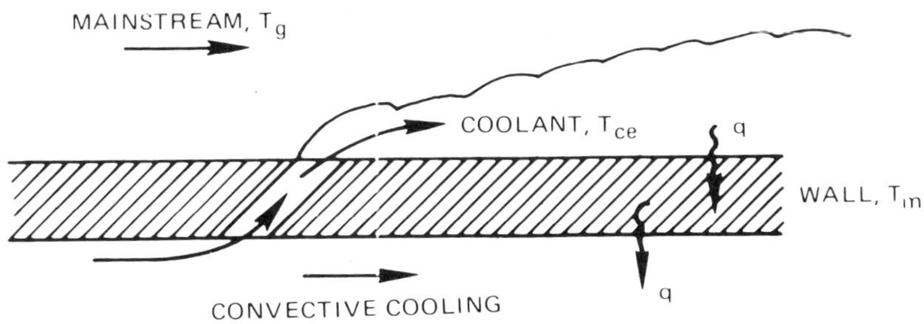

Fig. 5.16 Film cooling.

The general concept of film cooling is shown in Fig. 5.16. A layer of coolant air is placed over the airfoil, which serves to insulate the airfoil from the hot turbine gases. Because the coolant mixes with the turbine gases, the film is not a very effective method of cooling in itself. However, when combined with convective cooling from the inside of the airfoil, it is a very effective method of cooling airfoils.

In the following discussion, the problem of film cooling will be divided essentially into two parts. The first will be to determine the adiabatic wall temperature in the presence of a film, but without convective cooling. This will be called film temperature. Film temperature is correlated in terms of film effectiveness. In general, two types of film effectiveness definitions are useful; local and lateral average. Local film effectiveness is defined by

$$\eta_f \equiv \frac{T_g - T_f(z, x)}{T_g - T_c} \tag{5.10}$$

and lateral average film effectiveness by

$$\bar{\eta}_f \equiv \frac{T_g - \bar{T}_f(x)}{T_g - T_c} \tag{5.11}$$

The second part will be to determine the heat-transfer coefficient based on the film and airfoil temperature difference as the driving potential. Some researchers have chosen to utilize instead an overall heat-transfer coefficient with the difference between the mainstream gas temperature and the airfoil temperature as the driving potential. This method of describing film cooling will also be discussed, but only for one type of film cooling—multihole film cooling.

Slot cooling. Research in film cooling started at least 30 years ago. Wieghardt[10] measured film temperature from a two-dimensional slot on a flat plate during World War II. Many others have measured film temperature since then. Goldstein[11] gives an excellent review of the work through 1971.

An approach often used to correlate slot film cooling is the heat sink or entrainment model, shown in Fig. 5.17, in which the mainstream air is assumed to be entrained into a film layer. In its simplest form, the entrainment W_{ge} is assumed to progress as in a boundary layer without film cooling. Furthermore, the air in the boundary layer is assumed to be at a uniform temperature corresponding to the mixed mean temperature of the entrained air and film cooling air. A seventh-power velocity profile is normally used to calculate the entrained flow. In each of various versions of the model, a choice must be made for the origin of the boundary layer. This is an important variable near the slot, but it becomes less so further downstream. A typical result is given in Ref. 11 for a flat-plate low-velocity

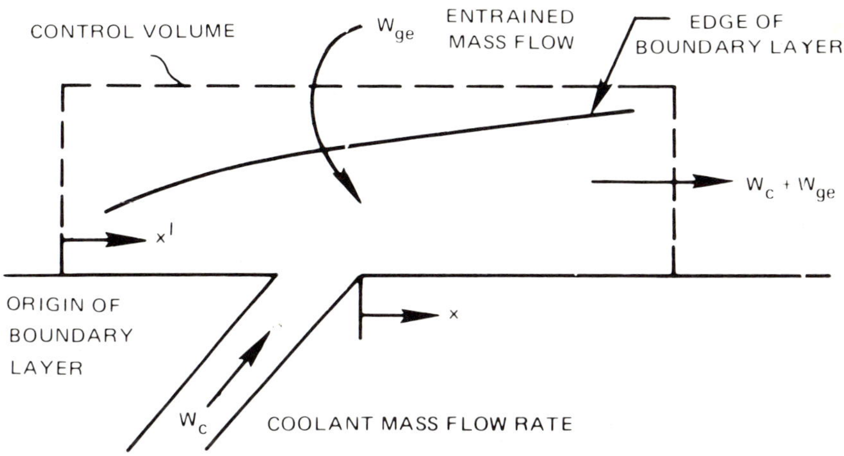

Fig. 5.17 Entrainment model for film cooling.

boundary layer starting at such a location that $W_{ge} = W_c$ at the injection point. The result is

$$\bar{\eta}_f = 1/(1 + 0.249\zeta^{0.8}) \tag{5.12}$$

where

$$\zeta = (x/Ms)\left[(\mu_c/\mu_g)Re\right]^{-0.25} \tag{5.13}$$

and Re is based on coolant mass velocity, viscosity, and slot height. This simple model compares favorably with a lot of data as is shown in Ref. 11. Other researchers such as Goldstein and Haji-Sheikh[12] have proposed more sophisticated models that account for the effects of blowing on the entrainment and distribution of the temperature within the boundary layer.

The heat-transfer coefficient under a slot film has been shown by Hartnett et al.[13] to be very close to that which would occur without blowing except near the slot at high blowing rates. This feature, which is found to be common in many applications of film cooling, is one of the reasons that dividing the problem into an adiabatic wall temperature and a heat-transfer coefficient is useful.

Rows of holes. Slots are not often used on airfoils because of thermal stress and mechanical design considerations. High thermal stresses are caused by the fact that the airfoil just upstream of a slot becomes very hot, while just downstream of a slot it becomes very cold. The mechanical design problem of slots is caused by the difficulty in holding the airfoil together at the slot, especially in the presence of the high thermal stresses. A commonly used approach is to substitute rows of holes for the slots. These tend to solve

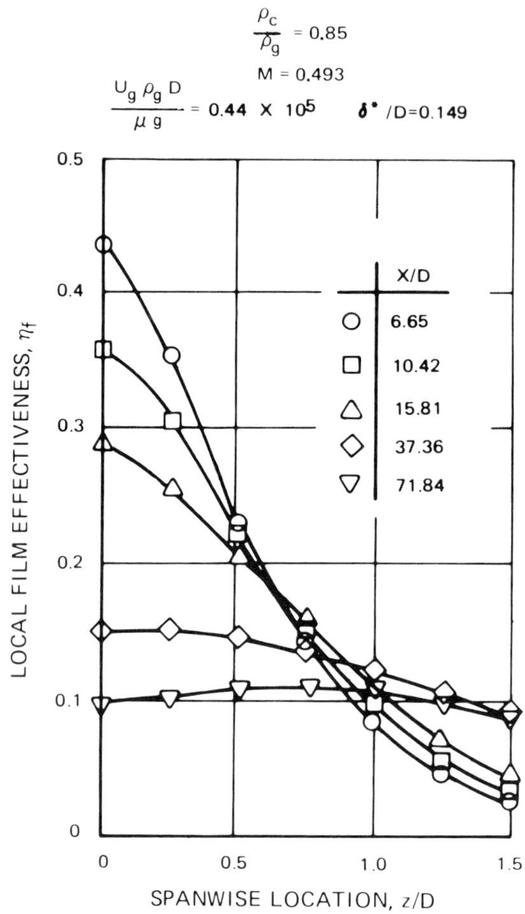

Fig. 5.18 Spanwise film effectiveness from a row of holes (from Ref. 14).

both problems because the metal between the holes relieves the thermal gradients and stresses and holds the airfoil together.

Rows of holes tend to solve some problems, but they do so at the expense of others. The film effectiveness is lower in general than from a slot and the flowfields are more difficult to predict or correlate. Much of the film cooling work in recent literature has addressed these problems. Some of this work will be discussed here.

The flow from a row of holes is very much three-dimensional as can be seen from film effectiveness data. Ericksen[14] obtained the film effectiveness data in Fig. 5.18 for the geometry shown in Fig. 5.19. It can be seen that the film does not become uniform until a distance somewhat greater than 40 diam downstream.

The three-dimensionality of the flowfield causes some questions in calculating heat fluxes to airfoils. Normally, one wishes to calculate the lateral

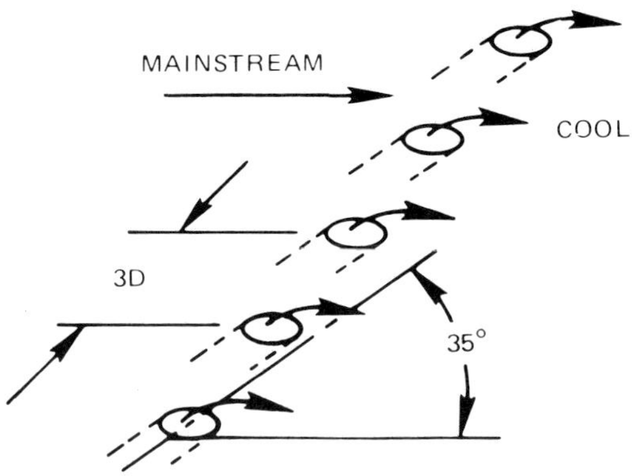

Fig. 5.19 Film cooling geometry.

average heat flux where the wall temperature is uniform in the lateral direction. To do this, one should calculate

$$\bar{q}(x) = \frac{1}{p}\int_0^p h(z,x)\left[T_f(z,x) - T_m(x)\right] \mathrm{d}z \tag{5.14}$$

where $h(z,x)$ and $T_f(z,x)$ are local values of the heat-transfer coefficient and film temperature. This would, of course, necessitate a rather extensive data base, which is expensive and time consuming to obtain. Fortunately, the problem is minimal in practice, because $h(z,x)$ is generally a weak function of z other than in the immediate vicinity of the holes and may be taken out of the integral. Then, Eq. (5.14) reduces to

$$\bar{q}(x) = h(x)\left[\bar{T}_f(x) - T_m(x)\right] \tag{5.15}$$

From Eq. (5.15) it can be seen that the lateral average heat flux may be calculated from correlations for lateral average rather than local film temperature.

A great deal of understanding of film cooling via rows of holes has been gained by Eckert and his co-workers[14–17] at the University of Minnesota. They worked extensively on the geometry shown in Fig. 5.19 on a flat plate at low velocities. Because of the large amount of information they generated on this geometry, rows-of-holes film cooling will be discussed largely from their work and that of Liess,[18] who tested the same geometry. This will eliminate the geometric variables of hole spacing and orientation and, hopefully, provide a clearer insight into the film cooling process. With the exception of Ref. 16, all of the work in Refs. 14–18 was performed at coolant-to-mainstream density ratios close to 1. For brevity, this will not be

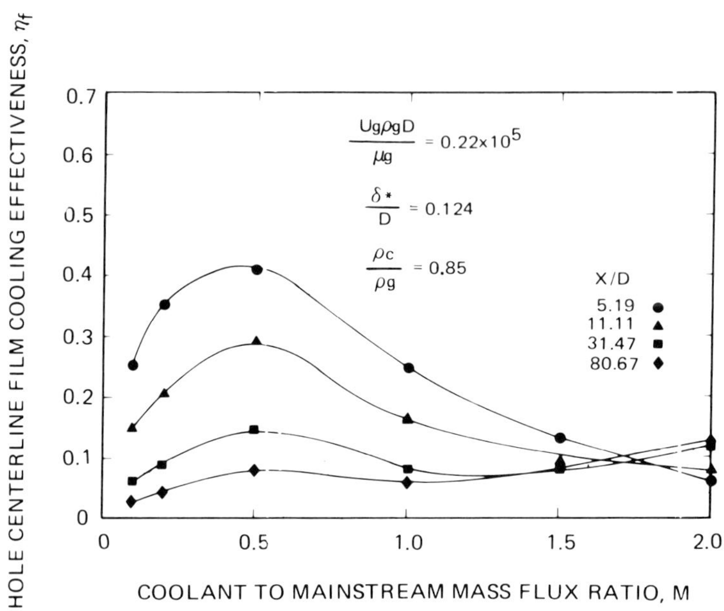

Fig. 5.20 Effect of blowing rate on centerline film cooling effectiveness (from Ref. 15).

mentioned again, but this parameter is an important one as is shown in Ref. 16.

Figure 5.20, showing data from Goldstein et al.,[15] illustrates one of the common features of film cooling from rows of holes. As the coolant-to-mainstream mass flux ratio M increases, the effectiveness first increases, reaches a maximum at a mass flux ratio of 0.5, and then decreases. This maximum is attributed to the penetration of the jet into the mainstream as opposed to its laying down on the surface. The effect is strongest near the row of holes and very weak far downstream.

Pedersen[16] investigated the effect of the coolant density in this same geometry. His data for lateral average effectiveness at a downstream location of 10 diam is shown in Fig. 5.21. The same general shape of the effectiveness curve in Fig. 5.20 can be seen here. As the coolant-to-mainstream density ratio increases, however, the maximum effectiveness increases. Reference 16 interprets the data to give the maximum effectiveness at a value of coolant-to-mainstream velocity ratio (U_c/U_g) of approximately 0.4. Ericksen et al.[17] proposed a model for film effectiveness based on a point heat sink located some distance above the surface. Reference 16 used the argument that the distance above the surface of the heat sink should be a function of the coolant-to-mainstream momentum flux ratio I and, using the above model, proposed a correlation of the form

$$\bar{\eta}_f/M = f(I) \tag{5.16}$$

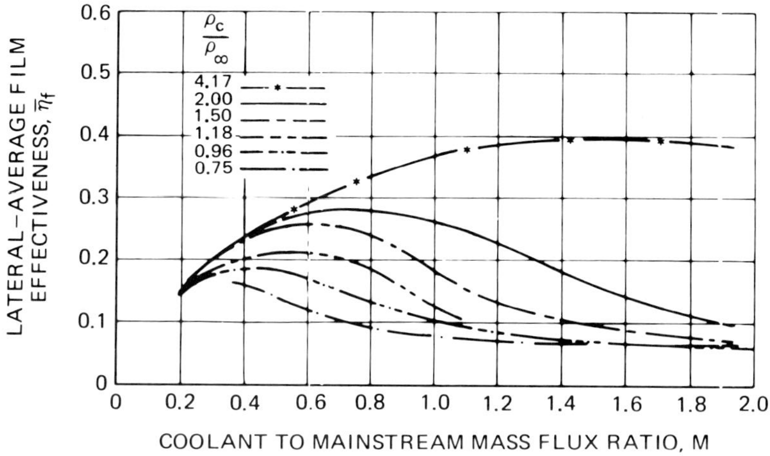

Fig. 5.21 Effect of coolant density on film effectiveness (from Ref. 16).

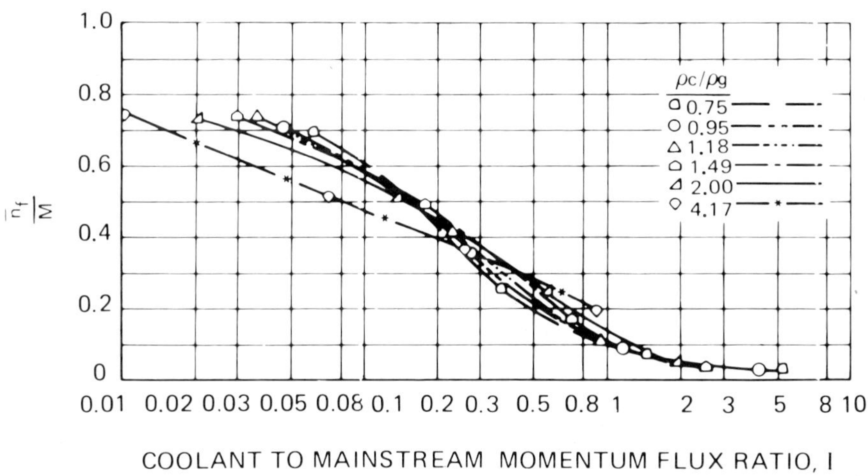

Fig. 5.22 Correlation of film effectiveness with coolant to mainstream momentum flux ratio (from Ref. 16).

for its data at constant U_g and hole diameter. The data from Ref. 16 at a distance 10 diam downstream are shown in Fig. 5.22. The correlation is reasonable except at large ρ_c/ρ_g. This may be related to the fact noted in Ref. 16 that, at values of U_c/U_g below 0.45, a change in the character of the flow seems to occur. At values of U_c/U_g below 0.45, the jet has the characteristic of appearing to remain attached to the wall and being independent of I, whereas above that value of U_c/U_g the jet appears to penetrate the mainstream further and to be controlled by I. This is about the same velocity ratio mentioned above as necessary for penetration to occur. It does appear somewhat surprising that the penetration condition should be controlled by velocity, but this effect also appears in other data. At this time, no further generalization on the effect of ρ_c/ρ_g has appeared in the literature.

The variables of coolant-to-mainstream mass flux, density, and momentum flux ratios appear to be the strongest parameter affecting film cooling. However, there are other variables of varying degrees of importance and levels of confirmation. Ericksen[14] and Liess[18] have measured the effect of the ratio of the displacement thickness of the inlet boundary layer to the hole diameter. They agree that there is a significant effect and that the

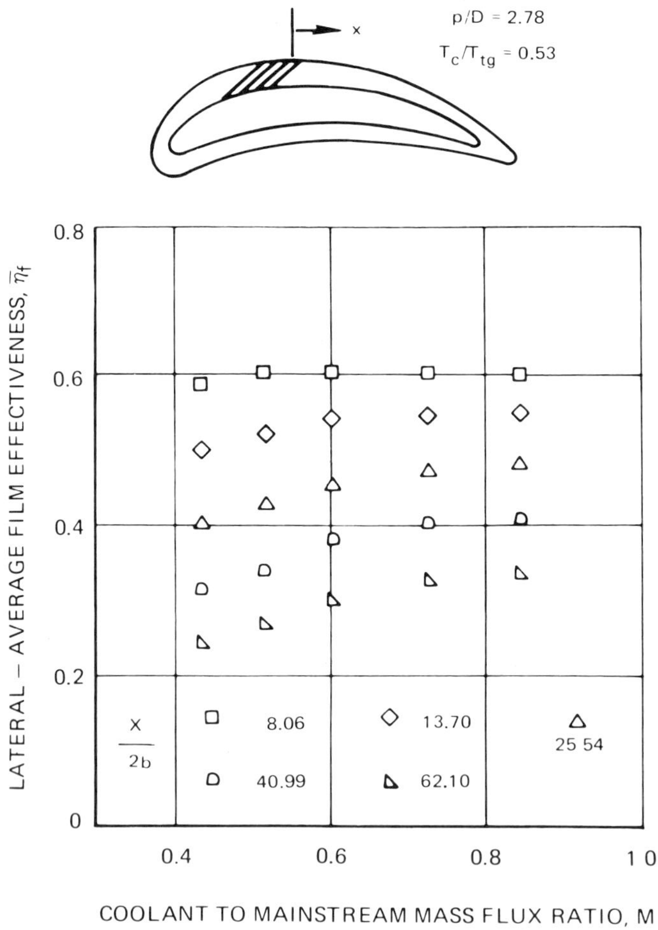

Fig. 5.23 Film cooling effectiveness on an airfoil with a double row of holes (from Ref. 20).

thinner boundary layer has a higher film effectiveness. They differ in that Ref. 14 finds that effectiveness on the hole centerline increases with decreasing boundary-layer thickness to a value of δ^*/D as low as 0.06, whereas Ref. 18 finds that the lateral average effectiveness reaches a maximum value at δ^*/D around 0.2. References 14 and 18 both indicate that the freestream Reynolds number $U_g D/\nu_g$ appears at most to be a weak variable. Reference 18 also indicates that the freestream Mach number is not an important variable. In addition, it finds that an accelerating pressure gradient reduces the film effectiveness at low coolant-to-mainstream mass flux ratios (< 1.0), but does not have as strong an effect at high mass flux ratios (> 1.0).

The heat-transfer coefficient does not differ greatly from that which would occur without film cooling. In the same geometry discussed above, Ericksen[14] shows that the heat-transfer coefficient varies hardly at all up to a coolant-to-mainstream mass flux ratio of 0.5 and at 2.0 differs by only 27% from the no-blowing case. Other than at the leading edge of an airfoil, the coolant-to-mainstream mass flux ratio is generally below 2.0.

The geometry discussed above is useful because it has demonstrated many features of film cooling. However, it is seldom, if ever, used in actual practice because the overall effectiveness is too low. In general, closer hole spacing and/or multiple rows of holes are utilized to obtain the high levels of film effectiveness required. Metzger and Fletcher[19] measured high effectiveness values with close spacing of the holes (pitch-to-diameter ratio of 1.55) on a flat plate at low velocities. Lander et al.[20] also measured high film effectiveness from a double row of holes on an airfoil in a cascade, as shown in Fig. 5.23. It should be noted that the coolant-to-mainstream total temperature ratio T_c/T_{tg} was 0.53 which gives a coolant-to-mainstream density ratio similarity with engines in the experiments of Ref. 20. Muska et al.[21] also showed very high effectiveness on a flat plate from double rows of holes at pitch-to-diameter ratios of 2.0 and low blowing rates. Goldstein et al.[22] showed that by flaring the holes of the single row of holes geometry of Fig. 5.19, a significant increase in the film effectiveness could be achieved.

In addition to film effectiveness data, Lander et al.[20] obtained heat-transfer coefficient data for the geometry of a double row of holes shown in Fig. 5.24. These data lead to two conclusions: (1) the presence of holes in the airfoil will likely lead to a rapid transition of the boundary layer to turbulent flow and (2) the effect of blowing on the heat-transfer coefficient is not large. This latter conclusion reinforces the results from flat plates.

Generally, total airfoils cannot be adequately cooled with just a single or double row of holes. Additional film cooling is required from another row or a double row of holes. An example of this is shown in Fig. 5.25, where there is film cooling at the leading edge plus additional film cooling from two additional sets of double rows of holes. Reference 21 shows that the effect of multiple rows of holes may be determined in an additive manner first suggested by Sellers[23] for slots. In this method, the mainstream temperature downstream of the first set of rows of holes is taken to be the film temperature from the film cooling at the leading edge. The mainstream temperature for all subsequent downstream rows is calculated in a similar

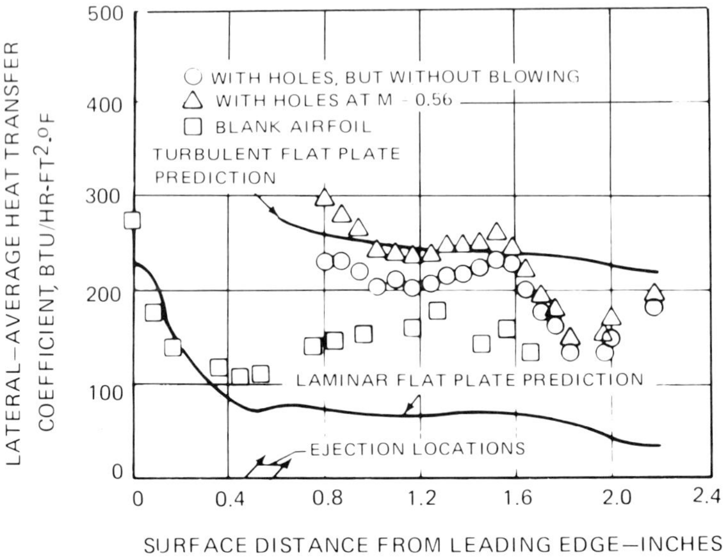

Fig. 5.24 Heat-transfer coefficient under film cooling (from Ref. 20).

Fig. 5.25 Film-cooled airfoil with multiple rows of holes.

manner. Then, behind the nth set of holes,

$$\bar{\eta}_f = \sum_{i=1}^{n} \bar{\eta}_{f_i} \prod_{j=0}^{i-1} \left(1 - \bar{\eta}_{f_j}\right) \tag{5.17}$$

where $\bar{\eta}_{f_0}$ is equal to zero and $\bar{\eta}_{f_i}$ is the lateral average film effectiveness from the ith set of film cooling holes. This has been shown to be valid down to spacings of 16.7 hole diam.

The thermal design of film-cooled airfoils must account for both the effects of the film cooling and the effects of internal convective cooling. Fortunately for the designer, the internal convective cooling information developed for purely convectively cooled airfoils is directly applicable and is, therefore, somewhat straightforward. However, the data in the literature for film cooling, some of which were discussed above, are *not* adequate. At the present time, in addition to the data in the literature, one must obtain data in cascades for geometries very close to that to be used. In performing these tests, one must simulate the Reynolds number, Mach number, coolant-to-mainstream density ratio, and coolant-to-mainstream mass flux ratio. The turbulence effects must also be simulated in some manner. The literature discussed above can be used for guidance in these tests.

This procedure to obtain information for film cooling assumes, as in convective cooling, that a two-dimensional solution is valid. The shortcomings of a two-dimensional approach have already been discussed for convectively cooled airfoils, arguments that can be equally applied to film cooling. There may be a difference in that the effects of rotation may be different in film-cooled airfoils because of the thickness of the film cooling layers. In spite of the shortcomings of the two-dimensional approach, it has been successfully applied to airfoil design. This is evidenced by the successful design of film-cooled turbines in the 2500°F class.

Multihole film cooling. In regions where very high effectiveness is required (> 0.60), multihole film cooling becomes desirable. Multihole film cooling differs from rows-of-holes film cooling in that the geometry is one of a uniform array of holes rather than of rows of holes and the holes tend to be smaller. An example of multihole film cooling can be seen in Fig. 5.26. This type of film cooling is treated separately from rows-of-holes film cooling because the analytical modeling for it is rather different and, in fact, is more similar to that for transpiration cooling discussed in a following subsection.

Both the film effectiveness and the overall heat-transfer coefficient approaches will be discussed for multihole film cooling. Both methods have their adherents and the available literature is somewhat evenly divided between them. The literature will first be discussed from the film effectiveness viewpoint.

As in all methods of film cooling, convective cooling is an important part of multihole film cooling. A simple one-dimensional model by Mayle and Camarata[24] has been found very useful to demonstrate this and to show the

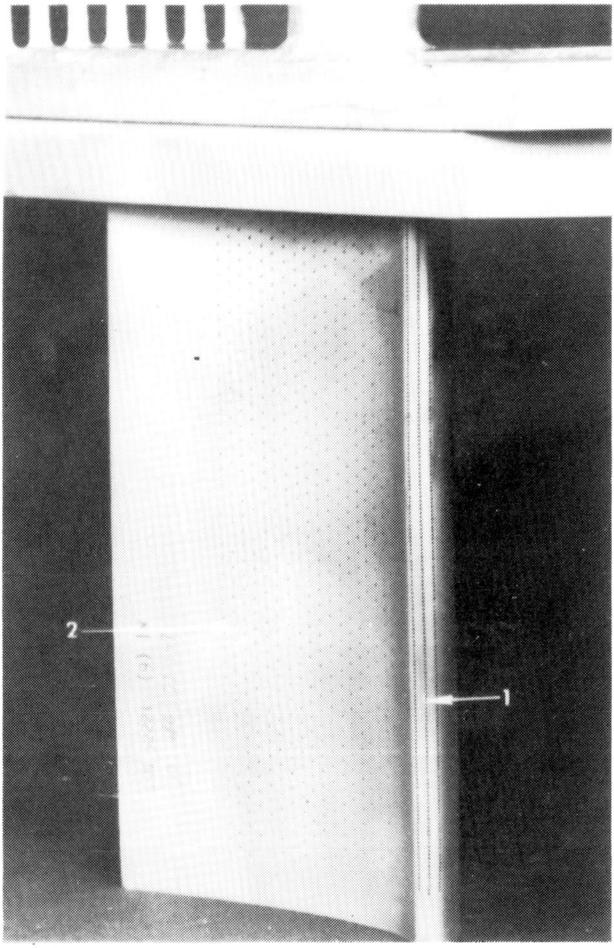

Fig. 5.26 Airfoil with multihole film cooling: (1) film-cooled leading edge, (2) multihole pattern on pressure side of vane.

relative effects of film and convective cooling for multihole film cooling. This model considers a small chordwise region where the film effectiveness, gas side heat-transfer coefficient, and metal temperature take on lateral average values. The model is shown in Fig. 5.27. It can be shown that a local cooling effectiveness Φ defined below can be calculated from

$$\Phi \equiv \frac{T_g - T_m}{T_g - T_c} = \frac{\bar{\eta}_f + \eta_c(F/St - \bar{\eta}_f)}{1 + \eta_c(F/St + \bar{\eta}_f)} \tag{5.18}$$

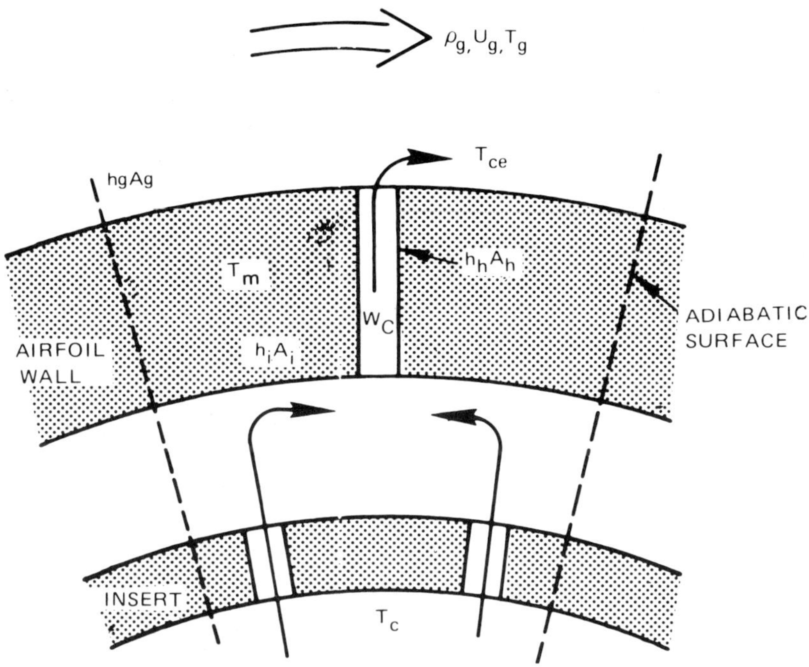

Fig. 5.27 One-dimensional model for multihole cooling.

where

$$\eta_c = 1 - \exp\left(\frac{A_h h_h + A_i h_i}{W_c c_{p_c}}\right) \quad \text{(convective effectiveness)} \tag{5.19}$$

$$\bar{\eta}_f = \frac{T_g - \bar{T}_f(x)}{T_g - T_{ce}} \quad \text{(lateral average film effectiveness)} \tag{5.20}$$

$$St = \frac{\bar{h}_g}{\rho_g U_g c_{p_g}} \quad \left(\text{external Stanton number based on } \bar{T}_f \text{ and } T_m\right) \tag{5.21}$$

$$F = \frac{W_c}{A_g \rho_g U_g}\left(\frac{c_{p_c}}{c_{p_g}}\right) \quad \text{(coolant mass flux parameter)} \tag{5.22}$$

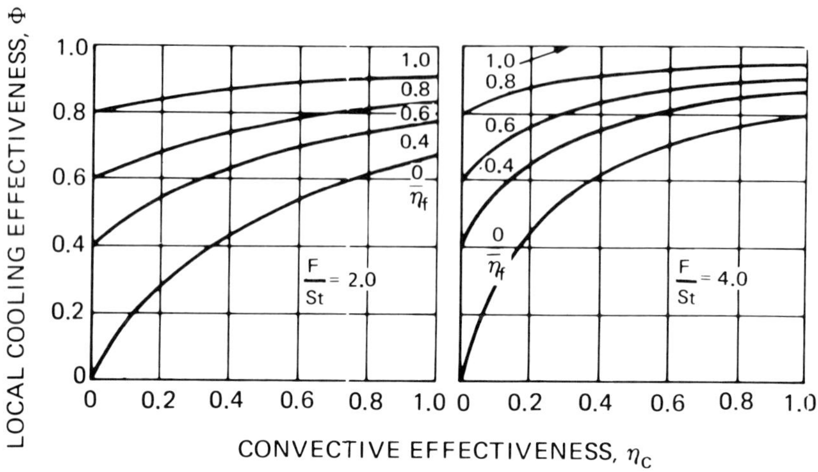

Fig. 5.28 Effects of convective and film effectiveness on overall effectiveness.

In airfoils it is usually found that

$$0.2 < \eta_c < 0.6$$

and where the unblown heat-transfer coefficient is used in the Stanton number

$$1 < F/St < 5$$

The results of Eq. (5.18) are shown in Fig. 5.28 for two values of F/St. The significant contribution of convective cooling in both these cases can be readily seen. Furthermore, it can be seen that convection is more important when F/St is larger.

As in rows-of-holes film cooling, the film effectiveness part of the model is the more difficult problem. Figure 5.29 shows a geometry tested at low speeds on a flat plate by Mayle and Camarata.[25] They have reported local as well as lateral average film effectiveness data for this geometry. LeBrocq et al.[26] and Launder and York[27] have also reported film effectiveness data for similar arrays of holes on a flat plate at low velocities, but in their work the holes were normal to the surface or angled directly to the rear. In Ref. 26 a nonstaggered array was also tested. The work in Ref. 25 was performed at a coolant-to-mainstream density ratio near 1, while the work in Refs. 26 and 27 was performed at several values of this density ratio.

Several interesting features of the flow for this type of film cooling are shown in Fig. 5.30. The top figure indicates a strong spanwise variation in film effectiveness even at a far downstream location. The figure on the bottom indicates a strong chordwise buildup in lateral average film effectiveness in the downstream direction. At a chordwise distance of 100, the

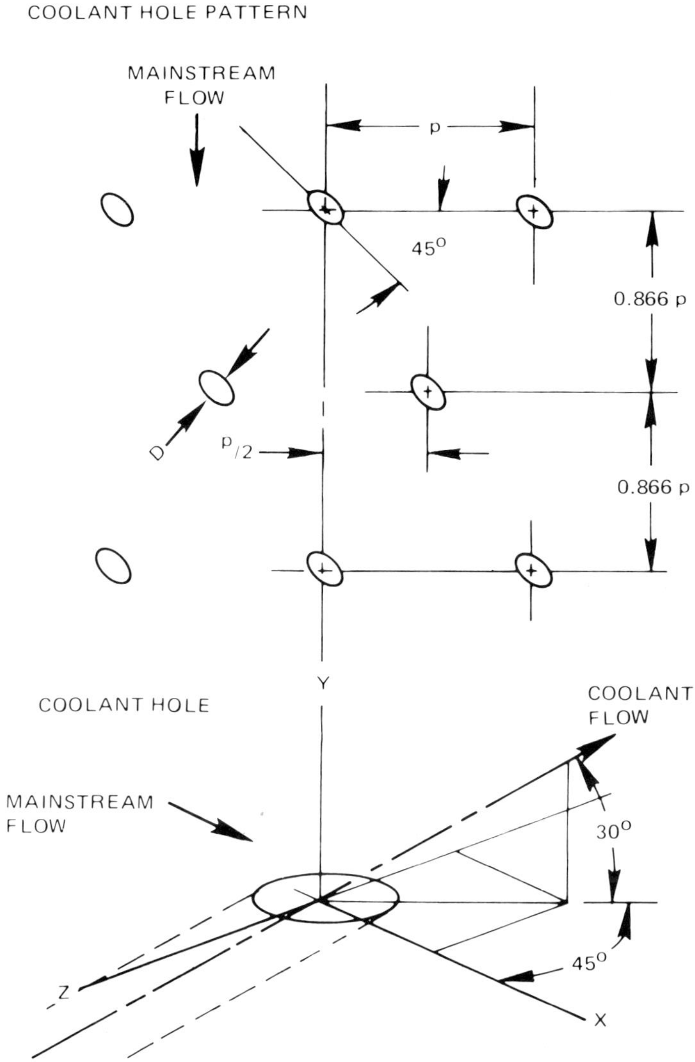

Fig. 5.29 Multihole cooling geometry (from Ref. 25).

buildup of film effectiveness flattens. This is interpreted in Ref. 25 as the region where the mixing of the jets with the mainstream just balances the replenishment of cooling air in the film. This feature is more prominent at lower values of M, becoming less so at higher values. Another feature is the slow drop-off in the effectiveness at the end of the hole array. This can add significantly to airfoil cooling, especially near the trailing edge where it is difficult to place film cooling holes.

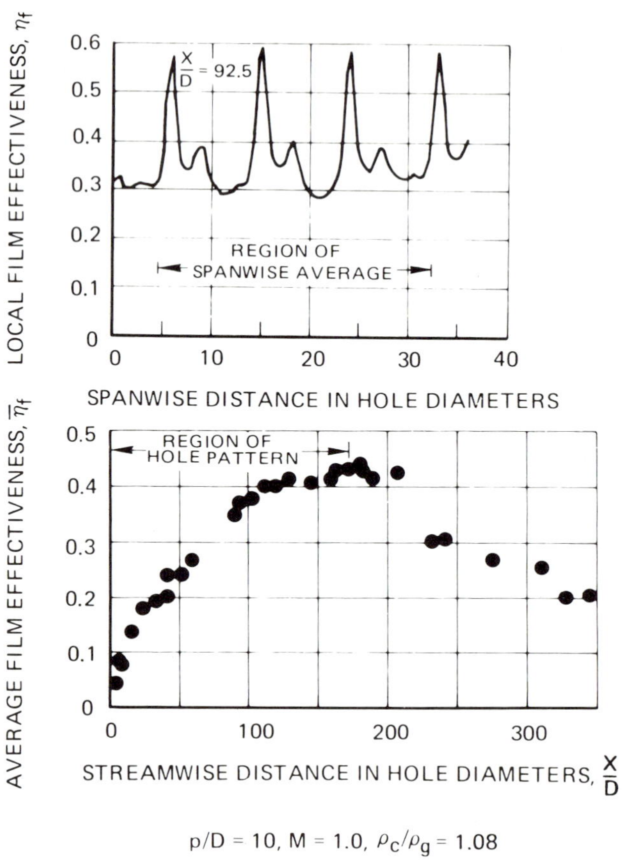

Fig. 5.30 Film effectiveness from multihole cooling (from Ref. 25).

Limited success was achieved in Ref. 25 in modeling the film effectiveness utilizing a heat sink model proposed by Ramsey et al.[28] for single holes and the additive film model of Ref. 23. In this case, the model of Ref. 23 was applied on a spanwise local basis, rather than on an average basis as described above for rows-of-holes film cooling.

It was also found in Ref. 25 that there appeared to be an optimum hole density. Decreases in the pitch-to-diameter ratio from 14 to 10 to 8 formed a small but definite optimum for film effectiveness at a pitch-to-diameter ratio of 10. This is very important to the manufacturers of turbines, who do not wish to drill a very large number of holes.

References 26 and 27 show similar results for holes normal to the surface and angled directly to the rear. In Ref. 26 it is argued from velocity profile measurements that staggered arrays of holes are superior to inline arrays and from direct data that holes angled to the rear are superior to normal holes. In Refs. 26 and 27 the effects of coolant density were also measured and found to be similar to that for rows-of-holes film cooling. The higher-density coolant caused higher film effectiveness, with the maximum effectiveness at levels of coolant-to-mainstream velocity ratio similar to those for rows-of-holes film cooling. In Ref. 27 it was found that accelerating the flow increased the film effectiveness locally behind the holes, but tended to decrease the effectiveness in the lateral direction. They also found that the freestream turbulence decreased the film effectiveness in accelerating flows, but not in constant-velocity flows. These latter two phenomena were explained in terms of transition of the coolant jet flow from laminar to turbulent flow. In practice, the Reynolds number of the coolant holes is such that the coolant jets are generally already turbulent. Thus, any effects of acceleration and turbulence that might be found on a turbine airfoil would require further rationalization.

Heat-transfer coefficients were also measured and reported in Ref. 25. It was found there that the effects of blowing were stronger than for a similar value of M in rows-of-holes film cooling. The effects were to increase the heat-transfer coefficient with the increasing mass flux of the coolant and the decreasing hole pitch. The levels were as much as 2.5 times greater than for a flow without blowing at a pitch-to-diameter ratio of 8 and a value of M of 1.5. Furthermore, the Stanton number for a given geometry reached a value independent of streamwise location when the blowing parameter F was greater than 0.007. In this region, they correlated the Stanton number of Eq. (5.21) with

$$St = 0.038(D/p)M^{0.28} \tag{5.23}$$

At lower values of F, the Stanton number monotonically approached the unblown Stanton number.

The heat-transfer data of Ref. 25 are for the lateral average case with the coolant and mainstream temperatures equal. The use of a lateral average may well be questioned when the effect of blowing on the heat-transfer coefficient is so strong, but at the present time there is little else in the literature.

Choe et al.[29] have taken the approach of a reduced heat-transfer coefficient concept. They have defined a parameter

$$\theta \equiv (T_g - T_{ce})/(T_g - T_m) \tag{5.24}$$

where T_m is a lateral average metal temperature. For the case of low-velocity flows and small temperature differences and thus small property variations, Ref. 29 shows that the linearity of the energy equation gives

$$St = St(\theta = 0) - \theta[St(\theta = 0) - St(\theta = 1)] \tag{5.25}$$

where St is based on $(T_g - T_m)$. Then, since the Stanton number is linear with θ, the Stanton number can be measured at two values of θ and all values of Stanton numbers can then be calculated. Furthermore, the adiabatic wall temperature can be found from the relationship

$$\bar{\eta}_f = 1/\theta(St = 0) \tag{5.26}$$

where $\bar{\eta}_f$ is the lateral average film effectiveness of a good conductor. Metzger et al.[30] have defined a similar parameter, but in their work the metal temperature is a lateral and chordwise average temperature.

This approach to the problem of multihole cooling could be quite useful because it allows the heat-transfer to an airfoil to be generalized from limited data. It has a major drawback, however, in the required linearity of the energy equation. In an engine, temperature differences are generally sufficiently large that linearity is not a valid assumption. Thus, application of this method is not straightforward.

Choe et al.[31] have shown how the reduced heat-transfer concept might reach greater application. They have shown that a two-dimensional boundary-layer type of model is capable of calculating the Stanton number in the presence of multihole cooling. Their model contains empirical input from the case where temperature differences are small, but presumably the analysis, with more appropriate empirical input, could be applied to conditions more closely related to those found in engines.

The current status of multihole film cooling in engines is somewhat behind rows-of-holes film cooling. There are much fundamental data available in the literature, but there is little application to service engine design. Many of the same recommendations must be made here as for rows-of-holes film cooling, if one wishes a multihole film-cooled airfoil. An airfoil close to the actual design must be tested in a cascade with Reynolds number, Mach number, coolant-to-mainstream density ratio, and coolant-to-mainstream flow ratio similarity to engine conditions. The effect of engine turbulence must also be modeled in some way. Questions raised by the rotating effects are the same as for rows-of-holes film cooling.

Special design problems appear on this type of airfoil because of the multiplicity of small holes in the surface. The large number of holes cause an especially difficult flow distribution problem that must be solved for each design. Also, there is a high cost associated with the large number of holes. Another problem is that the small holes are suspected to be more prone to plugging from dirt in the cooling air. In spite of these problems, the relatively high cooling effectiveness achievable with multihole cooling makes it an attractive cooling scheme for very-high-temperature engines.

Leading-Edge Film Cooling

Airfoils requiring film cooling generally need it over the whole airfoil, including the leading edge. The leading-edge region is especially difficult to cover with a film because the film must be established at the stagnation point, the exact location of which is not known. To cover the leading edge, large quantities of cooling air are introduced through rather densely packed holes. Examples of leading-edge film cooling holes are shown in Figs. 5.25 and 5.26. Because of the large quantities of cooling air, it might be thought

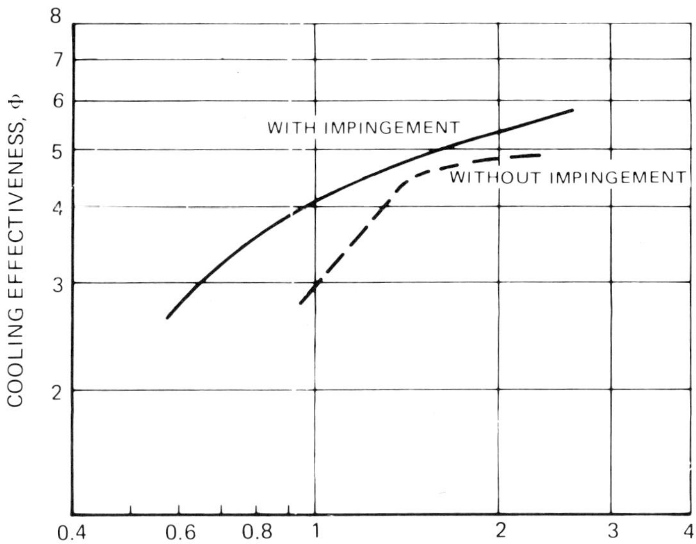

Fig. 5.31 Cooling effectiveness of leading-edge film cooling (from Ref. 32).

that the cooling mechanism is purely film effectiveness. However, this is not necessarily so. The one-dimensional model of multihole cooling is equally valid here and can be used to determine the relative importance of the film and convective effectiveness. At the leading edge, the parameter F/St is calculated to be large if $\bar{h}_g$ is assumed to be of the same order as without blowing. Then, from the arguments noted above, the convective effectiveness is seen to be very important. The exact levels of $\bar{h}_g$ in the region of leading-edge film cooling are not known, but some results reported in the literature, discussed in the next paragraph, confirm that convective cooling is very important in leading-edge film cooling.

Hiroki and Katsumata[32] obtained data on a four-row leading-edge geometry. They tested a leading edge with inside impingement as well as film cooling holes and a leading edge with only film cooling holes. They performed these tests on models made of conductive materials. Their results are shown in Fig. 5.31. It can be seen that the leading edge with impingement is definitely superior to that without impingement. Because of the use of conductive models, a large amount of convective cooling must have occurred in both models because of the convective cooling within the holes. The additional effect of impingement convective cooling, as shown in Fig. 5.31, is indicative of the importance of convective cooling in leading-edge film cooling. Unfortunately, there is very little other data published in the literature on this subject. Thus, a designer without unpublished information must rely again on testing to gain design data.

One other feature of leading-edge film cooling should be noted. The film formed at the leading edge persists to the trailing edge of the airfoil. This is shown in Ref. 21. The level of film effectiveness decreases from the leading to the trailing edge, as one might expect. However, its presence is important to airfoil design and should not be discounted.

Aerodynamic Losses of Film Cooling

The efficiency of a cooled turbine can be thought to be determined by a series of entropy production processes (called unit processes), as was pointed out in Chap. 4. In that chapter, it is shown that there are two contributions to entropy production in a unit process: the changes in the total temperatures and total pressures of the coolant and mainstream flows. In a unit process, each of these properties is defined in terms of the appropriate reference frame (rotating or stationary). If one considers a unit process to be across a row of airfoils, the total temperature changes can be readily calculated from the energy equation. However, the total pressure changes cannot be so readily calculated. Thus, to calculate cooled turbine performance, a method is required to calculate the total pressure changes. This is generally reported in terms of the decrease in the total pressure of the mainstream, which is herein called the total pressure loss.

There is a significant total pressure loss associated with injecting cooling air into the hot turbine flow. The present state-of-the-art is that the loss cannot be completely calculated from first principles. However, a simple mixing model has been shown to be reasonably accurate by Hartsel.[33] In this model, it is assumed that the coolant does not in any way affect the pressure distribution on the airfoil. Furthermore, it does not interact with the boundary layer, and thus the skin friction, on the airfoil. The model is one where the coolant mixes with some portion of the mainstream, called the mixing layer, in a one-dimensional manner. The losses are calculated within the mixing layer from the leading to the trailing edge, being accumulated toward the trailing edge. At the trailing-edge plane, the two mixing layers (suction and pressure side) are mixed with the mainstream unaffected flow at the average trailing-edge static pressure. This flow is then mixed with the viscous boundary layers and any trailing-edge coolant flow in a manner that also accounts for trailing-edge blockage. Each of the mixing calculations is performed using simple control volume arguments. In addition, the losses within the mixing layer for each coolant flow W_c are calculated using the one-dimensional linearized relationship

$$\frac{\Delta P_{tg}}{P_{tg}} = \frac{\gamma}{2}\mathscr{M}^2 \frac{W_c}{W_g}\left[1 + \frac{T_{ce}}{T_{tg}} - 2\frac{U_c}{U_g}\cos\beta\right] \tag{5.27}$$

where all quantities are evaluated at the point of injection. It is not at all obvious what fraction of the total mainstream flow W_g should be. However,

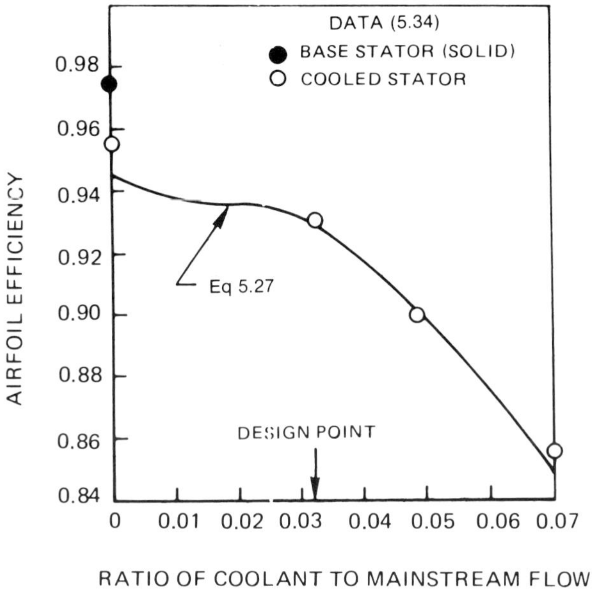

Fig. 5.32 Comparison of data for cooled cascade efficiency (from Ref. 34) with mixing analysis [Eq. (5.27)].

in Ref. 33, W_g was varied at 3–30% of the total mainstream flow and very little effect on the loss was found. Figure 5.32 from Ref. 33 shows the results of this analysis compared to the data of Prust et al.[34] The comparison is made on the basis of airfoil efficiency, which is defined in Ref. 33 as the ratio of the actual kinetic energy of the total flow to that which would occur if the coolant and the mainstream air each isentropically expanded to the same downstream static pressure. The corroboration is good, especially at coolant flow rates equal to or greater than design conditions. One might question the data and the analysis at coolant flow rates below design conditions, where the airfoil efficiency does not approach the uncooled airfoil efficiency. The data are exemplary of cascade data where the coolant flow is supplied from a single source to several regions of different static pressures on an airfoil. At flow rates and thus supply pressures below design conditions, mainstream flow enters into the airfoil in regions of high airfoil static pressures, while the coolant continues to flow out in regions of low airfoil static pressures. The outflow of coolant can cause total pressure loss even when the net coolant flow is zero.

The analysis of Ref. 33 is similar to others in the industry. At this point, it does not appear that the effect of film cooling on the skin friction can be properly taken into account, but the mixing effect seems to account for much of the loss. The important end result of this is that it gives guidance to the designer on the optimum location for film cooling holes. It is apparent from Eq. (5.27) that film cooling near the stagnation point and on the pressure side of the airfoils (where the Mach numbers are low) will give small total pressure losses, whereas film cooling on the suction side near the throat (where Mach numbers are high) will give high total pressure losses.

It is quite often argued that the trailing-edge ejection of the coolant flow will reduce the aerodynamic losses by filling in the wake. The works of Anderson et al.[35] and Anderson and Heiser[36] have shown analytically and experimentally that Eq. (5.27) can be used to approximate the behavior of the trailing-edge ejection in the absence of gross changes in the mainstream flow (e.g., base pressure) by the coolant flow. Close examination of Eq. (5.27) will show that with fixed geometry and temperatures the total pressure loss caused by the coolant flow will first rise with the coolant flow rate, reach a maximum, then decrease with the coolant flow rate to zero and become negative (total pressure gain). This occurs because the velocity (and thus the total pressure) of the coolant must continually increase with the coolant flow rate. The characteristic of reducing total pressure loss with high coolant flow rates is, unfortunately, of little practical value. Although the total pressure loss of the mainstream may be reduced to zero or negative values in this way, the entropy generation of the coolant and the mainstream considered together (see Chap. 4) causes losses in efficiency because of the high coolant total pressures required for high coolant flow rates. It is, of course, possible that the mainstream flow might be affected by the trailing-edge ejection, especially the base pressure. As discussed in Chap. 4, any changes in the base pressure would have a strong effect on total pressure losses. However, there have been no reports in the literature showing variations in the base pressure caused by trailing-edge ejection of the coolant flow. From these arguments, it appears at the present time that trailing-edge ejection of coolant flow causes a net loss in overall turbine performance. Similar arguments can be applied to all other film cooling flows.

Transpiration Cooling

Transpiration cooling is considered by many to be the ideal cooling method. In this method, coolant is introduced through a porous wall, so that it convectively cools the porous wall to the maximum theoretical limit (metal temperature equals coolant discharge temperature) and also film cools it with a well-distributed low-velocity coolant. This type of cooling is amenable to analysis and has received much attention from a theoretical boundary-layer approach. As a result, transpiration cooling can be considered calculable by the boundary-layer methods of Chap. 6.

Transpiration-cooled airfoils have reached the stage of having been extensively tested in turbines. Most of the tests on transpiration-cooled turbine airfoils have been conducted on airfoils fabricated from woven wire cloth that simulates a transpiration surface. Moskowitz and Lombardo[37] have run such airfoils for several hours at 2750°F turbine inlet temperature.

In spite of the desirability of the cooling method, transpiration-cooled airfoils have not received widespread acceptance and utilization, for reasons that are mostly structural in nature. Porous metals are generally not strong enough to make viable airfoils. The woven wire described above is stronger, but is still difficult to manufacture into good aerodynamic shapes with adequate smoothness. There is also a large amount of surface area on the metal that causes oxidation to be a problem, which must be offset with lower metal temperatures. In spite of these problems, there are still strong proponents of transpiration-cooled airfoils and, with effort, they may yet prove their case.

5.4 End Wall Cooling

The end wall regions shown in Fig. 5.33 are generally exposed to a lower gas temperature than the corresponding airfoils, as was discussed in Sec. 5.2. In the past when the turbine inlet temperature levels were low enough, the temperature profile could be so shaped that no end wall cooling was required. However, as gas temperatures have increased, cooling has become required on the end walls. This is a particularly difficult area to cool because of the adjacent complex flowfields.

The environment for each of the end wall regions is somewhat different. The stator platforms must accept a random temperature distribution, just as the stator vanes must. The stationary shroud around the blade tips is exposed to a similarly random temperature field, which is further disturbed by blade tip passing. In addition, the stationary shroud is exposed to a gas with a decreasing total temperature because of work being performed on the rotor. The rotor platform receives the benefit of an average temperature because of the rotation.

Much of the literature on turbine cooling has consisted of work on turbine airfoils. Literature on the end wall region is very limited. There are some data available from engine manufacturers, but again less so than on airfoils. At the present time, the state-of-the-art is such that an end wall

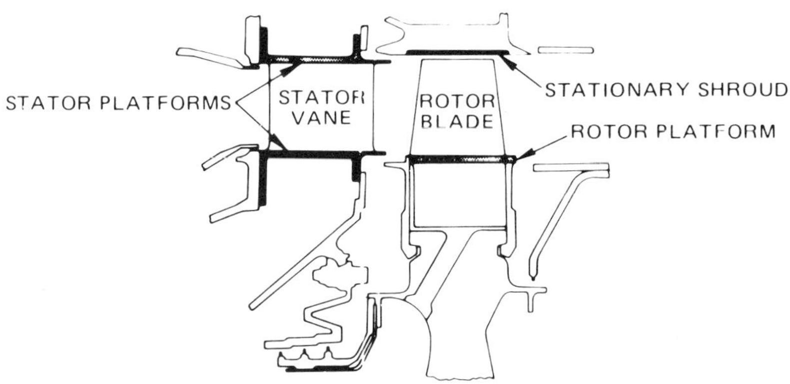

Fig. 5.33 End wall regions.

cooling configuration is designed by some relatively crude design process, after which the configuration must be tested and modified.

Stator Platforms

The stator platforms are subjected to strong secondary cross flows. These cross flows are caused by vorticity in the mainstream inlet flow and by viscous effects within the passage. Blair[38] has published what is probably the most detailed experimental data on stator end walls. He measured the film effectiveness and heat-transfer coefficients for a first stator vane passage. The tests were run on a single passage at large-scale low-velocity flow conditions. The results are shown in Figs. 5.34 and 5.35.

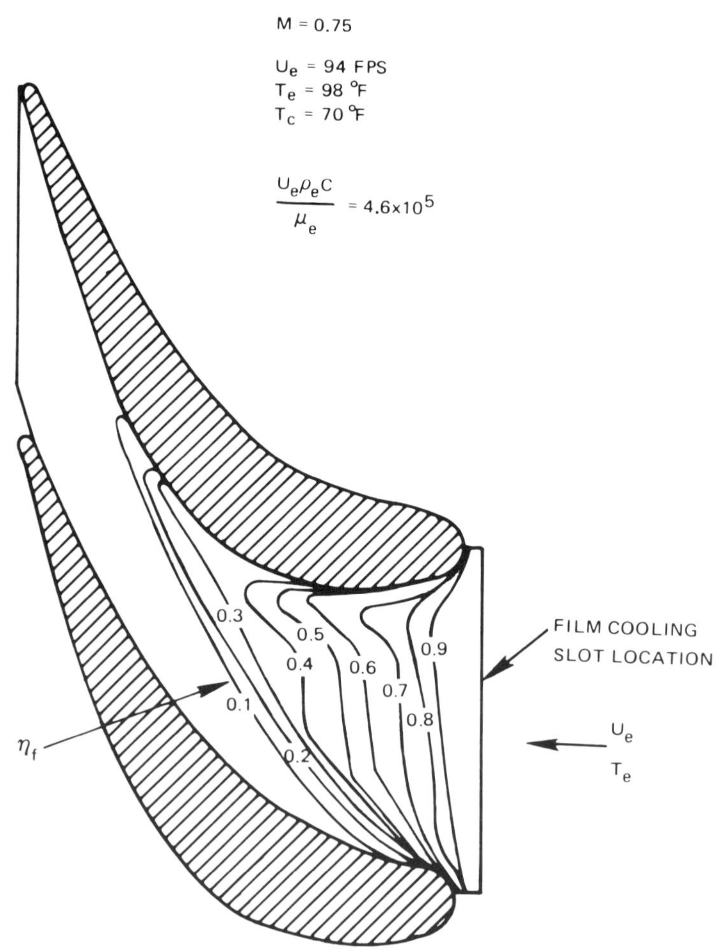

Fig. 5.34 End wall film cooling data (from Ref. 38).

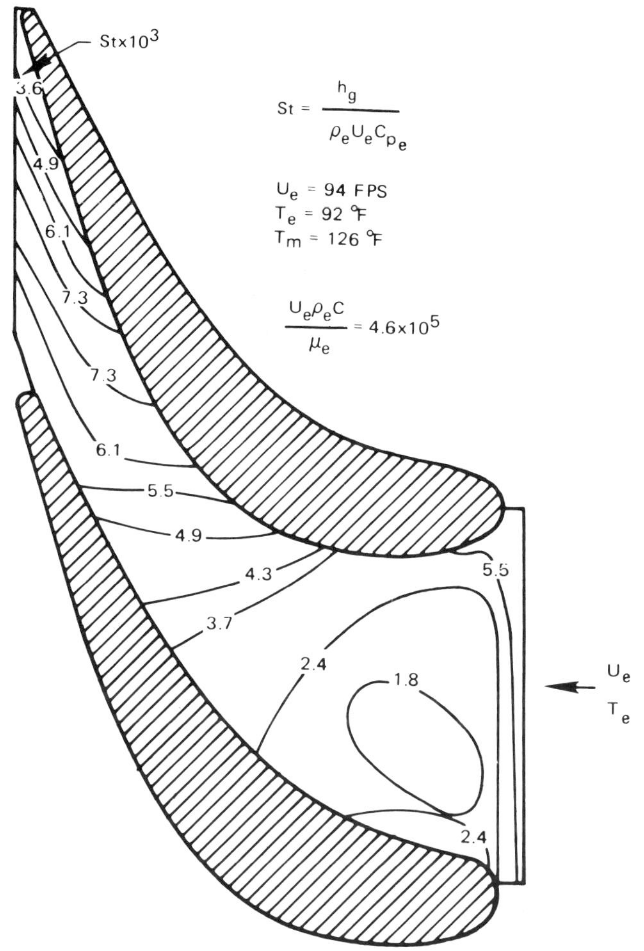

Fig. 5.35 End wall heat-transfer distribution (from Ref. 38).

The effect of film cooling from an upstream slot is shown on Fig. 5.34. The most interesting feature of this figure is that the cross-channel flow drives the film away from the pressure surface so strongly that little of the upstream slot film goes very far down the channel near the pressure surface. For this reason, film cooling of the stator platforms must utilize further replenishment within the channel.

The heat-transfer coefficient distribution is shown in Fig. 5.35. As one might expect, the heat-transfer coefficient is highest where the velocity is the highest. In Ref. 38 an attempt was made to rationalize the heat-transfer coefficient distribution in terms of the airfoil heat-transfer coefficient distribution. The results indicated that the level of the heat-transfer coefficient distribution on the end wall is close to that on the airfoil forming the passage. Louis[39] obtained heat-transfer coefficient data on stator platforms in a turbine operating under transient blowdown conditions. His data were less detailed than that in Ref. 38 because of the small size of his turbine, but he did measure heat-transfer coefficients at several locations on a stator platform. His data had similar trends to that of Ref. 38. He also made an attempt at an analysis of the heat-transfer coefficient. In spite of his efforts and those of Blair,[38] the problem of heat transfer to the end walls must be considered beyond analysis at this time.

Little other data exist in the literature on stator platforms. Hopefully, more will be forthcoming to clarify the heat-transfer picture in this complex flow region.

Stationary Shrouds

Very little data have been published on stationary shroud heat transfer. Reference 39 has reported data from the same blowdown turbine facility mentioned in the previous section. Very high heat-transfer coefficients near the leading edge of the blade row, which then tapered off rapidly in the axial direction, were measured. The data did not appear amenable to analysis. The reported heat-transfer coefficients were based on the temperature into the blade row. Therefore, at least some of the drop in the reported heat-transfer coefficients might be attributed to the work being performed by the blade.

Stationary shrouds may require film cooling at high turbine inlet temperatures. Little data are available in the literature on this type of cooling. The data required are very difficult to obtain because it must be obtained in a rotating turbine. Louis[39] has obtained a limited amount of data for this cooling problem, but it is difficult to ascertain the generality of the information.

Rotating Platforms and Shrouds

Heat transfer to rotating platforms and shrouds have received the least attention. At lower temperatures, most turbines have rotating shrouds for aerodynamic performance reasons. At higher temperatures where shroud cooling might be required, however, the designer generally chooses to relieve the blade stress problems caused by rotating shrouds by eliminating them. Thus, rotating shroud heat transfer is of only small interest. However, the platforms at the blade roots cannot be eliminated and may require some cooling. This problem has received little attention because several factors tend to diminish it: (1) the gas temperature has a radial profile such that the platform sees a lower temperature than the rotor blade; (2) the rotor blade platform tends to see an average temperature rather than a local temperature; and (3) the relative total temperatures incident on a given rotor platform are lower than those incident on the corresponding vane platform because of vane cooling air dilution and relative velocity effects. Thus, only at very high turbine inlet temperatures is rotor platform cooling required. At present, there appears to be no published data on this cooling problem. The only recourse appears to be to apply stator vane-type data and analyses to design rotor blade platforms. Since only meager data are available in the literature for stator vane platforms, this is not all that easy to do. There is, of course, the remaining question of whether the effects of rotation are significant on the end walls.

5.5 Conclusions

The state-of-the-art in turbine cooling is that turbines with inlet temperatures of the order of 2500°F can be successfully designed. This is seen most dramatically by observing that gas turbine engines in this temperature class are in successful commercial operation. In previous sections, the open literature data and analyses that form the background for the design of such engines have been discussed. A point mentioned several times but worth repeating is that, in addition to the open literature, much data and analyses exist which are proprietary to engine companies. Without this additional information, it is questionable whether these engines could have been successfully built.

The state-of-the-art in airfoil profile convective cooling is probably the most advanced with relatively accurate designs currently possible. Much of the information for this appears in the open literature. The state-of-the-art in film cooling is not as far advanced. Enough design information exists so that successful turbines can be built utilizing film cooling, but a certain amount of testing after design is required to assure adequate airfoil life. The actual design data for this are not in the open literature, although much background information is. Transpiration cooling, which intuitively appears to be the most efficient means of cooling, has not achieved great success. This appears to be more for structural reasons than for cooling problems. The end walls, which fortunately require less cooling, are several years behind airfoils in cooling technology. Because of the complexity of end wall flows, this will probably always be so.

5.6 Recent Advances

In the period since 1976 when the preceding material was written, turbine inlet temperatures have risen only about 200 °F. This is in contrast to what might have been expected from Fig. 5.1. The technology efforts have been directed toward improving existing designs and design concepts for greater durability and aerodynamic performance, rather than toward radically higher temperatures. Much of this effort has been in the form of design, fabricate, test, and redesign by the various engine companies. While this is a fascinating process in itself, it is not suitable for discussion here. On a more fundamental basis are parallel research efforts to provide an expanded data base and a better physical understanding of the turbine cooling process. These latter efforts have been well documented in the literature and are discussed here.

Airfoil Heat-Transfer Coefficients

Heat-transfer coefficients to turbine airfoil surfaces have received considerable attention. Data have been generated in cascades and rotating turbines and improved analytical procedures have been developed. Only the experimental results are discussed here, since the analytical procedures are extensively discussed in Chap. 6.

References 40–42 report heat-transfer coefficient data on airfoils in plane cascades. References 40 and 41 report the results of tests on airfoil contours appropriate for rotating blades. The effects of Reynolds number, turbulence level, and Mach number on the heat-transfer coefficients were measured. In these tests, the pressure distributions were measured and the values of the ratio of the total temperature of the freestream to the temperature of the wall were controlled to be approximately what would exist in an engine. Reference 42 reports the results of tests on airfoil contours that are appropriate for a first stator vane. The effects of the Reynolds and Mach numbers with only small variations in a high level of freestream turbulence were measured. As in Refs. 40 and 41, airfoil pressure distributions were measured and the values of the ratio of the total temperature of the freestream to that of the wall were controlled to be approximately what would exist in an engine. However, in addition, this temperature ratio was also varied to a small degree. Attempts were made in the above references to match the data with boundary-layer analyses. None of the procedures used were able to completely match the data, but some came reasonably close. The primary value of these data must be considered to be for calibrating future analyses.

Other recent works[43-45] report data for heat-transfer coefficients on turbine airfoils in a rotating turbine stage where interaction effects between stator vanes and rotor blades can be expected. Some interaction effects were found, but their significance to airfoil cooling design has not yet been completely determined.

The unshrouded blade tip cooling problem mentioned in Sec. 5.3 has proved to be a serious one. Some research is currently being conducted on this problem, but only a limited quantity of information has yet been published. One important piece of information is Ref. 46, which indicates that the flow over the tip creates heat-transfer coefficients similar to those which would exist without the relative motion of the blade tip with respect to a stationary wall.

Airfoil Internal Convective Cooling

Recent investigations of airfoil internal convective cooling have produced results in the literature that are very applicable to practical designs. The investigations were primarily experimental and were directed toward specific geometries, including impingement with cross flow, pin fins, and turbulators. Some work has also been reported on the effects of Coriolis and centrifugal forces.

Impingement with cross flow has been systematically and extensively investigated by Metzger and his co-workers and is reported in Refs. 47–50.

Heat-transfer coefficients were determined for rectangular and staggered array impingement hole patterns and were measured for the cases where all the cross flow was from spent impingement air and where there was an initial cross flow. The latter case examined the effects of the initial cross flow being at a temperature different from the impingement flow. Specific correlations may be obtained from the references, which should be very useful in improving the calculation of impingement heat transfer. Several features of the data and correlations are of interest:

(1) Rectangular array, or inline, holes appear to be superior in heat-transfer performance to staggered arrays. This is, of course, with all other parameters being the same, such as hole size, hole number, etc.

(2) The effects of an initial cross flow at a temperature different from the impingement flow was handled by defining an adiabatic wall temperature similar to that used in film cooling. The method appears to be a reasonable approach to this problem where heat transfer is governed by three temperatures—cross flow, impingement air, and metal surface.

(3) It was shown that a simple one-dimensional momentum model allows accurate calculation of the impingement flow distribution.

A significant quantity of data on the heat-transfer coefficient and pressure drops with the short pin fins used in airfoils has been published. References 51–54 report experimental results on pin fins with height-to-diameter ratios of 0.5–4.0. The data and resulting correlations are all similar to each other. One important feature of the correlations is that the heat-transfer coefficients are always significantly less than would have been predicted from the long pin fin data existing in the literature. It was noted in Ref. 51 that the distribution of heat-transfer coefficients in the streamwise direction is such that the coefficient initially increases from the first to about the fourth row of pins and then decreases slowly. In Ref. 55 it was shown that it may be worthwhile to put the pins only part way across the channel. Having the pins protruding from the surface, but not extending across the channel, decreases pressure drops and heat-transfer coefficients. However, the pressure drops are more strongly reduced than are the corresponding heat-transfer coefficients.

Additional information on turbulators has been reported in Ref. 56. There are several findings of interest to airfoil cooling designers. It was found that both friction factors and heat-transfer coefficients reach a maximum at a turbulator pitch-to-height ratio of 10. Furthermore, it was found that a trapezoidal cross-sectional turbulator gives about the same heat-transfer coefficients, but lower friction factors, than a square turbulator. Finally, it was found that turbulators oriented 45 deg to the flow, as opposed to the 90 deg usually used, give much reduced friction factors while maintaining virtually the same high levels of heat-transfer coefficients.

There is a growing awareness that centrifugal and Coriolis forces can cause significant secondary flows in rotating blade convective cooling systems. These secondary flows should in turn affect the heat-transfer coefficients. In Ref. 57, data have been reported for experiments where the rotational effects were present in flow in a tube rotating about an axis normal to the centerline of the tube. It was found that the average heat-transfer coefficients can be either increased or decreased by the effects of rotation.

Film Cooling

Film cooling from rows of holes has continued to receive a significant research effort. As in the past, the work at the University of Minnesota represents a logical sequence of research from which one may gain insight into film cooling. References 58–61 report some of their work on film cooling. In Ref. 58, film cooling data in an airfoil cascade are reported. Using the same film cooling geometry reported in Refs. 14–16, film effectiveness on the pressure and suction sides of airfoils was measured. It was found that at the coolant flow ranges over which gas turbines usually operate, the suction side has a higher film effectiveness and the pressure side a lower film effectiveness than a flat plate. The use of a double row of holes, as opposed to a single row, was investigated in Ref. 59. It was found that, for a given quantity of coolant flow, the double row is superior. In Refs. 60 and 61, the effects of freestream turbulence on film effectiveness were investigated. Both the turbulence intensity and scale were found to affect film effectiveness in a complex manner that depends on the blowing rate. At this point, it is not clear how this information on turbulence can be utilized in a cooling design.

Heat fluxes in the presence of film cooling have been measured in the same facilities used to measure the convective heat-transfer coefficients reported in Refs. 40 and 41. Typical results are given in Refs. 41 and 62. The film cooling results are reported in terms of the ratio of the heat flux in the presence of film cooling to that without film cooling. This ratio of heat flux values can be related to film effectiveness and overall cooling effectiveness, but the necessary additional information is generally not precisely known and must be assumed. Reference 62 discusses the problems involved in establishing this relationship.

Analytical methods have been applied to the film cooling problem. An example of this is given in Ref. 63 where the film effectiveness has been calculated downstream of a row of holes. The results appear promising but, as the authors point out, the numerical method deteriorates at the mass flux ratios of practical interest (> 0.5). With the current strong emphasis on numerical methods, it is expected that improvements in this direction will be forthcoming in the not too distant future.

Film cooling experiments on a rotating blade have been reported in Ref. 64. It was found that the suction surface film cooling behaved very much as in a plane cascade, while the pressure surface was quite different. On the pressure surface, the film cooling from a row of holes near the leading edge skewed very strongly outward. Another work[65] by the same investigators led them to believe that this skewing was caused by an inviscid effect of the freestream flowfield. These observations indicate that future designs should include considerations of the three-dimensional nature of the flow in the airfoil rows.

Multihole film cooling has not proved to be a widely used cooling method, except at the leading edges of airfoils. The problems of high cost and potential hole plugging mentioned in Sec. 5.3 have caused only limited acceptance of multihole film cooling. At the leading edge, where it is commonly used, larger holes ($\approx$ 0.020 in.) relieve these problems. The use of larger holes represents a heat-transfer compromise, one accepted because there are few alternatives at the leading edge.

Even though leading-edge film cooling is an important problem area, only a limited quantity of data has been published on it. Reference 66 reports data on a leading-edge type of configuration in terms of the ratio of heat fluxes with and without a film. Therefore, the data must be treated with the same caution discussed above. Also, the ratio of hole diameter to cylinder diameter suggests that the data are most applicable to leading edges with very small holes.

Aerodynamic Effects of Cooling

There have been several investigations to determine the effects of the trailing-edge discharge of coolant on the base pressure of turbine airfoils. Reference 67 is representative of these works. In Ref. 67, it was found that base pressure first increased, reached a maximum, and then decreased with coolant flow. In the range of coolant flow tested (and of interest to the cooling problem), the base pressure was always above that with no coolant flow. This increase in base pressure tends to improve the aerodynamic performance of turbine airfoils.

End Wall Cooling

End wall heat transfer has received considerable attention. Reference 68 reports experiments to measure the heat-transfer coefficients to the first stator-type end wall regions in plane cascades. The effects of Reynolds number, Mach number, inlet turbulence level, and boundary-layer thickness were all measured. A large quantity of data was generated and placed into a computerized data base. Reference 69 reports flow visualization studies of the same geometry tested in Ref. 68. Reference 70 reports data of a similar nature in a rotor blade configuration with incompressible conditions and a fixed Reynolds number and turbulence intensity. Only the inlet boundary-layer thickness was varied. These investigations identified the effects of an inlet horseshoe vortex, which were to create a local region of high heat-transfer coefficients near the leading edge. The effect of the horseshoe vortex as it passed downstream was not very strong in the rest of the passage, however. Another feature was a region of high heat-transfer coefficients just downstream of the trailing edge. It was also found in Ref. 70 that the heat transfer to the suction side of the airfoils near the end walls was strongly affected by the end wall secondary flows. Perhaps of equal importance, it was also observed that the pressure surface behaved very much like a two-dimensional flow all the way to the end wall.

The data of Refs. 68–70 were gathered for the purposes of gaining greater insight into the heat-transfer phenomena and for providing data to check calculation procedures. Some understanding has been gained, but numerical procedures to compute three-dimensional flowfields and the resulting heat-transfer coefficients have not yet been well developed. Some attempts have been made to make such computations, but without conclusive results.

Thermal Barrier Coatings

Airfoils with thermal barrier coatings are considered to have great potential for future applications. These are airfoils fabricated in a conventional manner with the exception of a thin layer (≈ 0.015 in.) of a high-temperature insulating material. Generally stabilized zirconia is used. This material can withstand very high temperatures and has a thermal conductivity less than one-tenth that of conventional superalloys. Airfoils coated with zirconia can run with much less cooling air at a given gas temperature or conversely can run at much higher gas temperatures at a given level of cooling air flow than uncoated airfoils. References 71 and 72 report some of the testing performed with zirconia-coated airfoils.

Designing airfoils with thermal barrier coatings presents some unique problems. Even when polished, the material has an inherent roughness, thereby increasing both the skin-friction and heat-transfer coefficients. References 71 and 72 are somewhat at odds with each other about the magnitude of these effects, with the former indicating only a small effect and the latter a large effect. The presence of a layer of thermal barrier coating also tends to thicken trailing edges, and there is less doubt that this is harmful to aerodynamic performance. One other factor is that the value of the insulating capability depends on the heat flux levels. At sea-level takeoff conditions where temperatures and pressures are high, the heat fluxes are also high and the temperature drop through the insulation is high. At high altitude, the temperatures may still be high, but the heat fluxes are lower because of low pressures. Therefore, the temperature drop through the thermal barrier coating is much less. Design of airfoils with thermal barrier coatings must weigh all of these factors.

Conclusions

The most recent advances in turbine cooling technology have been toward refinements of cooling methods already in use. Improved designs are resulting from these refinements, but much remains to be done before an adequate understanding of turbine cooling will exist. At least one new direction in turbine cooling has emerged in the concept of using a thin layer of insulating material. If the materials should prove to be sufficiently durable, they will allow great increases in turbine inlet temperature and engine performance.

References

[1] Holland, M. J., "Olympus 593 Turbine Cooling," AGARD CP 73, 1970.

[2] Huile, K. R., "A Theoretical Study of Three-Dimensional Compressible Turbulent Boundary Layers on Rotating Surfaces in Turbomachinery," PhD Thesis, Massachusetts Institute of Technology, Cambridge, 1976.

[3] Chupp, R. E., Helms, H. E., McFadden, R. W., and Brown, T. R., "Evaluation

of Internal Heat Transfer Coefficients for Impingement Cooled Turbine Airfoils," *Journal of Aircraft*, Vol. 16, May-June 1969, pp. 203–208.

[4]Metzger, D. E., Baltzer, R. T., and Jenkins, C. W., "Impingement Cooling Performance in Gas Turbine Airfoils Including Effects of Leading Edge Sharpness," *Journal of Engineering for Power*, *Transactions of ASME*, Ser. A, Vol. 94, July 1972, pp. 219–225.

[5]Kercher, D. M. and Tabakoff, W., "Heat Transfer by a Square Array of Round Air Jets Impinging Perpendicular to a Flat Surface Including the Effect of Spent Air," *Journal of Engineering for Power*, *Transactions of ASME*, Ser. A, Vol. 92, Jan. 1970, pp. 73–92.

[6]Colladay, R. S., *Turbine Cooling*, NASA SP-290, Vol. 3, 1975, Ch. 11.

[7]Webb, R. L., Eckert, E. R. G., and Goldstein, R. J., "Heat Transfer and Friction in Tubes with Repeated-Rib Roughness," *International Journal of Heat and Mass Transfer*, Vol. 14, 1971, pp. 601–617.

[8]Halls, G. A., "Nozzle Guide Vane Cooling—The State of the Art," AGARD CP 73, 1970.

[9]Grimison, E. C., "Correlation and Utilization of New Data on Flow Resistance and Heat Transfer for Cross Flow of Gases over Tube Banks," *Transactions of ASME*, Vol. 59, 1937, pp. 583–594.

[10]Wieghardt, K., "Hot Air Discharge for Deicing," Air Material Command, FTS-919-RE, 1946.

[11]Goldstein, R. J., *Film Cooling Advances in Heat Transfer*, Vol. 7, Academic Press, New York, 1970, pp. 321–379.

[12]Goldstein, R. J. and Haji-Sheikh, A., "Prediction of Film Cooling Effectiveness," *JSME 1967 Semi-International Symposium*, Tokyo, 1967, pp. 213–218.

[13]Hartnett, J. P., Birkebak, R. C., and Eckert, E. R. G., *International Developments in Heat Transfer*, Pt. IV, ASME, New York, 1961, p. 682.

[14]Ericksen, V. L., "Film Cooling Effectiveness and Heat Transfer with Injection Through Holes," NASA CR-72991, Aug. 1971.

[15]Goldstein, R. J., Eckert, E. R. G., Ericksen, V. L., and Ramsey, J. W., "Film Cooling Following Injection Through Inclined Circular Tubes," NASA CR-72612, 1969.

[16]Pedersen, D. R., "Effect of Density Ratio on Film Cooling Effectiveness for Injection Through a Row of Holes and for a Porous Slot," PhD. Thesis, University of Minnesota, Minneapolis, March 1972.

[17]Ericksen, V. L., Eckert, E. R. G., and Goldstein, R. J., "A Model for the Analysis of the Temperature Field Downstream of a Heated Jet Injected into an Isothermal Crossflow at an Angle of 90°," NASA CR-72990, 1971.

[18]Liess, C., "Film Cooling with Ejection from a Row of Inclined Circular Holes, An Experimental Study for the Application to Gas Turbine Blades," Tech. Note 97, von Kármán Institute for Fluid Dynamics, March 1973.

[19]Metzger, D. E. and Fletcher, D. D., "Surface Heat Transfer Immediately Downstream of Flush, Non-Tangential Injection Holes and Slots," *Journal of Aircraft*, Vol. 8, Jan. 1971, pp. 33–38.

[20]Lander, R. E., Fish, R. W., and Suo, M., "External Heat Transfer Distribution on Film Cooled Turbine Vanes," *Journal of Aircraft*, Vol. 9, Oct. 1972, pp. 707–714.

[21]Muska, J. W., Fish, R. W., and Suo, M, "The Additive Nature of Film Cooling," *Journal of Engineering for Power*, *Transactions of ASME*, Ser. A, Vol. 98, Oct. 1976, pp. 457–464.

[22]Goldstein, R. J., Eckert, E. R. G., and Burggraf, F., "Effects of Hole Geometry and Density on Three Dimensional Film Cooling," *International Journal of Heat and Mass Transfer*, Vol. 17, 1974, pp. 595–607.

[23]Sellers, J. P., "Gaseous Film Cooling with Multiple Injection Stations," *AIAA Journal*, Vol. 1, Dec. 1963, pp. 2154–2156.

[24]Mayle, R. E. and Camarata, F. J., "Heat Transfer Investigation for Multihole Aircraft Turbine Blade Cooling," AFAPL-TR-73-30, June 1973.

[25]Mayle, R. E. and Camarata, F. J., "Multihole Cooling Film Effectiveness and Heat Transfer," *Journal of Heat Transfer*, Vol. 97, Nov. 1975, pp. 534–538.

[26]LeBrocq, P., Launder, B. E., and Priddin, C. H., "Discrete Hole Injection as a Means of Transpiration Cooling," *Proceedings, Institution of Mechanical Engineers*, Vol. 187, 1973, pp 17–73.

[27]Launder, B. E. and York, J., "Discrete-Hole Cooling in the Presence of Free-Stream Turbulence and Strong Favorable Pressure Gradient," *International Journal of Heat and Mass Transfer*, Vol. 17, 1974, pp. 1403–1409.

[28]Ramsey, J. W., Goldstein, R. J., and Eckert, E. R. G., "A Model for Analysis of the Temperature Distribution with Injection of a Heated Jet into an Isothermal Flow," *Heat Transfer 1970*, Elsevier Publishing Co., Amsterdam, 1970.

[29]Choe, H., Kays, W. M., and Moffat, R. J., "The Superposition Approach to Film Cooling," ASME Paper 74-WA/HT-27, 1974.

[30]Metzger, D. E., Takeuchi, D. I., and Kuenstler, P. A., "Effectiveness and Heat Transfer with Full-Coverage Film Cooling," ASME Paper 73-GT-18, 1973.

[31]Choe, H., Kays, W. M., and Moffat, R. J., "The Turbulent Boundary Layer on a Full-Coverage Film-Cooled Surface: An Experimental Heat Transfer Study with Normal Injection," Thermosciences Div., Dept. of Mechanical Engineering, Stanford University, Stanford, Calif., Rept. HMT-22, May 1975.

[32]Hiroki, T. and Katsumata, I., "Design and Experimental Studies of Turbine Cooling," ASME Paper 74-GT-30, 1974.

[33]Hartsel, J. E., "Prediction of Effects of Mass Transfer Cooling on the Blade Row Efficiency of Turbine Airfoils," AIAA Paper 72-11, Jan. 1972.

[34]Prust, H. W., Schum, H. J., and Szanca, E. M., "Cold Air Investigation of a Turbine with Transpiration-Cooled Stator Blades, I: Performance of Stator with Discrete Hole Blading," NASA TM X-2094, 1970.

[35]Anderson, L. R., Heiser, W. H., and Jackson, J. C., "Axisymmetric One-Dimensional Compressible Flow Theory and Applications," ASME Paper 70-GT-82, May 1970.

[36]Anderson, L. R. and Heiser, W. H., "Systematic Evaluation of Cooled Turbine Efficiency," ASME Paper 69-GT-63, March 1969.

[37]Moskowitz, S. L. and Lombardo, S., "2750 Degree F Engine Test of a Transpiration Air-Cooled Turbine," *Journal of Engineering for Power, Transactions of ASME*, Ser. A, Vol. 93, April 1971, pp. 238–248.

[38]Blair, M. F., "An Experimental Study of Heat Transfer and Film Cooling on Large Scale Turbine Endwalls," *Journal of Heat Transfer*, Vol. 96, Nov. 1974, pp. 524–529.

[39]Louis, J. F., "Heat Transfer in Turbines," AFAPL-TR-75-107, Sept. 1975.

[40]Consigny, H. and Richards, B. E., "Short Duration Measurements of Heat Transfer Rate to a Gas Turbine Rotor Blade," ASME Paper 81-GT-140, March 1981.

[41]Daniels, L. C., "Film Cooling of Gas Turbine Blades," Dept. of Engineering Science, University of Oxford, Oxford, England, Rept. 1302/79, May 1978.

[42]Nealy, D. A., Mihelc, M. S., Hylton, L. D., and Gladden, H. J., "Measurements of Heat Transfer Distribution over the Surfaces of Highly Loaded Turbine Nozzle Guide Vanes," ASME Paper 83-GT-53, March 1983.

[43]Dring, R. P., Hardin, L. W., Joslyn, H. D., and Wagner, J. H., "Research on Turbine Rotor-Stator Aerodynamic Interaction and Rotor Negative Incidence Stall," AFWAL-TR-81-2114, Nov. 1981.

[44]Dunn, M. G. and Hause, A., "Measurement of Heat Flux and Pressure in a Turbine Stage," ASME Paper 81-GT-88, March 1981.

[45]Dunn, M. G., Rae, W. J., and Holt, J. L., "Measurement and Analysis of Heat Flux Data in a Turbine Stage, Part II: Discussion of Results and Comparison with Predictions," ASME Paper 83-GT-122, March 1983.

[46]Mayle, R. E. and Metzger, D. E., "Heat Transfer at the Tip of an Unshrouded Turbine Blade," *Proceedings, 7th International Heat Transfer Conference*, Munich, 1982, Vol. 3, pp. 87–92.

[47]Metzger, D. E., Florschuetz, L. W., Takeuchi, P. I., Behee, R. D., and Berry, R. A., "Heat Transfer Characteristics for Inline and Staggered Arrays of Circular Jets with Crossflow of Spent Air," *Journal of Heat Transfer*, Vol. 101, Aug. 1979, pp. 526–531.

[48]Florschuetz, L. W., Berry, R. A., and Metzger, D. E., "Periodic Streamwise Variations of Heat Transfer Coefficients for Inline and Staggered Arrays of Circular Jets with Crossflow of Spent Air," *Journal of Heat Transfer*, Vol. 102, Feb. 1980, pp. 132–137.

[49]Florschuetz, L. W., Truman, C. R., and Metzger, D. E., "Streamwise Flow and Heat Transfer Distributions for Jet Array Impingement with Crossflow," *Journal of Heat Transfer*, Vol. 103, May 1982, pp. 337–342.

[50]Florschuetz, L. W., Metzger, D. E., and Su, C. C., "Heat Transfer Characteristics for Jet Array Impingement with Initial Cross Flow," ASME Paper 83-GT-28, March 1983.

[51]Metzger, D. E., Berry, R. A., and Brown, J. P., "Developing Heat Transfer in Rectangular Ducts with Staggered Arrays of Short Pin Fins," *Journal of Heat Transfer*, Vol. 104, Nov. 1982, pp. 700–706.

[52]Metzger, D. E., Fan, Z. X., and Shepard, W. B., "Pressure Loss and Heat Transfer Through Multiple Rows of Short Pin Fins," *Proceedings, 7th International Heat Transfer Conference*, Munich, 1982, pp. 137–142.

[53]Van Fossen, G. J., "Heat Transfer Coefficients for Staggered Arrays of Short Pin Fins," *Journal of Engineering for Power, Transactions of ASME*, Ser. A, Vol. 104, April 1982, pp. 268–274.

[54]Brigham, B. A. and Van Fossen, G. J., "Length to Diameter Ratio and Row Number Effects in Short Pin Fin Heat Transfer," ASME Paper 83-GT-54, March 1983.

[55]Peng, Y., "Heat Transfer and Friction Loss Characteristics of Pin Fin Cooling Configuration," ASME Paper 83-GT-123, March 1983.

[56]Han, J. C., Glicksman, L. R., and Rohsenow, W. M., "An Investigation of Heat Transfer and Friction for Rib-Roughened Surfaces," *International Journal of Heat and Mass Transfer*, Vol. 21, 1978, pp. 1143–1156.

[57]Morris, W. D. and Ayhan, T., "Observations on the Influence of Rotation on

Heat Transfer in the Coolant Channels of Gas Turbine Rotor Blades," *Proceedings, Institution of Mechanical Engineers*, Vol. 193, pp. 303–311.

[58]Ito, S., Goldstein, R. J., and Eckert, E. R. G., "Film Cooling of a Gas Turbine Blade," *Journal of Engineering for Power, Transactions of ASME*, Ser. A, Vol. 100, 1978, pp. 476–481.

[59]Jabbari, M. Y. and Goldstein, R. J., "Adiabatic Wall Temperature and Heat Transfer Downstream of Ingestion Through Two Rows of Holes," *Journal of Engineering for Power, Transactions of ASME*, Ser. A, Vol. 100, April 1978, pp. 303–307.

[60]Kadotani, K. and Goldstein, R. J., "On the Nature of Jets Entering a Turbulent Flow, Part A—Jet-Mainstream Interaction," *Journal of Engineering for Power, Transactions of ASME*, Ser. A, Vol. 101, July 1979, pp. 459–465.

[61]Kadotani, K. and Goldstein, R. J., "On the Nature of Jets Entering a Turbulent Flow, Part B: Film Cooling Performance," *Journal of Engineering for Power, Transactions of ASME*, Ser. A, Vol. 101, July 1979, pp. 466–470.

[62]Richards, B. E., Ville, J. P., and Appels, C., "Film Cooled Small Turbine Blade Research—Film Cooling Effectiveness at Simulated Turbine Conditions," von Kármán Institute for Fluid Dynamics, Tech. Note 120, June 1976.

[63]Bergeles, G., Gossman, A. D., and Launder, B. E., "The Prediction of Three Dimensional Discrete-Hole Cooling Processes, Part 2: Turbulent Flow," *Journal of Heat Transfer*, Vol. 103, Feb. 1981, pp. 141–145.

[64]Dring, R. P., Blair, M. F., and Joslyn, H. D., "An Experimental Investigation of Film Cooling on a Turbine Rotor Blade," *Journal of Engineering for Power, Transactions of ASME*, Ser. A, Vol. 102, Jan. 1980, pp. 81–87.

[65]Dring, R. P. and Joslyn, H. D., "Measurement of Turbine Rotor Blade Flows," *Journal of Engineering for Power, Transactions of ASME*, Ser. A, Vol. 103, April 1981, pp. 400–405.

[66]Luckey, D. W. and L'Ecuyer, M. R., "Stagnation Region Gas Film Cooling: Spanwise Angled Injection from Multiple Rows of Holes," NASA CR 165333, April 1981.

[67]Sieverding, C. H., "The Influence of Trailing Edge Ejection on the Base Pressure in Transonic Turbine Cascades," ASME Paper 82-GT-50, April 1982.

[68]York, R. E., Hylton, L. D., and Mihelc, M. S., "An Experimental Investigation of Endwall Heat Transfer and Aerodynamics in a Linear Vane Cascade," ASME Paper 83-GT-52, March 1983.

[69]Gaugler, R. E. and Russell, L. M., "Comparison of Visualized Turbine Endwall Secondary Flows and Measured Heat Transfer Patterns," ASME Paper 83-GT-83, March 1983.

[70]Graziani, R. A., Blair, M. F., Taylor, J. R., and Mayle, R. E., "An Experimental Study of Endwall and Airfoil Surface Heat Transfer in a Large Scale Turbine Blade Cascade," *Journal of Engineering for Power, Transactions of ASME*, Ser. A, Vol. 102, April 1980, pp. 257–267.

[71]Liebert, C. H. and Stepka, F. S., "Ceramic Thermal Barrier Coatings for Cooled Turbines," *Journal of Aircraft*, Vol. 14, May 1977, pp. 487–493.

[72]Liang, G. P. and Fairbanks, J. W., "Heat Transfer Investigation of Laminated Turbine Airfoils," Paper presented at ASME Winter Annual Meeting, 1978, pp. 21–30.

CHAPTER 6. COMPUTATION OF TURBOMACHINERY BOUNDARY LAYERS

Henry McDonald
Scientific Research Associates, Inc., Glastonbury, Connecticut

6. COMPUTATION OF TURBOMACHINERY BOUNDARY LAYERS

6.1 Introduction

In this chapter, the term "boundary layer" is applied very loosely in the generic sense of a shear layer that is not necessarily thin, rather than the quite precise terminology usually implied in classical external aerodynamics. Excluded from this very broad category of flows are shear layers whose behavior can be adequately described by assuming that the transport of momentum or energy across the streamlines of the mean flow is quite negligible. The present chapter is devoted to describing the concepts and procedures available to allow the prediction of the behavior of these "boundary layers" in the diverse and special circumstances found in turbomachinery.

In view of the crucial role of boundary layers in setting loss levels, heat-transfer rates, and operating limits, there can be no doubting the desire of the turbomachinery designer to understand and predict the behavior of these boundary layers as they are influenced by flow and geometric changes. Despite this desire, boundary-layer prediction schemes are not, with minor exceptions, at present in extensive use in practical design systems. Real engines have been found to introduce complexities that cannot be allowed for in the prediction schemes and that subsequently dominate the flow behavior. Fortunately, it appears that in the near future this rather sad state of affairs will be changed, due to the combined development of efficient methods for solving multidimensional nonlinear systems of partial differential equations, together with recent advances in the modeling of turbulent transport processes. In the subsequent sections, a hierarchy of techniques will be evolved, all based on using modern computers to effect the solution. The particular turbomachinery application of the various techniques will be introduced, together with their current status and shortcomings and, since many of the techniques are still being evolved, a prognosis for their eventual success will be given.

In developing the solution hierarchy, it is first convenient to divide the problem into its two major constituents: (1) the problem of developing and then solving the set of equations purporting to describe the problem at hand, and (2) the problem of describing the turbulent transport of momentum and energy to some reasonable degree of accuracy for that problem. The first part of the problem can be broken down further on the basis that certain physical approximations greatly simplify the numerical process of solving the set of equations. In particular, the neglect of streamwise diffusion in the equations is usually a very good approximation. When this is followed by the assumption that the pressure forces can be approximated either locally by mass conservation (i.e., a blockage correction) and/or without an a priori knowledge of the boundary-layer behavior, the governing equations for time-averaged steady flow are greatly simplified and can be solved relatively easily by forward marching procedures. Thus, a solution hierarchy is developed based, first of all, on one-, two-, and three-dimensional marching schemes followed by techniques in one, two, and three space dimensions that cover problems which do not permit the aforementioned simplification and require both the upstream and downstream boundary conditions to be satisfied. Turbulence models are treated separately, although in the literature a given model is often associated with a particular scheme. As a practical matter, the various turbulence models are usually interchangeable.

Turning to the special problems that arise as a consequence of the operating mode of the gas turbine, it turns out that the inlet conditions to many components of interest are often unsteady and more often than not contain total pressure and total temperature variations, usually distributed in a highly three-dimensional manner. In a time-averaged frame (which can either be rotating with the blades or stationary on a stator or vane), such unsteadiness can be viewed as "inlet turbulence" with, in some instances, both ordered and random components in the motion. It is to be expected that the response of the flow to this "inlet turbulence" will vary, depending on the composition of the turbulence. Primary measures of turbulence such as rms intensity and/or some spatial scale would be expected to give only some sort of zeroth-order estimate of its effect. Within the device itself, blades and vanes are noticeably more cambered than is usual with conventional wing sections. This camber leads to the possibility of curvature effects on the turbulence structure within the boundary layer itself, but fortunately in all but film-cooled turbine blades the effect of curvature on the turbulence structure may probably be of secondary importance when compared to other unknowns. However, the camber and the consequent large amounts of flow turning, such as occurs in the turbine, can cause the annulus wall (hub and casing) boundary layers to overturn within the blade or vane passageway, leading to pronounced boundary-layer migrational effects that can eventually result in streamwise vortices of considerable strength. Streamwise vortices can also be produced in the region of vane or blade intersection with the annulus wall boundary layers—the so-called horseshoe vortices. The appearance and the subsequent fate of such streamwise vortices appear to have a considerable impact on the aerodynamics and heat-transfer rates within the device; their presence usually signifies that conventional boundary-layer concepts are no longer valid. These and related problems are discussed in Chaps. 4 and 5 of this volume.

Insofar as the operating conditions of the compressor and turbine are concerned, it appears that in modern compressor technology the aim is to obtain higher loadings, leading to the airfoils working closer and closer to a catastrophic flow breakdown (stall) somewhere in the machine. It turns out that the operating limits of highly loaded stages are very difficult to predict and the stage performance appears to be only loosely related to the two-dimensional airfoil boundary-layer characteristics. These facts have led compressor designers to place comparatively little reliance on boundary-layer calculations and to develop a correlation approach based on rig tests. On the other hand, turbines accelerate the flow and consequently the problem of operating close to stage stall is mitigated. However, as is pointed out in Chap. 4 of this volume, separation can still be a problem—even in turbines. Further, the local heat transfer on the vanes or blades is of considerable design interest and dominated by the boundary layer. In view of the relative success of two-dimensional boundary-layer analyses in predicting the heat-transfer rates in the turbine, turbine designers have come to place a greater reliance on boundary-layer analyses. More recent developments in turbine technology (such as even higher turbine inlet temperatures, film cooling, and higher work extraction levels per stage) have led to a commensurate interest by designers in developing and using appropriate analyses. In particular, the higher work extraction levels and the accompanying high levels of flow acceleration cause extended regions of transition from laminar to turbulent flow (and vice versa). The various practical film cooling schemes introduce three-dimensional effects and the higher turbine inlet temperatures make it necessary to know with some precision the intrablade hub and casing heat-transfer distributions.

The foregoing problems are in many ways much more difficult to treat than the boundary-layer problems encountered in external aircraft aerodynamics. The wide variety of possible configurations open to the turbomachinery designer, and the probable impact of the boundary layer on any selected configuration, leads one to suspect that the correlation (i.e., the extrapolation approach) would be both uncertain and expensive. Further development of boundary-layer analyses seems entirely justified, even if it results in an improvement in only the correlation approach to the design of turbomachinery.

In the subsequent sections, attention is first devoted to both integral and finite difference procedures for predicting the two-dimensional blade or vane boundary layers. Here the aim is to provide the means of estimating the airfoil section loss coefficient and the detailed distribution of heat-transfer coefficient around the airfoil. This is followed by a discussion of the axisymmetric pitch-averaged equations and the procedures available for solving this set of equations and their role in predicting the pitch-averaged, i.e., circumferentially-averaged hub and casing boundary-layer behavior. Next, the rather limited role played by conventional three-dimensional boundary layers in turbomachinery will be introduced and the available schemes, both integral and finite difference, discussed. The much enlarged capability of what is here termed the extended three-dimensional boundary-layer procedures, sometimes termed (rather euphemistically) the parabolized Navier-Stokes equations, will then be described and the problems of this concept discussed. The difficulties arising from boundary-layer separation are then introduced and the status of the very limited number of schemes for treating this problem as rigorously as possible reviewed. Finally, the problems of turbulence modeling and the current approaches to this most difficult of topics will be reviewed. Each topic is discussed from the point of view of the special problems of turbomachinery applications and the aim is to give the reader guidance in selecting an approach, concept, or procedure suitable for his particular problem.

6.2 Two-Dimensional or Axisymmetric Boundary Layers

The turbomachinery designer has a very profound interest in vane and blade boundary layers from the heat-transfer and loss point of view. Further, the flow in the various ducts and the resulting total pressure losses are of a similar considerable interest. In spite of the often very noticeable three-dimensional variations present in both of the foregoing instances and largely as a result of the absence up until now of better simulations, two-dimensional and axially symmetric boundary-layer analyses have been applied to these problems. In the present section, both integral and finite difference techniques for predicting the two-dimensional vane boundary-layer behavior will be discussed. Finally, the additions required to describe axisymmetric flow, both swirling and nonswirling, will be introduced.

Integral Methods for Predicting the Blade Boundary Layers

In view of the highly satisfactory status of methods which directly and numerically integrate the boundary-layer partial differential equations of motion, it seems at first sight incongruous at this point in time to dwell upon the so-called integral procedures for predicting boundary-layer development. The arguments against the further development or application of integral methods seem very persuasive. Direct numerical treatment of the governing partial differential equations of motion can now be performed routinely, and in view of the general availability and comparatively low cost of modern computers, quite cheaply compared to, say, the cost of engineering man-hours. The potential cost savings of integral procedures are dependent upon utilization and code construction and, in some applications, an order-of-magnitude reduction in an already small computing cost is not a significant factor. Ease of use is not usually a deciding factor either, since a number of the direct numerical procedures have been made to operate, upon user request, with the same identical input usually demanded by integral procedures, together with such features as optional automatic grid selection. The direct procedures, by virtue of their not requiring the *a priori* adoption of velocity, temperature, and turbulence profile families, certainly contain less empiricism than the integral methods and, thereby, focus attention upon the essential problem of the turbulent boundary layer, that is, adequate specification of the turbulent transport mechanism. Finally, it is claimed by their protagonists that the direct procedures are much more general and flexible with regard to such items as boundary conditions and inlet profiles, so that features such as heat transfer, wall transpiration, rough walls, and film cooling, for instance, may be readily incorporated into the procedure, subject solely to the accuracy of the boundary-layer approximations and the turbulent transport model.

In spite of the foregoing, however, a case for the continued development and application of integral methods for predicting boundary-layer behavior in some instances does exist. It is becoming apparent, for example, that a whole category of flow problems arise where a rapid and often iterative estimate of the boundary-layer growth is required as a subtask in a procedure for predicting, say, the pressure field in or around a body. In such instances, provided the required degree of accuracy is attainable, the potential cost savings of the integral procedure might be very considerable and hence desirable. Although the cost savings attributed to the integral procedures are usually thought of as arising from the reduced use of the computer, the engineering labor required to code and debug an integral procedure can be one or more orders of magnitude smaller than that required by the better direct numerical procedures. However, since detailed listings of a number of satisfactory direct procedures are available in the open literature, the code construction cost savings may not be realized. Also, it does not follow that the use of empirical information, such as velocity profile families, necessarily degrades a prediction; this is the case only when the empirical input is inaccurate or inappropriate and the parameters of interest depend upon the empirical input. It is true, however, that the necessity of supplying this additional empirical information does limit integral techniques to those problems where such empirical information exists and has been suitably correlated. Here, the degree of collapse to which the empirical correlations must adhere is dictated solely by the user's overall required predictive accuracy and this, of course, is the user's perogative to decide. However, it does seem clear that, for instance, the displacement thickness over a smooth shock-free two-dimensional unseparated airfoil without heat transfer at high Reynolds numbers may be predicted quite satisfactorily by a number of simple integral procedures. On the other hand, if the problem is changed to estimate the heat transfer to the same airfoil with a rapidly varying wall temperature distribution typical of those encountered in gas turbine operations, few, if any, of the currently available integral procedures could be relied upon to provide an acceptable prediction of the heat-transfer rate. The reason for the failure in the presence of heat transfer is the inadequacy of the presently available temperature profile families when the wall temperature varies rapidly.

In the subsequent discussion, an attempt will be made to delineate those areas where present integral methods might be expected to be inaccurate as a result of the inadequate additional empirical information required, relative to direct procedures. At the same time, some integral procedures possess characteristic features that are of considerable importance in the convenient application of the procedure. These desirable features will also be emphasized. Also, certain integral procedures can be fashioned to permit incorporation of turbulence models of the same type as are currently being developed for the direct numerical procedures. Such features are obviously attractive and they, too, will be emphasized in the subsequent development.

The momentum integral equation. As is well known, the basic technique of deriving an infinite family of integral momentum equations from the partial differential equations of motion consists of multiplying the partial differential equations by a factor $y^n u^m$ and integrating the equations in the coordinate direction normal to the wall. When $n = m = 0$, the von Kármán momentum integral equation is obtained and, if the external flow is varying slowing in time compared to the typical turbulent velocity fluctuations,[1] this equation can be written for compressible flow, neglecting the Reynolds normal stresses, as

$$\frac{\partial\theta}{\partial x} + \frac{\theta}{u_e}\frac{\partial u_e}{\partial x}(2+H) + \frac{\theta}{\rho_e}\frac{\partial\rho_e}{\partial x} + \frac{1}{u_e}\frac{\partial\delta^*}{\partial t} + \frac{\delta^*}{u_e^2}\frac{\partial u_e}{\partial t}$$

$$+ \frac{\delta^*}{\rho_e u_e}\frac{\partial\rho_e}{\partial t} - \frac{1}{\rho_e u_e}\frac{\partial\rho_e}{\partial t}\theta_\rho - \frac{1}{u_e}\frac{\partial\theta_\rho}{\partial t} = \frac{c_f}{2} + c_Q \tag{6.1}$$

where

$$\rho_e\frac{\partial u_e}{\partial t} + \rho_e u_e\frac{\partial u_e}{\partial x} = -\frac{\partial p}{\partial x}$$

$$\theta = \int_0^\delta \frac{\rho u}{\rho_e u_e}\left(1 - \frac{u}{u_e}\right)\mathrm{d}y$$

$$\delta^* = \int_0^\delta \left(1 - \frac{\rho u}{\rho_e u_e}\right)\mathrm{d}y$$

$$\theta_\rho = \int_0^\delta \left(1 - \frac{\rho}{\rho_e}\right)\mathrm{d}y \tag{6.2}$$

and H is the shape factor given by δ^*/θ and

$$c_f = \frac{\tau_w}{\frac{1}{2}\rho_e u_e^2} \qquad c_Q = \frac{(\overline{\rho v})_w}{\rho_e u_e} \tag{6.3}$$

Given the external velocity distribution $u_e(x,t)$ and external density field $\rho_e(x,t)$ from an inviscid calculation of the flow around the body displacement surface, the momentum integral equation relates the three thickness parameters and the skin friction. Obviously, additional relationships must be supplied to form a determinate system. Before proceeding to develop the required additional equations, some observations seem appropriate. First, the axial momentum integral equation is not particularly controversial and most investigators have made it a point to base their analysis upon this foundation. Some authors include the Reynolds normal stress terms, but this seems to be quite optional and to date has not proved to be a particularly crucial item (except possibly near separation where, in any event, the conventional boundary-layer analysis is in difficulties). Evans and Horlock[2] have pointed out that unless the integration is carried out far enough into the freestream, the fact must be taken into account that the skin-friction term on the right-hand side of the momentum integral equation (6.1) is actually the net result of the wall stress minus the apparent Reynolds shear stress $-\overline{u'v'}$ at the $y = \delta$ point where the integration is terminated. In principle, this does not cause any difficulty with the momentum integral equation, since it is evident from Eq. (6.2) that the upper limit on the integration can be arbitrarily large for the defect thicknesses defined there.

Problems can arise, however, with the auxiliary relationships if they involve integral thickness parameters depending on the location of the boundary-layer edge. The problem manifests itself principally in flows where the velocity profile tails off very gradually into the freestream, such that there might be a factor in excess of 1.25 between the point at which the local velocity equaled 0.99 of the external stream ($y = \delta_{0.99}$) and the point at which the local velocity equaled 0.995 of the freestream ($y = \delta_{0.995}$). This long tail seems to be a characteristic of flows with a significant amount of freestream turbulence present, i.e., $T_u = (\overline{u'^2})^{\frac{1}{2}}/u_e > 0.03$. The implication here is that boundary layers with this characteristic long tail should be integrated out to where the Reynolds apparent shear stress is negligible compared to the wall stress. This was the approach adopted by McDonald and Kreskovsky.[3] As an alternative, an estimate of the Reynolds stress at some convenient thickness can be made and this approach was adopted by Evans and Horlock.[2]

Finally, it is observed that the momentum integral equation is independent of any assumption about the form of the mean velocity or temperature profile and is equally valid for laminar, transitional, or fully turbulent flow. This fact is convenient in applying the momentum integral equation and leads one to seek, where possible, auxiliary equations with this same formal detail profile independence property. Detail profile independence permits the overall technique to be constructed so as to be valid for any type boundary-layer flow and places the flow distinction mechanism in the more easily isolated region of turbulence model and profile specification.

Skin-friction laws and the mean velocity profiles. Turning to the auxiliary relationships to be supplied, the great majority of methods specify, often explicitly, a skin-friction law relating the integral thickness parameters θ, δ^*, etc., to the skin-friction coefficient c_f. Typical forms of these explicit relationships are discussed by Nash[4] for incompressible turbulent flow and an apparently quite satisfactory relationship is derived by Nash and MacDonald[5] for turbulent compressible adiabatic flow. Several other similar quite satisfactory explicit relationships are available in the literature for compressible unseparated flow in a pressure gradient on a smooth wall without heat transfer. Less satisfactory, however, are the skin-friction laws for flows with heat transfer when either the freestream or the wall temperature is varying rapidly. As an alternative but equivalent process to assuming an explicit skin-friction law, and as a result of the near wall dependence of velocity on wall friction, a skin-friction law may be obtained as a by-product of the assumed velocity profile family, to be discussed in detail subsequently. This latter practice is much more consistent with the overall analysis, although not necessarily any more accurate, and, of course, it does demand that a velocity profile depending on the wall friction be adopted. Certainly, obtaining the skin friction from the assumed velocity profile makes a mechanical process of the extrapolation of the skin-friction law into compressible flows with pressure gradients, or any other flow where the skin-friction measurements to be correlated are sparse, and is the procedure recommended by the present author at this time.

Turning now to the question of mean velocity profile correlations, first of all, three broad classifications of approaches can be discerned in turbulent incompressible flow. In the first of these, now largely abandoned for turbulent flows, a simple polynomial representation of the mean velocity profile is proposed, usually of the type velocity proportional to $y^{1/n}$ where n is some exponent, perhaps even an integer, that can vary with the boundary-layer condition. Experience has shown that, for all but the crudest of estimates, the simple polynomial is inadequate. Fortunately, much better representations are possible and this leads to the second category, where the correlations are based on the law of wall with an allowance for a departure from the logarithmic region in the far wall region. The best-known formulation of this type is that due to Coles. Experience has shown his correlation to be remarkably accurate for a wide range of unseparated low-speed flows in both favorable and adverse pressure gradients.[6] Coles[7] suggests a profile of the following form for flows with surface transpiration:

$$\frac{2}{v_w^+}\left[(1+u^+v_w^+)^{\frac{1}{2}}-1\right]=\frac{1}{\kappa}\ell n\, y^+ + C + \frac{\Pi}{\kappa}2\left(1-\cos\frac{\pi y}{\delta}\right) \tag{6.4}$$

where

$$\rho u_\tau^2 = \tau_w, \qquad v_w^+ = v_w/u_\tau$$

$$u^+ = u/u_\tau, \qquad y^+ = yu_\tau/\nu$$

$$C = C_0 + (2/v_w^+)\left[(1+B_0 v_w^+)^{\frac{1}{2}}-1\right] - B_0 \tag{6.5}$$

B_0 is Simpson's[8] blowing intercept taken as 10.805, κ the well-known von Kármán constant, and C_0 the additive constant in the law of the wall, taken as 0.41 and 5.0, respectively, by Coles for smooth walls. Π is the wake strength parameter[9] and Eq. (6.4) may be integrated to give the desired integral profile thickness parameters. It is also clear that if the profile is evaluated at $y = \delta$ where $u = u_e$, a skin-friction law is obtained. The strength of the wake component Π may be eliminated from a knowledge of the integral thickness parameters. The skin-friction relationship is apparently quite an awkward transcendental formulation; however, if necessary, it is easily and very efficiently solved by a Newton-Raphson scheme. Another shortcoming of the Coles profile as presented is that it is not continuous all the way to wall and is valid only outside the viscous sublayer, say for $y^+ > 50$. Coles does, however, present corrections to the various integral formulas to account for the neglect of the sublayer and, if needed, Waltz[10] has constructed a version of Coles' profile that is continuous all the way down to the wall. Also, it should be noted that Coles' profile does not have zero gradient at the point where $y = \delta$, although this discrepancy is usually not significant in normal profile usage.

In the third and final category, considerable emphasis is placed on the local equilibrium hypothesis to obtain velocity profiles. To understand this concept, it is necessary to recall that an equilibrium boundary layer is one that in a normalized sense exhibits velocity profile similarity as it develops downstream. These normalized and similar (self-preserving) profiles are functions only of a nondimensional pressure gradient parameter β, where $\beta = \delta^*(\mathrm{d}p/\mathrm{d}x)/\tau_w$ and their existence in an equilibrium turbulent flow, where β is constant was demonstrated by Clauser.[11] (A slight Reynolds number effect is to be expected for these turbulent equilibrium profiles as a result of the viscous sublayer and superlayer, but it generally may be safely neglected.) The local equilibrium hypothesis simply asserts that nonequilibrium boundary layers, characterized by nonconstant values of the pressure gradient parameter β, have velocity profiles that may be described by some, as yet undetermined, equilibrium boundary-layer velocity profile, regardless of the streamwise rate of change in the pressure gradient parameter. To determine the appropriate equilibrium boundary layer, it is necessary to note that, as Clauser observed, equilibrium velocity defect profiles are of the form

$$(u_e - u)/u_\tau = f(y/\delta) \qquad \beta = \text{const} \tag{6.6}$$

Clauser further suggested useful integral parameters G and I, where

$$G = \int_0^1 f^2 \mathrm{d}\left(\frac{y}{\delta}\right) \Big/ \int_0^1 f \mathrm{d}\left(\frac{y}{\delta}\right) = \sqrt{\frac{2}{C_f}}\left(1 - \frac{1}{H}\right)$$

$$I = \int_0^1 f \mathrm{d}\left(\frac{y}{\delta}\right) \qquad = \left(\frac{2}{C_f}\right)^{\frac{1}{2}} \frac{\delta^*}{\delta} \tag{6.7}$$

and obviously an infinite sequence of integral shape parameters can be defined to relate the various commonly used integral parameters arising from the integral moment equations to equilibrium parameters. It follows that, in an equilibrium boundary-layer, specification of the pressure gradient parameter β immediately determined the profile shape $f(y/\delta)$ and the sequence of shape parameters G, I, etc. follow. Indeed, Nash,[12] for example, has correlated the available experimental and theoretical equilibrium profile information to obtain the relationship between the shape parameter G and the parameter β in the form

$$G = 6.1(\beta + 1.81)^{\frac{1}{2}} - 1.7 \tag{6.8}$$

and, of course, a similar relationship may be developed for the other parameters, I, etc. The local equilibrium hypothesis then suggests that given any one of the infinite sequence of shape parameters G, I, etc., all the others in the sequence are determined as belonging to that particular equilibrium boundary layer. The β parameter thus is no longer to be interpreted as the pressure gradient parameter, but merely as a characterizing independent parameter of the velocity profile. The actual shape of the profiles may be obtained from several sources, such as the calculations of Mellor and Gibson[13] or by examining the measured equilibrium flows. Usually, the detailed definition of the velocity profile is not required in the prediction scheme and the velocity profile serves only to provide a relationship between the various integral thickness parameters. Therefore, the user can evaluate the required integral relationships in advance and express the results in simple look-up tables or in polynomials with G or β as the independent variable.

In normal usage, it is much more convenient to omit the β parameter altogether and evaluate the required profile relationships with the shape parameter G as the dependent variable, as did Michel et al.,[14] for instance. Nash's skin-friction law,[4] previously mentioned, was evolved in the foregoing spirit from equilibrium concepts proposed initially by Clauser[11] and later developed by Rotta.[15] The general technique is closely related to the well-accepted technique for predicting laminar boundary development where the appropriate equilibrium information is obtained from the Falkner-Skan solutions.

In the practical matter of accuracy, there seems little to choose between the Coles' profiles and the local equilibrium profiles. Both representations have been used with considerable success by differing authors; however, on balance this author has found the Coles' representation more convenient in view of its analytic profile specification.

Insofar as the effect of compressibility is concerned, the satisfactory development of two-parameter incompressible velocity profile families has led to the search for a transformation technique that will reduce the compressible problem to an equivalent incompressible problem for which the existing correlations would be adequate. Two broad categories of approach have been pursued in this area, the first of which exploits the fact that, since the integral approaches usually demand only a relationship between integral thickness parameters, the detailed profiles can be ignored and essentially empirical correlation techniques developed to map integral thickness relationships from incompressible to compressible flow. The well-known reference temperature methods (see, for instance, Rubesin and Johnson[16] or the Appendix to the paper by Coles[17]) provide a very simple mapping of this type that might be useful in certain restricted applications. The second broad category of approach is much more ambitious, in which attempts are made to define a point-by-point mapping of the velocity profile and to develop the required integral thickness relationships and skin-friction law as a consequence of the pointwise mapping. Two techniques have emerged from these efforts, both of which seem to have been sufficiently successful to justify their consideration for use in calculation schemes. The first mapping is one based on the observation of a number of investigators (for instance, Waltz[10]) that, when expressed in coordinates not containing local density (or equivalently local temperature), the velocity profiles become fairly insensitive to Mach number. Winter et al.[18] show this to be true in the usual logarithmic region for their measurements at Mach 2.2, when they used the kinematic viscosity and density evaluated at the wall temperature in the usual law of the wall formulation for velocity. Subsequently, Winter and Gaudet[19] showed that the wake component for their conditions also had a shape independent of the Mach number. The results of this very simple mapping are quite encouraging, particularly for low Mach numbers, but at the present time caution must be advocated since to date little in the way of detailed evaluation of this concept in flows with pressure gradients has been carried out.

The second mapping has been more thoroughly evaluated and is based on a transformation originally developed by Van Driest[20] from mixing length arguments for the law of the wall region of the flow and subsequently found by Maise and McDonald,[21] following an observation by Coles,[22] to yield a surprisingly accurate collapse of a wide range of compressible adiabatic constant-pressure boundary-layer profiles, including the wake region of the boundary layer. More recently, Mathews et al.[23] demonstrated that this same profile formulation gave reasonable results in adiabatic compressible flow in adverse gradients. The transformation was derived by Van Driest from the mixing length hypothesis together with the Crocco temperature profile assumption. Using the usual assumption that, in a transpired boundary layer, the local shear stress τ is given by

$$\tau = \tau_w + \rho u v_w \tag{6.9}$$

the suggested form of the velocity profile for the compressible transpired boundary layer may be readily derived as

$$u^* = \frac{1}{\tilde{A}} \sin^{-1}\left[\frac{2\tilde{A}^2 u^+ - \tilde{B}}{(\tilde{B}^2 + 4\tilde{A}^2)^{\frac{1}{2}}}\right]$$

$$= \frac{1}{\kappa} \ell n y^+ + \tilde{C} + \frac{\Pi}{\kappa} 2\left(1 - \cos \pi \frac{y}{\delta}\right) \tag{6.10}$$

where Π is the conventional wake component and $\tilde{C}$ an additive constant. With negligible transpiration, i.e., $v_w^+ \ll \tilde{B}$, there is no ambiguity, so ensuring that at low Mach numbers the conventional law of the wall is returned requires that

$$\tilde{C} = -\frac{1}{\tilde{A}}\sin^{-1}\left[\frac{\tilde{B}}{(\tilde{B}^2 + 4\tilde{A}^2)^{\frac{1}{2}}}\right] + C_0 \qquad v_w^+ \ll \tilde{B}$$

where C_0 is as defined in Eq. (6.5). However, with significant transpiration present, determining the additive constant $\tilde{C}$ becomes quite involved and, in view of the uncertainty of the effect of compressibility on Simpson's blowing intercept B_0, the topic is best left for future investigation. Also, in Eq. (6.10),

$$\tilde{A}^2 = \frac{(\gamma - 1)}{2} M_e^2 \frac{C_f}{2}$$

$$\tilde{B} = \left[\left(1 + \frac{\gamma - 1}{2} M_e^2\right)\frac{T_e}{T_w} - 1\right]\left(\frac{C_f}{2}\frac{T_w}{T_e}\right)^{\frac{1}{2}} + v_w^+ \tag{6.11}$$

In the definition of y^+ and u_τ, the kinematic viscosity and density are evaluated at the wall temperature, which, although it gives reasonable results, it still is a fairly arbitrary choice of temperature. The consequent skin-friction law follows immediately by evaluating the profile at $y = \delta$, just as in incompressible flow.

Recently, a number of researchers have investigated various suggested forms for the compressible velocity profile and skin-friction law at hypersonic Mach numbers, where, of course, the compressibility effect is accentuated. For instance, Hopkins et al.[24] and Owen and Horstman[25] found the Van Driest relationship given above performed acceptably for adiabatic flow and better on average than the wall temperature procedure mentioned earlier. Finally, it is noted that quite satisfactory mean velocity profile predictions have been made by the numerical procedures using turbulence models that reduce to the same mixing length model Van Driest adopted in developing his transformation. It seems plausible that, provided the actual temperature profile is reasonably close to the Crocco distribution assumed by Van Driest, the Van Driest mapping will be satisfactory. In view of more recent confirmatory experimental evidence, at the present time the Van Driest mapping is the recommended procedure for obtaining the compressibility effect on the mean velocity profile and skin friction in turbulent adiabatic flow. In view of the poor predictions of the velocity profiles in the presence of heat transfer, noted by Maise and McDonald[21] and others, the application of the Van Driest transformation to flows with heat transfer must be viewed with caution for the present; here the temperature profile may well be the culprit.

More recently, Fernholz and Finley[26] have performed an extensive survey of compressible boundary-layer profiles and taken issue with the Van Driest mapping as applied by Maise and McDonald.[21] Fernholz and Finley object to the use of the conventionally defined boundary-layer thickness scale δ as a profile scaling parameter to compare the measured data to the assumed profile because of the experimental difficulty of determining δ with precision. This same objection holds as well for Coles' well-established incompressible profile family.[27] Here, the view taken is identical to that of incompressible flow, where the assumed profile family also defines a relationship between the various thickness scales (such as momentum, displacement, and velocity defect thickness, etc.) and the skin friction. Measured velocity profiles may be reduced using measured values of these more accurately determined defect scales. Indeed, the measured profiles may be curve fitted using the various scales as free parameters with an after-the-fact comparison between the measured scales and those determined from the curve-fitting operation. From this point of view, the previous comparisons obtained, for instance, by Coles in incompressible flow[27] and by Maise and McDonald in compressible flow[21] seem quite acceptable. Fernholz and Finley present an extensive compilation of measured compressible boundary-layer profiles data in Ref. 26 and in Ref. 28 an alternative profile family that appears to perform satisfactorily.

Before leaving the topic of the compressibility effects on the mean velocity profile, however, mention must be made of those transformations that have their roots in the Stewartson-Illingworth transformations which have proved so useful for laminar flow. This latter type of transformation seeks to obtain a rigorous mapping of the governing partial differential equations into an equivalent constant-density system. The Van Driest transformation introduced earlier is, of course, also a transformation, but not in the same sense as implied here, since no attempt is made in the Van Driest formulation to reduce the governing partial differential equations to an equivalent constant-density form. The velocity profiles in the Van Driest sense could also be regarded as simply empirical correlations. Coles[27] has reviewed the status of the rigorous transformation approaches for turbulent flow and suggested some improvements. However, even Coles' improved formulation suffers from observable defects.[21] The central problem is that these rigorous transformations imply a relationship between the apparent Reynolds stresses in the two fields. At one time, it was felt that when a transformation was obtained which could properly relate the mean profiles in the two fields, the resulting Reynolds stress mapping would, as a consequence, be valid. Unfortunately, for this approach it turned out, because of the apparent validity of Morkovin's hypothesis,[29] much easier to determine the effect of compressibility on the Reynolds stress than to obtain a rigorous reduction of the compressible equations of motion to an equivalent incompressible form. Indeed, a satisfactory rigorous transformation technique for the turbulent boundary layer apparently does not yet exist.

Turning now to wall roughness, it turns out that, with a little effort, profile families based on the law of the wall generalize readily to account for wall roughness effects. The generalization follows since the direct effect of surface roughness is felt only in the vicinity of the wall and that, further from the wall, the flow is independent of viscosity and scales on the wall stress, but is relatively independent of how that wall stress is generated. Indeed, as Coles[7] points out, it is virtually impossible to determine whether the wall is smooth or rough by examination of a velocity profile given in the defect form of $(u_e - u)/u_\tau = f(y/\delta)$. Similarity considerations lead to a roughness scale $k^+ = ku_\tau/\nu_w$ and the observed roughness effects are contained in the additive constant of the velocity profile C, which is now symbolically written $C(k^+) = C_0 - \Delta u_r^+$. The constant $C(k^+)$ is a function of the geometry of the roughness and data for various geometries has been correlated by a number of investigators. For the time being, it is necessary to restrict attention to the case of zero transpiration and hence following the suggestion of Clauser,[11,30] for instance, the incompressible Coles' profile may be written

$$u^+ = \frac{1}{\kappa}\ell n y^+ + C_0 + \frac{\Pi}{\kappa}2\left(1 - \cos\frac{\pi y}{\delta}\right) - \Delta u_r^+ \tag{6.12}$$

and as a result of the observation that the fully rough wall profile scales on u^+ vs $f(y^+/k^+)$, there results, for $k^+ > 70$ say,

$$\Delta u_r^+ + C_r + (1/\kappa)\,\ell n k^+ \tag{6.13}$$

where C_r is a constant for a particular geometry of roughness and, on the basis of experimental investigations, Dvorak[31] suggests the correlation

$$C_r = \alpha\left(\log_{10}\lambda^\beta - 1\right) \tag{6.14}$$

with

$$\begin{aligned} \alpha &= 17.35 \qquad \beta = 1.625 \qquad 1 < \lambda < 4.68 \\ &= 5.95 \qquad\;\; = 1.103 \qquad\;\; \lambda \geq 4.68 \end{aligned} \tag{6.15}$$

where λ is the ratio of the total surface area to the roughness (wetted) surface area. Typically, for Nikuradse's sand grain roughness, C_r is near -3.4. For roughness heights in the intermediate (transitional) range of $5 < k^+ < 70$, it is necessary to fair the Δu_r^+ parameter down to zero in some reasonable manner as k^+ is reduced below 70. Note that Δu_r^+ is always positive. As before in dealing with the additive constant C, it is supposed that the wall temperature effects are allowed for by evaluating any of the temperature-dependent parameters appearing in the roughness relations at the wall temperature. The assumption that the compressible defect profile remains unaltered by the wall roughness appears very logical based on the incompressible arguments. Indeed, Chen[32] has found this to be the case using Young's[33] measurements, as can be seen in Fig. 6.1. Lastly, it is observed that Simpson's[8] blowing intercept B_0, which determined the additive constant C for smooth wall transpiration, cannot be expected to hold for rough walls and that to date this problem has not been resolved.

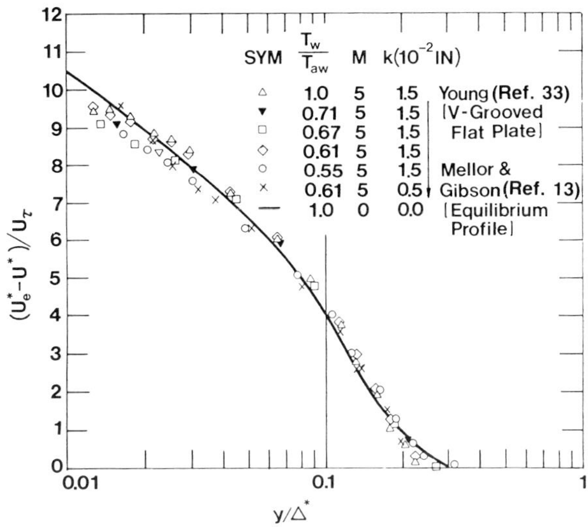

Fig. 6.1 Compressible turbulent velocity defect profiles on a rough surface with heat transfer [*M* is the freestream Mach number, *k* the roughness height, *u Van Driest's[20] transformed velocity, and Δ* Clauser's thickness scale δ*I** (Eq. 6.7)] (from Ref. 32).**

The temperature profile. In developing the governing integral equations, the local density appears under the integral sign. Excepting the case of incompressible flow with small temperature differences, it becomes necessary to determine the variation of the density across the boundary layer. Since for boundary-layer approximations the static pressure is constant across the boundary layer, the gas law gives the product of local density and local temperature as constant; thus, the problem becomes one of specifying the local temperature. There is little doubt, however, that the specification of the temperature profile across the boundary layer has been, and is, the least satisfactory area of the overall problem of predicting boundary-layer behavior by an integral technique. It seems clear that, if the mean velocity profile had been as poorly characterized as the mean temperature profile, investigators would have long since abandoned simple integral procedures in even greater numbers than have done so at present. The difficulty is implicit in the solution to the problem adopted by Dvorak and Head,[34] who, for heat transfer in low-speed flow, computed the development of the velocity profile sufficiently accurately for their purposes via the very compact simple integral scheme due to Head.[35] However, in order to obtain a commensurately accurate solution of the energy equation, Dvorak and Head felt obliged to resort to a finite difference technique with an assumed turbulent effective Prandtl number and they performed the integration of the energy equation with velocities and turbulent shear stress obtained from the momentum integral calculation. Clearly, they did not consider any temperature profile family known to them at that time adequate for the purpose of predicting heat transfer in the presence of a pressure gradient. However, the overall prognosis now is not quite as bleak as the preceding remarks might lead one to believe. For instance, Green[36] was able to successfully extend Head's scheme to compressible flow with adiabatic walls. Therefore, it follows that the required and actual accuracy of the temperature profile is dependent on the flow situation and that, in order to clarify the areas of application and the various computational strategies, it is necessary to review the entire question of determining the boundary-layer temperature.

A similar review is given by Fernholz and Finley,[26] who also compare the available data with several proposed temperature profile families.

Starting with the energy equation in the form

$$\bar{\rho}\frac{\partial \bar{H}}{\partial t} - \frac{\partial p}{\partial t} + \bar{\rho}\bar{u}\frac{\partial \bar{H}}{\partial x} + \overline{\rho v}\frac{\partial \bar{H}}{\partial y} = \frac{\partial}{\partial y}(Q + u\tau) \tag{6.16}$$

where H is the stagnation enthalpy defined in the usual manner as

$$\bar{H} = \bar{h} + \bar{u}^2/2 \qquad \bar{h} = \int C_p \, \mathrm{d}\bar{T} \simeq C_p \bar{T} \tag{6.17}$$

since for the usual gas turbine applications the specific heat at constant pressure may be taken as constant. The total apparent heat flux Q is defined as

$$Q = k\frac{\partial \bar{T}}{\partial y} - \bar{\rho}\overline{v'h'} = (k + k_t)\frac{\partial \bar{T}}{\partial y} \tag{6.18}$$

where, in spite of its shortcomings, an effective conductivity k_t times the temperature gradient has been used to represent the velocity enthalpy transport correlation. Recalling the definition of the effective viscosity ν_t, the energy equation may be rewritten as

$$\bar{\rho}\frac{\partial \bar{H}}{\partial t} - \frac{\partial p}{\partial t} + \bar{\rho}\bar{u}\frac{\partial \bar{H}}{\partial x} + \overline{\rho v}\frac{\partial \bar{H}}{\partial y}$$

$$= \frac{\partial}{\partial y}\bar{\rho}\left[\left(\frac{\nu_t}{P_{rt}} + \frac{\nu}{P_r}\right)\frac{\partial \bar{h}}{\partial y} + (\nu_t + \nu)\frac{\partial}{\partial y}\frac{\bar{u}^2}{2}\right] \tag{6.19}$$

and the laminar and turbulent effective Prandtl number have been introduced where

$$P_r = \bar{\rho} C_p \nu / k \qquad P_{rt} = \bar{\rho} C_p \nu_t / k_t \tag{6.20}$$

Note that the introduction of the turbulent effective Prandtl number is not in itself controversial if the relationship given above is regarded simply as a definition of the turbulent Prandtl number. As will become evident, the usual mode of operation is, of course, quite the inverse and in a calculation scheme a turbulent Prandtl number distribution is often specified and the above relationship interpreted as a definition of the turbulent effective conductivity k_t. Viewed as a method of determining the turbulent effective conductivity, the concept of a well-behaved turbulent Prandtl number is best regarded for the time being as an acceptable working hypothesis subject to the very similar limitations as the mixing length hypothesis undoubtedly is for determining the turbulent momentum transport. For instance, various definitions of the turbulent Prandtl number can be adopted based upon whether the transport of static or stagnation enthalpy is used. Here, static enthalpy has been adopted and Owen and Horstman[25] found that this Prandtl number was much less sensitive to Mach number than one based on stagnation enthalpy. The usual arguments for the similarity of the transport mechanism for the heat and momentum of a gas in both laminar the turbulent flow leads to the assumption of unit laminar and turbulent Prandtl numbers. If this is done, the energy equation further simplifies and, as Crocco pointed out for laminar flow and later Young for turbulent flow, by inspection it can be seen that the condition of constant stagnation enthalpy ($\overline{H} = \overline{H}_e = \overline{H}_w$) satisfies the energy equation. Constant stagnation enthalpy cannot allow for wall heat transfer, but it does give perhaps the simplest temperature profile for an insulated wall. Actually, as Crocco pointed out, a more general profile can readily be determined from the energy equation by assuming that the stagnation enthalpy $\overline{H}$ is some undetermined function of local velocity alone, i.e., $\overline{H} = f(u)$. If this (similarity) assumption is introduced into the energy equation and the result compared with the momentum equation, it can be seen that, if the axial pressure gradient is negligible, solutions of the assumed form can be obtained if $\mathrm{d}^2 f/\mathrm{d}u^2 = 0$. Integrating twice and introducing the wall and freestream boundary conditions gives the profile

$$\overline{H}^\circ = \overline{H}_w + \left(\overline{H}_e^\circ - \overline{H}_w\right)(\bar{u}/\bar{u}_e) = C_p\overline{T} + \bar{u}^2/2 \tag{6.21}$$

which is the well-known Crocco quadratic temperature profile. The Crocco profile allows for wall heat transfer; it has been widely used in calculation schemes and its limitations will be discussed in detail below. Note that if the Crocco profile is differentiated to give the temperature gradient at the wall, the skin friction C_f and Stanton numbers S_t can be introduced to give the relationship

$$2S_t/C_f = 1/P_r \tag{6.22}$$

where the Stanton number is defined as

$$S_t = \frac{Q_w}{\rho_e u_e\left(\overline{H}_e^\circ - \overline{H}_w\right)} \tag{6.23}$$

but, since it was assumed that $P_r = P_{rt} = 1$ in deriving the temperature profile, the Stanton number/skin-friction relationship given by Eq. (6.22) should have the Prandtl numbers set to unity. Normally, this is not done, but in any event the ratio $2S_t/C_f$ is termed the Reynolds analogy factor.

Having covered the preliminaries, the various commonly adopted profile strategies can now be reviewed. The first observation is that the assumption of constant stagnation temperature, although very convenient, can be valid only for an insulated wall—if at all. Second, the wall temperature is generally observed to fall below the freestream stagnation temperature, so that if a wall recovery factor r is defined as

$$r = (T_w - T_e)/(T_e^\circ - T_e) \tag{6.24}$$

experimentally the recovery factor in turbulent flow over an insulated wall is generally found to lie between 0.8 and 0.9. This fact obviously cannot be allowed for within the framework of a constant stagnation enthalpy.

The next level of sophistication is obviously the Crocco quadratic temperature profile, a profile used extensively in calculation schemes. The Crocco profile does at least allow the wall temperature to reach its recovery value, but it will be recalled that the derivation would appear to restrict its use to boundary layers in thermal equilibrium [$T = f(u)$ alone will suffice as a definition of thermal equilibrium at this time] in the absence of streamwise pressure gradients. In view of the lack of suitable alternatives, various authors have used the Crocco profile outside the region where it was observed to be a reasonable approximation of the temperature within the boundary layer. For insulated walls, the results were not too unreasonable, but there seems little justification for the use of the Crocco relationship in flows with heat transfer and pressure gradients. The observed results show the Reynolds analogy factor to vary widely from the value near unity deduced from the Crocco relationship and observed in constant-pressure boundary layers in thermal equilibrium. In light of this, there have been a number of attempts to derive modified Reynolds analogy factors, but none seem to be satisfactory at this point. The search for a modified Reynolds analogy factor is motivated to a considerable degree by its convenience in application. If such a factor could be found, it would allow the heat transfer to be computed after the fact from a knowledge of the skin friction. In this manner, the heat transfer may be computed without an overt solution of the energy equation, although the solution is implied in the statement that the enthalpy is a function of local velocity alone.

While one might overlook the measured turbulent Prandtl number being slightly different from unity, the requirement for a negligible axial pressure gradient gives cause for concern and the assumption of enthalpy-velocity similarity is positively alarming. The latter assumption is obviously violated when the wall temperature varies in an arbitrary manner, such as on a cooled turbine blade. The high accelerations present on the suction side of a turbine blade jeopardize the axial pressure gradient restriction. The more recent experimental evidence is quite unambivalent. Boundary layers developing in the absence of heat transfer or pressure gradients eventually reach an equilibrium state where the temperature distribution is adequately described by a Crocco-type relation. However, even in the absence of pressure gradients, boundary layers that are (or have been) subjected to severe wall temperature variations possess temperature profiles not agreeing very well with the Crocco relation until after recovery. The length of such recovery could take many boundary-layer thicknesses of development. The controversy over the temperature profile on a wind tunnel nozzle wall vis-à-vis the same profile measurement on a boundary layer grown in an essentially insulated flat plate (see Ref. 37 for discussions on this point) is a clear indication of the combined effect of favorable pressure gradients and wall temperature. The nozzle temperature profile is markedly different from the Crocco distribution and recovers very slowly, even when the wall is maintained at its adiabatic temperature.

Rotta[38] has reviewed a number of the measurements and suggested that, in spite of the measurement difficulties, there were many instances where he felt it could be concluded that the Crocco distribution performed poorly, particularly in the presence of heat transfer. As a corollary, obviously the simple Reynolds analogy factor in many cases failed to predict the heat transfer. Dvorak and Head[34] evidently agreed with Rotta and were led to the numerical scheme discussed earlier to obtain their temperature profile. The only positive counters that can be made to date to the preceding are the observations that, first of all, a Van Driest type of relationship obtained using mixing length arguments and a Crocco temperature relationship correlates well with the skin-friction measurements on insulated and hot or cold walls over a wide Mach number, wall temperature, and Reynolds number range.[39] Second, calculation methods using a Crocco temperature relationship usually perform quite as well on insulated nonhypersonic wall boundary layers growing in an arbitrary pressure gradient as those techniques solving the energy equation directly. Here again, the user's subjective judgment on acceptable accuracy enters and the existing evaluations have not normally proceeded much beyond the usual integral thickness parameter comparisons. Little in the way of evaluation has been done in similar flows with heat transfer but it seems abundantly clear that, while integral thickness parameters may or may not be adequately predicted by an integral technique, the heat transfer will not be acceptable if it is obtained from a simple Crocco temperature profile, especially for gas turbine applications.

As a result of the dissatisfaction with the Crocco distribution for certain applications (for which it really was never meant to apply), remedies have been sought by a number of investigators. Of particular note, Cousteix et al.[40] developed an analogous treatment for the temperature field to the local equilibrium concepts previously introduced for the mean velocity profile. In their treatment, Cousteix et al. developed profile families of temperature (enthalpy) and velocity from predicted equilibrium boundary layers over a range of Mach numbers, wall temperature ratios, and pressure gradient parameters. As with the velocity profiles in incompressible flow, the shape factor G is used to replace the pressure gradient parameter β. Although here, Cousteix et al. base the friction velocity u_τ appearing in the definition of G upon a transformed (i.e., "incompressible") skin-friction coefficient $\bar{C}_f$ and express $\bar{C}_f$ as a function of Mach number, wall temperature ratio, local Reynolds number based on a boundary-layer integral thickness, and shape parameter G. The formulation is, therefore, somewhat awkward but as with the Coles' incompressible skin-friction law, solutions can readily be obtained by a Newton-Raphson process. Reynolds analogy factors are similarly expressed and, consequently, the formulation of Cousteix et al. does allow for the direct effect of pressure gradient upon the Stanton number in a reasonable manner. The principle shortcoming of this procedure is that the equilibrium solutions are, of necessity, obtained with a specified variation of wall temperature. The pathological case of a rapid variation in wall temperature, such as might occur on a turbine blade in passing over a cooling labyrinth, is still not properly accounted for in this procedure. The interested reader can refer to the original work for the detailed formulas.

The last technique that will be discussed here, although not completely developed at the present time, is very promising and apparently works well in simple flows. The basic idea involved is to develop a temperature profile analogous to the law of the wall/law of the wake profile for velocity. The idea has considerable merit, both from similarity and/or mixing length arguments, if the concept of a well-behaved turbulent Prandtl number is tenable. A few preliminaries are involved and in view of the fact that this approach has probably the best chance of success, these preliminaries will be explained in detail. First of all, a temperature law of the wall can be obtained using the mixing length or similarity arguments using the previously introduced relationships

$$Q = k_T \frac{\partial T}{\partial y} \qquad P_{rt} = \frac{\bar{\rho} C_p \nu_t}{k_t} \tag{6.25}$$

and defining heat flux parameter Q_τ analogous to the friction velocity u_τ as

$$Q_w = \bar{\rho}_w C_p u_\tau Q_\tau \tag{6.26}$$

If it is now assumed that across the wall region of the flow the local total stress τ does not vary $(= \tau_w)$, the static temperature gradient outside the viscous sublayer but within the fully turbulent wall region of the flow is written after a little manipulation as

$$\frac{1}{P_{rt}} \left(\frac{\rho}{\rho_w} \right)^{\frac{1}{2}} \frac{\mathrm{d}T^+}{\mathrm{d}y^+} = \frac{1}{\kappa y^+} (1 - A_Q u^+) \tag{6.27}$$

$$Q = Q_w - u\tau_w$$

$$T^+ = T/Q_\tau$$

$$A_Q = \tau_w / (\bar{\rho}_w C_p Q_\tau) \tag{6.28}$$

and the relationship $\nu_t = lu_\tau(\bar{\rho}_w/\bar{\rho})^{\frac{1}{2}}$ has been introduced. Now using the definition of stagnation temperature and differentiating one obtains

$$\frac{\mathrm{d}T^+}{\mathrm{d}y^+} = \frac{\mathrm{d}T^{\circ +}}{\mathrm{d}y^+} - A_Q u^+ \frac{\mathrm{d}u^+}{\mathrm{d}y^+} \tag{6.29}$$

and recalling the compressible law of the wall, one can write

$$\left(\frac{\rho}{\rho_w}\right)^{\frac{1}{2}} \frac{\mathrm{d}u^+}{\mathrm{d}y^+} = \frac{1}{l^+} \tag{6.30}$$

and consequently the stagnation temperature can be used to reduce the thermal equation to

$$\left(\frac{\rho}{\rho_w}\right)^{\frac{1}{2}} \frac{\mathrm{d}T^{\circ +}}{\mathrm{d}y^+} = \frac{P_{rt}}{l^+}\left[1 - A_Q u^+\left(1 - \frac{1}{P_{rt}}\right)\right] \tag{6.31}$$

with the consequent simplification when $P_{rt} \simeq 1.0$ of

$$\left(\frac{\rho}{\rho_w}\right)^{\frac{1}{2}} \frac{\mathrm{d}T^{\circ +}}{\mathrm{d}y^+} = \frac{1}{l^+} \tag{6.32}$$

and by comparison with the compressible velocity law of the wall there results

$$T^{\circ +} = u^+ + C_0 \tag{6.33}$$

The linear stagnation temperature/velocity profile relationship thus derived is quite intriguing. It follows immediately that the Van Driest compressible velocity law of the wall can be used with a simple change of variable to describe the stagnation temperature. However, for the record, integrate to obtain the stagnation temperature law of the wall directly. First of all, the density ratio can be expressed in terms of the stagnation temperature using the linear velocity/stagnation temperature relationship developed previously together with the assumption of a negligible static pressure gradient normal to the wall and this results in

$$\frac{\rho}{\rho_w} = \frac{T_w}{a_0(T^{\circ +})^2 + a_1(T^{\circ +}) + a_2} \tag{6.34}$$

where

$$a_0 = -A_Q Q_\tau/2 = u_\tau^2/2C_p$$

$$a_1 = Q_\tau(1 + A_Q C_0)$$

$$a_2 = -C_0^2 A_Q Q_\tau/2 = -C_0^2 u_\tau^2/2C_p \tag{6.35}$$

Now using a mixing length $l^+ = \kappa y^+$, the temperature profiles integrate out to give

$$\frac{-T_w^{\frac{1}{2}}}{(-a_0)^{\frac{1}{2}}} \sin^{-1} \frac{2a_0 T^{\circ +} + a_1}{(a_1^2 - 4a_0 a_2)^{\frac{1}{2}}} = \frac{1}{\kappa} \ell n y^+ + B_Q \tag{6.36}$$

As with the compressible velocity law of the wall, the profile given above can be reinterpreted as a temperature transformation of the form

$$\frac{T^*}{Q_\tau} = \frac{-T_w^{\frac{1}{2}}}{(-a_0)^{\frac{1}{2}}} \sin^{-1} \frac{2a_0 T^{\circ +} + a_1}{(a_1^2 - 4a_0 a_2)^{\frac{1}{2}}} + \frac{T_w^{\frac{1}{2}}}{(-a_0)^{\frac{1}{2}}} \sin^{-1} \frac{a_1}{(a_1^2 - 4a_0 a_2)^{\frac{1}{2}}} \tag{6.37}$$

where T^* is the transformed, i.e., "incompressible" value of stagnation temperature. Insofar as the constants C_0 and B_Q are concerned, Rotta[38] previously suggested an intercept relationship, while not for precisely the same linear profile as is given by Eq. (6.33), nevertheless sufficiently close that it could be used herein, of the form

$$C_0 = \beta_q (1 + 3.4\beta_q - 0.2 M_\tau) \tag{6.38}$$

where

$$M_\tau = M_e \sqrt{C_f/2}$$

$$\beta_q = Q_w / (\rho_w C_p u_\tau T_w) = 1/T_w^+ \tag{6.39}$$

The remaining constant B_Q takes on values identical to the same constant in the mean velocity profile, that is about 5.0, according to McDonald and Owen.[41] A comparison of this temperature profile to data is given in Fig. 6.2; in this zero pressure gradient flow, the results are very satisfactory. The similarity of the transformed temperature profile to the mean velocity prompts the hypothesis that a wake component of temperature might exist such that

$$B_Q = \frac{2\Pi}{\kappa} \left(1 - \cos \frac{\pi y}{\delta_T}\right) \tag{6.40}$$

where δ_T is the thickness of the thermal boundary layer. Although the initial results are encouraging (as can be seen in Fig. 6.3), further evaluation of this hypothesized profile is required. Initial unpublished results by McDonald and Owen in pressure gradient flows indicate that, apart from an apparent variation of C_0 with the pressure gradient, the proposed log law holds up quite well.

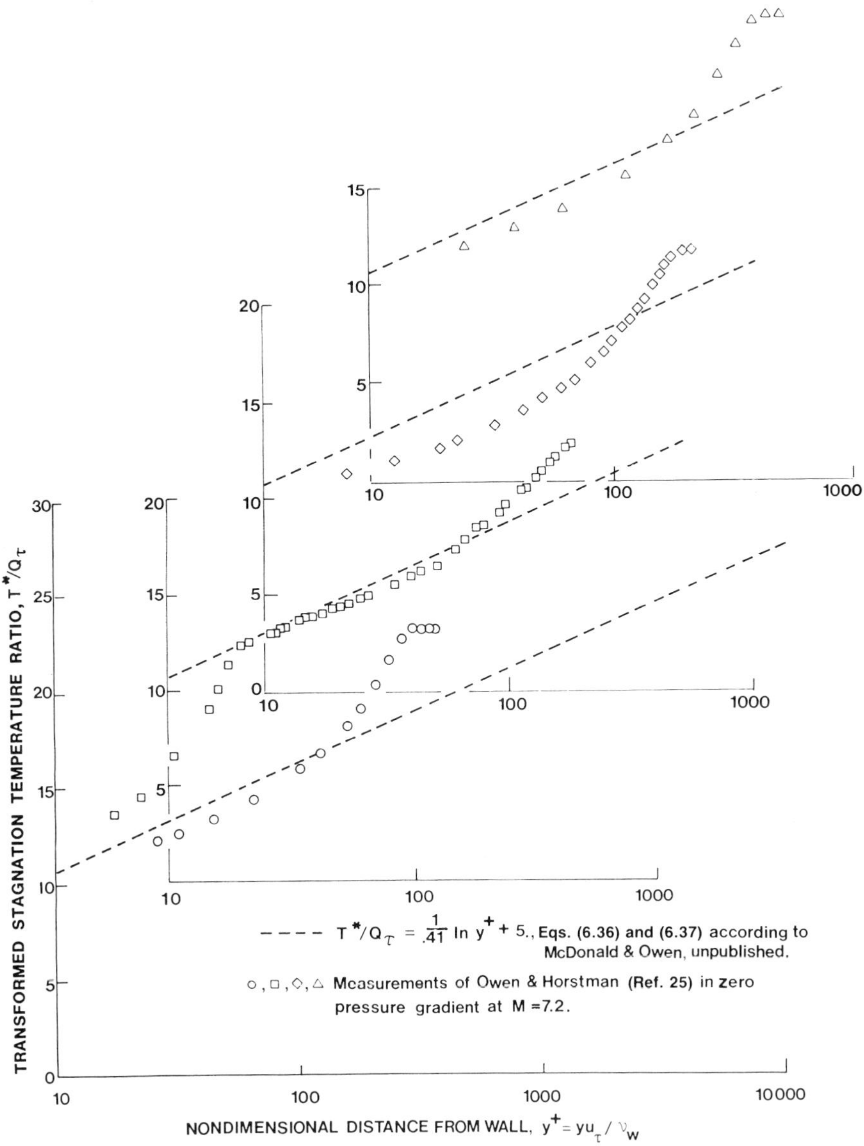

Fig. 6.2 Temperature law of the wall.

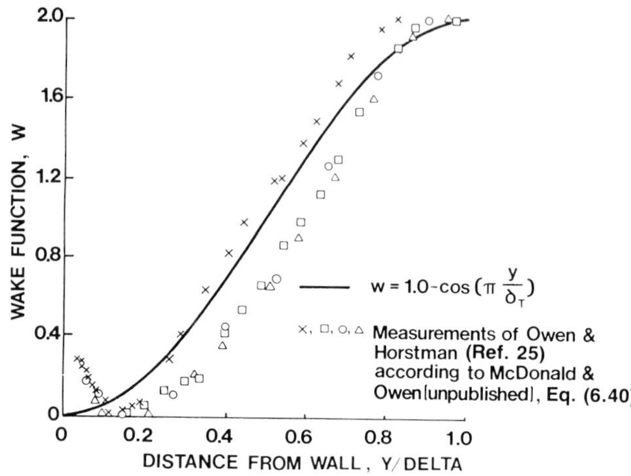

Fig. 6.3 Temperature wake component.

The compressible temperature law of the wall given by Eq. (6.36) is apparently not widely known or used. Before going on, it should be noted that the temperature profile given by Eq. (6.36) has been derived somewhat obliquely by assuming a unit turbulent Prandtl number. An alternate derivation can be made by neglecting A_Q in Eq. (6.27) and simplifying for small temperature differences. The central point is that in the alternate derivation the assumption is made that the local apparent heat flux Q is constant in the near-wall region and equal to the wall heat flux Q_w, whereas in the derivation given here leading to Eq. (6.36) the local apparent heat flux is given by $Q = Q_w - u\tau_w$. Certainly, the experimental evidence cited by Meir and Rotta[42] clearly favors the $Q = Q_w - u\tau_w$ assumption used here.

Rotta[43] developed a similarly motivated but slightly different thermal law of the wall, starting from the relationship

$$T^+ - T_w = \Delta T^+ = \frac{\kappa}{\kappa_T}(u^+ - C_0) + C_T \tag{6.41}$$

Taking the accepted values 0.4 for κ, 0.5 for κ_T, 5.0 for C_0 smooth wall, and 4.0 for C_T gives a net additive constant of zero, although for the moment this fact will not be used. If Rotta's previously used parameter β_q is now introduced, the static temperature profile at low speeds in the region of the wall can now be written

$$\frac{T}{T_w} = 1 + \beta_q u^+ \frac{\kappa}{\kappa_T} + \beta_q\left(C_T - C_0\frac{\kappa}{\kappa_T}\right) \tag{6.42}$$

In developing his thermal profiles, Rotta added a wake contribution to the above profile of strength B' and distribution $(e^{-3F}D)^{\frac{3}{4}}$ where

$$F_D = \frac{u_e - u}{u_\tau}\left(\frac{T_w}{T_e}\right)^{\frac{1}{2}} \tag{6.43}$$

The strength of the temperature wake component B' may be determined by evaluating the profile at the boundary-layer edge, that is

$$B' = \frac{T_e}{T_w} - 1 - \frac{\kappa}{\kappa_T}\beta_q\left(\frac{T_e}{T_w}\frac{2}{C_f}\right)^{\frac{1}{2}} - \beta_q\left(C_T - C_0\frac{\kappa}{\kappa_T}\right) \tag{6.44}$$

The resulting profile was shown by Rotta to be in very good agreement with some low-speed data. Actually, Rotta was a little more general in his derivation than implied by the above and he generated a temperature profile that did not necessarily require the velocity and temperature to have the same (logarithmic) functional dependence on y^+. Note that as given by Rotta the temperature profile thickness cannot exceed the velocity profile thickness, and this might prove an embarrassment in a highly accelerated flow.

Rotta has yet to apply this same concept to high-speed flow, but several points do seem appropriate. First, Eq. (6.42) does give rise to a logarithmic stagnation temperature profile for moderate-to-small stagnation temperature ratios, of the form

$$T^{\circ +} - T_w^+ = \frac{1}{\kappa_T}\ell n\, y^+ + C_T \tag{6.45}$$

or, if Rotta's wake contribution is added

$$\frac{T^\circ}{T_w} = 1 + \frac{\beta_q}{\kappa_T}\ell n\, y^+ + \beta_q C_T + B' e^{-3F_D^{\frac{3}{4}}} \tag{6.46}$$

and, as before, one could now replace the $\ell n\, y^+$ term in the temperature profile by the compressible law of the wall derived earlier. This poses an additional minor problem, however, since the compressible law of the wall, it will be recalled, was obtained by introducing the Crocco temperature profile to enable the velocity profile to be obtained by integration via the mixing length hypothesis. In more general terms, the mixing length analysis gives the velocity profile as

$$\int_{\text{ref}}\left(\frac{\rho}{\rho_w}\right)^{\frac{1}{2}} \mathrm{d}u^+ = \frac{1}{\kappa}\ell n\, y^+ + C \tag{6.47}$$

and the Van Driest law of the wall emerges when it is assumed that the static pressure is constant across the wall region and the temperature is given in accordance with the Crocco profile. For modest heat-transfer rates and Mach number, Rotta suggested that the density ratio would vary less than the velocity in the law of the wall region, leading him to suggest a compressible law of the wall of the form

$$\left(\frac{\rho}{\rho_w}\right)^{\frac{1}{2}} u^+ = \frac{1}{\kappa}\ell n\, y^+ + C \tag{6.48}$$

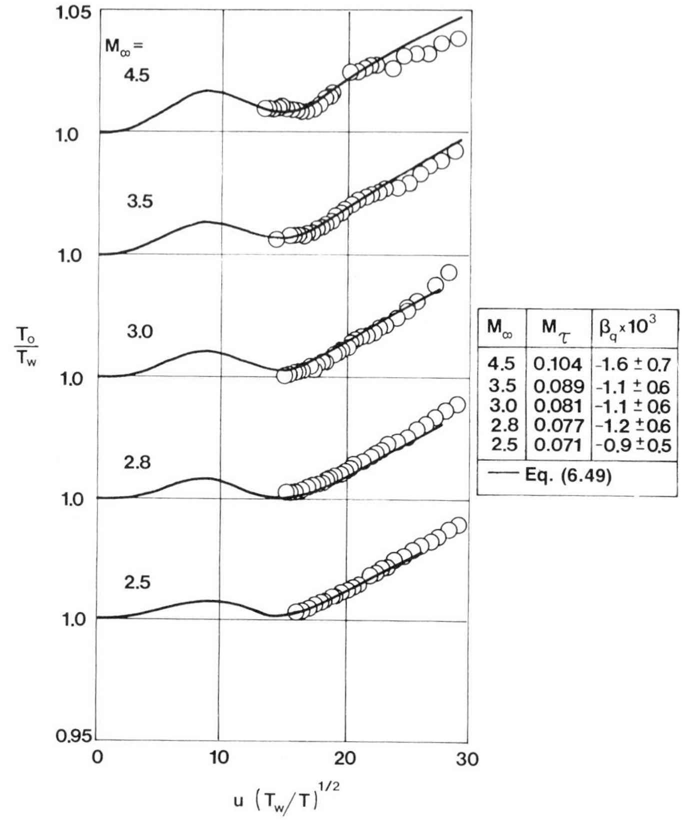

Fig. 6.4 Calculated and measured total temperature distributions.

and the experimental evidence leads one to believe that this is quite a reasonable approximation. If Rotta's form of the compressible law of the wall is inserted into the stagnation temperature law of the wall/law of the wake profile, there results the relationship

$$\frac{T^{\circ}}{T_w} = 1 + \beta_q \frac{\kappa}{\kappa_T}\left(\frac{\rho}{\rho_w}\right)^{\frac{1}{2}} u^{+} + \beta_q\left(C_T - \frac{\kappa}{\kappa_T} C\right) + B' e^{-3F_D^{\frac{3}{4}}} \tag{6.49}$$

and as with the low speed static temperature profile, the wake contribution B' is readily evaluated from the assumed profile value at the edge of the layer. Note that for an insulated wall the wake contribution B'' is related to the recovery factor r defined earlier. Meir and Rotta[42] present a number of measured stagnation temperature profiles plotted against $M_\tau = u^{+}(\rho/\rho_w)^{\frac{1}{2}}$ in the Mach number range 2.5 ~ 4.5 and the resulting linear behavior is quite remarkable, as can be seen in Fig. 6.4. A similar plot at low speed[43] also demonstrated the same linear behavior and, as mentioned previously, the intercept (additive constant) was close to zero. However, not surprisingly, in view of the kinematic viscosity temperature dependence, the additive constant appears to exhibit a local temperature dependence so that at high speeds

$$\beta_q\left(C_T - \frac{\kappa}{\kappa_T}C\right) \sim 0.03, \beta_q \sim -1 \times 10^{-3} \tag{6.50}$$

The implication of the additive constant being nonzero is twofold. First, the absolute magnitude is small but significant, it being apparent that, if the linear behavior were extrapolated back to zero velocity, the temperature intercept would give a temperature value of 0.97 as opposed to unity for zero additive constant. Second, the velocity profile evidence would lead one to believe that C_T is exhibiting the dependence on β_q as opposed to C, the velocity profile intercept, although the evidence is not at all conclusive. If it is the temperature intercept C_T that is varying, its variation is certainly very large, going from a value of 4.0 to a value of 35.0 in the Meir-Rotta experiments. Rotta's calculations at low speed indicate a modest variation of the temperature intercept with β_q, the heat-transfer parameter, giving a value of C_T near 9 for a value of β_q of -0.03. It would appear that C_T was also varying with some Mach number parameter to give rise to the value of 35.0 when β_q was in the region of -0.001 at a freestream Mach number of between 2.5 and 4.5. Obviously, this point must be clarified before extensive use of the temperature law of the wall/law of the wake is to be recommended.

Several additional points should also be made concerning the stagnation temperature law of the wall. The first of these points concerns the relationship to the Crocco distribution, which many investigators have shown to be valid in certain restricted classes of the boundary layer. A cursory examination of the stagnation temperature law of the wall shows that the replacement of $\ell n y^+$ by Rotta's suggested compressible law of the wall velocity gives rise to a term in $u^+(\rho/\rho_w)^{\frac{1}{2}}$ that is implicitly a nonlinear u velocity term and is in conflict with the simple u^+ dependence which would arise from the Crocco distribution. It should be recalled from the section of the compressible velocity profile, however, that Winter et al.[18] have shown a very simple law of the wall not containing the $(\rho/\rho_w)^{\frac{1}{2}}$ term gives a reasonable representation of the insulated wall compressible profiles in the low-to-moderate supersonic Mach number range. It is further recalled that Maise and McDonald[21] obtained a satisfactory collapse of the insulated wall velocity profile data using the Van Driest compressible coordinates, which again differs from both the Winter and Rotta suggestions. The foregoing would seem to indicate that the $(\rho/\rho_w)^{\frac{1}{2}}$ term was not causing a first-order effect in the insulated wall velocity profiles and hence, for nonhypersonic boundary layers with low heat-transfer rates, it could be expected that the stagnation temperature law of the wall profile would exhibit a near-linear velocity dependence, consistent both with observation and the Crocco relationship.

To conclude, although a logarithmic temperature profile is observed in many cases in both, the slope $1/\kappa_T$ and the additive constant C_T show considerable scatter even within a given experiment. Even though the experiment is not easy to perform, the question must arise as to the validity of the hypothesis on which the profile has been derived and/or the departure from equilibrium arising from the thermal history of the flow. To date neither of these questions have been satisfactorily explored or answered. It must be noted, however, that the success of direct numerical schemes using mixing length and turbulent Prandtl number concepts leads one to suspect the other approximations should the logarithmic profile not be observed. Obviously, a great deal of additional work remains to be performed so that the temperature profile may be placed on the same footing as the mean velocity profile. In the interim, the stagnation temperature law of the wall using wake component [Eq. (6.33)] would seem to be the procedure of choice, with perhaps a wake component added. Of course, the introduction of an additional parameter, the wall heat-transfer parameter β_q, in both this profile and that of Cousteix et al.,[40] requires the introduction of an additional determining relationship. For this, the integral thermal energy equation will be introduced.

The integral thermal energy equation. The integral thermal energy equation in determining heat transfer plays an analogous role to the von Kármán momentum integral equation and its role in determining the skin friction. In the skin-friction case, the von Kármán equation could be used to determine the magnitude of the skin friction after the fact if the streamwise behavior of the various integral thicknesses together with the axial pressure were known. In practice, the equation is not used in this manner, even if it were possible, in view of the known inaccuracies inherent in this approach. Instead, the various integral thicknesses are related one to another and to the skin friction by means of the assumed velocity profile or equivalently by the skin-friction law. In this fashion, the von Kármán equation can be recast either implicitly or explicitly as a differential equation for the skin friction and the local skin friction obtained accurately by streamwise integration. A precisely equivalent treatment of the integral thermal energy equation can be adopted and Cousteix et al.[40] provide a heat-transfer (Stanton number) relationship that depends on the local flow and the various integral thickness parameters. Use of this relationship reduces the integral thermal energy equation to an implicit differential equation for the wall heat transfer. On the other hand, a stagnation temperature law of the wall with a Rotta-type wake component allows the integral thermal energy equation to be reduced to an explicit differential equation for the heat transfer.

The integral thermal energy equation is readily derived by integrating the energy equation (6.16) from the wall to the freestream to give

$$\frac{1}{\rho_e u_e}\frac{\partial}{\partial t}\rho_e\delta_H^* + \frac{\delta}{\rho_e u_e}\frac{\partial p}{\partial t} + \frac{1}{\rho_e u_e}\frac{\partial}{\partial x}\rho_e u_e \theta_H = C_Q\frac{\bar{H}_e - \bar{H}_w}{\bar{H}_e - \bar{H}_{\text{ref}}} + S_t \qquad (6.51)$$

where

$$\delta_H^* = \int_0^\delta \frac{\bar{\rho}}{\rho_e} \frac{\overline{H}_e - \overline{H}}{\overline{H}_e - \overline{H}_{\text{ref}}} \, \mathrm{d}y$$

$$\theta_H = \int_0^\delta \frac{\bar{\rho}\bar{u}}{\rho_e u_e} \frac{\overline{H}_e - \overline{H}}{\overline{H}_e - \overline{H}_{\text{ref}}} \, \mathrm{d}y \tag{6.52}$$

where θ_H is termed the enthalpy thickness and the Stanton number definition has been generalized to

$$S_t = Q_w / \rho_e u_e (\overline{H}_e - \overline{H}_{\text{ref}}) \tag{6.53}$$

In the above derivation, the freestream stagnation enthalpy is assumed constant, although this assumption can readily be relaxed. Rotta's heat-transfer parameter[43] introduced earlier is related to the above defined Stanton number by

$$\beta_q = \frac{Q_w}{\rho_w \overline{H}_w u_\tau} = \left(\frac{\rho_e}{\rho_w} \frac{2}{C_f} \right)^{\frac{1}{2}} \frac{\overline{H}_e - \overline{H}_{\text{ref}}}{\overline{H}_w} S_t \tag{6.54}$$

The implementation of the energy equation is best visualized from a predictor-corrector point of view. At a given streamwise location with a steady external flow, the current values of the problem parameters (such as skin-friction coefficient, Stanton number, momentum thickness, etc.) can be used to integrate the momentum and energy integral equations, yielding the values of the dependent integral thicknesses at the next streamwise station. These new level thicknesses can be used in collaboration with the mean velocity and temperature profiles to yield a new level skin friction, Stanton number, and all the other integral thickness parameters. Thus, a corrector step can be taken where the required information can now be assumed to be the average of the new and current station values.

Moment of momentum equation. It is quite clear that the use of a two-parameter velocity profile, either that based on Coles' profile or that obtained from the local equilibrium hypothesis, demands one more relationship in addition to the momentum integral equation in order to specify the second profile parameter, provided of course that one of the velocity profile parameters contains the skin friction. Various forms of auxiliary relationships have been employed and a number are discussed by Rotta[15] and many examples are to be found in the Stanford Conference Proceedings. The entirely empirical auxiliary relationships will not be considered further herein, mainly because they have been largely abandoned by the research community because of their inherent limitations, which are liable to be very severe in gas turbine applications. The remaining auxiliary relationships are usually derived by taking integral moments of the axial momentum equation. As mentioned earlier, this process is formalized by multiplying the boundary-layer partial differential equations of motion by $y^n u^m$ and integrating, either partially or entirely, across the boundary layer. When $n = 1$ and $m = 0$, the y moment of momentum equation is obtained, which has been used quite successfully by a number of authors. When $n = 0$, $m = 1$ the kinetic energy integral equation is obtained, which has also found favor with a number of investigators. At one time, it was felt by a number of individuals that the y moment equation was to be preferred to the kinetic energy equation, in spite of the fact that the y moment equation does not integrate out neatly in terms of the accepted integral thickness parameters until a velocity profile is specified. Even then in incompressible flow with a two-parameter velocity profile such as Coles', the resulting differential equation is quite clumsy. The argument against the use of the kinetic energy integral equation was simply that, for many turbulent flows, the velocity varies slowly above the sublayer so that the u momentum equation has a tendency to approach a constant times the momentum equation, which ultimately leads to indeterminacy when the momentum and kinetic energy equations are solved simultaneously to predict boundary-layer behavior. In view of the apparent lack of difficulty experienced by users of the kinetic energy integral equation, it would appear that these fears are unfounded.

It turns out, however, that linear dependence within the system of equations can and does occur; that is, for some set of conditions, one of the equations turns out to be a simple linear combination of other equations and the determinant of the coefficients of the system goes to zero. The difficulty is not restricted to systems containing the kinetic energy integral equation, but, in fact, also arises with the y momentum equation under very similar conditions, notably near (but not at) separation or reattachment. The problem of linear dependence is explained in detail by Shamroth,[44] who points out that no physical significance should be ascribed to such singular points since they pose no difficulty to the direct numerical procedures. Shamroth also suggests ways of dealing with the problem, which has long gone unrecognized, almost certainly as a result of the problem occurring close to the region where the predictions would be expected to be in considerable error. Certainly, the most powerful remedy to the problem is Shamroth's suggested least squares technique, which is probably best viewed as an alternative system to be used only when the existing system is in the region of a singular point, as determined by monitoring the determinant of the coefficients of the system. When the determinant becomes small in a normalized sense, one can develop an N moment equation by integration from the wall to the point Ly/N, $L = 1, N$. A weighted least squares averaging can then be performed to reduce the N equations to one to replace the offending linearly dependent equation. Obviously, if N is taken to be large, there would be little chance of encountering a system singularity, but at the expense of greatly increasing the computer run time. Certainly, a less expensive answer and possibly quite an effective one would be to integrate the appropriate moment of the momentum equation out to, say, $y = \delta^*$ and use it to replace the linearly dependent $y = \delta$ moment equation. Other less expensive techniques could also be investigated.

One additional moment equation, the u^{-2} equation, also integrates out quite easily and after some manipulation can give a differential equation for the displacement thickness. Weinbaum and Garvine[45] derived this equation and obtained some very interesting conclusions regarding the viscous analogue of the sonic throat. Bradshaw and Ferriss[46] also present a version of this equation; however, they did not attempt to use it in a calculation scheme. Although not singular at the wall, care must be taken with the u^{-2} moment equation to ensure the proper limiting behavior as the wall is approached. Since it offers no clear advantage for normal boundary-layer computation, it cannot be recommended over the kinetic energy equation at this time.

In summary then, if for no reason other than the fact that the kinetic energy integral equation integrates out to a convenient form without the introduction of a specific velocity or temperature profile, the kinetic energy integral equation is to be recommended for use, bearing in mind that linear dependence may occur and must be guarded against. The kinetic energy integral equation can be written for a slowly varying time-dependent external flow as

$$\frac{\partial \theta_E}{\partial x} + 3\frac{\theta_E}{u_e}\frac{\partial u_e}{\partial x} + \frac{2}{u_e}\frac{\partial u_e}{\partial x}(\delta^* - \delta_u^*) + \frac{\theta_E}{\rho_e}\frac{\partial \rho_e}{\partial x} + \frac{1}{u_e}\frac{\partial}{\partial t}(\delta^* + \theta)$$

$$+ \frac{\theta}{\rho_e u_e^3}\frac{\partial \rho_e u_e^2}{\partial t} + \frac{2}{u_e^2}\frac{\partial u_e}{\partial t}(\delta^* - \delta_u^*) - \frac{1}{u_e}\frac{\partial \theta_\rho}{\partial t} - \frac{1}{\rho_e u_e}\frac{\partial \rho_e}{\partial t}\theta_\rho$$

$$+ \frac{\delta^*}{\rho_e u_e}\frac{\partial \rho_e}{\partial t} = \frac{2}{\rho_e u_e^3}\int_0^\delta \tau \frac{\partial u}{\partial y}\,\mathrm{d}y + C_Q \tag{6.55}$$

where the various thicknesses are as previously defined, but in addition

$$\delta_u^* = \int_0^\delta \left(1 - \frac{u}{u_e}\right)\mathrm{d}y \tag{6.56}$$

and

$$\theta_E = \int_0^\delta \frac{\rho u}{\rho_e u_e}\left[1 - \left(\frac{u}{u_e}\right)^2\right]\mathrm{d}y \tag{6.57}$$

The turbulence model. At this point, it is worth recalling that the appearance of turbulent correlation coefficients in the time-averaged momentum equations represents the contribution to the momentum transport by the turbulent motion. While not appearing directly in the von Kármán momentum integral equation, various integral moments of these turbulent correlation coefficients appear in the auxiliary equations introduced earlier. For example, in the kinetic energy integral equation, the term

C_D is called the dissipation integral (more properly, the production integral), where $\rho_e u_e^3 C_D = \int_0^\delta \tau\, \partial u/\partial y\, \mathrm{d}y$. In general, the specification of the dissipation integral, or some other integral moment of the turbulent stress, is crucial to the overall accuracy of the more general methods. However, there are specific flows where simplification is possible, for instance, when the turbulent transport is negligible compared to the inertial effects. In such flows, very poor estimates of the turbulent transport can still result in acceptable predictions and consequently a degree of overoptimism concerning the general predictive capabilities of a number of schemes. Also turbulence information is, in fact, introduced into the system of equations by two means: (1) via the various aforementioned integral moments of the Reynolds stress and (2) by means of the assumed velocity profile family. Indeed, if it were possible to define an accurate one-parameter velocity profile with skin friction as that parameter, the von Kármán equation could be integrated without introduction of any turbulence model. After the fact, the implied dissipation integral could be recovered directly from the computed solution and the kinetic energy integral equation. Higher moment equations would eventually allow the complete implied Reynolds stress distribution to be reconstructed. In fact, White[47] has suggested such a one-parameter velocity family and his suggestion may be recast slightly and the hypothesized profile interpreted as a Coles' profile with an assumed wake parameter Π/pressure gradient parameter β relationship. White justifies the one-parameter profile on the grounds of expediency in deriving a prediction scheme feasible for hand computation. The defects of the one-parameter profile can have major consequences and, in cases where the assumed $\pi \sim \beta$ relationship is inappropriate, White's method would be expected to give overall poor predictions. A case in point, for instance, is the constant-pressure recovery from separation or near separation. Here the observed wake component is initially very large, yet the pressure gradient is zero, a fact completely inconsistent with the assumed $\pi \sim \beta$ relationship. Such a recovering flow places a strong emphasis upon the adequacy of the turbulence information as it is clear that, without a pressure gradient to drive the flow, the rate of recovery is entirely dictated by the extent of the turbulent momentum transport and its ability to energize the flow near the wall. Thus, restricted procedures that rely either on an inertial effect dominating the turbulent transport in some region of the boundary layer (Stratford[48] develops this concept very clearly) or on the assumed one-parameter mean velocity profile to furnish a sufficient description of the turbulent transport (such as White's[47] procedure) will not be considered further as the restriction may, in fact, result in quite misleading predictions for general vane or blade boundary layers.

The simplest method of obtaining the necessary turbulent stress integrals is to correlate the measured distributions and in this manner, for example, Escudier and Nicoll[49] developed for incompressible flow the relationship

$$C_D = \int_0^\delta \frac{\tau}{\rho_e u_e^3} \frac{\partial \bar{u}}{\partial y}\, \mathrm{d}y = \frac{1}{6}(2\zeta + 1)C_f + C_1|1 - \zeta|^n \tag{6.58}$$

where

$$\zeta = (3 - H)/2H \tag{6.59}$$

and

$$C_1 = 0.00565 \qquad n = 2.715 \qquad \zeta < 1$$

$$C_1 = 0.01 \qquad n = 3 \qquad \zeta \geq 1 \tag{6.60}$$

which for many applications they found quite satisfactory. Interestingly, Escudier and Nicoll also showed that, over a fairly wide range of conditions pertaining to attached boundary layers, the dissipation integral could equally well be deduced by assuming that the turbulent Reynolds stress was related to the local mean velocity gradient by a simple mixing length hypothesis, so that, for instance, differentiation of the Coles' profile and specification of the mixing length could yield the dissipation integral directly

$$C_D \approx \int_0^\delta \frac{-\bar{\rho}\overline{u'v'}}{\rho_e u_e^3} \frac{\partial \bar{u}}{\partial y} \, \mathrm{d}y \tag{6.61}$$

where

$$-\overline{u'v'} = l^2 \left(\frac{\partial \bar{u}}{\partial y} \right) \left| \frac{\partial \bar{u}}{\partial y} \right| \tag{6.62}$$

and

$$l = \kappa y \qquad y < l_\infty / \kappa$$

$$l = l_\infty = 0.0758 \qquad y \geq l_\infty / \kappa \tag{6.63}$$

This latter means of obtaining the required turbulence structural information (by making the same detailed postulates as do the direct numerical schemes) goes some way to countering the claim that the direct numerical schemes are to be preferred since they introduce the turbulence information in a much more direct and easily identifiable manner. To continue, rather than perform the differentiation and integration demanded by the mixing length formulation, the correlation given in Eq. (6.58) is equivalent in incompressible flow and less time consuming in a calculation scheme. The mixing length formulation, however, provides a straightforward, if laborious, means of obtaining the effect of compressibility upon the dissipation integral. This extension to compressible flow follows immediately, since the compressible velocity profile is already supposedly available and Maise and McDonald[21] have shown that the mixing length profile is not sensitive to compressibility effects. Similarly, other features that can effect the turbulence structure (such as wall curvature), which have been expressed effectively as a change in the mixing length,[50] can readily be incorporated into the evaluation of the shear integrals.

The next stage in the development of a more general stress integral relationship follows from the observation that the simple mixing length profile used by Escudier and Nicoll[49] and the equivalent eddy viscosity profile used by Mellor and Gibson[13] is really valid only for equilibrium turbulent flows in the sense previously introduced by Clauser.[11,30] A number of investigators (for instance, Goldberg[51] and Bradshaw and Ferriss[46]) have measured in varying degrees the failure of the invariant normalized mixing length or eddy viscosity profile. The fact remains that equilibrium assumptions give reasonable results for a large number of measured nonequilibrium boundary layers. This observation testifies to the validity of the local equilibrium hypothesis (introduced earlier) for these particular flows, even to the extent of determining the various turbulent stress integrals solely on the basis of some profile shape parameter such as G [see Eq. (6.7)]. In order to decide when a local equilibrium procedure such as that presented by Escudier and Nicoll[49] might or might not suffice (and here the subjective criteria of the individual user must be acknowledged), it is necessary to obtain some yardstick to warn of the failure of the local equilibrium hypothesis. An appealing technique would, of course, be some parameter characterizing the state of the turbulent transport relative to the equilibrium transport, but obviously no local equilibrium prediction procedure could return this information. One plausible hypothesis fitting within the framework of a local equilibrium prediction scheme is the suggestion that the further the boundary layer is from actual equilibrium the less likely it is that the local equilibrium hypothesis will be valid. Thus, as the calculation proceeds, the computed value of, say, the shape factor $G(X)$ could be compared to the equilibrium value of G ($= \hat{G}$), which would result from the local pressure parameter $\beta(X)$. The value of $\hat{G}$ could readily be evaluated for correlations of equilibrium boundary layers such as that given by Nash[12] and reproduced here as Eq. (6.8). Again, it depends upon the user's criteria of accuracy, but based on the Kline et al.[52] results, if in a local equilibrium calculation $(G - \hat{G})/\hat{G}$ exceeded 0.5 at some location, as a rough guide, an improvement in the predictions might be expected in going to a better, nonequilibrium, turbulence model.

Early attempts to improve the turbulence transport description to allow for nonequilibrium effects for use in integral methods centered on developing empirical rate expressions for the integral moments of the turbulent stress, such as the dissipation or stress integral of the general form

$$\frac{\mathrm{d}C_D}{\mathrm{d}x} = k(C_D - C_{De}) \tag{6.64}$$

where C_{De} is the local equilibrium value of the dissipation integral that would result from a $C_D \sim G$ relationship such as that of Escudier and Nicoll[49] given by Eq. (6.59) and k is an empirically determined constant of approximate value $0.013/\delta^*$. Relationships of the foregoing type were suggested by Goldberg[51] for the dissipation integral and by Nash and Hicks[53] for the stress integral.* Such relationships are obviously very

*Goldberg also proposed that C_D be used rather than C_{De}. However, C_{De} would seem to be the appropriate choice.

convenient to use and add little in the way of computer logic or run time. Thus, in view of both the aesthetics and observed improvements, their use would appear almost mandatory with any of the existing incompressible local equilibrium procedures. With an entirely empirical expression such as Eq. (6.64), the extension to, say, compressible flow or flows with freestream turbulence does pose possible problems that to date have not been addressed.

As a further development, McDonald[54] was able to show that rate expressions quite similar to the type suggested intuitively by Goldberg and Nash could be developed from the turbulence kinetic energy equation using turbulence structural similarity concepts advanced by Townsend.[55] The turbulence kinetic energy equation is an exact conservation equation governing the kinetic energy residing in the turbulent fluctuations and its derivation from the Navier-Stokes equations is given by Favre for compressible flow. As a result of the exact formulation of the turbulence kinetic energy equation and the approximate validity of the turbulence similarity concepts invoked, it has become a popular method of developing nonequilibrium turbulence models with the potential of allowing for some of the observed effects particularly significant in gas turbine applications. To observe how the turbulence kinetic energy equation may be used in the integral boundary-layer procedures, this equation can be integrated across the boundary layer for nonhypersonic flows and for those flows where the mean flow varies slowly with time compared to the turbulent motion, yielding,

$$\frac{1}{2}\frac{\partial}{\partial t}\int_0^{\delta}\bar{\rho}\overline{q^2}\,\mathrm{d}y+\frac{1}{2}\frac{\partial}{\partial x}\int_0^{\delta}\bar{\rho}\bar{u}\overline{q^2}\,\mathrm{d}y=\int_0^{\delta}-\bar{\rho}\overline{u'v'}\frac{\partial\bar{u}}{\partial y}\,\mathrm{d}y-\int_0^{\delta}\bar{\rho}\varepsilon\,\mathrm{d}y$$

$$-\int_0^{\delta}\bar{\rho}\left(\overline{u'^2}-\overline{v'^2}\right)\frac{\partial\bar{u}}{\partial x}\,\mathrm{d}y+\int_0^{\delta}\overline{P'\frac{\partial u_1'}{\partial x_1}}\,\mathrm{d}y+E+w \tag{6.65}$$

where

$$w=\left(\overline{q^2\rho v}\right)_w$$

$$E=\left[\frac{1}{2}\overline{q^2}\left(\bar{\rho}\bar{u}\frac{\partial\delta}{\partial x}-\overline{\rho v}\right)-\overline{P'v'}+\frac{1}{2}\overline{(\rho v)'q^2}+\frac{1}{2}\rho q^2\frac{\partial\delta}{\partial t}\right]_e \tag{6.66}$$

The resulting equation is, like the von Kármán momentum equation, quite uncontroversial; however, it is also quite unhelpful in fashioning a calculation scheme without the introduction of further approximations. For instance, it can be seen that, although the previously introduced dissipation integral appears explicitly in the integral turbulence kinetic energy equation, it serves in part simply to determine the rate of change of yet another turbulence quantity—an integral of the turbulence kinetic energy flux.

There are, of course, many ways in which simplifications and approximations can be introduced into the turbulence kinetic energy equation to convert it into an equation that will control the development of the Reynolds shear stress, either implicitly or explicitly. As mentioned earlier, McDonald[54] and later Green et al.[56] applied the turbulence kinetic energy equation along the locus of the maximum stress occurring within the boundary layer. After the introduction of Townsend's structural similarity arguments, McDonald[54] developed a rate expression for the stress integral appearing in the y moment of the momentum equation; on the other hand, Green et al.[56] constructed an equation for the streamwise rate of change of an entrainment coefficient. The principle difficulty with these approaches, which follows the locus of the maximum shear stress, is that this locus may be difficult to determine with the necessary precision.

Hirst and Reynolds[57] introduce a series of intuitive suggestions to reduce the integral turbulence kinetic energy equation given above to a differential equation governing the streamwise development of the entrainment rate M, where M is defined as

$$M = \frac{\mathrm{d}}{\mathrm{d}x}\int_0^{\delta} u\,\mathrm{d}y = \frac{\mathrm{d}}{\mathrm{d}x} u_e(\delta - \delta^*) \tag{6.67}$$

but as with the entirely empirical rate equations of Goldberg[51] and Nash and Hicks,[53] although satisfactory for conventional low-speed boundary layers, difficulties arise in attempting to introduce compressibility and curvature effects, for instance, into the Hirst-Reynolds scheme.

A less intuitive analysis than those described above has been developed by McDonald and Camarata,[58] who followed Townsend[55] and Bradshaw et al.[59] and defined structural parameters a_n and L, together with a mixing length l, where

$$-\overline{u'v'} = a_1\overline{q^2}, \overline{u'^2} = a_2\overline{q^2}, \overline{v'^2} = a_3\overline{q^2}$$

$$\overline{w'^2} = (1 - a_2 - a_3)\overline{q^2}$$

$$\varepsilon = \left(-\overline{u'v'}\right)^{\frac{3}{2}}/L, \left(-\overline{u'v'}\right) = l^2\left|\frac{\partial \bar{u}}{\partial y}\right|\frac{\partial \bar{u}}{\partial y} \tag{6.68}$$

This permits the turbulence kinetic energy equation to be written as

$$\frac{1}{2}\frac{\partial}{\partial t}\int_0^{\delta}\bar{\rho}\overline{q^2}\,\mathrm{d}y + \frac{\partial}{\partial x}\left(\frac{\phi_1\rho_e u_e^3\delta^+}{2a_1}\right)$$

$$= \rho_e u_e^3\left(\phi_2 - \phi_3 + \frac{E}{\rho_e u_e^3}\right) + \int_0^{\delta}\overline{P'\frac{\partial u_1'}{\partial x_1}}\,\mathrm{d}y + W \tag{6.69}$$

where

$$\phi_1 = \int_0^{\delta/\delta^+} \frac{\bar{\rho}\bar{u}}{\rho_e u_e} \left(\frac{l}{\delta^+} \frac{\partial \bar{u}/u_e}{\partial \eta} \right)^2 \mathrm{d}\eta$$

$$\phi_2 = \int_0^{\delta/\delta^+} \frac{\bar{\rho}}{\rho_e} \left(\frac{l}{\delta^+} \right)^2 \left(\frac{\partial \bar{u}/u_e}{\partial \eta} \right)^3 \left(1 - \frac{l}{L} \right) \mathrm{d}\eta$$

$$\phi_3 = \int_0^{\delta/\delta^+} \frac{\bar{\rho}}{\rho_e} \left(\frac{a_2 - a_3}{a_1} \right) \left(\frac{l}{\delta^+} \frac{\partial \bar{u}/u_e}{\partial \eta} \right)^2 \frac{\delta^+}{u_e} \left(\frac{\partial \bar{u}}{\partial x} \right)_{y=\mathrm{const}} \mathrm{d}\eta \qquad (6.70)$$

and where η is a nondimensional transverse distance y/δ^+ and δ^+ is arbitrary. On the basis of the experimental evidence, a dissipation length profile and a one-parameter mixing length profile are assumed to be of the general form

$$L/L_\infty = \tanh[\kappa y/L_\infty]$$

$$l/l_\infty = \tanh[\kappa y/l_\infty] \qquad (6.71)$$

and the value of L_∞ can be taken to be 0.1δ on the basis of Bradshaw's measurements. For fully developed turbulence, the structural parameters a_1, a_2, and a_3 are assumed to have reached a condition of structural equilibrium characterized by constant values of 0.15, 0.5, and 0.2, respectively, chosen on the basis of the available evidence. In accordance with Morkovin's[29] hypothesis, the aforementioned structural parameters are assumed to be independent of the direct effects of compressibility for nonhypersonic boundary layers. If the normal stress terms are neglected, Eq. (6.69) becomes an integrodifferential equation for a wake value of the mixing length l_∞. This integrodifferential equation can be integrated in the streamwise direction along with the other momentum integral equation and the wake value of the mixing length derived by a Newton-Raphson iterative scheme once a value of ϕ_1 at the streamwise location in question is obtained. The process is described in detail by McDonald and Fish,[60] who have quite successfully used the scheme in conjunction with a direct numerical procedure that solved the boundary-layer partial differential equations of motion. Once a value of the wake mixing length l_∞ is obtained, the various stress integrals can be calculated directly from the assumed velocity profile family and the mixing length hypothesis.

As formulated above, the integral turbulence kinetic energy scheme has the disadvantage that the dependent variable l_∞ appears in the kernel of an integrodifferential equation. The latter disadvantage is quite troublesome in an integral procedure that must execute rapidly or lose to the generality of the direct numerical procedures. If implementation of this type of scheme is contemplated, some consideration should be given to ways of simplifying the integrodifferential equation even further. With this in mind, it is observed that the piecewise linear mixing length family used by Escudier and Nicoll[49] could also be used in the integral turbulence kinetic energy equation with a resulting major simplification. The join point between the inner region of the boundary layer where $l = y$ and the outer region where $l = l_\infty$ is designated by δ_j. Note that

$$\eta_j = \delta_j/\delta^+ = l_\infty/(\kappa\delta^+) \tag{6.72}$$

It is then further supposed that the dissipation length L can also be represented by a similar piecewise linear distribution as the mixing length. Since the mixing length is usually not too dissimilar from the dissipation length L in value, the turbulence integral thickness parameters appearing in the turbulence kinetic energy equation can be written to a reasonable degree of approximation as

$$\phi_1 = \left(\frac{l_\infty}{\delta^+}\right)^2 \int_{\eta_j}^{\eta_e} \frac{\bar{\rho}\bar{u}}{\rho_e u_e}\left(\frac{\partial \bar{u}/u_e}{\partial \eta}\right)^2 \mathrm{d}\eta + \kappa^2 \int_0^{\eta_j}\left(\eta\frac{\partial \bar{u}/u_e}{\partial \eta}\right)^2 \frac{\bar{\rho}\bar{u}}{\rho_e u_e}\,\mathrm{d}\eta$$

$$\phi_2 = \left(\frac{l_\infty}{\delta^+}\right)^2\left(1 - \frac{l_\infty}{L_\infty}\right)\int_{\eta_j}^{\eta_e}\left(\frac{\partial \bar{u}/u_e}{\partial \eta}\right)^3 \frac{\bar{\rho}}{\rho_e}\,\mathrm{d}\eta \tag{6.73}$$

In this simplified system, the wake mixing length no longer appears under the integral sign, so the interpretation of the integral turbulence kinetic energy equation as a differential equation for the wake mixing length greatly simplifies the implementation of the concept. The resulting scheme is still appreciably more complicated than the simple rate expressions of Goldberg[51] or Nash.[53] However, the effects of time dependency, compressibility, freestream turbulence, wall transpiration, and curvature are all accounted for within a relatively simple framework (presuming Bradshaw's[50] correction to the dissipation length L for the effect of curvature is accepted). In view of the demands of the turbomachinery environment, schemes such as the foregoing, based on the turbulence kinetic energy equation or other equations in the double velocity fluctuation set, have much to recommend them.

The disadvantages of schemes based on the turbulence kinetic energy equation are mainly that a profile family for the turbulent shear stress usually must be adopted; although if the assumed profile is reasonable, little is lost. In the foregoing scheme, by means of the mixing length assumption, the shear stress profile is related to the mean velocity profile in a manner that does appropriately recover the observed equilibrium state. However, at the same time, the assumed profile does suffer from the discrepancy that the Reynolds stress will disappear precisely at the same point the velocity gradient does. Boundary layers with velocity profile overshoots (such as film-cooled boundary layers) would therefore probably be subject to error, but almost certainly other factors will be involved such that the errors from the stress profile assumption might be insignificant. A problem also can arise with the assumed invariance of the relationship between the kinetic energy and the Reynolds shear stress, for instance when the shear stress changes sign but, of course, the energy does not. Such considerations give impetus to the development of better turbulence models; but, for the more conventional boundary layers, the assumptions given above are probably as accurate as the assumed mean velocity profile.

Additional moments of the turbulence kinetic energy equation could also be introduced to allow a more complex stress profile to be used, but it must be acknowledged that at the present time there is neither motivation nor information to proceed in this direction. A much more promising avenue for future development would be to integrate the equations governing the individual components of the turbulence kinetic energy, that is, the $\overline{u'^2}$, $\overline{v'^2}$, and $\overline{w'^2}$ equations, in conjunction with the Reynolds shear stress equation that governs the production of $-\overline{u'v'}$. Such a strategy alleviates some of the more restrictive aspects of the structural similarity hypothesis previously invoked. A preliminary evaluation by the present author of a four integral equation scheme using Rotta's hypothesis for the partitioning of dissipation yielded very favorable results.

As a final observation, it should be clear, if it has not become so earlier, that it should be possible to take integral moments of any of the current model partial differential equations that describe the spatial development of the time-averaged Reynolds stress tensor. In this manner, eventually it should be possible to construct versions of these model systems suited for use with the integral procedures for solving the boundary-layer equations. The additional capability attributable to these newer turbulence models would therefore become available within the framework of the integral procedure, subject of course to the limitations of the profile families, both of velocity and turbulence.

Conclusion on two-dimensional integral methods for blade boundary layers. In arriving at certain recommendations, it seems appropriate to mention briefly those particular methods that demonstrate the various recommended characteristic properties. To some degree, this has already been done in the previous discussion, but certainly it can bear repeating and re-emphasizing here. A great deal of the work done on integral methods for predicting turbulent two-dimensional or axially symmetric boundary layers is summarized in the 1967 Stanford Conference Proceedings[52] and only a comparatively small number of additional contributions have been made since that time. It is not the purpose of this section to exhaustively review these various contributions, but merely to point out certain desirable features and to cite examples where these desirable features have been employed.

A convenient starting point is probably the method due to Head,[35] since this particular method seems to mark the beginning of the present era of more accurate boundary-layer predictions. Head's procedure is, in the previously introduced terminology, a local equilibrium procedure in that the implied Reynolds apparent shear stress is uniquely determined by the local mean velocity. The technique is sufficiently simple as to be suitable for use with a programmable pocket calculator and its accuracy is quite acceptable for a surprisingly wide range of flows. The procedure was incorporated into a heat-transfer prediction scheme by Dvorak and Head,[34] although in this case the resulting scheme was a hybrid, since the energy equation was solved by a finite difference procedure after the mean flow was obtained by Head's integral scheme. Later Head's scheme was extended to compressible flow over insulated walls by Green[36] and for transpired flow by Thompson.[61] Perhaps with the exception of the heat-transfer scheme, any potential user would be well advised to inquire whether or not Head's scheme (or an existing development of it) might be adequate for their purposes. Insofar as the heat-transfer version of Head's scheme is concerned, it seems, at least to this author, that the hybrid heat-transfer scheme really does not have such a great deal to offer to warrant its use compared to a full finite difference scheme.

Most well-constructed local equilibrium procedures do perform, or could probably be made to perform, about as well as Head's method. However, methods using the integral kinetic energy equation are particularly convenient in that they have a simple form independent of the assumed form of the boundary layer mean profile. A good example of this type of method is to be found in the work of Escudier and Nicoll.[49] As an additional feature, methods such as that due to Escudier and Nicoll can permit the direct use of the same type of turbulence models as are being developed for the direct numerical procedures. It follows then that once the mean velocity and temperature profile families are specified, variables such as wall transpiration, wall roughness, freestream turbulence, and streamwise curvature can eventually be allowed for in this type of method. Compared to the direct numerical procedures, all the constraints of the integral procedures lie in the adopted velocity and temperature profile families. To continue, Lubard and Fernandez[62] developed a local equilibrium procedure that used the integral kinetic energy equation for transpired flows. Alber[63] has developed an impermeable wall procedure similar to the Lubard and Fernandez scheme but for compressible flow. Based on these precedents and the suggestions and development given earlier in the present work, it would be a simple matter to construct a compressible, transpired, rough wall, integral procedure based on local equilibrium concepts using the von Kármán momentum equation and the integral kinetic energy equation. In view of good results previously obtained with the various individual procedures, such a synthesis of methods would be expected to be quite successful for boundary layers in near equilibrium, with perhaps the compressible law of the wall additive constant with transpiration being the only questionable parameter over and above the mean velocity and temperature profiles at the present time.

The next step in the hierarchy would be to consider the departure from local equilibrium and here an expression for the rate of change of the appropriate integral of the turbulent stress, in the spirit of Goldberg[51] or Nash and Hicks[53] for instance, is very convenient and would result in a trivial amount of coding and increased computation. Turbulent lag equations of the foregoing type, although presently developed only for incompressible flow, do seem, at least conceptually, to result in an improvement over the local equilibrium procedures. Less empirical, but more laborious, are the schemes that use lag equations developed from the turbulence kinetic energy equation. The simplified version of the lag equation developed by McDonald and Camarata[58] from the integral form of the turbulence kinetic energy equation presented allows for compressibility, freestream turbulence and, using Bradshaw's[50] suggestion, for streamwise curvature. In collaboration with what could be termed the synthesized procedure described in the previous paragraph, the simplified integral turbulence kinetic energy equation would certainly typify the state-of-the-art of integral procedures for predicting turbulent boundary-layer development in 1975.

The final stage in the hierarchy of problems to be considered here is the problem of heat transfer. It is clear from the literature that, for insulated wall boundary layers, a Crocco-type temperature profile suffices to predict the effects of compressibility on the integral thickness parameters. It is also clear from work such as Rotta's[38] that really none of the convenient so-called Reynolds analogy factors, modified or otherwise, can adequately predict the heat transfer with a variable wall temperature in the presence of streamwise pressure gradients. These rather demanding conditions are unfortunately precisely those normally encountered in turbine blade design. The preliminary indications are that a stagnation temperature law of the wall with a wake component, such as that originally suggested by Rotta,[43] when used in conjunction with the integral thermal energy equation might improve the current poor status of heat-transfer predictions by integral methods. Finally, while it is very possible that heat-transfer prediction for internally convectively cooled turbine blades or vanes might be made acceptable relatively easily, it seems less likely that film-cooled components will be adequately treated within the framework of integral boundary-layer prediction schemes. Even for the idealized case of parallel slot injection of the coolant film, the resulting velocity profiles are not well described by the existing law of the wall/law of the wake concepts. At this point, the research effort required to further enlarge the mean velocity and temperature profile families to encompass film cooling, taken together with the estimated success probability, would seem to outweigh the increased cost of the present-day direct numerical schemes that already have shown their capability to handle at least simple film-cooling configurations. Thus, film-cooled boundary layers almost certainly lie outside the scope of present integral procedures and will probably remain there for some time to come.

Numerical Procedures for Solving the Boundary-Layer Partial Differential Equations

Motivation. In the light of the preceding discussion, caution must be advocated when using integral procedures for solving the boundary-layer equations whenever the profile families for the dependent variables are either inadequately verified or known to be inaccurate. It may well be that, in spite of the known profile shortcomings, as a consequence of the averaging out of the integral schemes, some of the flow properties will be predicted adequately for certain purposes. Unfortunately, heat transfer is usually a very demanding property to predict accurately and one of very keen interest to turbine designers. The problems that arise with two- and three-dimensional profile families in highly accelerated flows when the laminar and turbulent transport of heat and momentum are of a similar order (or in a film-cooled boundary layer developing on a curved surface) can be euphemistically termed "difficult." Numerical procedures that directly solve the partial differential equations and hence do not require this profile information therefore become much more attractive when these particular "difficult" problems must be treated.

Another very valid reason for using a numerical procedure for solving the partial differential equations lies in the flexibility of the resulting scheme. Changes in boundary conditions, higher-order terms, generalizations to include chemical reactions, and magnetohydrodynamic forces may all be incorporated with surprisingly little additional work. In some instances, such as the flow in a duct, the merging of shear layers causes perhaps only a reassessment and readjustment within the turbulence model being used within the numerical procedure; whereas, in the integral scheme, it calls for major reconstruction. Such developments are not simply mere possibilities, but even now clearly demonstrate the advantages of the numerical procedures.

General comments. In the course of the subsequent development, frequent reference will be made to existing or potential numerical procedures that can be used to obtain solutions to the problem under discussion. At this point, it is worth observing that in the present context a potential numerical procedure is considered to be a technique which has shown promise on simple problems, such as the transient heat conduction problem in one space dimension. Since the real problems are generally nonlinear, coupled, and multidimensional with initial and boundary conditions to be specified, there is a considerable difference between a potential and demonstrated capability and it is necessary to keep this distinction firmly in mind. In some instances, certain of the candidate procedures have clear advantages over others, provided the procedures can, in fact, return the solution to the degree of numerical accuracy required by the user. The potential advantages in practice might be quite unrelated to the relative cost of running the computer to obtain the solution. To evaluate the relative advantages it is first recognized that the engineer user has the choice of constructing a scheme based upon one of the more promising potential procedures to solve his problem or of adapting an existing available procedure for this purpose. Therefore, the engineer user must on an ad hoc basis evaluate the relative economics of the projected utilization of the possibly less than optimum existing procedures weighed against the engineering effort required to develop a new procedure embodying the better suited technique for his particular problem. Since costs even for the same computer vary from installation to installation dependent upon the computer configuration, the degree to which the computer is utilized, the accounting procedures used to obtain a charge rate, and the actual charge algorithm itself (which weights the various internal processes and resources within the computer), the relative computing economics can change dramatically from institution to institution. While computing costs at a given site may usually be projected with relative certainty, the task of the engineer user in making the economic evaluation is often plagued with major uncertainties. In particular, the engineer user, and more often than not, the computational expert as well, may not be sufficiently knowledgeable of the problem or of candidate techniques to accurately assess the labor required to cast them into a working procedure, or of their possible operational idiosyncrasies once ready for use. Since the more complex techniques can take many man-months of engineering effort to develop into operational procedures, a judgment error by either the computational expert or the engineer user in the evaluation phase can have a catastrophic economic impact.

In addition, sometimes both the existing and the prospective procedures might not behave in an acceptable manner for the particular application the user has in mind, and this fact might not be widely known or apparent to the user in advance. Similarly, certain procedures are user sensitive in operation and considerable skill might be necessary with these procedures to obtain satisfactory solutions. In seeking information upon which to base a reasonable evaluation of the risk factors involved, the diligent engineer is often further confounded by the conflicting claims and observations made by different authors regarding the operational characteristics of certain of the procedures. Several factors contribute to the generation of these conflicting claims. Principal among these is the user sensitivity mentioned earlier, which leads to a successful computation being dependent on such user-selected items as the computational mesh distribution and density or the initial and boundary conditions. Also a factor is a sensitivity of many of the procedures to computational detail, where the precise treatment of such items as nonlinearities, boundary conditions, or simply the ordering of the computation can have a major impact on the eventual outcome of the computation. Precise details such as this are often not given in archival journal publications for, among others, the obvious reason of space limitation.

Naturally, such real and prospective pathological difficulties demand that the engineer be thoroughly familiar with the existing candidate procedures and that very powerful reasons must exist to justify the development cost of any new and relatively untried procedure. There are, of course, valid reasons for embarking upon the prolonged development of a new procedure embodying more appropriate techniques for solving the particular problem. Unavailability or the general unsuitability of the existing procedures would be key factors in such a decision. Here again, however, conflicting claims of suitability or unsuitability are frequently to be found in the literature, often simply as a result of different authors having differing constraints and standards. The need to use the procedure on a computer with a very small high-speed memory might, for instance, sway judgment one way, whereas the need for a locally refined mesh, say to define the viscous sublayer of a turbulent boundary layer, might be a critical factor in another user's estimation. Less tangible, but not trivial, is the consideration that, should the engineer decide to develop the relatively untried procedure himself, he then is fully aware of the numerous approximations and prejudices incorporated into the procedure and is then well equipped to further extend the procedure into other problem areas.

The foregoing is in part designed to caution the engineer against a precipitous adoption of a promising but relatively untried numerical scheme for solving his particular problem. It is also aimed in part at answering the oft-heard question of the computational expert who feels that he has the most efficient algorithm for solving the particular problem under study and who questions why the engineering public continues to use what is to him a nonoptimum method.

Before going on to discuss the actual methods available for treating the boundary-layer equations, it is necessary to develop some basic concepts about numerical methods in general. However, it is not the purpose here to provide a broad introduction into general numerical methods for solving partial differential equations as there are some fine textbooks available on this subject. Only those topics that would be appropriate to aid a potential user in evaluating the numerical aspects of the candidate procedure will be discussed here. In this regard, three broad categories of methods exist for numerically solving the equations of fluid mechanics: the finite difference, finite element, and spectral methods. In the main, the subsequent discussion concerns finite difference methods simply because at the present time such methods are the furthest advanced in terms of fluid mechanics applications. Hybrid techniques also have made their appearance; for instance, finite difference methods may be viewed as a collocation technique with the dependent variable represented by a polynomial. If now the dependent variable in one or more coordinate directions is instead represented by some trigonometric expansion such as a Fourier series, the resulting scheme is termed a pseudospectral or hybrid pseudospectral method, depending on the degree of finite differencing employed. Indeed, other related methods can readily be formulated using orthogonal polynomials or piecewise polynomial splines to represent the dependent variables in one or more directions. In the usual spectral methods, the trigonometric series used to represent the dependent variable is introduced into the governing equations by means of the Galerkin technique, but at present the pseudospectral methods seem more efficient than the spectral methods by a factor of about two. Finite element methods have been very successful in structural problems, where they have proved especially useful with irregular boundaries. At the present time, the application of the finite element procedure to fluid mechanics problems is sparse and inconclusive. Certainly, in many applications the matrix inversion problem obtained from the finite element technique is identical to that arising from implicit finite difference formulations, so that such schemes must have comparable computational costs. The claimed attribute of finite element methods of ease of treatment of irregular boundaries is obtained at the expense of a nontrivial element construction process. To this must be added the observation that element shape and size variations can have a detrimental effect on accuracy.

In structural problems where finite element techniques have had remarkable success, most of the problems addressed appear to possess a single or at most two length scales. Viscous flow problems proliferate length scales with unfortunate ease. For instance, a turbulent boundary layer introduces three length scales normal to the wall. When the viscous sublayer is resolved, mesh aspect ratios of 10^3 or greater can be obtained based on the streamwise rate of change of physical processes. This type of mesh, and the rate of change of aspect ratio required to achieve reasonable solution with multiple scales present, does not cause a significant problem for many finite difference schemes, whereas it can cause major ill-conditioning for finite element schemes.

Thus, it does not at this point appear that the finite element approach will be quite the hoped-for panacea for geometric complexity. Finally, in the present work, the usual Eulerian body-fixed coordinate systems will be considered, as opposed to the Lagrangian system where the coordinates move with fluid. Lagrangian coordinate systems have considerable utility for relatively uncomplicated time-dependent flows containing discontinuities, but in the steady distorted flow the mesh interpolation problem appears at present to outweigh any benefit.

Order accuracy. The simplest way to discuss order accuracy is in terms of a truncated Taylor series. In setting up a finite difference molecule, the function value at adjacent grid points can be related to conditions at the grid point in question by means of a Taylor series. The Taylor series may then be appropriately truncated and an expression for the derivative at the point obtained. The truncation term in the Taylor series then determines the formal order accuracy of the difference scheme. For example, consider the three-point scheme with a mesh spacing of h and the continuous function f with single-valued finite derivatives

$$f(x+h)=f(x)+hf'(x)+h^2/2f''(x)+h^3/6f'''(x)+\cdots$$

$$f(x-h)=f(x)-hf'(x)+h^2/2f''(x)-h^3/6f'''(x)+\cdots$$

so by subtraction,

$$f'(x)=\frac{f(x+h)-f(x-h)}{2h}+\frac{h^2}{6}f'''(x)+\cdots$$

and hence the terminology that, if in the above equation the terms in h^2 and higher are neglected, the resulting finite difference representation of the derivative is termed second order. A fourth-order representation would be one where the first neglected term was h^4, and so on. The second-order finite difference representation given above is usually termed the first-derivative second-order central difference formula. A one-sided derivative is sometimes of use and from the foregoing in simplest form is written for a forward difference first derivative as

$$f'(x) = \frac{f(x+h) - f(x)}{h} - \frac{h}{2} f''(x) + \cdots$$

and clearly this is a first-order approximation.

Frequently and erroneously, the order accuracy of a finite difference formula for the derivative as defined above is appended to the solution procedure for the governing system of equations as a whole. Clearly, other factors can contribute to the errors in the solution to the system of equations. Three obvious sources of error that can swamp the errors arising from the derivative formula are: (1) linearization errors arising from the nonlinear nature of the governing equations; (2) errors arising from the treatment of the coupling between the equations within the system, including the boundary conditions; and (3) machine roundoff. The first two of these three sources of error will be discussed further later. In the subsequent development, order accuracy will be appended only to the difference formula and not to the overall procedure. Machine roundoff cannot be ignored as a possible source of error, but usually this problem manifests itself catastrophically during the initial development of the procedure, where appropriate remedies must be developed or the procedure abandoned.

It is observed that higher-order difference formulas are desirable in principle and, in some cases, represent the only reasonable alternative to reducing the mesh to an absurd degree to obtain accuracy. Several points must be noted, the first being that to obtain the increased order accuracy there has usually been an increase in labor. The question must now be asked if it is not more economic to use a lower-order difference with a more refined mesh. A number of authors have argued against methods of orders greater than fourth on this basis. Various ad hoc particularities can affect this judgment, however; for instance, in performing reacting boundary-layer calculations, the chemistry computation could greatly overshadow the fluid mechanics, thus driving one toward the use of as few grid points as possible, regardless of the cost per grid point of the fluid mechanics calculation.

Additional difficulties can be associated with higher-order methods, the obvious one being the application of boundary conditions since more extensive spatial differencing in the region of the boundaries would almost certainly be needed. Further, it has been observed that higher-order schemes tend to be difficult to work with in practice, due to their propensity to develop solutions with undesirable features such as "wiggles." Without doubt, improved higher-order schemes will appear in the future and will alleviate these problems.

It should be clear from the above that higher-order methods are not necessarily to be preferred over lower-order schemes. Accuracy can always be obtained from the consistent lower-order schemes, computer willing, by mesh refinement. The question is then simply the cost to the user of the required mesh. Should this cost be prohibitive, then a higher-order method or an alternative more efficient lower-order scheme might be the answer. In the subsequent discussion of the various methods, order accuracy, including the more subtle errors that might arise from linearization and decoupling, will be touched upon, but hopefully will not be overstressed.

Explicit and implicit methods. In the usual steady turbulent boundary-layer equations, if the laminar transport is not neglected with respect to the turbulent transport, the governing equations are everywhere parabolic and upon specification of a turbulence model form a well-posed initial boundary value problem. In certain cases, the neglect of the laminar transport can result in a hyperbolic set of equations, but this depends on the precise nature of the model of the turbulence transport. In either event, the equations may be solved by a forward marching technique, since one coordinate direction will possess a time-like property. In the subsequent discussion, this marching direction will frequently be termed the time coordinate for obvious reasons. The numerical methods for solving parabolic or hyperbolic equations that form a well-posed initial boundary value problem are usually termed either explicit or implicit methods. In an explicit technique, the unknown variables are expressed at the advanced "time" level entirely in terms of the known values at the current time level together with the appropriate boundary conditions. On the whole, explicit schemes tend to be conditionally stable. That is, a stable calculation can be obtained only by adhering to certain limitations on the time step. In an implicit procedure, the finite difference representation introduces unknown variables at both the advanced and the current time level. Generally speaking, a procedure of this type necessitates the solution of a set of simultaneous equations, which usually results in a complex and time-consuming procedure on a per grid point per time step basis. Implicit methods are usually stable for large time steps (although not necessarily unconditionally stable) and are therefore competitive with conditionally stable explicit procedures when the physical processes are changing on a time scale much greater than the marching step permissible with the conditionally stable explicit schemes. It turns out that the stability limits of the better known explicit schemes can be very restrictive in performing turbulent boundary-layer calculations; consequently, implicit schemes have found considerable favor. Unconditionally stable explicit schemes such as the DuFort-Frankel scheme are available, but to date these schemes have all been inconsistent. That is to say that, in the limit as the mesh is refined, the truncation error in the Taylor series expansion does not disappear for an arbitrary mesh spacing. In such a scheme, even given an unlimited computer, the user no longer has the ability to make the error as small as he wishes simply by mesh refinement. Therefore, inconsistent schemes have not been widely used, although Pletcher,[64] for example, has not had any apparent difficulty in using a DuFort-Frankel scheme on the boundary-layer equations.

Predictor-corrector procedures have been developed and these can be either explicit or implicit, or of some hybrid of the two types. Predictor-corrector schemes have become quite attractive for three-dimensional boundary layers and the prospect for treating aspects of the nonlinearities in the governing equations within the predictor-corrector framework is appealing.

Lastly, the so-called shooting methods defy description by the simple explicit-implicit terminology. In this and related techniques, the marching direction is discretized first, resulting in a nonlinear ordinary differential equation that poses a two-point boundary value problem. After linearization, the ordinary differential equation can be treated by a number of different schemes as an initial value problem. For example, in perhaps the most successful version of the shooting technique, a particular integral and a complimentary function are generated by numerical integration. The outer boundary condition then serves to determine the arbitrary constant in the complimentary function, and the solution of the linearized equation is obtained by adding the particular integral to the complimentary function. Unfortunately, iteration is required if more than one outer boundary condition is to be applied, as is usually the case in boundary-layer flows. Although successfully used for laminar flows and some turbulent flows, the convergence of the iteration involved in the shooting technique is not particularly reliable for arbitrary boundary-layer flows and has now been largely abandoned in favor of implicit methods. More recent developments of the shooting process, such as parallel shooting, have apparently not been evaluated on the boundary-layer equations.

In assessing the candidate methods for suitability to one's own particular problem, a definite screening process can be suggested based on the foregoing discussion. This process will now be discussed in detail in view of its importance in practice. Particular emphasis will, of course, be given to the special problems arising in turbomachinery applications.

Some factors influencing the choice of numerical method. On the basis of the preceding discussion, the choice of the most suitable numerical method for a given problem is an ad hoc one. Perhaps the most critical item is to estimate the rate at which the physical processes of interest can change in the marching (or iteration) direction. This leads to an estimate of the maximum step in the marching direction that the changing physical processes permit. This maximum physical step can then be compared to the maximum computational marching step that the various candidate procedures allow. The number of computational steps required to march the maximum physical step is thus estimated. To now compare the various schemes, an estimate must be made of the computational effort required to take one computational step. With this piece of information, the computational cost of marching the maximum physical step can be estimated and used to eliminate some of the candidate schemes. Finally, computer storage requirements and coding complexity, together with the estimated reliability and one's prior experience, may be factored in to complete the choice of method.

To see how this process works in practice, consider the simple incompressible turbulent boundary-layer momentum equation,

$$(\rho u)u_x + (\rho v)u_y = -p_x + (\mu_e u_y)_y \tag{6.74}$$

where the turbulent transport of momentum has been represented by an eddy viscosity coefficient μ_e. Now, based on experience with the simple transient heat conduction equation,

$$\frac{\partial T}{\partial t} = K\frac{\partial^2 T}{\partial y^2} \tag{6.75}$$

where t is the time (marching) coordinate, K the thermal conductivity, and T the temperature. It is well known that when applied to the simple heat conduction equation (6.75), most conditionally stable schemes exhibit a step size restriction of the form

$$R_{VSL} = \frac{K\Delta t}{\Delta y^2} \leq \frac{1}{2} \tag{6.76}$$

In the linearized boundary-layer problem, Eq. (6.76) approximately transforms into

$$\frac{\mu_e \Delta x}{\rho u \Delta y^2} \leq \frac{1}{2} \tag{6.77}$$

and it is found in practice that this type of restriction does indeed hold when these same methods are applied to the boundary equations. Now, Clauser[30] has suggested that in the outer region of an equilibrium incompressible turbulent boundary layer

$$\mu_e \simeq 0.016\rho_e u_e \delta^* \tag{6.78}$$

and this suggestion is in good accord with measurement. If now an equally spaced mesh normal to the wall is introduced

$$\Delta y = \delta/N \tag{6.79}$$

where N is as yet the undefined number of intervals, the viscous stability requirement reduces to

$$R_{VSL} \simeq 0.016N^2\delta^*\Delta x/\delta^2 \leq 1/2 \tag{6.80}$$

Typically in a turbulent incompressible boundary layer, $\delta/\delta^* \sim 6$, so that there results in round figures

$$\Delta x/\delta \leq 200/N^2 \tag{6.81}$$

Unfortunately, it is observed that the computational marching step decreases as the square of the number of grid points increases, which, of course, aggravates the problem of having to add grid points for increased accuracy. Further, based on observation, the physical processes change such that axial steps of greater than or equal to the boundary-layer thickness would be desirable in a computational scheme. It then follows that, for more than about 14 grid points normal to the wall in the boundary layer, the viscous stability restriction would force an axial step size of less than the desired boundary-layer thickness. This would not necessarily rule out schemes governed by a viscous stability criterion if the computational cost per axial step or accuracy were sufficiently better than competitive schemes taking a larger axial step. Now, a significant portion of the cost of marching one axial step goes into forming the coefficients for the integration scheme and is roughly independent of the type of integration, be it an explicit or implicit formulation. Further, very efficient matrix elimination schemes are available to treat the system of linear equations resulting from the implicit formulation. Thus, on balance, the implicit schemes generally tend in practice to be only a factor of two or three times more costly per axial step than the explicit schemes. Thus, superficially, the break-even point between conditionally stable explicit schemes and the (nominally) unconditionally stable implicit schemes for the usual boundary-layer equations occurs at roughly 20 grid points normal to the wall.

The axial step restriction can actually be a great deal more severe than the foregoing simplified arguments indicate. First, accuracy studies have indicated that 40 or more grid points normal to the wall might be required with a second-order difference scheme to accurately define the boundary layer in many circumstances, particularly when skin friction (or heat transfer) is important. (This figure could be reduced in some cases by fine tuning the mesh distribution.) Second, the previous arguments were based upon an equal mesh spacing normal to the wall and the viscous stability restriction is greatly aggravated if it becomes necessary to try to define the viscous sublayer of the turbulent boundary layer. The only alternative to defining the viscous sublayer by means of mesh points is to assume the validity of the law of the wall between the first grid point away from the wall and the wall proper. However, although the viscous sublayer is very thin, approximately 50% of the velocity change in the entire boundary layer occurs across it, so it is a comparatively important region. Thus, to date only the remarkably wide validity of the law of the wall has made the fitting of the profiles a viable alternative to adding more grid points. Unfortunately, in a number of instances of critical importance in gas turbine applications, the law of the wall is influenced by severe favorable pressure gradients and heat transfer (among other factors) and such influences cannot readily be accounted for as a simple correction to the law of the wall. Indeed, as was discussed in the section on integral prediction schemes, the whole topic of the thermal law of the wall is much less secure than the velocity law of the wall. Hence, in certain cases, much better success has been obtained using assumptions concerning the turbulence structure than assuming velocity or thermal profile similarity in the viscous sublayer. Thus, in gas turbine heat-transfer applications, there is a desire to place grid points deep within the viscous sublayer that virtually precludes the use of any conditionally stable procedure with a viscous stability limit near one-half. With implicit schemes if out of core storage is required, it usually can be obtained and the cost factored into the evaluation. Coding complexity, sometimes raised as a factor against implicit methods, turns out to be quite mythical, as that portion of the complete prediction procedure dealing with the actual integration can comprise less than 15% of the overall coding. For this reason, nominally unconditionally stable methods such as implicit schemes have found favor in boundary-layer prediction methods.

A second commonly encountered stability restriction is termed the Courant-Fredericks-Lewy (CFL) criterion. Considering the simple advection equation,

$$\frac{\partial u}{\partial t} = a\frac{\partial u}{\partial x} \tag{6.82}$$

the CFL criterion generally requires that

$$a\Delta t/\Delta x \le 1 \tag{6.83}$$

which when interpreted in the boundary-layer frame of reference requires

$$v\Delta x/u\Delta y \le 1 \tag{6.84}$$

and since v is small within a boundary layer, the CFL requirement generally does not pose as severe a restriction as the viscous stability condition discussed previously.

Nonlinear aspects. As has been observed earlier, the boundary-layer equations are nonlinear when expressed in terms of the velocity variables. Thus, some discussion of the particular problems arising from this complication is required. A change of variables can somewhat alleviate the nonlinearities, but with a temperature-dependent viscosity, compressible flow, or a turbulent effective viscosity, no completely linear system can be achieved by a change of variable. As a general rule, implicit methods with their eventual goal of a system of linear algebraic equations make a formal linearization at some point inevitable if the governing equations are nonlinear. The key point here is that inadequate treatment of the nonlinear terms can seriously degrade the overall accuracy of the method. Two popular methods of dealing with the nonlinearities are to be found in the literature. In the first method, finite difference operators are applied to the governing equations and a nonlinear set of algebraic equations obtained. The nonlinear set of algebraic equations can then be solved, for instance, by a Newton-Raphson technique. As an alternative, the governing equations can be linearized by some procedure and the finite difference operators then produce a linear algebraic system without further ado. Explicit procedures can, of course, be applied to exactly the same set of linearized governing equations; however, as a class they do permit a simpler process of linearization as a result of the extensive use of the known level solution. In the explicit schemes, it is usual to formulate the problem in the so-called conservative form of the governing equations, that is, as many as possible of the dependent variables are placed within the derivative operator. For example, by use of the continuity equation, the streamwise momentum equation given by Eq. (6.74) can be rewritten as

$$(\rho u^2)_x + (\rho uv)_y = -p_x + (\mu_e u_y)_y \tag{6.85}$$

and denoting X marching locations by the subscript j and y locations by the subscript i, the momentum equation can give, for instance, the explicit formula

$$(\rho u^2)_{j+1} = (\rho u^2)_j + (x_{j+1} - x_j)\left[-\left(\frac{(\rho uv)_{i+1} - (\rho uv)_{i-1}}{y_{i+1} - y_{i-1}}\right)_j - \left(\frac{p_{j+1} - p_j}{x_{j+1} - x_j}\right)_i + \left(\frac{(\mu_e u_y)_{i+1/2} - (\mu_e u_y)_{i-1/2}}{(y_{i+1} - y_{i-1})/2}\right)_j\right] \tag{6.86}$$

and clearly the only remaining problem, since everything at the j location is known and $p(x)$ given, is the determination of ρ and v, which is accomplished by similar integration of the continuity and energy equations. The linearization problem has been effectively circumvented by the explicit integration of the conservative form of the governing equations. However, the conditional stability problem of explicit methods remains.

There are, of course, other advantages to using a conservative form of the governing equations and its use is not restricted to explicit schemes (see Ref. 65). Here it is not proposed to discuss this particular aspect in depth. Clearly, if the two forms of the governing equations are equivalent, then for a properly resolved flow the same answer should be obtained from the conservative as from the nonconservative system, to within the truncation error. The preference for a particular form, usually the conservative form, stems either from the linearization simplification or from the case where the flow is only marginally or poorly resolved. In the latter case, the truncation errors are large and are often more palatably disguised within the conservative scheme. For this reason, when either form can be used without prejudice, the conservative form is usually selected. Some procedures, such as the Lax-Wendroff scheme, require governing equations written in a conservative form.

Returning to the question of linearization with implicit schemes, in addition to the two previously mentioned approaches at least one other strategy has been evolved. This third approach is an ad hoc process that examines an expression and in each term the highest derivative for the dependent variable is taken as the implicit term and all others evaluated as a known coefficient, using either a previous station's information or an estimate. For example, in the streamwise momentum equation, considering incompressible flow for the moment, the streamwise convective term would be linearized as

$$(\rho u)u_x = (\rho u)_{j,i}\left[\frac{u_{j+1}-u_j}{x_{j+1}-x_j}\right]_i \tag{6.87}$$

Obviously, such a linearization is quite crude and introduces a first-order error in the solution. A second and considerable improvement is to use a Newton-Raphson process to obtain

$$u_{n+1}^2 = 2u_n u_{n+1} - u_n^2 + O(u_{n+1}-u_n)^2 \tag{6.88}$$

and so obtain

$$uu_x = \tfrac{1}{2}(u^2)_x = \tfrac{1}{2}\left(2u_n u_{n+1} - u_n^2\right)_x \tag{6.89}$$

where n is an iteration index. The concept here is that starting from an initial guess, perhaps the previous x station solution, a sequence of linear problems can be solved until the nonlinear terms are adequately represented. The point to note with this process is that it can be time consuming and unrewarding, in fact, to iteratively reduce the errors arising from the linearization approximation below the basic truncation error of the difference scheme; for instance, why represent u^2 much better than the streamwise derivative? This point is the basis of noniterative linearization that will now be developed in detail in view of the importance of implicit methods in boundary-layer theory and the critical role of linearization in implicit methods.

Briley and McDonald[66,67] applied a noniterative linearization technique in developing implicit procedures for solving the multidimensional compressible Navier-Stokes and reduced Navier-Stokes equations. The linearization was based on a Taylor expansion of the nonlinear terms about the known level solution. In order to illustrate this technique and introduce the concepts of the relationship of truncation error arising from linearization to truncation errors arising from spatial discretization, consider the nonlinear ordinary differential equation

$$\frac{d\phi}{dx} = f(\phi) \tag{6.90}$$

with exact solution $\phi = \Phi(x)$. A centered difference between levels $n+1$ and n leads to

$$\frac{\phi^{n+1}-\phi^{n}}{\Delta x}=\frac{1}{2}\left(f^{n+1}+f^{n}\right)-\frac{\Delta x^{2}}{12}\Phi'''(x_i) \tag{6.91}$$

The central problem of the linearization now is to obtain a satisfactory representation of f^{n+1} that contains only linear contributions from ϕ^{n+1}. This can be done by expanding about the n level, as

$$f^{n+1}=f^{n}+\left(\frac{\mathrm{d}f}{\mathrm{d}\phi}\right)^{n}\Delta\phi+\left(\frac{\mathrm{d}^{2}f}{\mathrm{d}\phi^{2}}\right)^{n}\cdot\frac{\Delta\phi^{2}}{2}+O(\Delta\phi^{3}) \tag{6.92}$$

and again by expansion about the n level,

$$\phi^{n+1}=\phi^{n}+\Delta x\left(\frac{\mathrm{d}\phi}{\mathrm{d}x}\right)^{n}+\frac{\Delta x^{2}}{2}\left(\frac{\mathrm{d}^{2}\phi}{\mathrm{d}x^{2}}\right)^{n}+O(\Delta x^{3}) \tag{6.93}$$

Thus, from Eq. (6.93), the error in Eq. (6.91) arising from retaining only linear terms in the representation of f^{n+1} given by Eq. (6.92), termed the nonlinear truncation error (NLTE), is of the order

$$\frac{1}{4}\left(\frac{\mathrm{d}^{2}f}{\mathrm{d}\phi^{2}}\right)^{n}\left[\Delta x\left(\frac{\mathrm{d}\phi}{\mathrm{d}x}\right)^{n}+\frac{\Delta x^{2}}{2}\left(\frac{\mathrm{d}^{2}\phi}{\mathrm{d}x^{2}}\right)^{n}\right]^{2}$$

$$=\frac{1}{4}\left[\frac{\mathrm{d}}{\mathrm{d}x}\left(\frac{\mathrm{d}f}{\mathrm{d}x}\Big/\frac{\mathrm{d}\phi}{\mathrm{d}x}\right)\Big/\frac{\mathrm{d}\phi}{\mathrm{d}x}\right]^{n}\left[\Delta x^{2}\left(\frac{\mathrm{d}\phi}{\mathrm{d}x}\right)^{2n}+\cdots O(\Delta x^{3})\right]$$

$$\approx\frac{1}{4}\Delta x^{2}\left[\Phi'''(x_i)-\Phi''^{2}(x_i)/\Phi'(x_i)\right] \tag{6.94}$$

where $f=\Phi'(x_i)\neq 0$, whereas the spatial truncation error (STE) in Eq. (6.91) is of order $(\Delta x^{2}/12)\Phi'''(x_i)$. Both truncation errors are of the same formal order so that in this example it is obviously inappropriate in general to proceed to improve only the NLTE by, say, a Newton-Raphson technique or a predictor-corrector procedure, since the effort involved is more effectively expended by decreasing the step size Δx, which would, consequently, decrease *both* the NLTE and the STE.

Based on the results of the simple example shown above, several additional points are suggested, notably that the ad hoc linearizations mentioned previously could effectively reduce the order of the truncation error of the overall method. Furthermore, an optimum situation is where the NLTE and the STE are of the same order. Since the overall truncation error is determined by the lowest-order error, the effort required to obtain the higher-order truncation error, be it NLTE or STE, must be carefully weighed against the lack of improvement in the overall method truncation error. Should a case arise where the NLTE is the determining factor, this error can be reduced in the present linearization by taking additional terms into account in the expansion about the known time level. These additional terms may be evaluated in a number of different ways, i.e., by iteration across a step, but once again care must be taken to ensure that the result is worth the effort. In this regard, the degree of iteration consistent with the spatial truncation error in the above linearization is quite evident, unlike, for instance, some of the techniques for solving the nonlinear difference equations where careful establishment of the convergence criterion (or number of allowable iterations) is necessary.

The procedure may be formalized for a typical system with ϕ a function of x, y,

$$\frac{\partial}{\partial x} H(\phi) = F(\phi) \frac{\partial}{\partial y} G(\phi) \tag{6.95}$$

Centering between the $n+1$ and n pseudotime levels and proceeding as in Eqs. (6.90–6.93) yields

$$\frac{H(\phi_i^{n+1}) - H(\phi_i^n)}{\Delta x} = F(\phi_i^n)\delta_y G(\phi_i^n) + \frac{1}{2}\left\{F(\phi_i^n)\delta_y\left[\left(\frac{\mathrm{d}G(\phi)}{\mathrm{d}\phi}\right)_i^n (\phi_i^{n+1} - \phi_i^n)\right]\right.$$
$$\left. + \left(\frac{\mathrm{d}F(\phi)}{\mathrm{d}\phi}\right)_i^n (\phi_i^{n+1} - \phi_i^n)\delta_y G(\phi_i^n)\right\} + O(\Delta x^2, \Delta y^2) \tag{6.96}$$

If $H(\phi)$ is simply equal to ϕ, the expansion given by Eq. (6.96) is both linear and second order and can then be utilized in the implicit elimination without further manipulation. Further, $F(\phi)$ is unity and $G(\phi)$ equal to ϕ, the conventional Crank-Nicolson scheme is recovered. It is apparent from Eq. (6.96) that if the pseudotime derivatives are themselves linear, then second-order accuracy requires no more than a simple Crank-Nicolson averaging of the transverse spatial derivatives followed by linearization in the manner of Ref. 66. If the pseudotime derivatives are nonlinear, such as occur in the conventional boundary-layer equations, then care is required to maintain second-order accuracy. The problem becomes apparent if, for the moment, attention is devoted to the simple problem of Eq. (6.90). Here it can easily be shown that even with the simplest of linearizations, i.e., that f^{n+1} is equal to f^n, first-order accuracy in the pseudotime direction is obtained. Thus, it is intuitively reasonable to expect even a modest linearization (approximation) to f^{n+1} to do better than first-order accuracy. Turning now to Eq. (6.96), where the pseudotime derivatives contain nonlinearities, it is obvious here that the simplest of linearizations, $H(\phi^{n+1}) = H(\phi^n)$, completely destroys the solution. Thus, considerable care is required in the case of nonlinear pseudotime derivatives. In these circumstances, an advantage is obtained by basing the linearization on an expansion about the $(n+1/2)$ level, as outlined below, and this yields

$$\frac{H(\phi^{n+1})-H(\phi^{n})}{\Delta x}=\left(\frac{\mathrm{d}H}{\mathrm{d}\phi}\frac{\partial\phi}{\partial x}\right)^{n+\frac{1}{2}}+O(\Delta x^2) \tag{6.97}$$

further,

$$\left(\frac{\mathrm{d}H}{\mathrm{d}\phi}\right)^{n+\frac{1}{2}}=\left(\frac{\mathrm{d}H}{\mathrm{d}\phi}\right)^{n}+\left(\frac{\mathrm{d}^2H}{\mathrm{d}\phi^2}\frac{\partial\phi}{\partial x}\right)^{n}\frac{\Delta x}{2}+O(\Delta x^2) \tag{6.98}$$

and, of course, in Eq. (6.97), $(\partial\phi/\partial x)^{n+1/2}$ is readily represented by a second-order accurate finite difference operator. The critical term in the linearization is then $(\partial\phi/\partial x)^{n}$ appearing in the expression for $(\mathrm{d}H/\mathrm{d}\phi)^{n+1/2}$. This term must be represented *without* a $(n+1)$ level term in the finite difference operator, otherwise linearity will be lost. However, it is clear that in Eq. (6.98) a first-order representation of $(\partial\phi/\partial x)^{n}$ is adequate to preserve overall second-order accuracy and so, with a central difference for $(\partial\phi/\partial x)^{n+1/2}$ and a backward difference for $(\partial\phi/\partial x)^{n}$, there results

$$\frac{H^{n+1}(\phi)-H^{n}(\phi)}{\Delta x}=\frac{\phi^{n+1}-\phi^{n}}{\Delta x}\cdot\left[\left(\frac{\mathrm{d}H}{\mathrm{d}\phi}\right)^{n}\right.$$

$$\left.+\left(\frac{\mathrm{d}^2H}{\mathrm{d}\phi^2}\right)^{n}\cdot\left(\frac{\phi^{n}-\phi^{n-1}}{\Delta x_b}\right)\frac{\Delta x}{2}\right]+O(\Delta x^2) \tag{6.99}$$

which is the desired expansion. The combination of Eqs. (6.96) and (6.99) then provides the linear difference representation of the model equation (6.95) to second-order accuracy.

The foregoing linearization is readily applied to the boundary-layer equations using chain rule differentiation and interpreting H, for instance, as ρu^2 and ϕ as ρuv. In combination with either a multilevel difference scheme or by use of iteration, any order nonlinear truncation error can be obtained if desired. Again, it must be clearly stated that the marching linearization as presented above is conceptually quite different from the Newton-Raphson linearized system developed as part of the solution of the nonlinear equation. In the marching case, by linearizing about the previous station's known solution the spatial discretization error and nonlinear truncation errors go hand-in-hand and improvements in both are obtained by reducing the marching direction step. Iteration, which improves only the nonlinear error (and perhaps stability), costs roughly the same as one marching step and if possible should be avoided. Newton-Raphson (or quasilinearization, as it is sometimes called) is a technique for solving a nonlinear equation that as a byproduct produces a linear system for iteration. If one identified the first (known) iterant as the previous marching direction solution and the second (unknown) iterant as the current marching direction solution, then in many cases identical formulas to the marching linearization are obtained. As a result, many investigators have not discriminated between marching linearization and Newton-Raphson (quasilinearization), but without this discrimination the critical relationship between spatial and nonlinear error is lost.

The coupled system of equations. In the usual boundary-layer system of equations, the transverse momentum equation reduces to the requirement that the static pressure be constant across the boundary layer; this leaves the streamwise momentum, continuity, and thermal energy equations to determine the velocity components u and v and the density ρ. Further, it is clear that these three remaining equations are coupled together in that all three of the dependent variables appear in each of the governing equations. The inadequate treatment of this coupling can seriously degrade the quality of the solution.[68] Perhaps the simplest coupling treatment is a Picard type of iteration with the streamwise momentum equation generally being regarded as an equation for u, the energy equation giving ρ, and the continuity equation yielding v. The three equations are then solved sequentially using the most recent values of the dependent variables. Iteration then serves to couple the equations together. Blottner[68] was able to show that a very large number of such iterations (greater than 10) was often required to obtain overall second-order accuracy with second-order spatial difference formulas with iterative coupling.

A much more attractive method is to introduce a stream function ψ that identically satisfies the continuity equation and the coupling remains only between the energy and streamwise momentum equation. This residual coupling is often handled in the sequential Picard manner outlined previously; certainly, in cases of high heat-transfer rates or at high Mach numbers, one would expect that a similar large number of iterations would be necessary in this scheme for overall second-order accuracy. Finally, of course, the system can be treated directly as a coupled system and the matrix enlarged to form a block-banded system. See, for instance, the pioneering work of Flugge-Lotz and Blottner[69] (also, see Ref. 68 for a bibliography). The cost of solving the complete coupled system is quite high compared to solving each of the equations individually. However, considering the number of iterations required in the sequential process and the possible convergence problems, the direct coupled system becomes very attractive. In two dimensions, there does seem quite a lot to be gained from eliminating the continuity equation by some means or other, although the resulting complication of the streamwise momentum equation might be significant. The continuity equation is a first-order equation, with the momentum equations being at least second-order. Boundary conditions on the continuity equation can pose a problem, so that its elimination from the system can be advantageous from this point of view as well. Unfortunately, in three-space dimensions the utility of the stream function is greatly diminished, since at least two stream functions must be introduced to eliminate the continuity equation.

Coordinate systems and finite difference grids. For a numerical solution of the boundary-layer equations, the finite differencing could be applied directly to the governing equations in a primitive variable form via Cartesian coordinates. Clearly, however, difficulty would be experienced as the boundary layer grew in the streamwise direction and the stagnation points would require special treatment.

The so-called body-fitted coordinates s and n are much to be preferred from the viewpoint of accuracy. In this case, (s, o) is aligned with the body surface and for convenience n is usually taken as orthogonal to the body surface with the velocity components aligned in the coordinate directions. This coordinate system avoids boundary condition interpolation at the intersection of the Cartesian coordinates and the body surface. Still left untreated is the growth of the boundary layer within the mesh and how to determine the outer edge of the computational domain such that no active region is left untreated. However, efficiency is preserved by not taking a large normal extent for the domain. A further observation is that, because of the nature of the boundary-layer approximations and the requirements this places on the geometry, the form of the equations in the s, n system is to first order identical to the usual Cartesian form. In view of this identity of form, convention will be observed and x, y used as the streamwise s and surface normal n coordinates.

At first sight it might appear that stagnation regions could be dispersed with in descriptions of turbulent boundary layers. However, the stated aim here is to perform airfoil-type boundary-layer calculations, so some treatment of the stagnation region is called for. One major attribute of finite difference methods is that the discriminations between the boundary-layer states can all reside in the turbulence model and that one single scheme can (conceptually) integrate the governing equations from the stagnation point into the wake.

The problem is further compounded in turbulent flow by having at least two (and usually three) length scales in the direction normal to the wall, y^+ and y/δ, where y^+ is the previously defined law of the wall coordinate $y(\tau_w/\rho_w)^{\frac{1}{2}}/\nu_w$ and δ is some outer region thickness scale such as the boundary-layer thickness. The problems of boundary-layer growth and stagnation points are generally treated by means of a transformation of the governing equations, while the problem of multiple length scales is usually treated by means of a nonuniform finite difference grid. Both topics will be briefly reviewed.

Insofar as transformations of the governing equations are concerned, much of the early work in laminar flow used the Crocco variables x and u as independent variables with the shear stress τ as dependent variable. This scheme has several valuable properties, one of which is the combining of the continuity equation and streamwise momentum equation. This has distinct advantages from the viewpoint of equation coupling discussed previously. In addition, by choosing equal increments in u, the problem of the multilength scale boundary layer is treated without the need for a nonuniform mesh. Unfortunately, this transformation does not countenance double-valued velocities such as might occur with a velocity profile overshoot, a common occurrence in a film-cooled boundary layer. Consequently, this transformation is not widely used in turbulent boundary-layer procedures.

As mentioned previously, many authors introduce the stream function, which can be done as an independent variable in the von Mises form of the equations. Here the advantages are similar to the use of the Crocco variables, but the double-valued stream function problem is not permissible in any event in the steady boundary-layer system of equations as it implies reverse flow. In the von Mises system, a singularity at the wall requires careful handling. However, in many circumstances the governing equations need not be solved at the wall proper, but can be structured such that the first point away from the wall is the first location at which the governing equations are used. Thus, the von Mises form is quite attractive, although it does not remove the problem of leading-edge singularities nor grid growth. Patankar and Spalding,[70] for instance, use this formulation in their widely used turbulent boundary-layer prediction scheme.

Other authors have used transformations of the Dorodnitsyn-Howarth-Stewartson-Illingworth variety, which introduced a new normal coordinate stretched by the local density. Use of this transformation reduces the resolution problem with severe density gradients; however, it does nothing for the stagnation point problem. This transformation is, in fact, incorporated into the Levy-Lees transformation, which does have the property of removing the leading-edge starting point difficulty. The Levy-Lees transformation is quite straightforward and is particularly well suited to laminar flow due to its close relationship to the laminar flow similarity coordinates. Obviously, this relationship is of less importance in turbulent flow, where the laminar flow similarity coordinates are not relevant and such a formulation may even be at some disadvantage. Be this as it may, the Levy-Lees transformation has been used quite successfully by Cebeci and Smith[71] for both laminar and turbulent flow.

In all of the foregoing transformations, use may be made of the known edge quantities to nondimensionalize variables. As an extension of this process, some authors use a transformation where the distance normal to the wall is normalized by some local boundary-layer thickness parameter, usually the "incompressible" displacement thickness δ^*. Such a transformation has the advantage that it can remove the leading-edge starting problem and also ease the nonlinear problem since, for instance, the stream function ψ/ψ_e as a function of y/δ^* does not change rapidly in the marching direction. Thus, a prior station's solution provides a very good approximation for treating the nonlinear process. Grid growth is also more easily treated in this frame. The principle disadvantage of this transformation is that the unknown thickness δ^* at the location where the solution is being sought is required and this can be accomplished either iteratively (as part of a nonlinear iterative process) or directly as part of the elimination process in the solution of the banded matrix. This δ^* transformation has been used successfully by Herring and Mellor[72] and McDonald and Fish.[60]

In summary, then, given the basic desire to have one formulation that can encompass the entire turbine airfoil boundary layer from the leading-edge stagnation point and to the airfoil wake, the choice would seem to lie between the Levy-Lees or δ^* transformation (or indeed a combination of them both) with the use of a stream function in either case. Apart from the stagnation region problem (which has been circumvented by a number of authors who simply start from a very thin initial boundary layer of some assumed shape), the von Mises transformation is also very attractive.

Turning now to the problem of distributing the mesh points across the boundary layer, two complicating factors in what would otherwise be a straightforward process are observed. The aim here is to have a grid that allows roughly equal increments in the variables ρ, u, or ρu, whichever is the most demanding. As was mentioned earlier, use of the von Mises transformation greatly simplifies this process in all but a few instances. However, in general, the first complication is the a priori determination of what will be an adequate grid for a given problem. The second complication concerns the structure of the finite difference formulation with a nonuniform mesh. Each of these factors will be discussed in turn.

Taking the easier factor first, that of considering the effect of the nonuniform mesh on the finite difference structure itself, it is noted that this problem arises only when the finite difference molecule requires more than two grid points. The Keller box method,[73] which centers the solution between grid points, thus very neatly circumvents the unequal mesh problem, since it is essentially a two-point method. The Keller box method does have certain other less attractive features, so that, in general, other treatments of the unequal mesh problem might be required. The very straightforward method of simply devising unequally spaced difference formulas by Taylor expansion or equivalently by Lagrange interpolation does have the problem that, unless the mesh eccentricity (i.e., ratio of adjacent grid step size) is near unity, a reduction in order accuracy of the calculation will result. Blottner[74] has discussed this problem in detail and advocates use of a transformation to determine the unequal mesh difference formulas where

$$y = f(y') \tag{6.100}$$

and in the y' space an equally spaced mesh is used. Difference formulas are then constructed by the chain rule

$$\frac{\partial}{\partial y} = \frac{\partial}{\partial y'}\frac{\mathrm{d}y'}{\mathrm{d}y} = \left(\frac{\mathrm{d}f}{\mathrm{d}y'}\right)^{-1}\frac{\partial}{\partial y'}$$

$$\frac{\partial^2}{\partial y^2} = \left(\frac{\mathrm{d}f}{\mathrm{d}y'}\right)^{-2}\left[\frac{\partial^2}{\partial y'^2} - \frac{\mathrm{d}^2 f}{\mathrm{d}y'^2}\left(\frac{\mathrm{d}f}{\mathrm{d}y'}\right)^{-1}\frac{\partial}{\partial y'}\right] \tag{6.101}$$

and Blottner[74] advocates a finite difference approximation to the derivatives of the stretching function f, with the second derivative evaluated in conservative form, e.g.,

$$\left(\frac{\mathrm{d}f}{\mathrm{d}y'}\right)_i^{-1} = \frac{\Delta y(\Delta y^{+2} + \Delta y^{-2})}{\Delta y^{+}\Delta y^{-}(\Delta y^{+} + \Delta y^{-})}$$

and

$$\Delta y^+ = y_{i+1} - y_i \qquad \Delta y^- = y_i - y_{i-1} \tag{6.102}$$

whereas Briley and McDonald[66] have preferred with some success to differentiate an analytic stretching function, notably the one suggested by Roberts.[75] Similar analytic differentiation of the geometric progression form of the distribution function used by Cebeci and Smith[76] and McDonald and Fish[60] is readily accomplished. In practice, given the choice, the transformation process is to be preferred for determining unequally spaced difference formulas, if for no other reason than to ensure a smooth distribution of truncation error.

Turning now to the second factor, that of determining a suitable mesh distribution for a given problem, here the various approaches exhibit little unanimity. The problem is greatly simplified if it is not desired to define the sublayer, a course of action previously deemed unacceptable in regions of high heat transfer or highly accelerated flows. Above the sublayer, a very nearly equal mesh suffices for the turbulent boundary layer and the only problem is the growth of the boundary layer. The growth problem may be treated by periodic reinterpolation of the solution and is particularly simple with an explicit scheme.[77] Two stretching techniques in use to define the sublayer are the previously mentioned geometric progression and the Roberts' technique. In using either technique, a minimum step size normal to the wall is estimated, usually on the basis of requiring a $y^+_{\min}$ to be in the region of two to four. In the geometric progression scheme, the number of grid points felt appropriate to define the boundary-layer thickness serve to define the required constant grid eccentricity. This selection process usually assures a good grid for turbulent flow, provided the eccentricity is less than about 1.1. However, for laminar flow the grid is almost certainly too refined, but overall this rather conservative selection works quite well in practice in performing airfoil boundary-layer calculations. Even so, McDonald and Fish found it convenient to have a reinterpolation scheme available so that if necessary the grid could be restructured during the course of a calculation.

In using the Roberts' technique, two or more domains of uniform but differing mesh sizes are identified. Obviously, in the boundary layer the two regions are the wall and wake regions. The Roberts' scheme then distributes the grid points to smoothly blend the various regions. Again, the desire to place, say, three or four grid points below a y^+ of 15 determines the inner scale, while the total number of available grid points and the overall maximum boundary-layer thickness provide the necessary information to prescribe the transformation. Again, in practice, the laminar region of a typical airfoil is probably overly described, but the overall problem does not usually lend itself to optimizing the normal grid for the axial flow development. The most feasible approach at this point is to monitor the axial development and, should the mesh become too badly distributed, perform a reinterpolation of the grid. Excepting very high Mach number flows where the density gradients near the outer edge of the boundary layer can be appreciable, either the geometric progression or the Roberts' scheme can lead to an acceptable grid. Use of a nondimensional y coordinate with a localized boundary-layer thickness scale as the normalizing agent does greatly simplify the problem of having an adequate grid for the entire airfoil surface.

Finally, it remains to define the term "adequate grid" used previously and here only very general guidelines can be given. It should be clear at this point that the onus of determining whether or not a grid is acceptable must lie with the user, in view of the ad hoc nature of the grid selection process. It is quite unrealistic to expect the originators of the procedure to provide a default mesh selection scheme that will be adequate for all boundary-layer problems. The user must select a mesh, both normal and axial, and determine the variation of some critical parameter (such as skin friction) with the mesh refinement on a typical problem case. Only when the critical parameter does not change to some acceptable degree of tolerance (user defined) can the mesh be considered adequate. Subsequent cases must still be examined to see that the solution remains similarly distributed to the mesh refinement case. For airfoil calculation, an approximate guide might be to take axial step sizes in the region of the boundary-layer thickness; this translates roughly into 50–60 axial locations per surface. Normal to the wall prior experience suggests something in excess of 40 grid points in a second- or lower-order scheme with the first grid point placed at a y^+ of between 2 and 4. Subsequently, it does not seem advisable to allow the velocity to change from grid point to grid point by more than 0.05 u_e without redistributing the mesh. Given a mesh of this degree of refinement, it might be expected that changes in the heat-transfer rates could be held to less than approximately 2% upon mesh refinement.

Starting profiles. For an initial value problem of the boundary-layer type, information must be supplied that either directly or indirectly allows the starting profiles of the dependent variables to be prescribed. The velocity and temperature profile families for both laminar and turbulent flow discussed in the section on integral methods have been widely used in this regard. These profiles can be used without difficulty to specify the density and axial velocity at some initial location. Difficulty, however, can be experienced in determining the variation of transverse (or spanwise) velocity across the boundary layer. Examination of the continuity equation reveals that this transverse velocity is a function of the axial rate of change of the axial velocity. Thus, for consistency, this transverse velocity really should be obtained from some suggestion as to the axial variation of the axial velocity profile. The various methods used in this regard will now be reviewed briefly.

First, the problem can be resolved by setting the transverse velocity to zero at the initial station, since indeed boundary-layer theory requires the transverse velocity to be small in any event. Subsequently, quite rapid changes in the predicted boundary-layer behavior can occur and this starting technique cannot be recommended unless the predictions in the region of interest can be shown to be insensitive to the starting conditions. When predictions are being compared with measurements, it is often desired (and the information known) to match the initial rate of change of, say, the displacement thickness to the observed value. By integrating the continuity equation, one obtains

$$\left(\frac{v}{u}\right)_e = \frac{\mathrm{d}\delta^*}{\mathrm{d}x} - \frac{(\delta-\delta^*)}{\rho_e u_e}\frac{\mathrm{d}}{\mathrm{d}x}(\rho_e u_e) \tag{6.103}$$

so that the desired transverse velocity v_e at the edge of the boundary layer at the initial location is known. By a simple axial iteration, the transverse velocity profile shape at the first computed axial location in terms of, for instance, v/v_e vs y/δ^* could be used at the initial location, with the v_e obtained from Eq. (6.103). Further, in cases where the measured axial rate of change of displacement thickness is not known, some approximation (such as the flat plate value) for this rate of change could be adopted.

As a straightforward development of the foregoing technique, it is often assumed that, in terms of certain similarity coordinates, the axial derivatives are negligible, i.e., a local equilibrium is assumed. Special forms of the governing equations independent of the axial coordinate are then obtained and the equilibrium values of all the velocity components found, usually iteratively by integration. This is in essence the treatment used at stagnation points for, although the stream function ψ or the velocities u and v might be zero, ψ/ψ_e, u/u_e, and v/v_e are still well behaved and can be derived from assumptions of the type

$$\psi/\psi_e = f(y/h) \tag{6.104}$$

and

$$\frac{\partial f}{\partial x} = 0 \tag{6.105}$$

By use of the local equilibrium hypothesis, both the axial and transverse velocity profiles can be obtained at the starting location. The iterative solution of the similarity equations can be started from the laminar or turbulent profile families described earlier. In this mode of operation, no additional information is required by the finite difference procedures that would be required to start the integral methods. This local equilibrium starting technique has been used very successfully in both laminar and turbulent flows by Herring and Mellor,[72] Cebeci and Smith,[76] and McDonald and Fish.[60]

A powerful and simple technique used by Krause for two-dimensional flows is to replace the streamwise derivative of the streamwise velocity in the linearized momentum equation by the negative of the normal derivative of the normal velocity, as follows from the continuity equation. Thus, given the streamwise velocity profile and pressure gradient, the normal velocity can be deduced from solution of the momentum equation, which now has the form of an ordinary differential equation.

From continuity,

$$u_x = -v_y$$

and from momentum,

$$\rho u(-v_y) + \rho v u_y = \tau_y - p_x$$

Hence, given u as a function of y and the pressure gradient p_x together with a stress specification relating τ to u, this momentum equation may be solved for v as a function of y. In this way, u and v are obtained at the initial plane.

Last, some finite difference schemes, notably some of the Russian methods of splitting described by Yanenko[78] offer the prospect that, with careful structuring, an initial profile of transverse velocity will not be required when applied to the boundary-layer equations. Such methods are multilevel in that several intermediate level solutions are obtained in the process of marching one axial step. While the user might not wish to use such a scheme once the solutions are started, as a starting technique they are very intriguing.

The current status of finite difference boundary-layer procedures. Unlike the integral procedures for predicting boundary-layer development, the status of finite difference schemes for solving parabolic and hyperbolic equations is changing rapidly at the present time. For this reason, it is not proposed to review the available schemes with a view to recommending some "best" procedure for the turbine problem. Such a recommendation would doubtless soon be outdated. Nor is it intended to describe any one scheme in detail, for to do justice considerably more space would be required than is available here. Instead, some of the better known current schemes will be reviewed in the light of the topics discussed in earlier paragraphs. Hopefully, by following this same process, the reader will be better able to evaluate the suitability of the newer schemes that will become available in the future. Again, the pitfalls of the relatively untried scheme must be cautioned against and anyone contemplating such an application should read the charming little essay on this subject by Moretti.[79] Finally, little comment will be made at this point on the various turbulence models incorporated into the different procedures. Almost certainly, the various models are interchangeable and best reviewed separately as a class.

A start is made by ruling out any scheme that cannot treat both laminar *and* turbulent boundary layers, for the simple reason that extensive regions of both types are found on turbine blades, for instance, the scheme of Bradshaw-Ferriss. The original Bradshaw-Ferriss code[80] is probably best viewed as a vehicle for implementing a turbulence model and not as a general purpose boundary-layer code. Nash[77] uses a modified version of Bradshaw's turbulence model in an explicit three-dimensional turbulent procedure, but with the same scheme he has performed a number of laminar two-dimensional calculations without difficulty. The major objection to Nash's scheme for gas turbine applications is that the conditional stability arising from the explicit integration does allow the viscous sublayer to be defined efficiently. Using the DuFort-Frankel scheme, Pletcher[64] developed an apparently unconditionally stable explicit procedure (in the usual linear sense) in which the sublayer was defined without difficulty. Pletcher has successfully tested his procedure on a wide variety of flows and the usual objection to inconsistent schemes (that of not being able to reduce the truncation error beyond some point for an arbitrary mesh), although disquieting, apparently did greatly influence his results as far as can be determined. However, coupling and nonlinear truncation errors might be expected to affect the aforementioned procedures in gas turbine applications.

Turning to the implicit schemes, attention still is restricted to mature procedures with a demonstrated capability for treating turbulent as well as laminar flow. Thus, promising schemes such as the Davis coupled scheme[81] and the fourth-order scheme of Peters[82] will not be discussed. Such schemes may well be the fabric of the future, but further discussion at this point is probably premature. Rapidly becoming one of the most widely used schemes, the procedure of Spalding and Patankar has much to recommend it. In its original form, only the lack of a sublayer or stagnation point could seriously be held against the procedure for gas turbine applications. The lack of the sublayer has been removed by a number of people, for instance, Launder and Jones[83] and a highly developed version widely available from Kays and Moffat[84] in a form well suited for gas turbine use, with the exception of the stagnation point treatment. In this regard, it is claimed by Becko[85] that starting the Patankar and Spalding[70] scheme for some arbitrary and small boundary-layer thickness is an adequate treatment of the stagnation region. However, should this fail to be satisfactory in some instances, an acceptable alternative stagnation point treatment might be very costly to develop.

Three widely available, highly developed, and extensively tested schemes of quite similar construction have been developed by Herring and Mellor,[72] Cebeci and Smith,[76] and McDonald and Fish.[60] In all three cases, the procedure of Hartree and Womersly is followed, in which the streamwise direction is discretized first. Originally, Mellor and Herring then treated the resulting two-point boundary value problem by a shooting technique, but subsequently abandoned this technique in favor of an implicit Gaussian elimination process very similar to that developed by McDonald and Fish.[60] Cebeci and Smith[76] used the Choleski process to treat their implicit system of equations. All three methods treat both the sublayer and stagnation points and have optional equilibrium starting procedures. As a result of their structure, this type of method requires the nonlinear two-point boundary value system to be iterated out to a fair degree of convergence to retain spatial stability. During this nonlinear processing, grid growth and equation coupling can be taken care of.

More recently, in an effort to improve the efficiency of performing routine boundary-layer calculations, presumably with the three-dimensional problem in mind, Keller and Cebeci[86] have applied Keller's so-called box scheme. In this scheme, the boundary-layer equations are written as a nonlinear set of coupled first-order equations that are treated directly by a Newton-Raphson process, which gives rise to a block tridiagonal matrix for elimination. Only four grid points are used in this scheme, hence the box terminology. Although with the passing of time the first wave of enthusiasm for this technique has diminished somewhat,[68] it does seem to be a very neat scheme for the boundary-layer equations and is marred really only by its storage requirements. The technique has apparently no problem at stagnation points and can treat the sublayer. At the present time, it is being extensively tested and developed.

Although not as widely known as the foregoing schemes, Harris[87] has developed a turbulent version of the laminar scheme originally developed by Flugge-Lotz and Blottner,[69] which apparently worked well but is not widely used. Adams[88] developed a scheme of quite remarkable similarity to that constructed by Harris. Finally, the strip scheme of Kendall and Bartlett[89] has come into widespread use for performing reacting boundary-layer calculations, for which purpose it has been highly developed. Given the ready availability of other schemes with well-documented capabilities, there has not been any great incentive to pursue the application of Kendall's scheme to turbomachinery problems.

A number of the foregoing schemes are very widely available and of quite similar nominal capability. In certain cases, the choice would appear to reduce to evaluating the relative economics of the competing schemes, since the methods mentioned earlier all seem quite robust in operation. Order accuracy considerations, which might be expected to provide some guidance, are such that, in most of the schemes, formal second-order accuracy is relatively easy to attain by proper centering of the scheme. However, in many schemes, a first-order difference in the axial direction is routinely used to avoid spatial wiggles that can remain undamped in some cases. Thus, the demonstrated ability to routinely use a second-order axial scheme and, consequently, use fewer axial steps in a computation would be an attractive feature of the selected scheme, but one that has not yet been clearly attributed to any developed procedure. Hence, given that the schemes all require about the same number of mesh points, the point can be made that a large portion of the cost of marching one axial step is made up of forming coefficients of the discretized governing equations and is roughly independent of the precise details of the integration scheme. Thus, the great variations in quoted run times among the various procedures, even given the difficulty of establishing a common base for comparison, must reside in the treatment of the nonlinearities and the coupling between equations. Rigorous treatments such as that in the Keller box scheme or the one used in the Hartree-Womersly-based schemes are obviously time consuming and very conservative in comparison to say, the scheme of Patankar and Spalding[70] with its almost pathological avoidance of iteration. The question of adequacy and accuracy can be answered only on an ad hoc basis and it must therefore be the onus of the potential user to satisfy himself that the solutions to his problem are mesh independent to his required degree of accuracy. Clearly, in the eyes of some experts, it will be necessary to use a much more refined axial mesh—either that or iterate at given axial locations—for some demanding problems using a method like Patankar and Spalding with its cursory treatment of nonlinearities and coupling, than with the Keller or Hartree-Womersly schemes. The ideal situation is one in which the treatment of the coupling, nonlinearities, and spatial differencing are all accurate to the same degree; Anderson[90] has tried to implement this concept in developing a modified version of the Keller box scheme. However, the key point is that as obtained "off the shelf," so to speak, the various suitable and similar schemes might exhibit markedly differing run times for a given problem. This run time variation would be related to the degree of conservatism (or lack of it) of the originator. Further, many of the quoted run times are those required to achieve a level of comparison with experiment, not necessarily some degree of mesh independence of solution.

At this point, discussion of the finite difference procedures for predicting the boundary-layer behavior is concluded. Several good robust schemes are widely available and the prospects are excellent for the development of even more efficient schemes.

Axisymmetric (Pitch-Averaged) Boundary Layers

Motivation. In the turbomachine, the hub and casing boundary layers (commonly called the "annulus wall boundary layers") can, in a very simple minded view, be thought of as swirling axially symmetric boundary layers. As such, the study of this type of boundary layer would obviously seem worthwhile. That such a simplified viewpoint necessitates a high degree of compromise is well known. The hub and casing surfaces are often not continuous and frequently stationary surfaces lie adjacent to the rotating components. Leakage and coolant flow often occur through gaps in the surface, but all these difficulties shrink to insignificance when compared to the effect of the blade or stator rows. Here the overturning of the annulus wall boundary layers within the blade or stator row, caused by the mainstream deflection and the periodic shedding of these large wakes into the subsequent stages, causes major effects that cannot be rigorously taken into account in any axially symmetric boundary-layer analyses.

The difficulties involved in any more rigorous approach are, of course, quite daunting and this has led to the development of the concept of "pitch averaging." In the pitch averaging approach, the azimuthal variations are expressed as a mean value plus a small perturbation and the equations of motion averaged in the azimuthal direction. The resulting equations are quite as rigorous as the original unaveraged set and, as in the similarly treated turbulence problem, the unknowns are now the mean values and the correlation of the perturbations or "fluctuation." The pitch-averaged approach has obvious value if the mean values and correlations can be approximated such that the resulting averaged flow is reasonably well predicted. The current indications are that, indeed, pitch averaging can be a very useful tool and this concept will now be developed. Much of the original work on this topic has been carried out by Horlock and his co-workers, with independent contributions by Mellor. Its application in an integral formulation is described in detail by Horlock and Perkins[91] and Mellor and Wood.[92]

The governing equations. Implicit in the form of the governing equations is a coordinate system. In developing integral moments of the governing equations, it has long been recognized that certain coordinate systems possess major advantages over others. In particular, streamline coordinates based on the local direction of the inviscid flow at the edge of the boundary layer greatly simplify the momentum integral equations. This advantage disappears when the partial differential equations themselves are treated directly by numerical methods, in which case the effort has been to develop coordinate systems that allow the solution to be obtained in some logical manner. Anderson[90] has pointed out that use of streamline coordinates or approximations to them on a local basis allows the governing equations to be simplified, since the velocities normal to the approximate streamlines may be small, particularly in internal flow where the flow is constrained between two boundary surfaces. For his finite difference scheme, Anderson used the potential flow streamlines as an approximation to the real flow. Most authors have also used orthogonal coordinate systems, although this is by no means necessary and in the numerical schemes orthogonality can readily be abandoned if a more convenient coordinate system results. Starting then from the orthogonal system of curvilinear coordinates originally due to Hayes[93] for the three-dimensional boundary layer, one has

Continuity:

$$\frac{\partial \rho}{\partial t} + \frac{1}{h_1 h_2 h_3}\left[\frac{\partial}{\partial x} h_2 h_3 \rho u + \frac{\partial}{\partial y} h_1 h_3 \rho v + \frac{\partial}{\partial z} h_1 h_2 \rho w\right] = 0 \qquad (6.106)$$

Streamwise momentum:

$$\frac{\partial}{\partial t}\rho u + \frac{1}{h_1 h_2 h_3}\left[\frac{\partial}{\partial x} h_2 h_3 \rho u^2 + \frac{\partial}{\partial y} h_1 h_3 \rho u v + \frac{\partial}{\partial z} h_1 h_2 \rho u w\right]$$

$$+ \frac{1}{h_1 h_2 h_3}\left[\rho u v h_3 \frac{\partial h_1}{\partial y} + \rho u w h_2 \frac{\partial h_1}{\partial z} - \rho v^2 h_3 \frac{\partial h_2}{\partial x} - \rho w^2 h_2 \frac{\partial h_3}{\partial x}\right]$$

$$= \frac{1}{h_2}\left[\frac{\partial}{\partial y}\frac{\mu}{h_2}\frac{\partial u}{\partial y}\right] - \frac{1}{h_1}\frac{\partial p}{\partial x} \qquad (6.107)$$

Normal momentum:

$$\frac{\partial}{\partial t}\rho v+\frac{1}{h_1h_2h_3}\left[\frac{\partial}{\partial x}h_2h_3\rho uv+\frac{\partial}{\partial y}h_1h_3\rho v^2+\frac{\partial}{\partial z}h_1h_2\rho vw\right]$$

$$+\frac{1}{h_1h_2h_3}\left[\rho vwh_1\frac{\partial h_2}{\partial z}+\rho uvh_3\frac{\partial h_2}{\partial x}-\rho w^2h_1\frac{\partial h_3}{\partial y}-\rho u^2h_3\frac{\partial h_1}{\partial y}\right]=-\frac{1}{h_2}\frac{\partial p}{\partial y} \tag{6.108}$$

Spanwise momentum:

$$\frac{\partial}{\partial t}\rho w+\frac{1}{h_1h_2h_3}\left[\frac{\partial}{\partial x}h_2h_3\rho uw+\frac{\partial}{\partial y}h_1h_3\rho vw+\frac{\partial}{\partial z}h_1h_2\rho w^2\right]$$

$$+\frac{1}{h_1h_2h_3}\left[\rho uwh_2\frac{\partial h_3}{\partial x}+\rho vwh_1\frac{\partial h_3}{\partial y}-\rho u^2h_2\frac{\partial h_1}{\partial z}-\rho v^2h_1\frac{\partial h_2}{\partial z}\right]$$

$$=\frac{1}{h_2}\left[\frac{\partial}{\partial y}\frac{\mu}{h_2}\frac{\partial w}{\partial y}\right]-\frac{1}{h_3}\frac{\partial p}{\partial z} \tag{6.109}$$

with the metric scale factors h_1, h_2, and h_3 giving the arc length ds by

$$\mathrm{d}s^2=h_1^2\,\mathrm{d}x^2+h_2^2\,\mathrm{d}y^2+h_3^2\,\mathrm{d}z^2 \tag{6.110}$$

and it is, in essence, assuming that the local boundary-layer thickness is small compared with the principle radius of curvature of the surface. Of the two possibilities, either to Reynolds time average first and then pitch average, or vice versa, the usual precedent of time averaging first is adopted. With the accepted boundary-layer approximations, the Reynolds time averaging of the equations results in

Continuity:

$$\frac{\partial\bar\rho}{\partial t}+\frac{1}{h_1h_2h_3}\left[\frac{\partial}{\partial x}h_2h_3\bar\rho\bar u+\frac{\partial}{\partial y}h_1h_3\overline{\rho v}+\frac{\partial}{\partial z}h_2h_1\overline{\rho w}\right] \tag{6.111}$$

Streamwise momentum:

$$\frac{\partial}{\partial t}\bar\rho\bar u+\frac{1}{h_1h_2h_3}\left[\frac{\partial}{\partial x}h_2h_3\bar\rho\bar u^2+\frac{\partial}{\partial y}h_1h_3\overline{\rho v}\bar u+\frac{\partial}{\partial z}h_2h_1\bar\rho\bar u\bar w\right]$$

$$+\frac{\bar\rho\bar u\bar w}{h_1h_3}\frac{\partial h_1}{\partial z}-\frac{\bar\rho\bar w^2}{h_1h_3}\frac{\partial h_3}{\partial x}=-\frac{1}{h_1}\frac{\partial\bar p}{\partial x}+\frac{1}{h_2}\frac{\partial}{\partial y}\left(\frac{\bar\mu}{h_2}\frac{\partial\bar u}{\partial y}-\bar\rho\overline{u'v'}\right) \tag{6.112}$$

Normal momentum:

$$-\frac{1}{h_2}\frac{\partial \bar{p}}{\partial y}=\frac{1}{h_1 h_2 h_3}\frac{\partial}{\partial y}h_1 h_3 \overline{\rho v^2} \tag{6.113}$$

Spanwise momentum:

$$\frac{\partial}{\partial t}\bar{\rho}\bar{w}+\frac{1}{h_1 h_2 h_3}\left[\frac{\partial}{\partial x}h_2 h_3 \bar{\rho}\bar{u}\bar{w}+\frac{\partial}{\partial y}h_1 h_3 \overline{\rho v}\bar{w}+\frac{\partial}{\partial z}h_2 h_1 \bar{\rho}\bar{w}^2\right]$$

$$-\frac{\bar{\rho}\bar{u}^2}{h_1 h_3}\frac{\partial h_1}{\partial z}+\frac{\bar{\rho}\bar{u}\bar{w}}{h_1 h_3}\frac{\partial h_3}{\partial x}=-\frac{1}{h_3}\frac{\partial \bar{p}}{\partial z}+\frac{1}{h_2}\frac{\partial}{\partial y}\left(\frac{\bar{\mu}}{h_2}\frac{\partial \bar{w}}{\partial y}-\bar{\rho}\overline{v'w'}\right) \tag{6.114}$$

where the Reynolds normal stresses have been neglected. Subsequently the stress terms $\bar{\mu}\,\partial\bar{u}/\partial y-\bar{\rho}\overline{u'v'}$ are written τ_x, τ_y, etc. The time derivatives are retained on the supposition that the mean motion can still be allowed to vary in time, provided the variation does not occur on a time scale approaching that of the turbulence. It has also been assumed that $\bar{\rho}\bar{w} \gg \overline{\rho'w'}$.

Anderson[90] retains the h_2 metric scale factor in his derivation of the averaged equations. Such a development might well be important in certain applications, but for the moment the usual precedent of setting the h_2 metric to unity is followed (the thin shear layer approximation). The pitch-averaged equations can now be derived by introducing the dependent variable f as being the sum of some pitch-averaged quantity $\bar{f}$ and a perturbation f'. Subsequently, f' will be assumed small relative to $\bar{f}$. By definition

$$\bar{f}=\frac{1}{z_a-z_b}\int_{z_a}^{z_b} f\,\mathrm{d}z \quad \int_{z_a}^{z_b} f'\,\mathrm{d}z=0 \tag{6.115}$$

Now for the typical variable f, an integral average of the derivative of f is

$$\int_{z_a}^{z_b}\frac{\partial f}{\partial \tilde{x}}\,\mathrm{d}z=\frac{\partial}{\partial \tilde{x}}(z_a-z_b)\bar{f}-f_b\frac{\partial z_a}{\partial \tilde{x}}+f_a\frac{\partial z_b}{\partial \tilde{x}} \tag{6.116}$$

where $\tilde{x}$ could be any coordinate direction except z. Now the governing equations are pitch averaged, dropping the Reynolds averaged overbars on all but the Reynolds shear stress term. The single overbar signifies the pitch-averaged value of some time mean quantity. The double overbar denotes the pitch-averaged value of a turbulent correlation. The single overbar over a primed quantity denotes the correlation between perturbations from the pitch-averaged mean of some time-averaged quantity. Taking the streamwise momentum equation as an example and writing $g=z_a-z_b$, there results

$$\frac{\partial}{\partial t}g\overline{\rho u}+\frac{1}{h_1h_3}\frac{\partial}{\partial x}h_3g\overline{\rho u^2}+\frac{1}{h_1h_3}\frac{\partial}{\partial y}h_1h_3\overline{\rho uvg}$$

$$+g\overline{\rho uw}K_1-g\overline{\rho w^2}K_3=-\frac{1}{h_1}\frac{\partial\bar{p}}{\partial x}g+p_b\frac{\partial z_b}{\partial x}-p_a\frac{\partial z_a}{\partial x}$$

$$+\frac{\partial}{\partial y}\bar{\tau}_x g-\tau_{xb}\frac{\partial z_b}{\partial x}+\tau_{xa}\frac{\partial z_b}{\partial x} \tag{6.117}$$

and in the usual manner the pitch-averaged terms can be expressed as

$$\overline{\rho u}=\bar{\rho}\bar{u}$$

$$\overline{\rho u^2}=\bar{\rho}\bar{u}^2+2\bar{\rho}\bar{u}+\overline{\bar{\rho}u'^2}$$

$$\frac{\overline{\rho uv}}{\bar{u}\overline{\overline{\rho v}}}=\bar{\rho}\bar{u}\bar{v}+\bar{u}+\bar{v}+\overline{\bar{\rho}u'v'}+\bar{u}\overline{\overline{\rho v}}+\overline{u'\rho v'} \tag{6.118}$$

and so on. Clearly, for low-speed flow it seems reasonable to neglect the density perturbation and certainly the triple correlations will normally be negligible in comparison with the double correlations. This leaves the pitch-averaged equivalent of the usual Reynolds shear and normal stresses. At this point in the development, little is known of the behavior of these correlations and in applications these terms are usually neglected or crudely approximated. It is to be hoped that the description of these terms will be improved upon in the future since it is by no means clear at this stage that they can be neglected. The remaining additional terms due to pitch averaging arise from the forces on the blade surface and these too, although perhaps less controversial, are still the subject of some debate. The inviscid (but not necessarily irrotational) pressures could be used for the blade surfaces. The stresses on the blades τ_a and τ_b emerge as part of a separate blade boundary-layer calculation, but in many instances their difference may be quite negligible. Finally, of course, the gradients of the pitch-averaged quantities in the azimuthal direction are dropped. Additional simplifications (such as assuming the blades are thin) allow the gap g to cancel everywhere except in the effective blade force terms. The assumption of the thin blade also allows the upper and lower surface gap coordinate rates of change to be set equal; however, for the moment this simplification is not applied. The governing equations are now expressed as

Continuity:

$$\frac{\partial\bar{\rho}}{\partial t}+\frac{1}{gh_1h_3}\left[\frac{\partial}{\partial x}gh_3\bar{\rho}\bar{u}+\frac{\partial}{\partial y}gh_1h_3\overline{\overline{\rho v}}\right]=0 \tag{6.119}$$

Streamwise momentum:

$$\frac{\partial}{\partial t}\bar{\rho}\bar{u}+\frac{1}{gh_1h_3}\frac{\partial}{\partial x}gh_3\bar{\rho}\left(\bar{u}^2+\overline{u'^2}\right)+\frac{1}{gh_1h_3}\frac{\partial}{\partial y}gh_1h_3\left(\bar{u}\overline{\overline{\rho v}}+\overline{u'\overline{\overline{\rho v'}}}\right)$$

$$+K_1\bar{\rho}\left(\bar{u}\bar{w}+\overline{u'w'}\right)-K_3\bar{\rho}\left(\bar{w}^2+\overline{w'^2}\right)=-\frac{1}{gh_1}\frac{\partial}{\partial x}g\bar{p}+\frac{1}{g}\frac{\partial}{\partial y}\bar{\tau}_x g$$

$$+p_b\frac{\partial z_b}{\partial x}-p_a\frac{\partial z_a}{\partial x}-\tau_{xb}\frac{\partial z_b}{\partial y}+\tau_{xa}\frac{\partial z_b}{\partial y} \qquad (6.120)$$

Normal momentum:

$$-\frac{1}{h_2}\frac{\partial\bar{p}}{\partial y}=-\overline{\overline{\rho v^2}}K_2 \qquad (6.121)$$

Spanwise momentum:

$$\frac{\partial}{\partial t}\bar{\rho}\bar{w}+\frac{1}{gh_1h_3}\frac{\partial}{\partial x}gh_3\bar{\rho}\left(\bar{u}\bar{w}+\overline{u'w'}\right)+\frac{1}{gh_1h_3}\frac{\partial}{\partial y}gh_1h_3\bar{w}\overline{\overline{\rho v}}$$

$$+\overline{w'\overline{\overline{\rho v'}}}-K_1\bar{\rho}\left(\bar{u}^2-\overline{u'^2}\right)+K_3\bar{\rho}\left(\bar{u}\bar{w}+\overline{u'w'}\right)+\frac{1}{g}\frac{\partial}{\partial y}\bar{\tau}_z g$$

$$-\frac{1}{h_3}(p_b-p_a)-\tau_{zb}\frac{\partial z_b}{\partial y}+\tau_{za}\frac{\partial z_a}{\partial y} \qquad (6.122)$$

and, for example,

$$\bar{\tau}_x=\bar{\mu}\frac{\partial\bar{w}}{\partial y}-\overline{\overline{\bar{\rho}v'w'}} \qquad (6.123)$$

To recap, the single overbar signifies the pitch-averaged value of some time mean quantity and the double overbar denotes the pitch-averaged value of a turbulent correlation. The single overbar over a primed quantity denotes correlation between perturbations from the pitch-averaged mean of some time-averaged quantity.

Having derived the pitch-averaged equations, as usual the choice of further integrating the governing equations or of treating the partial differential equations directly remains open. During the initial stages of the development of this concept, the various authors wisely chose to use the simple integral approach (see Horlock and Perkins[91] for a discussion of the development of this approach). Encouraged by their success, Anderson[90] developed a finite difference scheme for solving the basic partial differential equations of motion. Both approaches will be briefly reviewed. Again, it is not the purpose here to present in detail the various approaches and the reader intent in performing calculations should consult the original references. Here the aim is to outline the approach and detail its strengths and weaknesses.

The momentum integral approach. A straightforward integral normal to the annulus wall of Eqs. (6.120) and (6.122) gives the pitch-averaged momentum integral equations. The derivation is given in detail by Daneshyar[94] for instance. Various integral thickness parameters can be introduced and the equations recast and, with certain coordinate systems, simplified somewhat. Here the analysis will not be specialized to that extent, but it serves the present purposes sufficiently well to present the integral equation in a comparatively primitive form, as

Continuity:

$$\overline{\overline{\rho v_e}} = \overline{\overline{\rho v_w}} - \frac{1}{g h_1 h_3}\left[\frac{\partial}{\partial x}\int_0^\delta \bar{\rho}\bar{u}\,\mathrm{d}y\right]\cdot g h_3 + \frac{\rho_e u_e}{h_1}\frac{\partial \delta}{\partial x}$$
$$-\frac{\partial}{\partial t}\int_0^\delta \bar{\rho}\,\mathrm{d}y + \bar{\rho}_e\frac{\partial \delta}{\partial t} \qquad (6.124)$$

Streamwise momentum:

$$\frac{\partial}{\partial t}\int_0^\delta \bar{\rho}\bar{u}\,\mathrm{d}y - \overline{\rho_e u_e}\frac{\partial \delta}{\partial t} + \frac{1}{g h_1 h_3}\frac{\partial}{\partial x} g h_3\int_0^\delta \bar{\rho}\left(\bar{u}^2 + \overline{u'^2}\right)\mathrm{d}y$$
$$-\frac{\rho_e}{h_1}\left(\bar{u}^2 + \overline{u_e'^2}\right)\frac{\partial \delta}{\partial x} + \bar{u}_e\overline{\overline{\rho v_e}} + \overline{u'\overline{\rho v_e'}} + K_1\int_0^\delta \bar{\rho}\left(\bar{w}\bar{u} + \overline{u'w'}\right)\mathrm{d}y$$
$$-K_3\int_0^\delta \bar{\rho}\left(\bar{w}^2 + \overline{w'^2}\right)\mathrm{d}y = \frac{1}{g h_1}\int_0^\delta \frac{\partial}{\partial x} g\bar{p}\,\mathrm{d}y + \bar{\tau}_{xw} + \tilde{D}_{x_1} \qquad (6.125)$$

Spanwise momentum:

$$\frac{\partial}{\partial t}\int_0^\delta \bar{\rho}\bar{w}\,\mathrm{d}y - \bar{\rho}_e\bar{w}_e\frac{\partial \delta}{\partial t} + \frac{1}{g h_1 h_3}\frac{\partial}{\partial x} g h_3\int_0^\delta \bar{\rho}\left(\bar{u}\bar{w}' + \overline{u'w'}\right)\mathrm{d}y$$
$$-\frac{\rho_e}{h_1}\left(\bar{u}_e\bar{w}_e + \overline{u'w_e'}\right)\frac{\partial \delta}{\partial x} + \bar{w}_e\overline{\overline{\rho v_e}} + \overline{w'\overline{\rho v_e'}}$$
$$-K_1\int_0^\delta \bar{\rho}\left(\bar{u}^2 + \overline{u'^2}\right)\mathrm{d}y + K_3\int_0^\delta \bar{\rho}\left(\bar{u}\bar{w} + \overline{u'w'}\right)\mathrm{d}y = \tau_{zw} + \tilde{D}_{z_1} \qquad (6.126)$$

and

$$\tilde{D}_{x_1} = \int_0^\delta\left(\frac{p_b}{h_1}\frac{\partial z_b}{\partial x} - \frac{p_a}{h_1}\frac{\partial z_a}{\partial x} - \tau_{xb}\frac{\partial z_b}{\partial y} + \tau_{xa}\frac{\partial z_a'}{\partial y}\right)\mathrm{d}y$$
$$\tilde{D}_{z_1} = -\int_0^\delta\left(\frac{p_b}{h_3} - \frac{p_a}{h_3} - \tau_{zb}\frac{\partial z_b}{\partial y} + \tau_{za}\frac{\partial z_a}{\partial y}\right)\mathrm{d}y \qquad (6.127)$$

If now the original governing equations are evaluated at the edge of the boundary layer and then nominally integrated normal to the wall, the resulting set of equations can, by subtraction, be used to recast the integral equations in integral defect form. Following Horlock and Perkins[91] and Mellor and Wood,[92] all of the contributions from pitch averaging can be lumped together into force defects of the type,

$$D_x = \sum D_{xn}$$

$$D_{x_1} = \tilde{D}_{x_1 e} - \tilde{D}_{x_1}$$

$$D_{x_2} = \frac{\partial}{\partial x} g h_3 \int_0^{\delta} \bar{\rho}\left(\overline{\overline{u_e'^2}} - \overline{u'^2}\right) \mathrm{d}y$$

$$D_{x_3} = \int_0^{\delta} \left[\frac{\partial \overline{p_e}}{\partial x} - \frac{\partial \bar{p}}{\partial x} \right] \mathrm{d}y, \text{ etc.}$$

$$D_z = \sum D_{zn}$$

$$D_{z_1} = \tilde{D}_{z_1 e} - \tilde{D}_{z_1}$$

$$D_{z_2} = \frac{\partial}{\partial x} \int_0^{\delta} \bar{\rho}\left(\overline{u'w_e'} - \overline{u'w'}\right) \mathrm{d}y, \text{ etc.} \qquad (6.128)$$

In this form, the pitch-averaged momentum equations can now be made to look exactly like the usual axisymmetric swirling momentum integral equations with the addition of force defect terms D_x and D_z. Adequate specifications of the force defects is therefore the key to the eventual success of this type of analysis. This point will be returned later. For the time being, it is observed that, in addition to the problem of determining the streamwise velocity $\bar{u}$, there now arises the problem of determining the swirl component $\bar{w}$ (commonly termed the cross-flow velocity), for which one additional equation, that for the spanwise momentum, has already become available (the continuity equation can be thought of as eliminating $\overline{\overline{\rho v_e}}$ from the system). In the usual integral manner, profiles of $\bar{u}$ and $\bar{w}$ must be specified and additional momentum equations constructed until the number of equations is equal to the number of parameters describing the profiles.

Considering now the velocity profile family problem, the strategy commonly adopted has been to consider a streamwise velocity component aligned with the direction of the edge inviscid streamline and a cross-flow component normal to this inviscid streamline direction. For small cross flows, it is argued that any of the accepted two-dimensional profiles, such as that due to Coles,[9] suffices for a streamwise profile, thereby also specifying the wall stress. As discussed at length earlier, Coles' profile is a two-parameter family, so that immediately an additional integral equation is required.

In the related problem of the infinite yawed wing, for instance, Cumpsty and Head[95] generalized Head's very simple entrainment concept into three dimensions and, subsequently, Horlock[96] applied an entrainment equation to the present problem. For the cross-flow profile, a number of suggestions have been made, such as the variation of Prandtl's profile made by Mager[97] or the well-known polar profile due to Johnson.[98] The Prandtl-Mager profile is a very simple one-parameter family of the form

$$w/u = \tan \beta_w (1 - y/\delta)^2 \tag{6.129}$$

where β_w is nominally the angle between the wall and the edge streamlines. Being a one-parameter family (β_w), the Prandtl-Mager profile requires that *no additional equations need be generated* other than the spanwise momentum equation, which probably explains its widespread use. Johnson's two-parameter polar profile has been shown to be an improvement over the Prandtl-Mager profile. Wheeler and Johnson,[99] however, noted in a review of this cross-flow profile problem that none of the cross-flow profiles available at that time (including some developments of the Johnson profile) were entirely satisfactory for their purposes. For this reason, Wheeler and Johnson went on to develop a direct numerical scheme for the three-dimensional turbulent boundary-layer equation. Horlock and Perkins[91] argue against Wheeler and Johnson's pessimism concerning the cross-flow profile for the present problem, citing that in the passage a reasonable form for the cross flow can be obtained from secondary flow theory.[96] Horlock's profile is similar in many respects to Johnson's and the necessary integrals are given by Lindsay.[100] For flows where the force defects D_x and D_z are absent, the set of equations including Head's entrainment equation can be solved without particular difficulty. For instance, Lindsay has compared predictions made using the Johnson profiles with the swirling axisymmetric flow without blades or body forces measured by Hoadley[101] with moderate success. Earlier calculations by Hoadley using the Prandtl-Mager profile, probably fortuitously, achieved slightly better agreement.

Turning now to flows with azimuthal variations, the critical problem of the force defect integrals D_x and D_z arises. The pressure on the blade surfaces can obviously be determined from some inviscid (but not necessarily irrotational) analyses. Estimates of the gapwise variations in u and w can be made from the same type of analyses and correlation coefficients constructed using the mean flow-averaged variables as the normalizing scale. The problem with this approach is that the variations within the end wall boundary layers are likely to be significant and possibly not well predicted by the preceding type of approach. By trial and error, Horlock and his co-workers have found that fair estimates of axial blockage can be made by setting $D_x = 0$, presumably as a result of the individual components cancelling out.[102] Horlock and Perkins[91] cite two different methods for determining D_z. The first is based on a secondary flow analysis (which also results in $D_x = 0$) and the second, due to Daneshyar,[102] is semiempirical. At this point, it seems clear that a very careful evaluation of the accuracy and generality of the presently available suggestions for the force defects D_x and D_z must be made. If satisfactory forms for D_x and D_z can or have been developed, it would appear that a very useful tool will exist where none is presently available, in spite of the known profile limitations. Obviously, end wall loss levels are important and any means of estimating their level and behavior are of keen interest to designers.

The finite difference approach. Given the generally accepted difficulty of adequately specifying the cross-flow velocity profile, a number of investigators (e.g., Nash,[77] Wheeler and Johnson,[99] Harris and Morris,[103] and Cebeci et al.[104]) have been led to develop finite difference techniques for solving the three-dimensional turbulent boundary-layer equations. The governing equations in the pitch-averaged problem are, in fact, the simplest nontrivial form of these same three-dimensional boundary-layer equations (that is, the axisymmetric swirling boundary-layer equations) with the addition of (given) source terms D_x and D_z. Therefore, no new numerical difficulties over and above those normally encountered with the three-dimensional boundary-layer equations would be expected with the pitch-averaged equations of motion. Furthermore, the presence of the large cross flow, which is often found in practice, leads to potential difficulty with an adequate specification of even the streamwise velocity profile family. Finally, the features of interest to the designer—compressibility and heat transfer—lead to a general concern about all of the profile assumptions, thus paving the way for a numerical approach. Naturally, the direct numerical approach would not have been embarked upon had not the simpler integral methods been encouragingly successful within their limited sphere of applicability.

It follows immediately that all of the available schemes for solving the three-dimensional boundary-layer equations could be used in the simplified situation considered here. Alternatively, the two-dimensional schemes discussed earlier could be extended to incorporate a swirl equation. In either event, almost all of the previous discussion on numerical methods for the boundary-layer equations applies in the present situation, although there is probably less interest in laminar flow or the regions of extended transition (both forward and reverse).

Anderson[90] appears to have been the first to apply a finite difference scheme to the pitch-averaged equations of motion. In Anderson's case, the approach was an extension of a two-dimensional scheme rather than a cut-down version of a three-dimensional scheme. Three-dimensional schemes will be discussed later in a separate section, so no discussion of the merits and demerits of the various three-dimensional schemes will be given here. Initially, Anderson used a straightforward explicit integration scheme with a law of the wall fit to cover the region below the first grid point. Difficulty with the law of the wall in swirling flow, together with the desire to have a large number of grid points for accuracy, led to the abandonment of the explicit scheme and the adoption of an implicit formulation. In view of its many attractive features, the Keller box scheme[73] was finally selected.

However, the inclusion of the swirl and energy equations led to a very large demand for storage with this technique, the problem mentioned earlier in connection with the Keller scheme. Also, Anderson followed the linearization concepts developed by McDonald and Briley[67] and linearized the governing equations to second order (consistent with spatial differencing) without any iteration such as that in the scheme used by Keller and Cebeci,[86] thus effecting a substantial savings in already modest computer run times. The Richardson extrapolation, also recommended by the originators, was not employed in Anderson's modified version of the box scheme since, in general, accuracy in skin friction below 1% was not sought. With the box scheme, Blottner[68] recently demonstrated the inadvisability of using the Richardson extrapolation unless extreme accuracy was being sought.

Turning to the question of the turbulence model, the penalty for the use of a detailed numerical approach is the requirement of specifying the $x-y$ variation of the turbulent correlations $\overline{\overline{u'v'}}$ and $\overline{\overline{u'w'}}$. As was mentioned earlier, Horlock and co-workers were able to circumvent this problem by the use of a generalized version of Head's entrainment law. The problem of three-dimensional turbulence models will be examined in detail in later sections of this chapter. For the moment, it is noted that for $\overline{u'v'}$ Anderson[90] used scalar eddy viscosity with Van Driest's suggestion for the sublayer, followed by Prandtl's mixing length for the wall region, and finally Clauser's suggestion for the outer wake-like region. In this concept, the extension to three dimensions of an essentially two-dimensional model is achieved by defining the eddy viscosity as

$$\varepsilon = l^2 \left| \frac{\partial \mathbf{Q}}{\partial y} \right| \tag{6.130}$$

where $\mathbf{Q}$ is the velocity parallel to the wall and l is the scalar mixing length. Consequently,

$$\left| \frac{\partial \mathbf{Q}}{\partial y} \right| = \left[\left(\frac{\partial u}{\partial y} \right)^2 + \left(\frac{\partial w}{\partial y} \right)^2 \right]^{\frac{1}{2}} \tag{6.131}$$

The stresses $\bar{\tau}_x$ and $\bar{\tau}_z$ incorporate $\overline{\overline{u'v'}}$ and $\overline{\overline{v'w'}}$ and are defined by

$$\bar{\tau}_x = \rho\varepsilon \frac{\partial \bar{u}}{\partial y} \qquad \bar{\tau}_z = \rho\varepsilon \frac{\partial \bar{w}}{\partial y} \tag{6.132}$$

where the scalar mixing length takes on the appropriate two-dimensional form in accordance with the Van Driest-Prandtl-Clauser suggestion discussed previously. Aside from a need to take a four-point y-averaged value of the eddy viscosity ε, the resulting scheme worked quite well in practice. Averaging of the eddy viscosity compromises either the claimed second-order accuracy of the scheme or the turbulence model, but is necessary to obtain stability with the scheme. A similar problem was treated in a like manner by Keller and Cebeci.[86]

In dealing with the force defects, Anderson now had to contend with the detailed specification of these quantities across the annulus wall boundary layers. Anderson utilized the strip theory, with each blade or stator element treated two-dimensionally as a plane cascade. The cascade performance was then estimated either theoretically or from one of the accepted correlations, both methods being available in the code. In this way, the hub-to-tip variations could be obtained by stacking the two-dimensional strips. Obviously, more complex alternative schemes to obtain the hub-to-tip variations can be developed and incorporated into Anderson's scheme. Given the encouraging results already obtained, such developments might be very rewarding and at the present time a thorough evaluation of the force defect problem is appropriate.

Conclusions on the pitch-averaged procedures. Insofar as formal pitch averaging is concerned, it must be emphasized that the resulting equations are quite as rigorous as the three-dimensional equations from which they were obtained. The simplification resulting from the decrease in dimensionality of the problem has been obtained at the expense of the loss of information on the details of the flow in the eliminated dimension. No approximations have as yet been introduced concerning the nature of this flow. The situation is identical to the Reynolds-averaged equations of motion, but almost certainly much more tractable due to the less random nature of the flow in the azimuth compared to the complex nature of the time variations in a turbulent flow. Therefore, the expectations are that, given reasonable estimates of the azimuthal variations and hence the correlation terms in the equations, accurate predictions of the pitch-averaged flow can be made if the vane or blade forces are known.

In spite of the foregoing difficulties, the pitch-averaged approach has a great deal to recommend it, particularly when the alternative of either giving up or treating the complete three-dimensional problem is considered. Much work remains to be done at this point; however, the design needs are pressing. The approach clearly provides a bridge between the simple throughflow analyses and the more exotic, but at present incomplete, three-dimensional intravane or blade analyses.

6.3 Three-Dimensional Boundary Layers

Within turbomachinery, three-dimensional flows are encountered with unfortunate regularity, both within the boundary layers and the freestream. The term "freestream" is used here to denote a region where little diffusion, dissipation, or production of turbulence occurs but also a region in which vorticity and turbulence might still exist by convection. Simple three-dimensional flows such as that encountered on conventional moderate and high aspect ratio wings (the so-called boundary sheets) are not usually found in turbomachinery applications. End effects, such as the strut or rotor end wall interaction, are the flows of critical concern. Such flows cannot be treated properly within the framework of conventional three-dimensional boundary-layer analyses. This failure of the conventional analyses may be seen in the previously introduced three-dimensional boundary-layer equations, due originally to Hayes.[93] Obviously in the derivation, it has been assumed that the diffusion of the momentum in one direction only is important (in this notation, in the y direction). Clearly, this assumption breaks down in the vicinity of a right-angled corner where the diffusion normal to both walls is significant. Fortunately, the extra diffusion terms do not change the basic nature of the problem from that of an initial boundary value treatable by means of a forward marching procedure, providing the diffusion in the marching direction is still negligible and the pressure can still be largely determined a priori. The extra complexity of the corner flow does, however, make solution of the system by an integral moment technique an exercise with a very low probability of success. Here, the equations embodying diffusion in both coordinate directions are termed the extended boundary-layer equations.

An additional feature of the extended boundary-layer equations concerns the absence of zones of influence. To discuss this problem, it is necessary to recall the work of Raetz,[105] who first investigated the properties of the conventional three-dimensional laminar boundary-layer equation. Raetz pointed out that diffusion occurred only in the direction normal to the wall in the conventional system and that in the other directions the equations were essentially hyperbolic and that, therefore, characteristic surfaces would exist. As a result, Raetz suggested that an "influence principle" would hold and that the influence of the solution at any point would be transferred to other points first by diffusion in the direction of the surface normal through the point and then by convection along all the streamlines passing through this surface normal. The zone of influence is then the region bounded by the body surface, the boundary-layer edge, and a pair of characteristic surfaces: one the surface normal envelope passing through the external streamline and the other the surface normal envelope passing through the limiting surface streamline. As a companion to the zone of influence, Raetz also suggested a "zone of dependence" that is bounded by the upstream edges of the normal to the wall passing through the freestream and the limiting wall streamline which intersect at the point in question. Only conditions within the zone of dependence influence the solution at the point in question and in turn only the flow in the subsequent zone of influence would be affected by the flow at the point.

Wesseling[106] considered the zone of influence problem from the point of view of the conventional three-dimensional turbulent boundary-layer equations and concluded that within this system Raetz's influence principle would hold, but the possibility of a smaller zone in turbulent flow than in laminar flow could not be ruled out. The smaller zone would arise from the nearly hyperbolic nature of turbulent transport normal to surface outside of the viscous sublayer.[59] The zone of influence problem has also been discussed by Wang.[107]

The zone of influence has obvious attributes when boundary conditions are being applied. For the conventional system, conditions on the initial surface and subsequently on the edges of the domain must be prescribed. The initial surface conditions are discussed by Ting[108] and it follows that only a limited number of dependent variables can be arbitrarily assigned starting values in view of the need to have starting values of the dependent variables that also satisfy the governing equations. The usual course of action is to specify the streamwise velocity and, either by implication or directly, its axial rate of change. The Reynolds shear stress, the cross-flow velocity and the corresponding cross-flow Reynolds stress, and the density must also be specified: one the surface normal to the wall passing through the external flow streamline, the other the surface normal to the wall passing through the limiting wall streamline. The transverse velocity is derived from the continuity equation and the variables specified everywhere on the inlet surface. Edge conditions pose a particular problem; but intuitively, outflow along the edge requires no boundary conditions as it is completely determined by the known interior flowfield. Wall surface and freestream edge condition are assumed given. The central problem rests with the unknown inflow edge conditions. Here the zone of dependence principle becomes invaluable for it clearly requires the effect of the unknown inflow boundary information to remain within the zone of influence of the inlet plane edges. Thus, if the region of interest is outside the inlet edge zone of influence, little concern for the unknown inflow boundary information need be expressed. The conventional three-dimensional boundary-layer equations therefore can be very useful for describing the boundary-layer flow on a wing surface, where the inflow regions near the wing root or tip have a limited zone of influence. Unfortunately, the converse is also true, that is, if the region of interest lies within the zone of influence of the edge, then the appropriate boundary condition information must be supplied. By the usual law of perversity of circumstances, the root and tip regions of wings, for instance, are regions of great boundry-layer interest in view of their major contribution to losses and to stall. Physically then, within these important regions the zone of influence principle is of little immediate help and it is of no great concern that, when diffusion in the spanwise direction is allowed for, as it is within the extended boundary-layer equations, the dependence principle is no longer obtained. The extended boundary-layer equations can also have difficulty with inflow boundary conditions, but they do possess the major attribute of being able to "turn the corner" and proceed to close the domain (as in a duct) or to proceed to a plane of symmetry where known information is available for the boundary conditions. Given the type of problem normally encountered in turbomachinery applications, the attributes of the extended boundary-layer equations are very impressive.

In the subsequent section, a very brief discussion of some of the better known methods of predicting the conventional three-dimensional boundary-layer equations will be given, mainly from the point of view of introducing some of the major difficulties involved with performing three-dimensional calculations. A more complete discussion of three-dimensional boundary layers has been given by Nash and Patel[109] and from the computational point of view by Blottner.[81] In view of the limited utility of the conventional boundary-layer equations, much more emphasis will be placed upon the development of the extended three-dimensional boundary-layer equations and the techniques for their solution. Although major subjects in their own right, the problem of determining the freestream and the geometric information will not be discussed here.

The Momentum Integral Approach

The conventional three-dimensional boundary-layer equations have been presented earlier during the development of the pitch-averaged equations. These boundary sheet equations were also integrated in the effectively normal direction y and can be cast into a number of different forms, depending on the integral defect thickness parameters introduced. Regardless of the precise form of the equations, if an n parameter profile family for the streamwise velocity u and an m parameter profile for the cross-flow velocity w are introduced, then, since the transverse mass flux ρv can be determined from the continuity equation, at least $m + n + 3$ independent equations must be provided to render the system determinate. The number three in this expression arises from the need to define the density ρ and the skin-friction components τ_{wx} and τ_{wz}. The problem of compressibility will not be directly addressed at this point and it will be assumed that some energy relationship exists to specify the local density. The skin-friction relationships will be addressed later. Of the required remaining $m + n$ equations arising from the adopted mean profile families, the first integral moment of the momentum equations provide two equations and, as has already been discussed at length for two dimensions, additional moment equations are readily obtained at the expense of introducing more information on the turbulent transport mechanisms. The problem of generating the additional moments of the turbulence will be deferred until turbulence models are discussed. Alternatively, additional equations may be derived by introducing plausible physical arguments such as Head's entrainment hypothesis. With the foregoing as background, the various procedures will be evaluated.

First of all, consider the streamwise profile. Although not mandatory, frequently in integral methods the direction for this velocity has been chosen to be that of the external streamline in a plane locally parallel to the surface. In this Cartesian coordinate system, the cross-flow component at the edge of the boundary layer w_e disappears, thus simplifying matters. If now the simple power law profile is supposed valid for the streamwise direction of the cross flow, the Prandtl-Mager profile is used. To recap, these are written by

$$u = u_e(y/\delta)^{1/n} \qquad w = u(1 - y/\delta)^2 \tan\beta \tag{6.133}$$

Now since u_e is given, these profile families introduce three parameters n, δ, and β, the angle between the wall and freestream streamline directions. Note that, since the limiting streamline angle β is a problem parameter, the skin-friction determination is simplified since

$$\tau_w = (\cos\beta)^{-1}\tau_{wx} \qquad \tau_{wz} = \tau_{wx}\tan\beta \tag{6.134}$$

and thus, only one skin-friction relationship need be specified, usually the streamwise component τ_{wx}. Perhaps the simplest use of the foregoing set was by Mager,[97] who assumed that the streamwise component of wall shear was given by the very simple Prandtl-von Kármán law for two-dimensional flow. Since the formulation now uses only three independent profile parameters, one more than the number of equations already on hand, Mager closed the set by specifying that the power law exponent n was a constant for some particular flow and usually ascribed a value around 7. For a simple flow, Mager achieved a very creditable degree of accuracy with his prediction, due almost certainly to his use of a value of n taken from the experiment.

The assumption of a constant power law exponent is now widely regarded as being an oversimplification and, as information on two-dimensional boundary-layer behavior increased over the years, a number of investigators have incorporated these developments into the three-dimensional framework following Mager's lead. Dring[110] later refined and developed Mager's technique and applied his scheme to the turbine end wall problem. Perhaps the next major development occurred when Cumpsty and Head[95] introduced Thompson's two-parameter profile family[111] for the streamwise velocity profile. Thompson's profile is in effect very similar to Coles' law of the wall/law of the wake profile family introduced earlier. Retaining the Prandtl-Mager profile for the cross-flow component left Cumpsty and Head with one auxiliary equation to specify, since the streamwise skin friction was one of the parameters of the streamwise velocity profile. For the auxiliary equation, Cumpsty and Head chose to postulate that the rate of entrainment of freestream fluid into the boundary layer was uniquely related to the shape of the streamwise velocity profile in the same way as in two-dimensional flow. To obtain the actual auxiliary equation, the continuity equation is integrated [Eq. (6.124)] and the terms representing the freestream entrainment rate collected and replaced by the assumed relationship to the streamwise profile. As a general observation, it is to be expected that Cumpsty and Head's procedure would perform much more satisfactory with (vanishingly) small cross flow than would Mager's procedure, in view of the close relationship of the Cumpsty-Head scheme to Head's simple and very effective two-dimensional boundary-layer prediction scheme. The relative behavior at higher cross flows is not so apparent, however, since it is by no means certain that for these flows the power law profile and other assumptions are inferior to law of the wall/law of the wake concepts for the streamwise velocity profile under these particular conditions. In any event, the predictions shown by Cumpsty and Head agree moderately well with their measurements for certain integral parameters on an infinite yawed wing.

In spite of the well-understood techniques for generating integral moment equations and the tolerance of the concept of an isotropic eddy viscosity even in some finite difference schemes (to generate the integral moments of the Reynolds stresses), there has been little attempt so far to generate three-dimensional multiparameter integral schemes. Eichelbrenner and Penke[112] used a polynomial cross-flow profile and essentially the u moment of the streamwise momentum equation, simplified somewhat. Horlock[96] and Lindsay[100] used developments of Johnson's profiles together with Head's entrainment concept. Smith[113] followed a course similar to Cumpsty and Head and used either the Prandtl-Mager profile or a simplified version of Johnson's profile (the simplification being that the apex of the Johnson's triangle is assumed to fall at some fixed value of u^+). Smith treated compressibility by assuming that the incompressible profile families hold in a y coordinate stretched by the local density,

$$\frac{u}{u_e} = f\left(\frac{y'}{\delta}\right) \qquad y' = \int_0^y \frac{\rho}{\rho_e \, \mathrm{d}y} \tag{6.135}$$

and the density evaluated by a quadratic temperature profile and, following Myring's[114] lead, abandoned external streamline coordinates. A very similar study was subsequently undertaken by Thompson and MacDonald.[115]

While not as demanding as the more complete numerical approaches, solution of the coupled nonlinear integral equations is not trivial. Most people have tended to use a variant of the scheme used by Mager,[97] who solved the coupled integral momentum equations iteratively and sequentially, starting with the streamwise momentum equation and lagging the cross-flow information. Once again, unless the number of iterations is small, the principle attribute of integral methods (that of speed) is being squandered. Thompson and MacDonald[115] encountered severe numerical problems with the sequential iterative solution technique near separation and in some regions of favorable pressure gradient. Thompson and MacDonald[115] thereupon solved the problem by means of a coupled implicit technique, treating the nonlinear coefficients by iteration. While no details of the Thompson-MacDonald algorithm were given, it would seem on the surface to be a very appropriate solution technique.

Two observations can be made on the work done to date. The first is that, in general, the level of agreement between prediction and measurement is poorer than that achieved by integral techniques in two dimensions. The second is that none of the existing three-dimensional schemes, when applied in two dimensions, are more advanced than say Green's versions[116] of Head's simple entrainment method. Clearly then, for three dimensions, there is much room for improvement since the existing level of agreement is in general not adequate, even for loss level estimates. In restricted classes of three-dimensional problems, it would seem feasible to modify the various assumptions such that for that particular class of problem good agreement with measurement could be obtained. The question that any user must answer is whether, for his particular projected use, the investment of effort is best spent in developing an integral scheme further or are the mean profile problems such that the effort is better invested in a finite difference scheme. Insofar as the profile assumptions are concerned, Wheeler and Johnson[99] and Nash and Patel[109] show the limitations of the existing suggestions and, while improvements can be made, it must be remembered that each additional profile parameter introduced to improve the profile description requires that an additional equation be added to the system. Only if the number of profile parameters is kept small do integral schemes retain an attractive speed advantage over the completely numerical treatment of the basic three-dimensional boundary-layer equations of motion. It must also be recalled that with the existing integral schemes compressibility is at best given a rather cursory treatment, the application envisaged for these schemes usually being the subsonic wing problem where compressibility effects are not large. Significant heat-transfer effects, such as might occur on turbine blades, are as yet not allowed for within the existing integral schemes. As a consequence of these problems with integral procedures, a number of investigators have been led to develop finite difference methods for solving the three-dimensional boundary-layer equations. These finite difference schemes would appear better suited for general turbomachinery applications than existing integral schemes, although they are still limited by the boundary-sheet approximations.

The Finite Difference Approach

Largely evaded in the treatment by existing integral methods is the question of the two Reynolds stress components. This problem is inescapable when the finite difference approach is considered, and the problem will be discussed and current solution techniques described in the section on turbulence models. The remaining problems are the variety of choices for the form of the equations and the coordinate system, the stability and accuracy of the numerical methods, the starting problem, and the previously mentioned inflow boundary conditions. As in the discussion of integral schemes, both the boundary-layer edge conditions and the metrical information are assumed known. In actuality, determination of both of these latter items might well involve more work than the determination of the boundary-layer behavior itself, but discussion of them is inappropriate here.

Historically, work on the three-dimensional boundary-layer equations has been concerned either with the development of some innovative numerical scheme or with the description of the turbulent transport in three dimensions. In the former case much progress has already been made and in the latter instance fairly simple numerical schemes were adopted to expedite the description and evaluation of the critical features of the turbulent flow, again with an encouraging degree of success. Presently, second-generation turbulent boundary-layer schemes are being evolved embodying aspects of the more recent numerical developments.

In discussing the numerical problem of treating the three-dimensional boundary-layer equations, the previous section on numerical methods for the boundary-layer partial differential equations is relevant. In the present section, the additional special features arising in the case of the three-dimensional boundary layer will be discussed in the same order of presentation as was adopted in the earlier numerical section. Starting with the choice of numerical method, yet another stability parameter arises. Frequently, due to the cross-flow velocity w, it is of the form

$$\gamma = \frac{h_1 w \Delta x}{h_3 u \Delta z} \tag{6.136}$$

where Δx and Δz are the marching and spanwise directions and h_1 and h_3 the appropriate metric scale factor. Krause et al.[117] have shown the stability parameter γ must satisfy certain restrictions for many schemes. Perhaps the most common restriction is that for a number of schemes γ must never be negative. The implication is that for these schemes the cross-flow velocity w cannot be allowed to change sign (the spanwise marching direction chosen to give w positive). Certainly the cross-flow velocity can change sign in flows of real interest, as evidenced by the measurements of Klinksiek and Pierce[118] in an S-shaped duct. Indeed, much of the criticism of Johnson's profile family[98] was directed at its inability to treat crossover profiles where the cross flow changed sign. It would seem then that the cross-flow stability parameter could prove catastrophically restrictive in some applications. Much less dangerous is the other restriction sometimes encountered—that $-1 < \gamma$, for here at least the prospect exists of reducing the axial step to satisfy the stability requirement and, on average, -1 is not a particularly small value of the parameter. Difficulties could be expected if a small spanwise mesh was desired simultaneously with a large axial step, a not uncommon set of circumstances. Also, it is possible to have local flow angles in excess of 45 deg so that w/u can exceed unity and this again might require that the axial step be reduced. Unfortunately, there is as yet no unconditionally stable scheme available for the usual three-dimensional boundary-layer equations (this observation will be qualified in a moment).

In the extended boundary-layer equations, diffusion in the spanwise direction is retained and several unconditionally stable schemes for the linearized system of equations are available. The converse is also true in that one of the unconditionally stable implicit techniques for treating the extended boundary-layer equations could be reduced by neglecting the spanwise diffusion. When applied to the conventional boundary-layer equations, it is indeed probably best to think of the procedure of Kendall et al.[119] in this light. This brings into focus the essential difference of the extended implicit schemes and the conventional implicit schemes. In the former, both the spanwise and the normal directions are considered implicitly and boundary conditions at either end of the spanwise sweep are satisfied. In the latter case of the conventional boundary-layer equations, only the normal direction is considered implicitly—an embarrassment if the initial value behavior of the spanwise sweep is not obtained. In line with the foregoing, it seems appropriate to term the case where only the normal direction is considered implicitly, a semi-implicit method, and when both directions are taken implicitly, a fully implicit method.

Insofar as the linearization process is concerned, no additional problems arise in three dimensions. The problems of ad hoc linearizations and coupling are emphasized, however, since non-negligible product terms ρuw, for instance, appear in the equations, calling attention to the coupled nonlinear nature of the problem. The cross-flow velocity w is, of course, not necessarily small and, indeed, a great deal of the difficulty in obtaining even integral solutions has been traced by Thompson and MacDonald[115] to the linearized sequential treatment of the equations. Aside from the explicit formulations (which retain all of their two-dimensional problems), the process of a formal linearization followed by a coupled solution of the governing equation would seem to be a very powerful method of treating the problem. Indeed, a number of techniques of this type, usually embodying a Newton-Raphson treatment of the nonlinearities, are described in the review of Blottner.[81]

In the early article on the turbulent equations, Nash[77] worked with the primitive variables *uvw*, $u'v'$, and $v'w'$ and treated the growth of the boundary layer within the computational domain by periodic reinterpolation. Stream function approaches lose some of their attractiveness in three dimensions in view of the need to introduce two such functions; however, a number of investigators (i.e., Cebeci[120]) have favored this approach to achieve the simplification that arises from the elimination of the continuity equation. In another approach, Harris and Morris[103] used the square of the normalized velocity defect to replace the normal coordinate y. While obtaining some advantage from this type of transformation, clearly the double value problem of streamwise velocity profile overshoot is a major difficulty for gas turbine applications. Apparently, the addition of the third dimension has not introduced new major problems (or remedies) and, insofar as the dependent variables are concerned, the problem is much the same as in two dimensions. Grid stretching functions also have a major role to play, although they have not yet been widely used since a great deal of the effort to date has been on laminar flows, where their use is not nearly as critical as it is in turbulent flow. Again, in three-dimensional turbulent problems, the desire to place grid points in the sublayer still exists, particularly for highly accelerated flows with heat transfer. The viscous sublayer problem becomes even more accute, since the three-dimensional version of the law of the wall is not nearly as secure as in two dimensions. Counterbalancing the desire to place grid points in the sublayer is the increased cost of doing so in three dimensions but, be that as it may, an inadequate sublayer treatment is castle building on a bed of sand.

Finally, to date, coordinate systems have been quite specifically tuned to the problem at hand. As a rule, the body surface is taken as one coordinate with the other two taken as convenient. The body surface not lying on a coordinate can lead to difficult computer logic and interpolation inaccuracies and hence is not recommended. In the past, orthogonal systems have often been selected, but nonorthogonal systems are quite readily accommodated in finite difference schemes and can be recommended should their use simplify the coordinates significantly. More general coordinate systems, not so specifically tuned to one particular geometry, are being developed. The overall problem is reviewed in detail by Eiseman and McDonald.[121] When constructing or modifying a computer code, the least that can be done at present is to leave the metric information arbitrary in the main body of the code and to set in some easily accessible subroutine.

The Reduced Navier-Stokes Equations

As was mentioned earlier, the conventional "boundary sheet" form of the three-dimensional boundary-layer equations takes into account only diffusive effects in one coordinate direction. The thin boundary sheet approximations have proved very valuable, but in a number of critical turbomachinery problems some of the boundary sheet approximations are not valid. For instance, the assumption that spanwise rates of change are negligible when compared to gradients normal to a wall is often violated.

It is worth recalling at this juncture that the reason for using the boundary-layer equations is to achieve a solution for the viscous flow effects more easily and hence more efficiently than solving the full Navier-Stokes equations. The success of this approach has been considerable, but it is conditional upon the validity of the boundary-layer approximations. In seeking a replacement for the boundary-layer equations for the internal flows of interest, two issues arise. The first is the adequacy of the physical approximations introduced into the Navier-Stokes equations to produce the simpler system. The second is the labor required to solve the simpler system. Given the physical approximations, it must be asked if the simpler, reduced system can be solved much more efficiently than the original Navier-Stokes equations. If not, then the approximations are probably not worth introducing. Both these issues will be addressed subsequently.

In looking at the three-dimensional flows involved it is important to recognize the fundamental differences that can arise compared to two-dimensional circumstances. First of all, note that, like many other high Reynolds number flows, a predominant direction of the flow can be identified, often arising from the ducted nature of many of the internal flows of interest. Many of these flows are inherently three-dimensional because of geometry, upstream distortions, or other factors. When viscous effects or other sources of vorticity are present, three-dimensional flows differ fundamentally from their two-dimensional counterparts in that large secondary flows are easily generated by a deflection of the primary flow and/or by other mechanisms. Secondary flow theory (reviewed by Horlock and Lakshminarayana[122]) affords considerable insight into the generation of secondary flow and establishes that streamwise vorticity on the order of 50% of the transverse vorticity (shear) can be generated by a lateral flow deflection of only 15 deg. The large secondary flows thus generated often exert an appreciable influence on the primary flow; thus, aerodynamic performance, viscous flow losses, and heat transfer can be affected significantly. To date, secondary flow theory has been concerned with inviscid flow. However, for the flows of interest, allowance must be made for the viscous effects that arise from vorticity production at the no-slip walls. From the earlier discussion, it is assumed that three-dimensional boundary-layer theory will be inadequate for this purpose. In seeking to replace the (invalid) boundary-layer equations and allow the generation of secondary flows in general accord with secondary flow theory, a number of authors have considered what are termed here "reduced forms of the Navier-Stokes equations" and this concept will now be discussed.

The present discussion will focus on physical approximations and their general effect on computational labor, but will not address numerical approximations or the error and labor of specific algorithms. Although the solution of the Navier-Stokes equations is feasible and has been demonstrated in three dimensions (e.g., see Refs. 123–126), the labor of solution using presently available algorithms is considerable for high Reynolds number flows with multiple length scales requiring locally refined three-dimensional meshes. For this reason, there is an extensive literature that addresses simplification of the steady Navier-Stokes equations for the purpose of numerical solution, often for application to high Reynolds number flows without streamwise flow reversal. A portion of this literature is reviewed by Davis and Rubin[127]; other discussion can be found in the references given here. All of the approaches considered here neglect the streamwise diffusion in some manner.

The viscous developing flow in a rectangular duct whose centerline has appreciable curvature is representative of the primary/large secondary flow structure of interest here and provides an orientation and physical basis for the flow approximations. The corresponding (incompressible) entry flow in a straight square duct has been considered in some detail by Rubin.[128] Excluding a small region of axial extent Re^{-1} near the leading edge, the flow structure in the straight duct consists of boundary layers, corner layer regions, and a potential core region with displacement interaction (blockage) effects. Further downstream, the flow structure evolves into a fully viscous region where the boundary layer, corner layer, and potential core regions merge and become less distinct. Eventually, the flow becomes fully developed and independent of the axial distance.

If the duct centerline is given an appreciable curvature in the region where thin boundary and corner layers are present, a different flow structure evolves. The potential flow for this curved geometry has radial pressure gradients beginning about one to two hydraulic diameters upstream of the bend. Three-dimensional boundary layers thus form on the end walls, with large radial cross flow toward the inside corners of the bend. A schematic of the cross-flow behavior near the change in curvature is shown in Fig. 6.5, as deduced from flow visualization experiments and other sources. Since the flow is confined, the cross flow is deflected in the corner and returns to the center region of the duct. A corner vortex structure may be formed, depending on the flow parameters present. As indicated in Fig. 6.5, the outer portion of the boundary layer and the corner flow regions away from the walls behaves as a rotational inviscid flow region in which secondary flow is generated (in the terminology of secondary flow theory) by the turning of the transverse vorticity (shear). The viscous effects are limited to the region very near the wall. This initial behavior of the end wall boundary layer near the change in duct curvature is consistent with empirical observations of boundary-layer velocity profiles having triangular polar diagrams (e.g., see Ref. 129). The rotational inviscid behavior is also reminiscent of the middle rotational inviscid layer of triple-deck theory for a two-dimensional boundary layer subjected to a streamwise disturbance, although the latter case does not involve streamwise vorticity. As the flow proceeds downstream, the secondary flow convects the streamwise vorticity and distorts the primary flow in what amounts to a strong interaction in the transverse directions between the viscous and inviscid (rotational or potential) flow regions. Significant regions of rotational, but essentially inviscid,

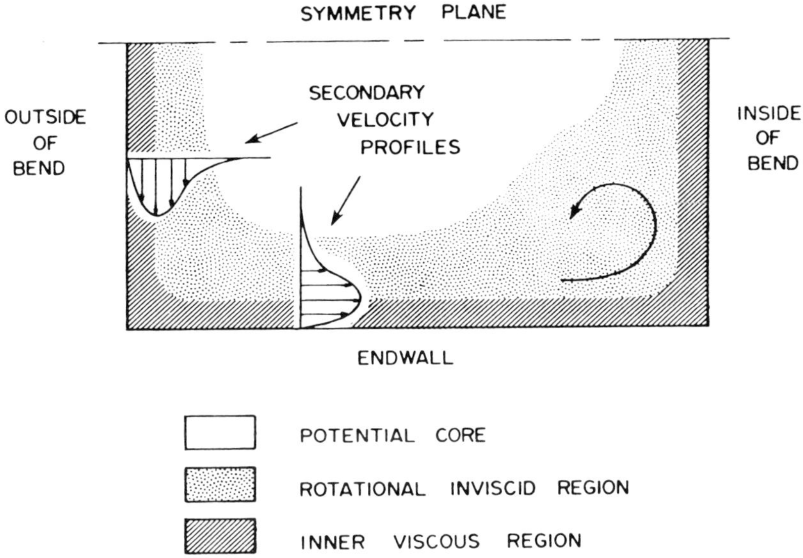

Fig. 6.5 Schematic of secondary-flow structures near start of bend.

flow occur as a result of this process and the potential core region will eventually disappear. If the duct curvature is constant, an incompressible flow can eventually reach a state of fully developed curved flow.

Finally, it has been known for some time that measured static pressure distributions tend to depart significantly from the potential flow pressure when large secondary flows are present. This is clear, for example, from the measurements of Stanitz et al.[130] in a strongly curved duct, where the measured static pressure is three-dimensional near the inside corners of the duct, even though the potential flow for this particular geometry is two-dimensional. This behavior is in contrast with a two-dimensional flow with small transverse velocity, where the modification of the potential flow pressure due to viscous interaction is often slight unless separation is present.

The various duct analyses usually attempt to introduce approximations that are consistent with an evolving flow structure of the type just described. Because of the strong interrelation of the potential core, rotational inviscid, and viscous regions near the walls, a single set of approximate equations are usually derived for simultaneous application to all of these flow regions at each transverse plane.

The goal of these form approaches has been to achieve a level of approximation that would yield accurate flow predictions, while reducing the labor of solution below that of the Navier-Stokes equations. The computational labor is greatly influenced by the type of governing equations solved, and here it is important to note that terminology such as "parabolic" and "elliptic" has been interpreted differently by different authors. The variations in terminology seem to derive from a scalar viewpoint that associates each momentum equation with a given velocity component, instead of considering the system of equations as a whole. For the present purposes, the term "elliptic" is applied to systems of equations whose characteristic polynomial has one or more imaginary roots, and systems for which no imaginary roots occur are referred to as "nonelliptic." Computationally, the distinction is very important, since elliptic systems are generally ill-posed for solution as an initial/boundary value problem, whereas nonelliptic systems are solvable by much more economical forward marching algorithms. Clearly, there is ample motivation to seek approximations of reasonable accuracy that offer the computational economy associated with nonelliptic approximating equations.

Systems of partial differential equations have been classified as elliptic, hyperbolic, or parabolic according to the roots of the characteristic polynomial equation (e.g., Courant and Hilbert[131]). Systems of an intermediate type that do not fit any one of these categories are encountered here and elsewhere. The present concern, however, is only whether the equations permit or preclude solution as a well-posed initial/boundary value problem, by integration in a "time-like" coordinate direction. Garabedian[132] considers when an independent variable may be identified as a time-like variable and, citing an example problem for a linear homogeneous system with constant coefficients, concludes that it is natural to require that every root of the characteristic equation be real, as this excludes solutions that may grow exponentially with the time-like variable. This criterion is employed here to test systems of equations for time-like behavior in a chosen coordinate variable. In view of the exponential growth that may occur in linear systems with constant coefficients, systems with one or more imaginary roots are presumed to be ill-posed for solution as an initial/boundary value problem and are referred to here as elliptic (with respect to the chosen coordinate). Systems with no imaginary root for the time-like coordinate are presumed to be well-posed and are referred to as nonelliptic.

The various systems of equations are usually of mixed order in that the velocity components appear in second-order derivatives, while the pressure appears only in first-order terms. To classify this mixed-order system, it is first rewritten as an equivalent first-order system and then examined following standard procedures for classification of n-dimensional systems of first-order partial differential equations (e.g., see Ref. 131). The equations can be written as the following quasilinear system of m three-dimensional partial differential equations:

$$A_1 \frac{\partial \mathbf{\Phi}}{\partial x} + A_2 \frac{\partial \mathbf{\Phi}}{\partial y} + A_3 \frac{\partial \mathbf{\Phi}}{\partial z} = \mathbf{B} \tag{6.137}$$

for the m-dimensional column vector $\mathbf{\Phi}$ of dependent variables, with x, y, and z the independent variables (spatial coordinates). The square matrices A_i and the column vector $\mathbf{B}$ contain coefficients depending only on x, y, z, and $\mathbf{\Phi}$. The derivative operators $h_1^{-1}\partial/\partial x$, $h_2^{-1}\partial/\partial y$, and $h_3^{-}\partial/\partial z$ are replaced by λ_x, λ_y, and λ_z, respectively, and the mth-order characteristic polynomial equation is given by

$$|A_1\lambda_x + A_2\lambda_y + A_3\lambda_z| = 0 \tag{6.138}$$

Since x has been identified here as the time-like coordinate, it is appropriate to examine Eq. (6.138) for imaginary roots λ_x, assuming that arbitrary real values are assigned for λ_y and λ_z. The presence of an imaginary root for λ_x would imply that the system is not suited for solution by forward marching as an initial value problem in space, according to Briley and McDonald.[133]

Since the viscous terms will be small over much of the flow region, the approximating equations must be nonelliptic both with viscous terms present *and* in the inviscid limit of small viscosity. The presence of second-order viscous terms suppresses the influence of first-order terms on the characteristic equation (selectively if the system is of mixed order). Since all second-order terms disappear in the limit of small viscosity, the viscous characteristic equation is degenerate in this limit and, consequently, the viscous and inviscid cases must be considered separately.

An assumption that one or both transverse velocity components is small (in some coordinate system) has often led to nonelliptic approximating equations. The two-dimensional boundary layer is a well-known example, where the usual order-of-magnitude estimates justify neglecting all convective and viscous terms in the normal y momentum equation. If coordinate curvature terms are absent or neglected, then this equation reduces to $\partial p/\partial y = 0$, which can be integrated to obtain $p = p(x)$ and introduced as a pressure approximation in the x-momentum equation. If significant curvature terms are present, as in the slender (but curved) channel approximations of Blottner[134] and Anderson,[135] then the y-momentum equation can no longer be integrated to approximate the streamwise pressure gradient term and must be solved by treating pressure as a dependent variable. Either approximation produces nonelliptic equations, however, since the coordinate curvature terms do not affect the analysis of characteristics. The same is true for three-dimensional boundary layers; nonelliptic equations are obtained when convective and viscous terms are neglected in the y-momentum equation, whether or not the streamwise pressure gradient term is approximated using $p = p(x, z)$.

If both transverse velocity components are small in a three-dimensional flow and if curvature terms are negligible, then the transverse momentum equations reduce to $\partial p/\partial y = 0$ and $\partial p/\partial z = 0$, justifying an approximation $p = p(x)$ in the x-momentum equation, which produces nonelliptic equations. The analysis of Rubin et al.[128] for entry flow in straight ducts contains an example of this type. Other approaches that assume that both transverse velocity components are small are discussed by Baker and Orzechowski[136] and Anderson and Hankins.[137]

In the turbomachinery flows of interest, both transverse velocity components may be large, significant curvature terms may be present, and the

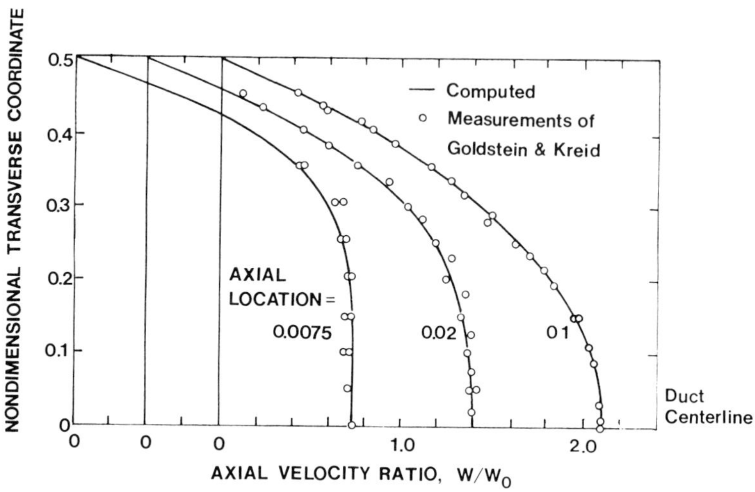

Fig. 6.6 Comparison between computed and measured axial velocity profiles for square duct on the midplane with laminar flow (from Ref. 140).

geometry may contain streamwise corners in which the surface normal changes direction. Conventional boundary-layer arguments based on a small normal velocity component are clearly not applicable in these circumstances. For this type of flow, many authors have instead used an approximation based on a multiple definition of pressure, in which the streamwise pressure gradient has a known form, but is completely independent of the transverse pressure gradient terms. Although not based on order-of-magnitude estimates, this approximation has been found to produce nonelliptic equations and has been used, for example, by Patankar and Spalding,[138] Carreto et al.,[139] Briley,[140] Ghia and Sokhey,[141] and Kreskovsky et al.[142] Experience shows that this particular approximation can provide reasonable accuracy for many problems if the streamwise pressure gradient is represented by a potential flow pressure with a mean pressure correction and with no approximation introduced for transverse pressure gradients. Nevertheless, the multiple definition of pressure does not allow the streamwise momentum balance to be influenced by experimentally observed distortions of the static pressure field, which are induced by large secondary flows. An example of this type of prediction is taken from the work of Briley[140] and shown in Fig. 6.6.

One approach that avoids approximations in the pressure gradient terms neglects only the streamwise diffusion terms in the Navier-Stokes equations. This approximation is not very restrictive and requires only a coordinate system aligned with the shear layers; these equations are also suitable for problems with reversed flow. The diverse approaches of Patankar and Spalding,[138] Moore and Moore,[143] and Pulliam and Steger[144] are all based on these approximating equations, which are often termed the parabolized or thin-layer Navier-Stokes equations. Although the neglect of streamwise diffusion is often described as a parabolizing approximation, it is generally agreed that in subsonic flow regions these governing equations are in some sense "elliptic" unless further approximation is introduced, usually in the streamwise pressure gradient. This is discussed, for example, by Davis and Rubin,[127] and the case of mixed supersonic/subsonic flow has been studied by Schiff and Steger[145] among others. If forward marching algorithms are used when only streamwise diffusion is to be neglected, then these algorithms are normally used within a context of global iteration analogous to that employed in time-like or other global iteration algorithms for the Navier-Stokes equations.

In a recent paper, Briley and McDonald[133] have sought to avoid the multiple definition of the pressure and yet retain the efficiency of obtaining a solution following a single pass through the duct. In the Briley-McDonald approach, boundary-layer order of magnitude estimates are re-examined in terms of a transverse velocity that is decomposed into potential (v_ϕ, w_ϕ) components which are themselves vectors. This is in contrast to the usual representation in terms of the velocity components v and w. It is found that the potential velocity vector must be small when either transverse coordinate can be the surface normal of a boundary layer, whereas the rotational velocity vector can be of order unity and parallels the growth in streamwise vorticity. The potential and rotational velocity vectors thus have the same role in deriving approximations for a viscous secondary flow as the "normal" and "cross-flow" velocity components in a three-dimensional boundary layer. An assumption of small scalar potential then leads to viscous approximations neglecting streamwise diffusion and inviscid approximations affecting the convective terms in the transverse momentum equations and stagnation enthalpy. On examination of the characteristic polynomial, it is found that the viscous approximation alone is sufficient to produce a parabolic system of equations. However, the presence of essentially inviscid flow regions at high Reynolds numbers suggests that the approximations used should also reduce the Euler equations to a nonelliptic form. The inviscid approximations introduced by Briley and McDonald[133] reduce the compressible Euler equations to a nonelliptic form for arbitrary Mach numbers. The resulting nonelliptic approximating equations can then be solved as an initial/boundary value problem, with a consequent saving in computer resources. Although the approximations introduced appear to have a wider range of applicability, which includes both internal and external flow problems, the investigation reported by Briley and McDonald is restricted to laminar subsonic flow in a square duct having appreciable streamwise curvature.

The foregoing methods may be extended to rotating coordinates and thus may be applied at least in concept to rotors. An example of this extension is given, for instance, by Govindan et al.[146] A major problem for these approaches in both the stationary and the rotating frames is concerned with the stagnation region. Here, either the approach breaks down or the approximations lose validity due to the occurrence of a stagnation point. Various additional approximations can be introduced, including, for example, a cusped leading model that avoids the stagnation point by removing it.

A further problem arises with obtaining starting conditions, but this will not be discussed further here, other than to suggest that where possible solutions ought to be started upstream where the cross-flow velocities are small. At the present time, a rigorous evaluation of these various reduced Navier-Stokes approaches proceeding from simple turning ducts all the way through to a rotor passage has not been performed. Such an extensive evaluation is a prerequisite to any recommendation for use, since a number of major simplifications have been introduced and there is a divergence of opinion on the efficiency of the approximations used. It is expected that in the next few years the necessary high-quality experimental data will become generally available and the rigorous evaluations will be published. In the event of a favorable outcome, these reduced Navier-Stokes analyses could become very significant tools for the turbomachinery designer.

6.4 Boundary-Layer Separation

Separation is the rather loose term given to the flow when it forms a region within itself that is inaccessible by convection to the flow coming from upstream. Usually, a wall forms one of the bounding surfaces of this region, but the recirculating region formed by the release of a swirling axial jet is an example of separation where a wall is not involved.

Motivation and Background

In several instances of very great concern to turbomachinery designers, the occurrence of significant regions of separated flow is termed "stall" and determines the boundary of the operating envelope of the device. Compressor stall is a very obvious example that springs to mind, diffuser stall is another. In both these examples, the significant regions of flow separation cause major changes in the pressure distribution around the body, as well as great increases in the drag of the system. Stall is often accompanied by an increased flow unsteadiness. The separation prediction problem is compounded by the observation that catastrophic flow separation (stall) is a very sensitive phenomenon and that small local separation regions can occur without significant deterioration in the flow. For a number of subtle reasons arising from both upstream and downstream conditions, these small and well-behaved local separations can suddenly degenerate into the large separation zones characterizing stall. It follows that it is often not enough simply to predict the onset of separation, difficult enough as that may be, but it can also be important to be able to predict at what point the small separation undergoes its metamorphosis into stall. The problem is quite acute in both compressor and diffuser stall since, to obtain optimum performance, it is often desired to operate the device on the verge of stall. Frequently, in this optimum condition, small local separation regions occur in the flow and a knowledge of the factors influencing this behavior would be very valuable.

In the absence of any reliable theoretical stall prediction procedure, the empirical approach has had to be adopted by the design engineers. In this empirical approach, maps of stall zones are built up by observation as the device is operated over a range of critical performance parameters. The resulting correlations are then attributed a more general significance and applied to other similar devices. Unfortunately, the sensitivity of stall to a myriad of factors has caused a great deal of concern, difficulty, and expense with the empirical approach in view of the ad hoc nature of the stall correlations. In an effort to improve the above type of stall correlation, the stall has been supposed to be determined by boundary-layer separation and the criteria for boundary-layer separation occurring within the components of the devices developed. Sanborn,[147] for instance, has discussed some of these separation criteria; generally, they relate the condition of the boundary layer at separation to the applied pressure gradient, in the general form

$$(\delta^*/\theta)_{\text{sep}} = f\left(-\frac{\theta^2}{\nu}\frac{du_e}{dx} \right) \tag{6.139}$$

where f is some function determined by observation and δ^* and θ the usual boundary-layer displacement and momentum thicknesses. The ratio of displacement to momentum thickness is usually termed the shape parameter H and the problem of determining the streamwise variation of the shape parameter H is left open at this point. The obvious suggestion of predicting the entire boundary-layer behavior and thereby determining the variation of shape parameter H is somewhat self-defeating at this point, since such a procedure would also either directly predict the separation point or allow an estimate of the separation point location to be made independently of the empirical correlation function f. Sanborn's view was that none of the available procedures (at that time) could give a reliable prediction of the separation location and that these empirical correlations would serve as "correctors" to the theoretical predictions. Valid as that observation may have been, it is to be expected that in the light of rapid developments efficient and accurate means for theoretically predicting the onset of separation will become available (this remark will be qualified subsequently). The crucial problem of determining at what point the separating boundary layer degenerates into stall can be solved by tracing the subsequent "boundary-layer" behavior. At this point, difficulties (conceptual and/or numerical) are encountered with spatial forward marching solutions to the governing equations. Downstream effects are most evidently propagating upstream and clearly both the governing equations and the integration schemes must allow this upstream propagation. The pressure requires special consideration since the thickening of the boundary layers as separation is approached can cause changes in the very pressure distribution that is driving the boundary layer to separate in the first place.

One solution approach is to resolve all these important length scales in one analysis that contains the necessary generality and then to apply it throughout the flow—even to those regions of the flow where the generality is not required and some simplification is possible. In an effort to be more efficient, alternative approaches have been explored. Here the multiscale problem of interest involves small regions of separated flow. For this problem, physical models of the various flow regions have been established and linked together by means of the boundary conditions. The simplified governing equation systems appropriate to each flow region are then solved efficiently making use of the (valid) approximations and linked together (interacted) through the appropriate boundary conditions. This approach encompasses three distinct areas: model formulation including interface boundary conditions, validation of the model, and computation of the solutions.

Considering the model formulation, it is known that a potential outer flow gives valid high Reynolds number force and moment distributions over bodies operating in unseparated flow conditions. Viscous effects dominate loss levels and set the operational limits of the device by determining separation and stall. Conventional boundary-layer theory provides the first level at which viscous effects can be allowed for, given the pressure distribution, and has in the past been limited to unseparated flow. The question addressed by a number of investigators was, given the possible utility of such an analysis, would it be possible to extend boundary-layer theory into limited regions of separated flow? Considerable encouragement for these efforts came from experimental studies. These experimental studies have indicated that in many small but important regions of separated flow, the boundary-layer-invoked approximations were valid, notably that v (a normal velocity) remained small compared to a streamwise velocity u and that across these separation regions the normal static pressure gradient was negligible. Since away from the small separation region the outer flow remained inviscid, it was natural to retain these approximations and attempt to construct a flow model to adequately predict these small separation regions, thereby enlarging the region wherein inviscid and boundary-layer calculations can provide accurate predictions of the flow. Having constructed such a flow model, the next question would be its validation and finally ascertaining that the resulting system of approximate governing equations could be solved more efficiently for the entire domain than, say, the Navier-Stokes equations.

Construction of a suitable model for small regions of separated flow has its roots in the work of Crocco and Lees,[148] who investigated the interaction of a laminar wake with an inviscid supersonic outer flow. A boundary-layer momentum integral technique was used to describe the shear layer and a simple local Prandtl-Meyer wave relationship to determine the freestream pressure at the edge of the shear layer. This approach was relatively successful, although a singularity caused a great deal of difficulty with the solution implementation. In this approach, the development of the boundary layer or shear layer and the freestream are independently governed by spatially parabolic and hyperbolic systems of equations, respectively. Rather remarkably, Garvine[149] later showed that a model of this problem was improperly posed as an initial value problem in space and should be approached as a boundary value problem. The boundary value nature was present even though each individual flow regime was governed by a system that could be solved by itself as an initial value problem in space; however, the boundary condition representing interaction of the two regions produced a boundary value problem. Garvine's observations explained much of the difficulty experienced in solving the problem as formulated by Crocco and Lees.

The next series of developments occurred almost simultaneously with Stewartson and Williams,[150] Neiland,[151] and Messiter[152] developing very similar asymptotic theories of shear layers interacting with external inviscid flows. In these analyses, valid for $R \to \infty$, Stratford's concept of a two-layer boundary layer[48] was used in conjunction with an analysis by Lighthill[153] to produce a rational, self-consistent structure for these interacting shear flows. Solved now as a boundary value problem, much of the previously elusive structure of this class of interacting flows emerged. These important results led to further developments and widespread application of what has become known as the triple-deck theory (e.g., see Ref. 154).

On a parallel development, work continued on the finite Reynolds number behavior of interacting shear layers. For a supersonic external stream, Werle and Vatsa[155] successfully obtained solutions to the laminar shock-wave/boundary-layer interaction problem, essentially as formulated by Crocco and Lees, by using an iteration procedure and solving the boundary-layer equations by an implicit finite difference scheme. Subsequently, very favorable comparison between triple-deck results and those of Werle and Vatsa were obtained as $Re \to \infty$ by Burggraf et al.[156] However, Davis,[157] Burggraf et al., and Davis and Rubin[158] pointed out the rather slow approach of these composite solutions at finite Reynolds numbers to the asymptotic limits of the triple deck, at least for this particular problem of a supersonic corner flow. This slow approach to the asymptotic limit has provided one justification for continued development of the "composite" interacted boundary-layer theories. As further justification, the interacted boundary-layer theory is less restrictive than the asymptotic theory and requires about the same computational labor.

In view of the elliptic nature of the external flow, in the case of a subsonic flow the boundary value nature of the interaction problem is not surprising. Important to the development of a computationally efficient composite approach was the recognition that, at finite Reynolds numbers in subsonic flow for small separation regions, the interaction was localized and could be represented as a perturbation of the exterior flow and determined by the particularly simple displacement body of linearized thin airfoil theory (the Hilbert integral). Briley and McDonald[159] used linearized subsonic thin airfoil theory in this manner to provide the change to the external flow resulting from the boundary-layer interaction. With this formulation, they were able to predict the development of transitional separation bubbles on airfoils and obtained a favorable comparison with experimental data. Subsequently, Kwon and Pletcher[160] and Cebeci and Schimke[161] used this same formulation, but with quite different numerical solution algorithms and turbulent transition models, and obtained good agreement with the data and with the earlier predictions of Briley and McDonald. A related study has recently been performed by Carter and Vatsa[162] for leading-edge bubbles.

While the interaction with both the subsonic and supersonic external flows can be well represented very simply for many problems using linearized flow theory, the mixed elliptic-hyperbolic nature of the transonic problem raises obvious problems. LeBalleur[163] and Carter[164,165] eliminated these problems in a straightforward manner by iteratively computing the external flow using an efficient transonic potential flow solver. Thus, the interacting boundary-layer analysis has been formalized and applied for localized separations in incompressible, transonic, supersonic, and hypersonic flows. Attention will now be given to results that validate the general approximations introduced in this analysis.

As mentioned earlier, order-of-magnitude estimates, experimental measurements, and comparisons between predictions and measurements all gave encouragement to the view that for small separation regions boundary-layer approximations would remain valid. Werle and Vatsa[155] compared favorably with both data and the Navier-Stokes solution obtained by Carter[166] for a compression ramp. Using triple-deck theory, Rizzetta et al.[167] compared favorably with data and with the Werle-Vatsa solutions for corner flow. Very direct tests of the validity of the boundary-layer approximations for subsonic flow were made by Briley and McDonald[159] and Ghia et al.[168] These comparisons are of special significance since the same algorithms, codes, and grids were used. In these instances, flow solutions were obtained using the Navier-Stokes equations and then the terms present in the Navier-Stokes equations (but not in the conventional boundary-layer equations) were deleted and the results were compared. Briley and McDonald examined the previously mentioned transitional separation bubbles and found only a small effect of the terms neglected in the incompressible interacted boundary-layer formulation. Ghia et al. examined the effect of a wide range of approximations upon the incompressible flow over a slab with varying degrees of nose bluntness and upon the flow over a circular cylinder with a rear splitter plate. They concluded that interacted boundary-layer-type approximations are valid in many instances of separated flow. Murphy et al.[169] compared solutions from the Navier-Stokes to those obtained from a scheme using boundary-layer approximations in both laminar and turbulent supersonic shock-wave/boundary-layer interaction. They concluded that, in these cases, as long as the flow displacement remained small, the neglect of the appropriate terms did not produce a significant discrepancy between the boundary layer and Navier-Stokes predictions. Vatsa et al.[170] compared interacted boundary-layer solutions with solutions obtained by a Navier-Stokes solver for turbulent flow over a boattail with a sting, where a small separation zone occurred at the boattail-sting juncture. Relatively good agreement was obtained between these two sets of predictions. Thus, it has been demonstrated that, for the flow problems examined above, use of the boundary-layer approximations does not lead to major errors compared to the Navier-Stokes solutions.

Structure of Proposed Solution Procedures

Now details of the numerical schemes must be introduced, since it can be concluded that a validated flow model for small separated flow regions interacting with an essentially inviscid outer flow region has been firmly established. The basic structure of this flow model leads one to expect considerable economy of computation relative to solving, for instance, the full Navier-Stokes equations. In the simplest of formats, the steady inviscid external flow around the geometry in question is computed once. If the external flow can be assumed irrotational, a single scalar potential equation may be used with obvious computational savings. The correction perturbation to this inviscid outer flow may be obtained from linearized flow theory and, at worst, in transonic flow with small separations, some perturbation potential would have to be solved throughout the flow. For subsonic flow, the correction can be obtained from the simple Hilbert line integral of thin airfoil theory. For supersonic flow, the correction can be obtained on a pointwise basis using, for instance, the linearized wave theory of Prandtl and Glauert. The remaining issue is the computation of the interacted boundary layer. The conventional boundary-layer approach leads one to think in terms of a parabolic system of equations in space and prior experience indicates that this system can be solved very rapidly and efficiently. Unfortunately, the boundary value nature of the interaction problem described earlier indicates that the interacted boundary layer is not well posed as an initial value problem in space and, if a forward marching solution algorithm is to be used for solving this boundary value problem, then multiple iterations through the parabolic system are required. Unless a sufficiently economical approach to the boundary value problem is adopted, the resulting effort of solution will seriously degrade the overall apparent computational efficiency of the formulation. Hence, the numerical approach adopted is of some importance.

Two subsidiary points emerge when the boundary value nature of the problem is addressed, particularly via iterated spatial forward marching approaches. The first point is the Goldstein singularity and the second is the reversed flow streamwise velocities that are encountered after the separation. These points will be addressed in more detail subsequently. For the present, it is noted that it is now widely accepted that the Navier-Stokes equations are regular at separation, while the Goldstein singularity arises in the steady (uninteracted) boundary-layer equations when the streamwise pressure gradient is specified as a boundary condition and the equations are solved as an initial value problem by forward marching in the streamwise direction. Convincing numerical evidence for the regularity of the Navier-Stokes equations at separation has been presented for instance by Briley,[171] Leal,[172] and Ghia and Davis.[173] Clear numerical evidence of the presence of the separation singularity in the steady boundary-layer equations when the streamwise pressure gradients are prescribed has been given by Werle and Davis[174] and Klineberg and Steger,[175] for instance. It has been established that the steady interacted boundary-layer/inviscid flow solutions, from both the composite equations and the triple deck agree with the corresponding Navier-Stokes solutions, and hence are regular at separation. Thus, the Goldstein singularity in interacting flows may have a very considerable impact on the calculation efficiency and strategy, but it is a matter for the solution algorithm, not of the formulation. Insofar as the difficulties associated with reverse flow velocities are concerned, these too are features of the solution algorithm and not of the problem formulation, as can be seen for instance from the work of Klineberg and Steger,[175] Carter,[166] and Williams.[176] The problems here arise from forward marching in space against the mean flow velocity direction. Such problems clearly affect certain types of solution algorithms, but again the elliptic boundary value nature of the interaction and its recognition in the formulation support the contention that the presence of reverse flows is not a difficulty of formulation, but of solution technique. Some aspects of the numerical algorithms that have been and are being used to solve the steady composite equations of interacting boundary-layer theory are examined in the next section.

Development of the Numerical Methodology

The emphasis here will be elucidating some of the numerical aspects of the various approaches adopted to solve the composite interacting boundary-layer equations. The governing equations can be written in the form,

$$(\rho u)_x + (\rho v)_y = 0 \tag{6.140}$$

$$\rho u u_x + \rho v u_y = -p_x - (\mu_T u_y)_y \tag{6.141}$$

where μ_T is the sum of the laminar and turbulent effective viscosity and an x or y subscript denotes differentiation. For turbulent compressible flow, the dependent variables will be assumed to have been mass averaged following Favre. The governing equations are subject to the impermeable wall and freestream boundary conditions

$$y = 0,\ u = v = 0,\ y = \delta,\ p = p_c(x) \tag{6.142}$$

The imposed pressure gradient is related to the streamwise velocity U_c

$$p_x = -\rho_c U_c (U_c)_x \tag{6.143}$$

and the streamwise velocity is decomposed into the imposed and perturbation velocity $U_c = U_c^\circ + U_c$, where U_c is the perturbation to the imposed inviscid velocity field U_c°, due to the interaction. For incompressible flow, thin airfoil theory defines the perturbation velocity by the so-called Hilbert integral wherein

$$U_c = \frac{1}{\pi} \oint_{-\infty}^{+\infty} \frac{(U_c \delta^*)\xi}{x - \xi}\, d\xi \tag{6.144}$$

where the symbol $\oint$ denotes that the Cauchy principle part of this singular integral is taken and δ^* the usual boundary-layer displacement thickness. An even simpler relationship between the local rate of change of displacement thickness and the perturbation velocity U_c can be derived in supersonic flow relative to a perturbation flow angle α,

$$(U_c)_x/U_c = (M_e^2 - 1)^{-\frac{1}{2}}\alpha_x \tag{6.145}$$

$$\alpha = (v/u)_{y=\delta^*} = \delta_x^* \tag{6.146}$$

In the supersonic flow literature, some minor variation in this type of expression can occur, depending on where the flow angle match point is taken and on the specifics of the linearization approximation. With this slight exception, the formulation given above is remarkably standard and, in addition to being part of the basic structure of the triple deck,[154] has been used as the appropriate system of governing equations describing composite interacting boundary-layer flows at finite Reynolds numbers. For instance, in subsonic flows, precisely this formulation has been used by Briley and McDonald,[159] Carter,[164] Carter and Wornom,[177] Kwon and Pletcher,[160] Veldman,[178] Cebeci and Schimke,[161] and Carter and Vatsa.[162] In supersonic flows, the composite formulation is essentially that used by Crocco and Lees[148] and Baum and Denison[179] for a wake and Werle and Vatsa[155] for boundary layers. In view of the wide acceptance of the formulation, the distinction between the various approaches can be made on this basis of the numerical solution algorithm and/or the turbulence model adopted.

In discussing the interaction formulation in subsonic flow, it was assumed that for efficiency the Hilbert integral perturbation would be the method of choice to obtain the interacted potential flow. In transonic flow, this no longer seems adequate and a more extensive inviscid flow calculation seems mandatory. For separated flow, LeBalleur[163] and Carter[164,165] have initiated such approaches. It is noted that a number of efficient iterative transonic potential flow solvers have become available and that it is possible the boundary-layer interaction could be incorporated into the iteration which would be required in any event to produce the basic (uninteracted) potential flow. Such a prospect is interesting and much remains to be explored in this regard. The point remains that deleting the Hilbert integral in and of itself should not affect in any substantial way the various stability problems mentioned earlier. The stability problems arise from the interactive response of the numerical system representing the potential flow and boundary layer, the potential flow being approximated by a basic flow and a linearized correction. Improving the inviscid flow representation, even to the extent of using the Euler equations, should not fundamentally alter the interaction numerical stability unless the numerical interaction algorithm is changed. The same is true for the boundary-layer calculation. Treating the boundary layer by a momentum integral spatial marching approach does not remove the numerical problems concerned with branching and, although the Goldstein singularity and the reverse flow problems are not apparent, other singularity and the reverse flow problems are not apparent, other singularities arise, as reported in Refs. 44 and 180. To this is added the need to characterize the mean velocity profile in some suitably general manner. In addition, for turbulent flow inherently the same basic information on the turbulence structure must be supplied, even if only in some moment form. Thus, while boundary-layer momentum integral calculation schemes have much to recommend them by way of computational efficiency, they do not by themselves present a solution to the interaction computational problems referred to earlier.

It is possible to preserve the interaction concept and solve alternative equations in the two regions. As was noted above, investigators are already using more complete descriptions of the interacted inviscid flow than can be obtained from a potential flow and a linearized correction. The same is true for the viscous flow and, for instance, there is no conceptual problem in solving the Navier-Stokes equations for this region in place of the boundary-layer equations. In some instances, for example, if the recirculating separation region were not thin, it would in fact be more appropriate to interact the inviscid flow with a solution of the Navier-Stokes equations. By limiting the extent of the flow domain over which the Navier-Stokes equations are solved to that region in which the equations are required to describe the fluid mechanics, an economy should be realized relative to solving such equations throughout the entire domain. Some of the problems that can arise with this approach are discussed by Levy et al.,[181] who present some demonstrative calculations for the interactive flow over the trailing-edge region of an elliptical airfoil.

Flowfield predictions made using the interaction concept for a subsonic transitional separation bubble taken from the work of Briley and McDonald[159] are given in Fig. 6.7. Very similar interactive predictions have also been made by Kwon and Pletcher[160] and Cebeci and Schimke.[161]

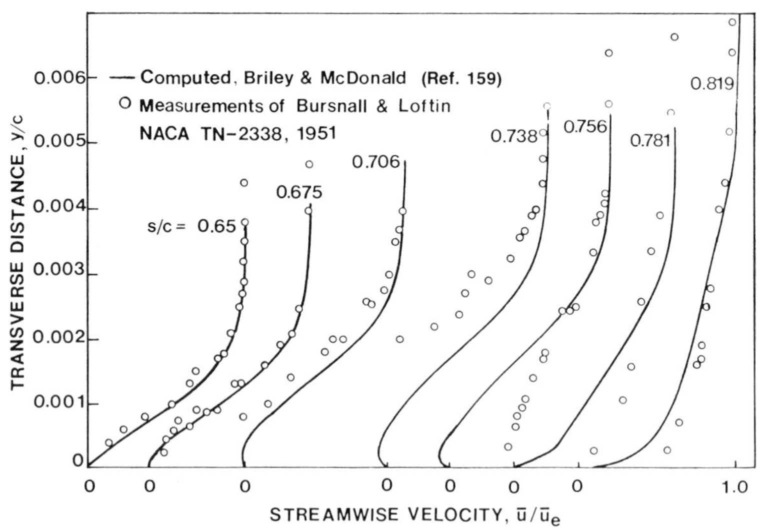

Fig. 6.7 Velocity profiles at selected streamwise locations in a separation bubble.

Conclusions on Boundary-Layer Separation

The prediction of separated flows in three dimensions is in its infancy. Given the developments that have occurred in two dimensions, time or pseudotime marching either the three-dimensional extended boundary-layer equations or the full Navier-Stokes equations would appear to have the best chance of success if zone embedding were feasible using double- or triple-deck concepts. Naturally, in view of the time-consuming nature of current three-dimensional calculation schemes, zone embedding becomes even more attractive in three dimensions. It seems also that the conventional three-dimensional boundary-layer equations, even if time marched, would not be adequate in view of the possible importance of the spanwise convective and diffusive effects in separated flows. Finally, in three space dimensions, the

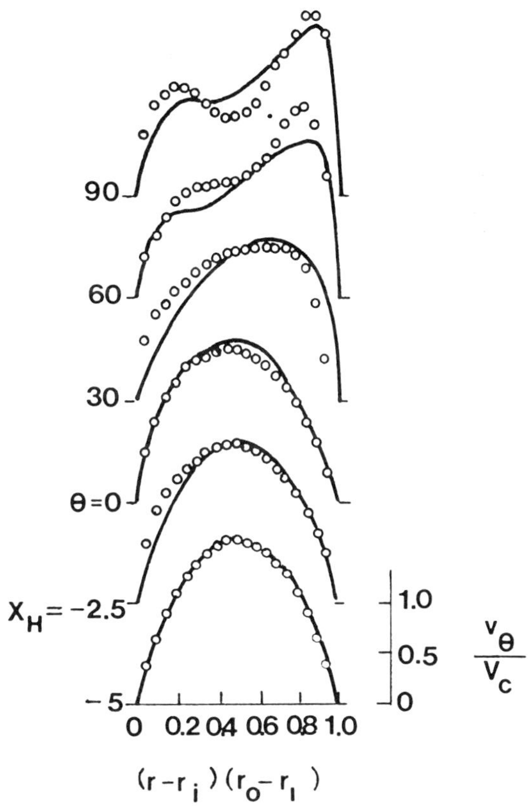

Fig. 6.8 Comparison between measured (circles) and computed velocity profiles along the midplane of a rectangular duct turning through 90 deg (*X* is the streamwise distance from the start of the bend, θ the duct centerline turning angle, r_i the inner wall radius, r_0 the outer wall radius) (from Ref. 227).

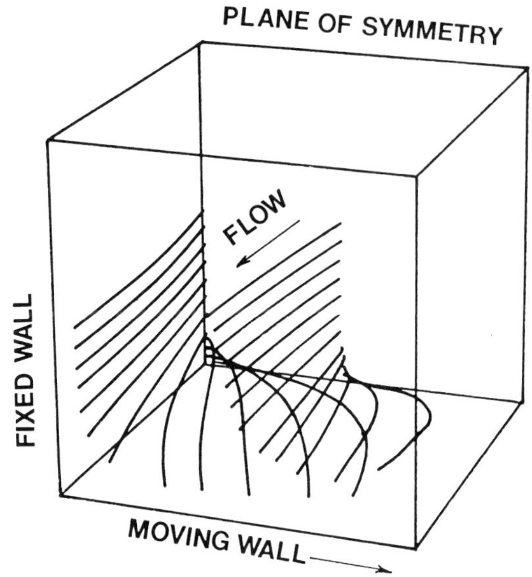

Fig. 6.9 Selected streamlines for duct flow with moving walls, $M = 0.44$, $Re = 60$. Predictions made using the compressible time-dependent Navier-Stokes equations (from Ref. 66).

problem of ensuring a gain in computational efficiency by using acceptable flow approximations becomes even more difficult. Further extensive work on these and similar schemes will be necessary before they can be used in turbomachinery problems. In particular, the capability to perform calculations in reasonable geometries is lacking and even the evaluation of the status of the presently available procedures is bedeviled by the high computing costs of performing the calculations. A demonstration of the capabilities of these two schemes is given in Figs. 6.8 and 6.9. Work in the area of representing the more realistic geometries is going forward.

In any event, it has not yet been established that the turbulence models proposed for use in this problem are either adequate or inadequate, even for the predictions of the gross features of the flow. Obviously, a great deal of work remains to be done, but at least a plan of attack on this most difficult of problems has been formulated and the initial stage implemented.

6.5 Turbulence Models

A physical description of turbulent motion was given in an earlier section. Here the emphasis is on the various prescriptions for the quantitative evaluation of the extent of the transport of heat, mass, and momentum that results from this turbulent motion. The turbulent transport mechanism is evidenced by the appearance of nonzero correlations of fluctuating quantities in the averaged equations describing the conservation of heat, mass, and momentum. These correlations are usually grouped together and for the momentum equations are termed the Reynolds stress tensor. The aim of the turbulence model is to provide an adequate description of the dominant terms in this Reynolds stress tensor for the flow under consideration. At the present time, the relationship between the Reynolds stresses and the mean motion is poorly understood, so it is necessary to resort to empiricism at some point in order to proceed. In principle, resorting to empiricism is not necessarily evil—indeed, the coefficient of laminar viscosity is an empirical quantity. The problem arises due to the lack of generality in the various turbulent transport models and, as attempts are made to formulate more general models, the complexity of application has increased as well as the speculative nature of the various closure hypothesis. In describing turbulence models then, it is still necessary to include the simple prescriptions, since they will continue to be used in the foreseeable future for those flows where their predictions are known to be reasonable. It is also necessary to outline where possible the areas in which the various models can, or do, fail to give reasonable predictions. As before, the emphasis will be on the special problems of turbomachinery. Several treatments giving the various points of view on turbulence models in general are to be found in the works of, for example, Launder and Spalding[182] and Bradshaw.[183] Since detailed turbulence models are normally thought of in terms of direct numerical solutions to the governing equation, the present discussion will reflect this bias. As previously mentioned, however, the detailed prescription can usually be incorporated into integral methods, at least with conceptual ease. Last, an extensive catalog of the more complex multiequation turbulence models is given in the thesis of Ng[184] and critical reviews of the topic by Bradshaw[183] and Reynolds.[185] Further discussion and categorization is also given by Launder and Spalding[182] and Mellor and Herring.[186] Interested readers are encouraged to read these works to get the full flavor of both the divergent and convergent views held by the various authors.

Special Problems of Flow in Turbomachinery

Here those aspects of the turbulent flow in turbomachinery giving rise to special problems are discussed. The order in which the problems are introduced does not imply their relative importance as this will vary from application to application. No consideration is given specifically to the very difficult problems of centrifugal compressors.

One of the most striking features of the flow in turbomachinery is its very pronounced three-dimensional nature. Annular ducts almost invariably have supporting struts whose intersection with the annulus wall boundary layers introduce regions of flow separation and merging shear layers, the "clash regions." Of course, this same problem occurs with compressors and turbines as well, with the added complication that the time-dependent nature of the flow within a stage introduces a further variable into the problem. Within the turbine high heat-transfer rates can be obtained in conjunction with highly cambered blades. The resulting overturning of the end wall boundary layers can result in the formation of vane or rotor passage vortices and, of course, the direct effect of curvature on the turbulence structure is possible. The strong accelerations and low Reynolds numbers occurring in the turbine give rise to extensive regions of transition from laminar to turbulent flow and vice versa. All of this occurs in a region where large and/or intense velocity fluctuations and pressure (acoustic) waves can develop in the nominal freestream and interact with the shear layers. Obviously, at the present time many of these effects cannot be taken into account in making flow predictions. Therefore, it follows that predictions of internal turbulent shear flows must be viewed more tolerantly (and yet more skeptically) than those in external flow.

On the positive side, two aspects of turbulence—compressibility and three-dimensionality—have been found to be less of a problem in external flow than thought at one time and, of course, the hope is that this fortunate state will also occur for internal flow problems. In the case of compressibility, Morkovin's hypothesis[29] that turbulence should be unaffected by fluctuations in density, provided the fluctuations are small compared to the mean density, has been supported by a number of investigators, e.g., Maise and McDonald.[21] Indeed, Morkovin's suggestion would seem to hold over a much wider range of density fluctuations than has hitherto been suspected,[25] leading one to believe that, even in the presence of the high heat-transfer rates found in turbomachinery, Morkovin's hypothesis might be adequate. In the case of three-dimensionality, it has been argued by a number of investigators (e.g., Bradshaw[187] and Nash[77]) that, since turbulence is a three-dimensional phenomenon even when the mean flow is two-dimensional, modest three-dimensionality would not affect turbulence "scalar" quantities. Application of this suggestion has led to quite acceptable three-dimensional boundary-layer predictions in some instances. Clearly, however, the strong three-dimensional effects occurring in turbomachinery stretch this suggestion to its limit and probably beyond. For the present, there is little in the way of substantiated working alternatives unfortunately.

In addition, rotational effects are encountered in internal flows in both compressors and turbines. According to Johnson[188] the rotating external blade boundary layers are sufficiently thin that the effects of the Coriolis and centrifugal forces both on the mean flow and turbulent structure are probably negligible. Internal or duct flows, such as might occur within a rotating cooling labyrinth, do, however, according to Johnson, exhibit significant rotational effects. At present little detailed quantitative information on duct rotational effects is available and insofar as performing calculations are concerned only the preliminary suggestions of Bradshaw[189] are available and extensive evaluation and/or development will be necessary before calculations can be performed with any confidence. As a first step in this long process, Hughes and Horlock[190] discuss the effect of centrifugal and Coriolis forces upon the turbulent shear stress and how these changes will affect the growth of boundary layers in turbomachines. The Hughes-Horlock paper is therefore mandatory reading for those interested in rotational effects on the boundary layer.

Categories of Turbulence Models

Turbulence models serve to define the unknown turbulent correlations appearing in the averaged conservation equations of heat, mass, and momentum ultimately in terms of the mean flow variables. In recent years, the practice has grown up of categorizing the turbulence model by the degree of complexity of the relationship between the correlations and the mean flow, usually by means of the number of partial differential equations that must be solved in order to obtain the Reynolds stresses. At the most elementary level, the previously introduced Prandtl's mixing length requires no partial differential equation to be solved in order to determine the Reynolds stress level and, consequently, this model might be termed a "zero equation" model or, more accurately, an algebraic stress model. In a similar vein, for example, after the empirical modeling of certain turbulent correlations, Donaldson[191] solves a partial differential equation for each of the three components of the Reynolds normal stress and the Reynolds shear stress; consequently, this suggestion would be termed a four-equation model.

The differential equations used can be either exact equations derived rigorously by taking moments of the Navier-Stokes equations or entirely empirical equations. If exact equations are employed, empirical relationships must be introduced for certain of the turbulence properties to "close" the system. A further discrimination between the turbulence models can be made on the basis of whether or not the differential equations are for one- or two-point turbulence properties, that is, the correlation at a point or the correlation between fluctuating quantities separated in linear space.

In a separate category, the use of differential equations for the one- or two-point turbulence properties may be replaced by the schemes that solve the time-dependent Navier-Stokes equations. In view of the wide range of scales normally encountered in turbulence, it is not feasible to resolve the flow at sufficiently small scale such that all of the motion contributing to the dynamics can be computed. Instead, a turbulence model can be postulated for the unresolved subgrid motion. Examples of this approach are to be found in the work of Deardorff[192]; but even with subgrid modeling, the computational cost of this approach is at present prohibitive for normal engineering applications.

In addition to the foregoing categorization, particularly in dealing with algebraic stress models and one- or two-equation turbulence models, the terminology equilibrium or nonequilibrium, historic or (with apologies) prehistoric is sometimes used. Here care has to be taken to discriminate between the various uses of the term "equilibrium." In particular, "equilibrium" is often used as a synonym for self-preserving or similarly in the sense of the very special boundary layers whose mean profile does not change in the axial direction when expressed in certain coordinates. Clauser[11] demonstrated that such boundary layers, previously well known in laminar flow, also exist in turbulent flow. Further, he also deduced that in such equilibrium flows the eddy viscosity [Eq. (6.153)] in the outer region of the boundary layer would have to be essentially constant when made nondimensional by a scale of velocity and length, in his case the freestream velocity u_e and the boundary-layer displacement thickness δ^*. This constant nondimensional eddy viscosity concept belongs in the algebraic stress model category and, in view of this, is sometimes referred to as equilibrium turbulence models.

Algebraic stress models all possess the feature that the local turbulent shear stress is given directly in terms of the local mean velocity profile. Therefore, such models cannot allow the observed lag in changes in the Reynold's shear stress when the mean motion is perturbed rapidly. Multiequation turbulence models do allow a lag and consequently these models are sometimes termed "nonequilibrium," "lag," or "historic" turbulence models. In view of the observation that equilibrium turbulence models do at least reflect the mean flow history, the claims and counterclaims about turbulence history or lack of it have been abandoned in favor of the less controversial categorizations of equilibrium and nonequilibrium turbulence model.

The equilibrium terminology also arises in part due to the interpretation of mixing length and eddy viscosity advanced by Batchelor[193] and Townsend.[194] According to this interpretation, the validity of the simple mixing length hypothesis and closely related well-behaved eddy viscosity is an expression of the local equilibrium between the production and dissipation of turbulence kinetic energy. This local equilibrium arises when the convection and diffusion of the turbulence kinetic energy is negligibly small. Care must therefore be taken in discriminating between the various uses of the term "equilibrium."

Algebraic Stress Models

For thin unseparated turbulent boundary layers, it is generally excepted that only the Reynolds shear stress $\overline{u'v'}$ need be retained in the mean axial momentum equation. Prandtl[195] proposed that in general the turbulent diffusional flux of some property $\phi = \bar{\phi} + \phi'$ could be expressed by

$$-\overline{v'\phi'} = \frac{1}{\sigma_\phi}\left(\overline{q^2}\right)^{\frac{1}{2}} l \frac{\partial \bar{\phi}}{\partial y} \tag{6.147}$$

where v is the velocity in the y direction normal to the mean flow and $\overline{q^2}$ the turbulence kinetic energy,

$$\overline{q^2} = \overline{u'^2} + \overline{v'^2} + \overline{w'^2} \tag{6.148}$$

the prime denotes a fluctuation, σ_ϕ a constant, and l a turbulent length scale. Note here the explicit postulate of a gradient transport mechanism. (See Ref. 196 for a discussion of the limitations of gradient transport in turbulence.) Attention is now restricted to the transport of axial velocity u and the constant σ_ϕ is incorporated into the definition of the length scale l.

At this point, two additional relationships must be provided before the shear stress can be determined: (1) some means to obtain the turbulence kinetic energy must be prescribed and (2) the length scale l must be specified. Indeed, a number of the two-equation models of turbulence to be discussed later attempt to provide partial differential equations for precisely these quantities. For the algebraic stress model, the relationship between shear stress and kinetic energy may be obtained by invoking the structural similarity concepts noted by Townsend,[194] among others. In the structural similarity concept, the hypothesis is made that after prolonged application of a mean strain the turbulence approaches a near-equilibrium condition, where the correlation coefficients are independent of the turbulent intensities and strain rate that produced them. Both theory and experiment have since indicated that structural similarity is only approximately obtained, even after prolonged straining, but it does provide a convenient relationship of the form

$$-\overline{u'v'} = a\overline{q^2} \tag{6.149}$$

where a is a coefficient that would not be expected to vary by an appreciable amount. Again, if a is incorporated into the definition of the length scale, Prandtl's formulation for the Reynolds stress then gives

$$\left(-\overline{u'v'}\right)^{\frac{1}{2}} = l\frac{\partial \bar{u}}{\partial y} \tag{6.150}$$

which is consistent with Prandtl's earlier mixing length formulation. In squaring both sides, it is established practice to write the formula

$$-\overline{u'v'} = l^2\frac{\partial \bar{u}}{\partial y}\left|\frac{\partial \bar{u}}{\partial y}\right| \tag{6.151}$$

so that the shear stress always has the opposite sign to the mean gradient. The definition of eddy viscosity is given as

$$-\overline{u'v'} = l^2\frac{\partial \bar{u}}{\partial y}\left|\frac{\partial \bar{u}}{\partial y}\right| = \nu_T\frac{\partial \bar{u}}{\partial y} \tag{6.152}$$

so that from a formal viewpoint mixing length l and eddy viscosity ν_T may be viewed interchangeably. Returning now to the specification of the mixing length in a boundary layer, a three-layer formulation is now widely favored. In this three-layer formulation, a viscous sublayer, a near-wall region, and a wake-like region of the boundary layer is identified and various suggestions for the mixing length (or equivalently for the eddy viscosity) made for each region.

The outer wake-like region. In the wake-like region of the flow, two suggestions are extensively used. The first is the constant eddy viscosity formulation deduced by Clauser[11] for equilibrium boundary layers, which can be written

$$\nu_T/\bar{u}_e\delta_K^* = 0.016 \tag{6.153}$$

and slight variations in the value of the constant 0.016 are observed in the literature. Herring and Mellor[72] were apparently the first to extensively use this definition of eddy viscosity for nonequilibrium and compressible boundary layers. In compressible flow, they adopted a kinematic definition of the displacement thickness, i.e.,

$$\delta_K^* = \int_0^\delta (1 - \bar{u}/\bar{u}_e)\,\mathrm{d}y \tag{6.154}$$

omitting density from the usual definition of δ^*. Later Maise and McDonald[21] found that use of the kinematic displacement thickness much improved the collapse of the eddy viscosities inferred from the correlated behavior of compressible constant-pressure boundary layers. Subsequently, Cebeci and Smith[71] adopted the Mellor-Herring suggestion and used the kinematic displacement thickness in Clauser's formula for compressible boundary layers, while adding an intermittency factor to give a variation of the eddy viscosity normal to the wall more in keeping with that observed.

As an alternative and conceptually equivalent process to specifying the eddy viscosity in the outer region, a number of people have suggested that when expressed as some fraction of the boundary-layer thickness δ the mixing length would be constant in the outer wake-like region of the boundary layer, i.e.,

$$l = 0.09\delta \tag{6.155}$$

and again minor variations in the constant 0.09 are observed in the literature. For instance, Patankar and Spalding[70] made effective and extensive use of this hypothesis following the lead of Escudier[197] who demonstrated that Eq. (6.155) was a reasonable approximation in many incompressible turbulent boundary layers. Maise and McDonald[21] later found that Eq. (6.155) appeared to be a very reasonable approximation and quite insensitive to Mach number in a constant-pressure compressible boundary layer.

Three major effects on the values adopted for the nondimensional outer layer mixing length or eddy viscosity have been identified: those of a low Reynolds number, freestream turbulence, and streamline curvature. Herring and Mellor[72] found, for instance, that in order to satisfactorily predict the behavior of a number of low Reynolds number boundary layers they had to introduce a low Reynolds number correction into the outer layer eddy viscosity and progressively increase its value as R_θ dropped below 5×10^3. These findings were, in fact, consistent with the observation of Coles,[17] who noted below R_θ of 5×10^3 the so-called wake component of the constant-pressure boundary layer eventually disappeared. Taking Coles' data correlation and velocity profile family, McDonald[198] deduced from the boundary-layer axial momentum equation the consistent Reynolds stress profiles and the corresponding nondimensional eddy viscosity and mixing lengths. The results were in broad agreement with the Herring-Mellor[72] suggestion and could be expressed as

$$(l/\delta) = (l/\delta)_\infty [1 + \exp(-1.63 \ln R_\theta) + 9.7] \qquad (6.156)$$

where $(l/\delta)_\infty$ is the high Reynolds number value of the mixing length constant. Subsequently, a very similar modification was proposed by Cebeci and Mosinskis.[199]

Consistent with the increase in mixing length and eddy viscosity noted at low Reynolds is Head's visual observation in his smoke tunnel that the turbulent interface with the freestream was much more contorted at low Reynolds numbers, presumably resulting in an increased entrainment of fluid into the boundary layer and the increased values of mixing length and eddy viscosity. Simpson[200] suggested that these low Reynold's number effects might be explained by an increase in the near-wall region turbulent transport, but the data analysis of Coles[17] does not support this idea and, subsequently, Huffman and Bradshaw[201] argued against Simpson's suggestion.

Turning now to the second major effect, that of freestream turbulence, Kline et al.[202] demonstrated some time ago the very large effect this parameter could have on a turbulent boundary layer. More recently, this effect has been experimentally studied in detail by Charnay et al.[203] and Huffman et al.,[204] among others. These results are indeed all quite consistent and show a large increase in the outer layer mixing length or eddy viscosity as the freestream turbulence is increased. The physical mechanism at play here is thought, once again, to be the increased contortion of the turbulent freestream interface due to the action of the freestream turbulence on the interface. It is also intuitively obvious that the scale of the freestream turbulence should have an effect on the interaction, since, if the scale is large and this freestream disturbance lifetime is much longer than the boundary-layer turbulence, the boundary layer will tend to respond more as it would to a slowly varying time-dependent freestream. Now, time-dependent freestreams can be treated if the freestream time scales involved are much longer than that of the energy-containing boundary-layer turbulence (which usually does not exceed the convection time to travel 10 boundary-layer thicknesses). For the usual concepts of freestream turbulence to be valid, the motion has to be characterized by scales less than 10 boundary-layer thicknesses. Following convention, the freestream turbulence is defined by

$$T_u = \frac{\left(\overline{q^2}/3\right)^{\frac{1}{2}}}{\bar{u}_e} \qquad (6.157)$$

which reduces to

$$T_u = \left(\overline{u'^2}\right)^{\frac{1}{2}} / \bar{u}_e \tag{6.158}$$

in the often assumed case of isotropic freestream turbulence.

A straightforward correlation of the Huffman et al.[204] data would suggest the outer region mixing length would vary very roughly as

$$l/\delta = (l/\delta)_{T_u=0} + T_u \tag{6.159}$$

Bayley et al.[205] successfully used a very simple relationship derived by trial and error in calculating the effect of freestream turbulence on the heat transfer to a turbine aerofoil. Within the mixing length framework, formulas of this type, which are (like the low Reynolds number correction) nothing more than data fits, probably suffice and doubtless as more data are acquired it will be possible to improve the data curve fit.

The third major effect on the outer layer mixing length arises from streamline curvature and this effect also has been clearly noted in a number of experimental studies. Bradshaw[189] discusses this curvature effect in detail and suggests on the basis of the Monin-Obeukhov meterological formula a first-order formula

$$l/\delta = (l/\delta)_{R_i=0}(1 - \beta R_i) \tag{6.160}$$

where R_i is an equivalent to the Richardson number and defined as

$$R_i = \frac{2u}{R}\left(\frac{\partial u}{\partial y}\right)^{-1} \tag{6.161}$$

where R is the radius of curvature taken positive on a convex surface. The constant β takes on a value about 7 for convex (blade upper surface) curvature and takes a value of about 4 on a concave surface. Typically, on a convex surface, βR_i might reach a value near 30 δ/R or 40 δ/R so that for ratios of boundary-layer thickness to surface radius of curvature greater than, say, 1/100, very substantial effects of curvature on the outer layer mixing length are to be expected. Fortunately, highly cambered blades are typical of turbines rather than compressors and the high acceleration levels of the turbine ensure that the naturally occurring surface boundary layers are very thin, so that except for film-cooled boundary layers the ratio δ/R does not usually approach the troublesome magnitudes. Nonetheless, surface (i.e., streamline) curvature must be observed as a potential source of difficulty. Hughes and Horlock[190] provide a discussion of this problem.

The near-wall region. Turning now to what has been termed the near-wall region of the flow, here almost complete unanimity in the choice of relationship is to be found among the various investigators, notably that

Prandtl's mixing length should be proportional to distance from the wall, viz.,

$$l = \kappa y \tag{6.162}$$

where κ is termed von Kármán's constant and is usually attributed a value close to 0.40. The crossover to the outer layer is achieved by selecting the smaller value of eddy viscosity or mixing length as computed by both the near-wall formula and the outer wake-like region formula. McDonald and Camarata[58] suggested the continuous relationship

$$l = l_\infty \tanh \kappa y / l_\infty \tag{6.163}$$

to cover both regions without the sharp break that follows from the use of two piecewise, discontinuous relationships. More recently, Galbraith and Head[206] have suggested that the von Kármán constant κ does in fact vary with pressure gradient, and this effect has also been noted by Glowacki and Chi.[207] While the effects of this variation in the von Kármán constant are negligibly small on the integral thickness parameters, the effect on skin friction and presumably heat transfer is significant. Glowacki and Chi deduced the turbulent shear stress from the axial momentum equation and the mean velocity profile development and found that

$$\kappa = 0.4 + 0.1823[1.0 - \exp(-0.3207\beta)] \tag{6.164}$$

where

$$\beta = \frac{\delta^*}{\tau_w} \frac{\mathrm{d}p_e}{\mathrm{d}x} \tag{6.165}$$

which when used in conjunction with the continuous distribution of mixing length, Eq. (6.163), and a Van Driest sublayer model (which will be described shortly) gave very good predictions when compared with measured equilibrium (self-preserving) boundary layers. Unfortunately, the use of a local pressure gradient parameter (constant for equilibrium boundary layers) in the correlation for κ gives rise to doubts about the ability to properly represent nonequilibrium boundary layers. A more significant parameter for generalization to nonequilibrium layers would be the Clauser shape parameter G. A replacement of β with G in Eq. (6.164) can be made using Nash's[12] relationship in Eq. (6.8). The implication at this point is that κ is exhibiting a dependence on the local turbulence structure, which on the basis of the local equilibrium hypothesis is related to the profile shape G that in turn is caused to change by the imposition of the pressure gradient β. An even more direct link in the argument follows from the observation, implicit in all of Coles' profile development and subsequently noted by Reeves,[208] that the fact the logarithmic law of the wall is observed to hold in pressure gradients is inconsistent with the linear mixing length assumption. It follows that if the logarithmic law of the wall holds in a pressure gradient, i.e., a flow where $\tau \neq \tau_w$ and $y^+ > 2A^+$, then the mixing length must be given by

$$l = \kappa y (\tau/\tau_w)^{\frac{1}{2}} \tag{6.166}$$

to recover the logarithmic law of the wall. This observation has, of course, been well known for some time and the predictors have usually responded by pointing out the good agreement between prediction and measurement that results from ignoring this dependence on the ratio of local to wall stress. Further, the departures from the log law as it fairs into the defect or wake formulation are subject to minor fitting variations that can have considerable impact upon the local velocity gradients crucial to the derived mixing length variations. Thus, predictors have ignored the strict interpretation of the log law in the presence of a pressure gradient and view the log law as an approximate relationship arising from $l = \kappa y$ and $\tau \simeq \tau_w$ and not vice versa, as Eq. (6.166) suggests. Indeed, a number of people have generalized the log law to account for pressure gradients (e.g., Townsend[209]) and the resulting profile certainly has a more realistic asymptotic behavior than the log law as τ_w decreases to zero. The inference of the generalizations of the law of the wall to account for pressure gradients is that indeed Eq. (6.166) does not hold. The truth probably lies somewhere in between as evidenced by the fact that Glowacki and Chi did not assume a log law in their data reduction, yet their results can be conveniently expressed as

$$l = \kappa_0 y (\tau/\tau_w)^a = \kappa_{\text{eff}} y \tag{6.167}$$

where the exponent a is nearer to $\frac{1}{4}$ than $\frac{1}{2}$ as in Eq. (6.166). The exponent of $\frac{1}{4}$ is of course a middle ground between the two viewpoints and would certainly result in more of a log law being evidenced by the predictions. The half-power variation given by Eq. (6.166) is not at all suitable for implementation in a finite difference scheme to determine the local stress τ, since the local stress to be obtained from the mixing length relationship would then cancel out the shear stress/mixing length formula. A caution also to the use of a more general (τ/τ_w) type of correction for the von Kármán constant lies in the asymptotic condition of vanishing wall shear and here, obviously, it is necessary to provide a limit of the type

$$\begin{aligned} \kappa_{\text{eff}} &= \kappa_0 (\tau/\tau_w)^{\frac{1}{4}} && 0.5 < \tau/\tau_w \leq 15.0 \\ &= 1.968\kappa_0 && \tau/\tau_w > 15.0 \\ &= 0.84\kappa_0 && \tau/\tau_w < 0.5 \end{aligned} \tag{6.168}$$

Finally, it is observed that a κ_{eff} relationship, such as in Eqs. (6.167) and (6.168), is of course inconsistent with the notion of a linear variation of mixing length, i.e., $\kappa_{\text{eff}} = \text{const}$, in the wall region except in some very special cases. If a linear dependence of mixing length on wall height is demanded (and here the experimental evidence in pressure gradient is not all that definitive), then it becomes necessary to postulate a dependence of mixing length upon the stress gradient $\partial\tau/\partial y$, which in turn possesses the ability to be nearly constant across the wall layer. Further, in equilibrium boundary layers near the wall, but outside the viscous sublayer, one observes that the stress gradient $\partial\tau/\partial y$ is nearly constant and given by

$$\frac{\delta^*}{\tau_w}\frac{\partial\tau}{\partial y} \simeq 0.7\beta \tag{6.169}$$

so that Glowacki and Chi's relationship in Eq. (6.164) can easily be converted to a stress gradient rather than a pressure gradient or shape parameter relationship.

Thus, in attempting to use the information deduced by Galbraith and Head and Glowacki and Chi, there are four obvious alternatives. The first is to use Glowacki and Chi's equilibrium correlation of the effective value for the von Kármán constant as presented on the basis of pressure gradient for nonequilibrium flows. It is suggested that this proposal should be rejected. To follow the local equilibrium hypothesis, it would be more reasonable to base the extension to nonequilibrium flows on a local mean velocity profile shape parameter G rather than the local pressure gradient parameter β. This second alternative seems quite reasonable for relatively unsophisticated turbulence models in view of the global nature of the mean velocity profile characterization. The third alternative is to take note of the widespread appearance of a logarithmic region in the mean velocity profile near the wall and characterize the effective von Kármán constant by some function of the local value of the ratio of the turbulent shear to wall shear stress. This latter practice, however, is inconsistent with the observed near-linear dependence of the mixing length on distance from the wall in the near-wall region (and poses difficulty near separation). The only way to recover a much wider region of linear dependence is to base the correlation of effective von Kármán constant on the local stress gradient as opposed to the local stress. Either of the latter two suggestions seems reasonable and presumably the correct choice will become obvious as more experimental evidence becomes available.

The viscous sublayer. Finally, a variation of eddy viscosity or mixing length across the viscous sublayer must be hypothesized. Here, however, the option is open of specifying this variation once and for all and integrating the so-called Couette flow approximation to the axial momentum equation

$$\begin{aligned}\tau &= -\overline{\rho u'v'} + \mu\frac{\partial u}{\partial y} = +\left[l^2(y)\left|\frac{\partial u}{\partial y}\right| + \mu\right]\frac{\partial u}{\partial y}\\ &= \tau_w + y\frac{\mathrm{d}p}{\mathrm{d}x}\end{aligned} \tag{6.170}$$

i.e., neglecting the convective terms in the boundary-layer axial momentum equation to obtain u at the edge of the viscous sublayer as a function of τ_w, ν, and $\mathrm{d}p/\mathrm{d}x$ and, of course, any of the parameters upon which $l(y)$ is supposed to depend on in the sublayer. This functional dependence of u at the edge of the sublayer can be parameterized and, consequently, the integration of the governing equations within the sublayer need not be performed so long as the Couette flow approximation is valid. This same parametric approach can be viewed alternatively as determining from observation the variation of the additive constant B in the law of the wall mean velocity profile Eq. (6.10). Patankar and Spalding[70] in their scheme chose to implement the once and for all integration of Jayatelleke,[210] while Bradshaw et al.[59] prefer to view the sublayer problem as an empirical problem of determining the variation of the law of the wall additive constant. In either case, the result is the same in that the governing equations are not integrated within the viscous sublayer; rather the solution is matched to some prescribed velocity outside the sublayer. A variety of rather neat matching techniques have been devised, but will not be discussed here. A similar process is followed for the energy and species equations.

There are two attributes to the previously outlined wall function procedure for dealing with the sublayer, aside from the obvious one of saving computer time by reducing the required number of grid points. The first is that if the sublayer profile is viewed simply as a problem of determining the law of the wall additive constants, the conceptual difficulty of explicitly postulating a turbulent transport mechanism for the very low Reynolds number asymptotic wall region is avoided. For instance, Bradshaw and Ferriss[80] cite this as one reason for selecting the law of the wall approach (in spite of the apparent ease with which reasonable sublayer profiles can be obtained from postulated transport models). The second attribute has been previously mentioned and derives from the absence of grid points very close to the wall, which results from matching the computed and analytic variables at some point removed from the wall. Clearly, stability-restricted prediction methods derive a considerable benefit from increasing the allowable axial step that follows from the increased step size normal to the wall. Indeed, it is doubtful that current stability-restricted methods would be at all competitive in terms of computer run times unless the wall function approach were adopted.

The problem with the wall function approach in general lies, first of all, with the limitations on the Couette flow approximation, for here it is observed that frequently the stress near the wall but outside the sublayer may be represented by the linear relationship

$$\tau = \tau_w + Ay\frac{\mathrm{d}p}{\mathrm{d}x} \tag{6.171}$$

where A can reach values as low as 0.2 and as high as 2.0 with relative ease. At the wall itself, a value of 1.0 for A must be obtained so that a variation in A across the sublayer, not taken into account in the Couette flow assumption, is to be expected. The value of velocity at the edge of the sublayer arrived at by a "once and for all" integration of the Couette flow approximation, therefore, cannot be relied upon in the case of strong pressure gradients or in the vicinity of flow separation. Further, there arises the question of the temperature variation in the sublayer—aside from the interest in this quantity for heat transfer, the temperature is required in the sublayer integration for the velocity profile. As was mentioned in the earlier discussion on the law of the wall, the temperature variation in the sublayer seems much more sensitive to variables such as pressure gradient than the velocity field. Consequently, the adoption of the Couette flow approximation for the temperature field or equivalently assuming an additive constant for the thermal law of the wall is more suspect than for the velocity field.

In view of the foregoing, it would seem reasonable that, if the wall function approach were to be selected, it is probably better to proceed as Bradshaw and Ferriss[80] do and view the sublayer as the problem of determining the law of the wall additive constants and correlate the variation of these constants with parameters such as pressure gradient, transpiration rate, or wall roughness, etc. On the other hand, viewed pragmatically, considerable success has been obtained by specifying a turbulent transport mechanism for the sublayer and not adopting the Couette flow simplification. This approach has particular merit in gas turbine applications where low Reynolds numbers combine with high acceleration levels to give extensive blending regions between laminar and turbulent flow. In such regions, the law of the wall does not hold, so the question of the value of the additive constant does not arise. Further, since in some turbine problems the sublayer may comprise the whole boundary layer, the neglect of the convective terms in the Couette approximation is clearly unacceptable.

Of the sublayer transport models proposed, the most commonly used in calculation schemes is that due to Van Driest,[211] who suggested that the mixing length in the wall region be written

$$l = \kappa y \cdot \mathscr{D} \tag{6.172}$$

where $\mathscr{D}$ is a damping factor defined by

$$\mathscr{D} = 1.0 - \exp(-y^{+}/A^{+}) \tag{6.173}$$

and A^{+} is a coefficient that could be given the physical interpretation as the thickness of the sublayer when expressed in law of the wall coordinates

$$y^{+} = y\sqrt{\tau/\rho}\,/\nu_{w} \tag{6.174}$$

It turns out that the sublayer damping factor is a critical parameter in obtaining accurate heat-transfer and wall friction predictions. Other suggestions for the sublayer damping factor have been given by Mellor[212] and

McDonald and Fish.[60] The differences in the various suggestions are in matters of detail and are not worth dwelling on here. Probably the most significant items in the damping factor are the definitions of the normal distance y^+ and the value ascribed to the Van Driest coefficient A^+. In the definition of y^+ given above, the local shear stress τ has been used, as seems proper. Frequently, however, this local stress is approximated by the wall value τ_w and, aside from being unnecessary in a direct numerical scheme, the approximation of the local stress by the wall stress leads to obvious difficulty near separation where $\tau_w \to 0$. Some authors, for example Cebeci[213] have proposed to remedy the situation by reverting to the Couette flow approximation replacing τ by

$$\tau = \tau_w + y\frac{\mathrm{d}p}{\mathrm{d}x} \tag{6.175}$$

As was pointed out above, there is considerable experimental evidence that Eq. (6.175) is not particularly accurate near a wall and there is little ground (except numerical instability and perhaps increased nonlinearity) for not using the total stress in the definition of y^+.

The second item of importance is the Van Driest coefficient A^+ (equivalent parameters arise in the formulations of Mellor[212] and McDonald and Fish.[60]) For boundary-layer flow over a smooth impermeable wall, the coefficient is usually given a value around 27. Unfortunately, the Van Driest coefficient is known to vary. For instance, Van Driest suggested that the effect of surface rougness could be modeled by varying A^+ such that, when k^+, defined by

$$k^+ = k\sqrt{\tau/\rho}\,/\nu_w \tag{6.176}$$

where k is the rms roughness height, exceeds A^+, the sublayer would be eradicated. A simple preliminary formulation embodying this feature has been implemented by McDonald and Fish[60] and Crawford and Kays.[214]

It is noted, however, that Van Driest's original suggestion of eradicating the sublayer does not account for the observed continued effect on the flow for roughness heights k^+ in excess of 27. In terms of the law of the wall additive constant B, this continues to decrease by very significant amounts as the roughness height k^+ increases beyond 27. In order to model this effect, either a shift in the effective wall location or a capacity of the roughness to enhance the turbulence production (or both) must be postulated. Although incorporated into the formula of McDonald and Fish, the turbulence production capability of wall roughness clearly merits much additional study. As given by Crawford and Kays,[214] there does not seem to be any mechanism to reduce B as k^+ increases beyond 27 in their formulation.

Insofar as pressure gradients and wall transpiration are concerned, numerous workers have observed the need to vary A^+ in order to match the experimental results (e.g., Cebeci[213]). However, it is noted that if the local stress τ is used in the definition of y^+ as opposed to the wall value τ_w, the required variation of A^+ is lessened. Indeed, both Herring and Mellor[72] and Kreskovsky et al.[215] in published and unpublished studies were able to satisfactorily predict the behavior of a wide range of boundary layers in strong variable pressure gradients for both transpired and impermeable walls, without resorting to varying the equivalent of A^+ in their formulations. This capability was attributed to the use of the local stress in the definition of y^+. Huffman and Bradshaw[201] found that even when the local stress was used, A^+ was observed to vary when the stress gradient $\nu_w/\rho_w u_\tau^3 \partial\tau/\partial y$ exceeded -5×10^{-3}. In any event, if the wall stress is used to define y^+, then A^+ must be allowed to vary. On the basis of extensive studies Kays and Moffat[84] suggest the empirical relationship

$$A^+ = \frac{24}{a\left[v_w^+ + b\dfrac{p^+}{1+cv_w^+}\right]+1} \tag{6.177}$$

where

$$\begin{aligned} a &= 7.1 \quad v_w^+ \geq 0.0 \quad \text{otherwise } a = 9.0 \\ b &= 4.25 \quad p^+ \leq 0.0 \quad \text{otherwise } b = 2.0 \\ c &= 10.0 \quad p^+ \leq 0.0 \quad \text{otherwise } c = 0.0 \end{aligned} \tag{6.178}$$

with a constant pressure zero transpiration value of 24 for A^+ is adopted, and where

$$p^+ = \frac{\nu_w}{\rho_w u_\tau^2}\frac{\partial p}{\partial x}$$

$$v_w^+ = \frac{v_w}{u_\tau} \qquad u_\tau = \sqrt{\frac{\tau_w}{\rho_w}} \tag{6.179}$$

Kay and Moffat have incorporated the foregoing suggestion into an extensively modified version of the Patankar-Spalding finite difference boundary-layer scheme[214] and achieved comparable results to Herring and Mellor.[72]

The question of relaminarizing boundary layers is also treatable within the mixing length framework, if the suggestion of Launder and Jones[83] is followed and A^+ obtained by integration of the differential equation

$$\frac{dA^+}{dx^+} = \left(A^+ - A_{eq}^+\right)/C \tag{6.180}$$

where C is a constant whose value has been found by trial and error to be about 4. A_{eq}^+ is the local equilibrium value of A^+ that would be obtained from a relationship such as Eq. (6.177). Launder and Jones subsequently abandoned this formulation in favor of a multiequation model of turbulence to be discussed subsequently.

The heat-transfer coefficient. Turning now to the turbulent transport of heat with the usual thin boundary-layer approximations, only the turbulent heat-transfer correlation $-\overline{v'T'}$ assumes importance. If Prandtl's formulation Eq. (6.147) is retained for consistency with the momentum transfer analysis, then the ratio of the turbulent heat-transfer correlation to the Reynolds shear stress is simply

$$\frac{-\overline{v'T'}}{-\overline{v'u'}} = \frac{\sigma_u}{\sigma_h} \frac{\partial \bar{T}/\partial y}{\partial \bar{u}/\partial y} \tag{6.181}$$

which from the definition of the eddy viscosity ν_T may be written

$$-\overline{v'T'} = \frac{\sigma_u}{\sigma_h} \nu_T \frac{\partial \bar{T}}{\partial y} \tag{6.182}$$

so that if an eddy diffusivity of heat is introduced

$$-\overline{v'T'} = \nu_h \frac{\partial \bar{T}}{\partial y} = \frac{\sigma_u}{\sigma_h} \nu_T \frac{\partial \bar{T}}{\partial y} \tag{6.183}$$

the turbulent Prandtl number is arrived at as

$$P_{rt} = \frac{\nu_T}{\nu_h} = \frac{\sigma_u}{\sigma_h} \tag{6.184}$$

which in Prandtl's formulation would be a constant.

Although it is quite feasible to deduce a model with three (or more) layers for the heat-transport correlation in an analogous fashion to the algebraic Reynolds shear stress models, it has long been felt that the mechanism resulting in the turbulent transport of momentum would be the same mechanisms at play in the turbulent transport of heat, so that the turbulent Prandtl number would indeed take on values close to unity. Initial boundary-layer calculations performed using this assumption of unit turbulent Prandtl number resulted in very good agreement with available data. Subsequent measurements by a number of different authors (e.g., Meir and Rotta[42] and Simpson et al.[216]) have all confirmed that, indeed, the turbulent Prandtl number takes on values close to unity in the turbulent boundary layer. Nonetheless, in spite of the experimental uncertainties involved in making such measurements at the present time, the results of the differing investigations are fairly consistent in showing values of the turbulent Prandtl number in excess of 1 near the wall and falling below 1 in the wake-like region of the boundary layer. This has led a number of authors to suggest distributions of turbulent Prandtl number for use in performing turbulent boundary calculations. For instance, Pai and Whitelaw[217] used

$$P_{r_t} = 1.75 - 1.25y/\delta \tag{6.185}$$

However, they found that by a large margin the best predictions of a film-cooled boundary layer were obtained with the old assumption of unit turbulent Prandtl number. On the other hand, Kays and Moffat[84] obtained reasonable predictions with a constant turbulent Prandtl number, but recommended a variation with y^+ if an accurate detailed temperature profile were to be desired. Hopefully, in the future improved techniques of making time-resolved temperature measurements in turbulent boundary layers will clarify this situation.

Extensions to three dimensions. The usual boundary sheet approximations in three dimensions result in two Reynolds stresses making their appearance in the axial and cross-stream momentum equation (see, e.g., Nash and Patel[109])

$$\frac{\tau_x}{\rho} = -\overline{u'v'} + \nu\frac{\partial \bar{u}}{\partial y}$$

$$\frac{\tau_z}{\rho} = -\overline{w'v'} + \nu\frac{\partial \bar{w}}{\partial y} \tag{6.186}$$

The natural extension of the previously described algebraic stress models to three dimensions is to take the eddy viscosity in three dimensions as

$$\nu_T = l^2\left|\frac{\partial \mathbf{v}}{\partial y}\right| \tag{6.187}$$

and

$$\left|\frac{\partial \mathbf{v}}{\partial y}\right| = \left[\left(\frac{\partial \bar{u}}{\partial y}\right)^2 + \left(\frac{\partial \bar{w}}{\partial y}\right)^2\right]^{\frac{1}{2}} \tag{6.188}$$

where l is assumed to be a scalar and have exactly the same form as that used in two dimensions. The closing assumption is now that the eddy viscosity is a simple isotropic scalar, i.e., its magnitude is independent of direction, so that one obtains for the stresses

$$\frac{\tau_x}{\rho} = (\nu_T + \nu)\frac{\partial \bar{u}}{\partial y}$$

$$\frac{\tau_z}{\rho} = (\nu_T + \nu)\frac{\partial \bar{w}}{\partial y} \tag{6.189}$$

Thus, as in two dimensions, the description of the three-dimensional Reynolds stresses can be poor when the mixing length assumptions are inadequate. However, in addition, in three dimensions the assumption that the eddy viscosity is an isotropic scalar results in the shear stress direction being that of the mean rate of strain. Experimental evidence shows that in many cases this is at best a crude assumption, but probably no worse than the assumption of a well-behaved mixing length or eddy viscosity; also, according to Nash and Patel, the calculations are not overly sensitive to the direction of the shear stress. In the foregoing, the generalization of mixing length formula to three dimensions has been given and the extension of the conventional two-dimensional eddy viscosity models, which usually differ only in the outer wake-like region in any event, are obtained similarly, i.e.,

$$\nu_T/\mathbf{v}_e\delta^* = 0.016 \tag{6.190}$$

where the freestream velocity u_e in two dimensions has simply been replaced by the freestream velocity vector parallel to the wall $\mathbf{v}_e$. As in two dimensions, the wall similarity scale is based upon the resultant stress, i.e.,

$$\frac{\tau}{\rho_w} = (\nu_T + \nu)\left[\left(\frac{\partial \bar{u}}{\partial y}\right)^2 + \left(\frac{\partial \bar{w}}{\partial y}\right)^2\right]^{\frac{1}{2}} \tag{6.191}$$

and the three-dimensional displacement thickness is defined

$$\delta^* = \int_0^\infty \left(1 - \frac{\mathbf{v}}{\bar{\mathbf{v}}_e}\right) \mathrm{d}y \tag{6.192}$$

Cebeci et al.[104] have in fact constructed a three-dimensional boundary-layer prediction scheme using an eddy viscosity formulation that has been generalized in the aforementioned manner from the two-dimensional model used by Cebeci and Smith.[71,76] Cebeci's initial results are quite encouraging. Wheeler and Johnson[99] examined three different three-dimensional shear stress models, including an eddy viscosity model similar in many respects to that used by Cebeci (only the viscous sublayer differed, as Wheeler and Johnson elected to fit a three-dimensional law of the wall profile). In general, Wheeler and Johnson concluded that there appeared to be a fairly large class of three-dimensional flows that could be quite well predicted by a direct numerical method, using even the very simple description of the Reynolds stresses obtained from the isotropic eddy viscosity model.

Summary of algebraic stress models. Even a cursory glance through the literature will show the quite remarkable accuracy and range of predictions possible with these very simple models of turbulence. Notable objections to mixing length and eddy viscosity, such as the fact that different outer layer prescriptions of the mixing length are required for differing types of flow operationally, causes little difficulty since the required typical value for that category of flow is known in advance. Difficulties with the transverse velocity gradient disappearing at some point other than where the shear stress goes through zero also can be circumvented, as was done in the film-cooling studies of, for example, Pai and Whitelaw.[217] Real problems arise with these models when substantial variations in the outer layer values of mixing length or eddy viscosity are required during the course of development of one particular shear layer and/or no convenient scale of length for the flow can readily be determined.

One-Equation Turbulence Models

There are two major defects in algebraic stress models, one being that practical flows can arise where the changes in the turbulent structure cannot keep pace with the changes in mean velocity and the other that in many turbulent flows no convenient turbulent scale of length is present. One-equation turbulence models are aimed at the turbulence lag problem. In using one-equation turbulence models, it is usual to suppose that a well-defined length scale such as the boundary-layer thickness δ is present; therefore, such models are of necessity restricted to flows where such a scale exists. Turning once again to Prandtl's constitutive relationship [Eq. (6.147)], it will be seen that the turbulent diffusioned flux is assumed to be related to some turbulent scale of length l and a turbulent velocity scale $\overline{q^2}$, where $\overline{q^2}$ is the turbulence kinetic energy. A number of one-equation models have then concentrated on solving a transport equation for the turbulence kinetic energy, which can be derived by manipulating the Navier-Stokes equations, while retaining the existing Prandtl formulation so that effectively a transport equation for the eddy diffusivity is constructed. Alternatively, an entirely empirical equation for the transport of the eddy diffusivity can be formulated such as was done by Nee and Kovasznay.[218] Finally, as an alternative to both of the preceding, the structural equilibrium hypothesis of Townsend[209] can be recalled and in this framework it is then necessary to predict only the turbulence kinetic energy to obtain the turbulent diffusional flux of interest, since with this hypothesis all correlations scale—however, a "dissipation length" now enters the problem. Prandtl's constitutive relationship is bypassed in the structural equilibrium approach, as is the local gradient transport hypothesis it embodies. An example of the structural equilibrium approach is to be found in the work of Bradshaw and Ferriss.[80] If the entirely empirical transport equation is omitted from further consideration (and most investigators prefer, where possible, to work with a rigorously based equation), the concensus seems to be that the turbulence kinetic energy equation is the most suitable single equation characterizing the Reynolds stress. Mellor and Herring[186] dispute this, however, arguing that the Reynolds shear stress equation (also derived by straightforward manipulation of the Navier-Stokes equations) is the "correct" equation, embodying as it does the transport of the Reynolds shear stress directly. However, at the outset of this particular development in turbulence modeling, the turbulence kinetic energy equation on a term-by-term basis was much better understood than the Reynolds shear stress equation and hence the selection of the energy equation for development.

In general, the various rigorous "transport" equations are formed from the instantaneous Navier-Stokes equations written in the form of fluctuating components by subtracting out the averaged Navier-Stokes equations. The resulting set may be regarded as equations governing the fluctuating components of velocity. The various single-point correlations may then be formed by taking appropriate moments of the fluctuation equations and averaging. Boundary-layer approximations, if valid, may then be introduced. The turbulence kinetic energy equation derived in the foregoing manner can be written in a three-dimensional Cartesian coordinate system in incompressible "boundary sheet" flow, (see Ref. 109) as

$$\frac{\partial}{\partial x}\frac{\overline{\bar{u}q^2}}{2}+\frac{\partial}{\partial y}\frac{\overline{\bar{v}q^2}}{2}+\frac{\partial}{\partial z}\frac{\overline{\bar{w}q^2}}{2}=-\overline{u'v'}\frac{\partial\bar{u}}{\partial y}-\overline{v'w'}\frac{\partial\bar{w}}{\partial y}$$

$$-\frac{\partial}{\partial y}\left(\frac{\overline{v'q^2}}{2}+\frac{\overline{v'p'}}{\bar{\rho}}\right)-\nu\left(\overline{u'\nabla^2u'}+\overline{v'\nabla^2v'}+\overline{w'\nabla^2w'}\right) \tag{6.193}$$

where the viscous term is the sum of the dissipative and diffusive effects of viscosity and often is rewritten as such, i.e.,

$$\nu\left(\overline{u'\nabla^2u'+v'\nabla^2v'+w'\nabla^2w'}\right)=\nu\sum_{i=1,3}\sum_{k=1,3}u_i'\frac{\partial^2u_i'}{\partial x_k^2}$$

$$=-\nu\sum_{i=1,3}\sum_{k=1,3}\left(\frac{\overline{\partial u_i'}}{\partial x_k}\right)^2+\frac{\partial^2\overline{u_i'^2}}{\partial x_k^2}=-\nu\sum_{k=1}\sum_{i=1,3}\left[\left(\frac{\overline{\partial u_i'}}{\partial x_k}\right)^2-\frac{1}{2}\frac{\partial^2}{\partial x_k^2}\overline{u_i'^2}\right] \tag{6.194}$$

where the gradients in kinetic energy are normally negligible, except close to a wall when the second derivative with respect to y may be important and can readily be retained without difficulty. Now, in order to proceed further, it is necessary to relate the higher-order correlations appearing in Eq. (6.193) to the lower-order correlations and/or the mean flow. This process is termed the "closure problem." Subsequently, the more commonly used strategies to effect this closure will be discussed.

Common to most approaches based on Eq. (6.193) is the treatment of the viscous energy dissipation terms in the energy equation and the appeal here is to the well-known observation that at high Reynolds numbers the rate of energy dissipation is controlled by the essentially inviscid turbulent motion supplying the energy which cascades to the smaller dissipative scales of the motion. Dimensional analysis then suggests that the viscous terms, represented by ε, can be written

$$\varepsilon=C_\varepsilon\left[\left(\overline{q^2}\right)^{\frac{3}{2}}/L\right] \tag{6.195}$$

where L is once again a turbulence length scale (prescribed) and C_ε some constant. At this point, if an appeal is made to Prandtl's constitutive relationship and the length scale L is prescribed as for the algebraic stress models, it remains only to set values for the various constants to close the system (normally the pressure velocity correlation gradient is neglected). Usually these constants are derived by forcing the system to return the mixing length values when the convective and diffusive terms in the kinetic energy equation can be neglected. Detailed values of the various constants are given in Ref. 182.

As an alternative to adopting Prandtl's constitutive relationship and the gradient transport hypothesis it embodies, Townsend's structural similarity arguments can be cited to yield

$$-\overline{u'v'} = a\overline{q^2} \tag{6.196}$$

where a could also be some prescribed function of the turbulent length scale L and hence y/δ. However, in application, Bradshaw and Ferriss,[80] Nash,[77] and McDonald and Camarata[58] all chose the simple expedient of setting a to a constant value of 0.15. At this point, two further quantities, the turbulent diffusive term and the turbulent length scale appearing in the expression for the dissipation remain to be specified. Insofar as the former is concerned, Bradshaw and Ferriss suggested

$$\frac{\overline{v'q^2}}{2} + \frac{\overline{v'p'}}{\rho} = G \cdot \overline{q^2}\left(\overline{q^2_{\max}}\right)^{\frac{1}{2}} \tag{6.197}$$

where $\overline{q^2_{\max}}$ is the maximum value of the turbulence kinetic energy in the boundary layer at that particular streamwise location. This quantity $\overline{q^2_{\max}}$ is introduced as a scale of the velocity responsible for the bulk transport (as opposed to gradient transport). G is a dimensionless function of normal distance in the boundary layer. Townsend[209] postulates a slightly different form, used by Nash,[77]

$$\frac{\overline{v'q^2}}{2} + \frac{\overline{v'p'}}{\rho} = -k\left(\overline{q^2}\right)^{\frac{3}{2}} \operatorname{sgn}\left(\frac{\partial \overline{q^2}}{\partial y}\right) \tag{6.198}$$

thus ensuring the transport is always down the gradient of $\overline{q^2}$. It is likely that both gradient and bulk diffusion of the kinetic energy are present in turbulent boundary layers and one or the other may dominate in various situations. At the present time, the situation is unclear and the arguments for both forms are given by their respective advocates in the appropriate references. Future study will doubtless clarify the matter, but for the moment the issue is sidestepped with the observation that boundary-layer predictions are not unduly sensitive to the assumptions made concerning diffusion of turbulence kinetic energy.

Turning to the problem of specifying the turbulent dissipation length scale: again, the proposition is that this scale is a unique function of position in the boundary layer. As with the other Prandtl-type formulations, a convenient method of estimating this length scale is given by considering equilibrium flows where convection and diffusion are negligible in comparison to production and dissipation of turbulence kinetic energy, leading to the observation that the dissipation length should have a very similar form to the mixing length distribution in an equilibrium boundary layer. This led McDonald and Camarata[58] to fit the dissipation length by a relationship similar to Eq. (6.163), notably incorporating the constant C_ε and defining the dissipation scale as $L = l/C_\varepsilon$, where

$$\frac{L}{\delta} = \left[0.1 \tanh\left(\frac{\kappa}{0.1}\frac{y}{\delta}\right)\right]\mathscr{D} \tag{6.199}$$

and the damping factor $\mathscr{D}$ is defined as before. McDonald and Camarata further chose to integrate the turbulence kinetic energy equation normal to the wall to remove the necessity for modeling the turbulent diffusion process. The closing assumption in this case was that it was then assumed that the usually defined mixing length could be described by formulas such as Eq. (6.163), but now it was possible to regard the outer layer value of the mixing length (the 0.09 δ referred to previously) as a free parameter whose value is determined by the turbulence kinetic energy equation.

All in all, the one-equation models as outlined above have performed at least as well—and usually better by varying degrees—than the algebraic stress models described earlier. A key ingredient to their success has been the fact that they represent a very modest and, in the main, physically plausible extension (indeed perturbation might be better) to the algebraic stress models. The one-equation models possess two powerful capabilities over the algebraic stress models of particular note in turbomachinery applications: (1) the ability to reflect the observed lag in turbulent boundary-layer response to severe pressure gradients being applied or removed and (2) the use of the turbulent kinetic energy equation with a nonzero freestream value allows the freestream turbulence to enter the prediction in a natural manner [see Fig. 6.10 and Eq. (6.66)]. Examples of the range and power of one-equation models are to be found in the predictions presented in Refs. 3, 77, 80, 159, and 219. The ability of the Townsend structural equilibrium approach to avoid zero shear stress when the mean velocity gradient goes through zero (which is not possible using the Prandtl constitutive relationship) is certainly attractive, but in most of the boundary-layer studies performed to date it does not seem to have been a crucial feature. An exception could arise in the film-cooling studies where the Townsend approach would eliminate the need for the simple ad hoc modification to the algebraic stress model predictions, such as that used by Pai and Whitelaw.[217]

One notable feature of the one-equation models that has not yet been mentioned concerns the effect of low Reynolds numbers. Glushko,[220]

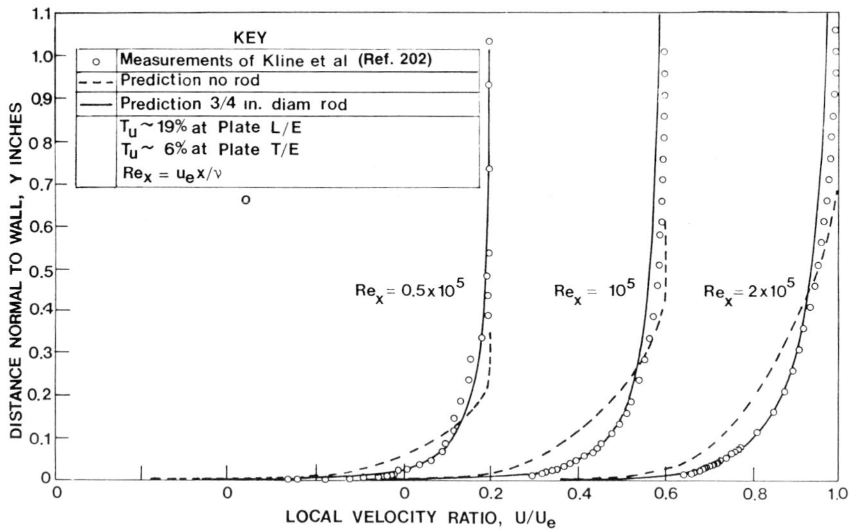

Fig. 6.10 Effect of freestream turbulence on the boundary-layer mean velocity profile.

McDonald and Fish,[60] and Jones and Launder[221] suggested that the structural coefficient used to close the turbulence kinetic energy equation would exhibit dependence on a turbulence Reynolds number

$$R_T = \frac{|-\overline{u'v'}|^{\frac{1}{2}} l}{\nu} \tag{6.200}$$

where l is the previously introduced turbulence length scale. A very similar alternative Reynolds number can be obtained by using $(\overline{q^2})^{\frac{1}{2}}$ in place of $(-\overline{u'v'})^{\frac{1}{2}}$. If this length scale is supposedly related to the conventional mixing length, then the turbulence Reynolds number is simply the ratio of the eddy viscosity to the actual viscosity. Using only the assumption that the local conditions obtained in Townsend's structural equilibrium state would also depend on the turbulence Reynolds number, i.e.,

$$-\overline{u'v'} = a \cdot f(R_T)\overline{q^2} \tag{6.201}$$

McDonald and Fish[60] were able to model both forward and reverse transition as it is influenced by the freestream turbulence wall roughness and pressure gradients. They argued that the mean strain effect on a would still be negligible for the flows of interest compared to the R_T effect. The functional dependence of the structural coefficient a on the turbulence Reynolds number R_T is given in Ref. 60 and was somewhat contrived, being chosen to have the correct asymptotic limits of 0 and 1—but with the

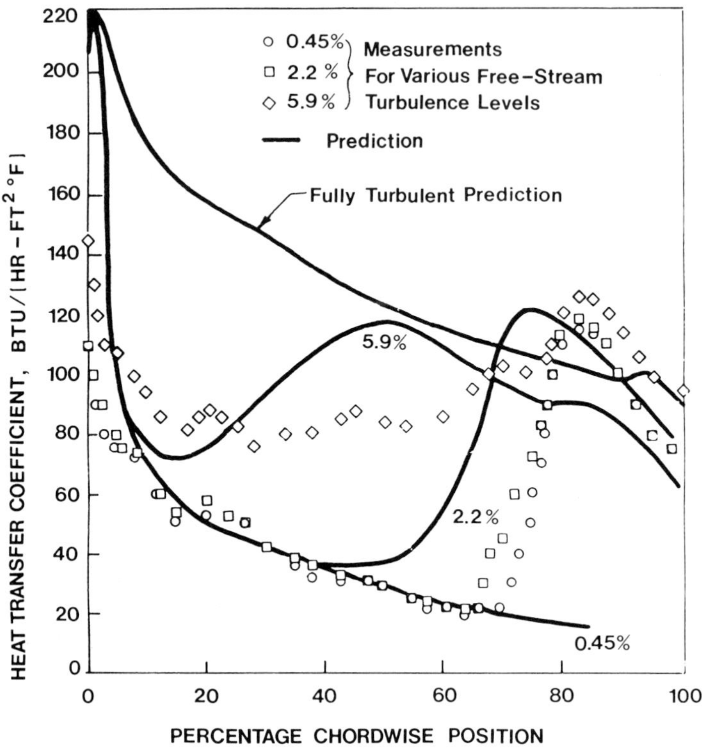

Fig. 6.11 Heat-transfer distribution on the suction side of a turbine airfoil (measurements of predictions according to Ref. 60).

variation in between selected judiciously to result in good agreement with the observed dependence of forward transition Reynolds number on freestream turbulence intensity. In spite of this, the selected dependence survived intact to give very reasonable predictions of boundary-layer relaminarization as a function of pressure gradient (e.g., see Ref. 60). Subsequently, it was found that the effect of convex curvature in inhibiting the forward transition process was reasonably well accounted for in this model, if Bradshaw's suggestion[189] for the effect of curvature on the dissipation length L were adopted. However, the destabilizing effect of concave curvature was not well predicted with this model.

Some predictions showing the effect of freestream turbulence on a fully turbulent boundary layer and on the heat-transfer to a turbine airfoil using the McDonald-Fish model are shown in Figs. 6.10 and 6.11.

Three observations are probably appropriate at this point. The first is that the heat transfer to a turbine airfoil is one of the major turbomachine boundary-layer problems, one dominated by low Reynolds number effects such as forward and reverse transition. Consequently, unless the boundary-layer analysis can account in some plausible manner for these phenomena as they are influenced by freestream turbulence and streamwise acceleration, the entire analysis is of little use to design engineers. Second, the alternatives to the adoption of the turbulence Reynolds number effect on the structural coefficient are either to abandon the problem as too difficult or to attempt to correlate the location of the initiation and extent of the transition region as it is observed to vary with Reynolds number pressure gradient and freestream disturbances. There is evidence that a unique correlation of this type does exist, since it is observed that, for the moderate-to-high levels of freestream turbulence typical of turbomachinery, the overall energy level of the broadband freestream disturbances seems to characterize the transition process at a given Reynolds number. If desired, the suggested $a - R_T$ relationship can be thought of as a particularly simple means of implementing precisely this type of empirical correlation. Third, a number of authors have suggested that a parameter such as R_T might at least characterize the relaminarization phenomenon such that, in general, when R_T fell below about 33 the turbulent motion would die out (e.g., see Ref. 84). The suggested variation of the structural coefficient a with R_T reflects this concept but, rather than have an on-off switch, the functional relationship adopted assures a smooth transition and a rate-controlled change in the structure that is quantitatively in accordance with observations. Bearing these three points in mind, it is felt that a contrived $a - R_T$ relationship (such as that suggested in Ref. 60), although at present unsupported by direct experimental evidence and doubtlessly a gross oversimplification of the physical phenomena at play, does introduce a degree of realism into the predictions—without which the analysis would be almost worthless for detailed turbine blade design purposes. Some idea of the complexities involved can be seen in Fig. 6.11.

In the context of the multiequation models to be discussed subsequently, a number of authors (e.g., Jones and Launder[221] and Donaldson[191]) have postulated a turbulence Reynolds number dependence of various parameters. In his one-equation model, Glushko[220] proposed a series of parameters varying with the turbulence Reynolds number, again with little direct experimental support. In any event, Glushko's model has not been as extensively tested or developed as that of McDonald and Fish,[60] which from the results of Kreskovsky et al.[215] can be seen to perform quite well over a very wide range of flows. It is to be expected that the dissipation process will exhibit a very marked effect on turbulence Reynolds number such that the dissipation length L will also depend on the turbulence Reynolds number. Indeed, Glushko[220] and Hanjalic and Launder[222] make specific suggestions as to the form of this Reynolds number dependence. McDonald and Fish,[60] while recognizing that such an effect is probably present, point out that in their one-equation formulation the ratio of turbulence length scale to dissipation length appears, so that, unless the dissipation length scale L decreased with Reynolds number much faster than the turbulence length scale l (an unlikely event), the low Reynolds number effect on dissipation would not have a significant effect on their predictions, provided that $l/L \ll 1.0$ at low Reynolds numbers. No matter which approach is favored—that of postulating an effect only on the structural coefficient a or an effect only on the dissipative process, or both —the postulated dependence can, at present, be obtained only by the indirect process of attempting to duplicate the measured flow development. Clearly, experimental guidance is required.

Two-Equation Turbulence Models

There is one major defect in the one-equation models in current use—the need to specify a turbulent length scale. If continued use of Prandtl's constitutive relationship is made in the one-equation models, the argument can also be valid that the use of the mean gradient further preserves the defect of zero correlation at the point of zero mean gradient. This point will not be pursued here, but it is noted that in a number of flows of practical interest no convenient shear layer thickness can be identified, for example, in can or annular combustors. In the two-equation system, the additional equation is aimed at providing a differential equation for some turbulence quantity that can be related to a turbulent scale of length, thus eliminating the need to directly specify some shear layer thickness upon which the turbulence length scale can be based. The idea has a great deal of merit, but unfortunately at present much of this work is still relatively preliminary and cannot be confidently used by design engineers. It is to be expected that this state of affairs will be improved upon in the near future, however. In view of the potential of two-equation models, some outline of the present status will be given here.

With few exceptions, the choice of one of the two equations has been the turbulence kinetic energy equation that performed so well in the single-equation system. The choice of the second equation has exhibited a great deal of variety, although the chosen dependent variables have generally been associated more or less directly with some turbulent length scale. Heuristic transport equations also have made their reappearance, justified persuasively on the grounds that so little is known about how to model the various correlations appearing in the rigorous equations that one might as well use a heuristic transport equation. Such arguments have led Bradshaw[183] and Saffman[223] to propose a heuristically dissipative length scale transport equation on the one hand and a pseudovorticity transport equation on the other. Criticism of the speculative nature of the various hypothesis required to convert the various rigorous equations into length scale transport equations or their equivalent is certainly justified. At this point, little in the way of experimental support, or even insight, is available to justify many of the suggestions. Be that as it may, it seems probable that, as the store of experimental information on the structure of turbulence is increased, an improvement in the modeling of the terms in the rigorous equations will result. Heuristic transport equations that cannot be eventually related to some rigorous transport equation seem destined to a limited lifetime. In seeking to remedy the lack of experimental support on the detailed modeling assumptions required to close the rigorous equations, an effort is usually made to insure that certain well-known results, usually concerning the decay of isotropic turbulence, are reproduced by the modeled equations. Again, this process on the surface seems very reasonable but, as argued by Bradshaw, the wide disparity between isotropic and shear flow turbulence confuses this issue dramatically.

Of the two-equation models proposed to date, only the Jones-Launder model[221] and the Saffman-Wilcox model[223] have been developed to any great extent. Earlier or alternative two-equation models, such as those developed by Ng[184] based on an even earlier proposal of Rotta[224] or the suggestion of Spalding,[225] have not as yet demonstrated any clear superiority and according to Launder are slightly more complicated. In his critical review, Reynolds[185] cautions against the use of the Saffman-Wilcox model as an engineering tool in view of the extreme sensitivity of the predictions to both the wall and freestream boundary conditions. To date the most success, measured in terms of satisfactory predictions obtained without changing the empirical constants, has been obtained using a modeled form of the equation governing the dissipation of turbulence kinetic energy. The reader then would be advised to choose from the two-equation model shopping list described by Launder and Spalding,[182] i.e., one model based upon the turbulence kinetic energy and the dissipation of turbulence kinetic energy. However, caution must still be advocated since clearly the use of an eddy viscosity, albeit derived from Prandtl's constitutive relationship, still suffers from the observed discrepancy of zero stress not occurring at zero mean gradient. Further, the known anisotropy of the eddy viscosity, so important in three-dimensional flows, is at present missing from these models. Clearly, much remains to be done on these multiequation models before they can be used with confidence by design engineers. The extension of the capability that these models offer, however, does lend encouragement to their continued development. Further discussion of turbulence models will be discontinued, since beyond this point the topic becomes even more controversial.

6.6 Conclusion

At the request of the editor, I would like to conclude this chapter with some observations about the audience that I had hoped to reach. I would also like to add some personal comments on the contents now that it has been completed; completed that is, in the sense that there is no longer any time to include additional material!

First, insofar as the hoped for audience was concerned, the model I had in mind grew out of my rather long and fruitful (at least on my part!) association with the engineers at Pratt & Whitney (Aircraft). The educational level of these engineers was generally at the masters or doctoral level in mechanical engineering. Often they were faced with the very difficult task of evaluating new or advanced fluid mechanics technology. If following the evaluation they deemed it attractive, they then had to further develop, validate, and ultimately transfer the technology into the design system, that it might usefully impact the end product. The understanding and prediction of general turbulent "boundary layers" has the potential of greatly assisting in the turbomachinery design process and, consequently, was and is a topic of considerable interest to these engineers. The literature on the subject is not turbomachinery biased, however, and sometimes methods or approaches were explored that had hereditary traits which either precluded successful application to turbomachinery problems or preordained certain very unfortunate limitations. It was in part with the goal of helping the industrially based engineers to become "selective shoppers" in the evaluation phase of their work that this chapter was initiated. It was further hoped that it would be of assistance to new recruits to the subject to help bridge the gap between graduate level courses and the current areas of interest of those involved in evaluation, development, validation, and implementation of viscous flow prediction schemes.

Concerning the actual material presented, in retrospect, it seems clear that too much effort was spent on the two-dimensional integral boundary-layer procedures. This occurred partly because the subject is relatively secure and partly because I poorly budgeted the time I had available to write the whole chapter. A much better allocation would have been to reduce the emphasis on integral procedures and include some discussion on wake flows. Although I did not intend to interpret the material I was requested to prepare as one dealing exclusively with boundary layers, over the years I had not detected any great interest by the engine manufacturers in wake flows. Hence their omission from the section. More recently, it appears that there is interest in wake-controlled problems and this area will grow in importance. A very encouraging contribution to this near-wake problem has recently been made by Pope and Whitelaw.[226] Insofar as three-dimensional flows are concerned, I regret not having more time available to develop the relationship between what my colleagues and I term "classical secondary flow theory," as described by Hawthorn and Horlock, and the extended layer equations discussed in the text. Such a relationship does exist and its exploration and further development may provide the key to understanding the observed complex intrablade or stator boundary-layer flows.

Acknowledgments

A major portion of the writing of this section was carried out while the author was on sabbatical leave from United Technologies Research Center. During this period, the author was appointed Senior Research Fellow at the University of Glasgow in the Department of Aeronautics. The author is deeply indebted to both institutions for their generosity and help.

The author is also indebted to his colleagues at Scientific Research Associates, Inc., for their help in the final preparation of this manuscript.

References

[1]McDonald, H. and Shamroth, S. J., "An Analysis and Application of the Time-Dependent Turbulent Boundary-Layer Equations," *AIAA Journal*, Vol. 9, Aug. 1971, pp. 1553–1560.

[2]Evans, R. L. and Horlock, J. H., "Calculations of the Development of Turbulent Boundary-Layers with a Turbulent Free-Stream," ASME Paper 74-FE-24.

[3]McDonald, H. and Kreskovsky, J. P., "Effect of Free Stream Turbulence on the Turbulent Boundary Layer," *International Journal of Heat and Mass Transfer*, Vol. 17, 1974, pp. 705–716.

[4]Nash, J. F., "A Note on Skin-Friction Laws for the Incompressible Turbulent Boundary Layer," NPL Aero Rept. 1135, Dec. 1964.

[5]Nash, J. F. and MacDonald, A. G. J., "A Turbulent Skin Friction Law for Use at Subsonic and Transonic Speeds," NPL Aero Rept. 1206; also British Aeronautical Research Council, CP 948, July 1966.

[6]Coles, D. E. and Hirst, E. A., *Proceedings: Computation of Turbulent Boundary Layers*, AFOSR-IFP Stanford Conference, Vol. II, 1968, Compiled Data, 1968.

[7]Coles, D. E., "A Survey of Data for Turbulent Boundary Layers with Mass Transfer," *Turbulent Shear Flows*, AGARD CP 93, Sept. 1971.

[8]Simpson, R. L., Ph.D. Thesis, Stanford University, Stanford, Calif., 1967.

[9]Coles, D. E., "The Law of the Wake in the Turbulent Boundary Layer," *Journal of Fluid Mechanics*, Vol. 1, 1956, pp. 191–226.

[10]Waltz, A., *Boundary Layers of Flow and Temperature*, MIT Press, Cambridge, Mass., 1969.

[11]Clauser, F. H., "Turbulent Boundary Layers in Adverse Pressure Gradients," *Journal of the Aeronautical Sciences*, Vol. 21, 1954, pp. 91–108.

[12]Nash, J. F., "Turbulent-Boundary-Layer Behavior and the Auxiliary Equation," AGARDograph 97, Pt. 1, 1965.

[13]Mellor, G. L. and Gibson, D. M., "Equilibrium Turbulent Boundary Layers," *Journal of Fluid Mechanics*, Vol. 24, Pt. 2, 1966, pp. 225–253.

[14]Michel, R., Quemard, C., and Durant, R., "Hypothesis on the Mixing Length and Application to the Calculation of the Turbulent Boundary Layer," *Proceedings: Computation of Turbulent Boundary Layers*, AFOSR-IFP Stanford Conference, Vol. I, 1968.

[15]Rotta, J. C., "Turbulent Boundary Layers in Incompressible Flow," *Progress in Aeronautical Sciences*, Vol. 2, Pergamon Press, New York, 1962.

[16]Rubesin, M. W. and Johnson, H., "A Summary of Skin Friction and Heat Transfer Solutions of the Laminar Boundary Layer on a Flat Plate," *Transactions of ASME*, Vol. 71, 1949, p. 383.

[17]Coles, D., "The Turbulent Boundary Layer in a Compressible Fluid," Rand Corp., Rept. R-403-PR, 1962.

[18]Winter, K. G., Smith, K. G., and Gaudet, L., "Measurements of Turbulent Skin Friction at High Reynolds Numbers at Mach Numbers of .2 and 2.2," AGARDograph 97, Pt. I, 1965.

[19]Winter, K. G. and Gaudet, L., "Some Recent Work on Compressible Turbulent Boundary Layers and Excrescence Drag," *Compressible Turbulent Boundary Layers*, NASA SP-216, 1968.

[20]Van Driest, E. R., "Turbulent Boundary Layer in Compressible Fluids," *Journal of the Aeronautical Sciences*, Vol. 18, 1951, pp. 631–638.

[21]Maise, G. and McDonald, H., "Mixing Length and Kinematic Eddy Viscosity in a Compressible Boundary Layer," *AIAA Journal*, Vol. 6, 1968, p. 73.

[22]Coles, D. E., "Measurements in the Boundary Layer on a Smooth Flat-Plate in Supersonic Flow," Ph.D. Thesis, California Institute of Technology, Pasadena, 1953.

[23]Mathews, D. C., Paynter, G. C., and Child, M. E., "On the Use of Coles' Universal Wake Function for Compressible Turbulent Boundary Layers," *Journal of*

Aircraft, Vol. 7, Feb. 1970, pp. 137–140.

[24]Hopkins, E. J., Keener, E. R., and Polek, T. E., "Hypersonic Turbulent Skin-Friction and Boundary-Layer Profiles on Non-Adiabatic Flat Plates," *AIAA Journal*, Vol. 10, 1972, pp. 40–48.

[25]Owen, F. K. and Horstman, C. C., "On the Structure of Hypersonic Turbulent Boundary Layers," *Journal of Fluid Mechanics*, Vol. 53, Pt. 4, 1972, pp. 611–636.

[26]Fernholz, H. H. and Finley, P. J., "A Critical Commentary on Mean Flow Data for Two-Dimensional Compressible Turbulent Boundary Layers," AGARDograph No. 253, May 1980.

[27]Coles, D. E., "The Turbulent Boundary Layer in a Compressible Fluid," *Physics of Fluids*, Vol. 7, No. 9, Sept. 1964, pp. 1403–1423.

[28]Fernholz, H. H. and Finley, P. J., "A Critical Compilation of Compressible Turbulent Boundary Layer Data," AGARDograph 223, June 1977.

[29]Morkovin, M. V., "Effects of Compressibility on Turbulent Flows," Collogues Internationaux du Centre National de Recherche Scientifique, Rept. 108, 1962, pp. 367–380.

[30]Clauser, F. H., "The Turbulent Boundary Layer," *Advances in Applied Mechanics*, Vol. 4, Academic Press, New York, 1956.

[31]Dvorak, F. A., "Calculation of Turbulent Boundary Layers on Rough Surfaces in Pressure Gradient," *AIAA Journal*, Vol. 7, Sept. 1969, pp. 1752–1759.

[32]Chen, K. K., "Compressible Turbulent Boundary Layer Heat Transfer to Rough Surfaces under Arbitrary Pressure Gradient," AIAA Paper 71-166, 1971.

[33]Young, F. L., "Experimental Investigation of the Effects of Surface Roughness on Compressible Turbulent Boundary Layer Skin Friction and Heat Transfer," University of Texas, Austin, Rept. DRL 532 (AD621085), 1965.

[34]Dvorak, F. A. and Head, M. R., "Heat Transfer in the Constant Properly Turbulent Boundary Layer," *International Journal of Heat and Mass Transfer*, Vol. 10, 1967, pp. 61–81.

[35]Head, M. R., "Entrainment in the Turbulent Boundary Layer," British Aeronautical Research Council, R&M 3152, 1958.

[36]Green, J. E., "Application of Head's Entrainment Method to the Prediction of Turbulent Boundary Layers and Wakes in Compressible Flow," Royal Aircraft Establishment, TR 72231, 1973.

[37]*Compressible Turbulent Boundary Layers*, NASA SP 216, Dec. 1968.

[38]Rotta, J. C., "Heat Transfer and Temperature Distribution in Turbulent Boundary Layers at Supersonic and Hypersonic Flow," AGARDograph 97, Pt. 1, 1965.

[39]Hopkins, E. J., Rubesin, M. W., Inouye, M., Keener, E. R., Mather, G. C., and Polek, T. E., "Summary and Correlation of Skin Friction and Heat Transfer Data for a Hypersonic Turbulent Boundary Layer on Simple Shapes," NASA TN D 5089.

[40]Cousteix, J., Houdeville, R., and Michel, R., "Couches Limites Turbulentes Avec Transport de Chaleur," *La Recherche Aerospatiale*, No. 6, 1974, pp. 327–337.

[41]McDonald, H. and Owen, F. K., Unpublished work.

[42]Meir, H. U. and Rotta, J. C., "Experimental and Theoretical Investigation of Temperature Distributions in Supersonic Boundary Layers," AIAA Paper 70-744, 1970.

[43]Rotta, J. C., "Die Turbulente Grenzschicht an Einer Stark Geheizten Ebenen Platte Bei Unterschattströmung," *Waerme- und Stoffuebertragung*, 1974, pp. 133–144.

[44]Shamroth, S. J., "On Integral Methods for Predicting Shear Layer Behavior," *Journal of Applied Mechanics*, Vol. 36, Dec. 1969.

[45]Weinbaum, S. and Garvine, R. W., "On the Two Dimensional Viscous Counterpart of the One Dimensional Sonic Throat," *Journal of Fluid Mechanics*, Vol. 39, Pt. I, 1969, pp. 57–85.

[46]Bradshaw, P. and Ferriss, D. H., "The Response of a Retarded Equilibrium Turbulent Boundary Layer to the Sudden Removal of Pressure Gradient," NPL Aero Rept. 1145, 1965.

[47]White, F. M., "A New Integral Method for Analyzing the Turbulent Boundary Layer with Arbitrary Pressure Gradient," ASME Paper 68-WA/FE 22, 1968.

[48]Stratford, B. S., "The Prediction of Separation of the Turbulent Boundary Layer," *Journal of Fluid Mechanics*, Vol. 5, Pt. I, 1959, pp. 1–16.

[49]Escudier, M. P. and Nicoll, W. B., "A Shear Work Integral Method for the Calculation of Turbulent Boundary Layer Development," *Proceedings: Computation of Turbulent Boundary Layers*, AFOSR-IFP Stanford Conference, Vol. I, 1968.

[50]Bradshaw, P., "The Analogy Between Streamline Curvature and Buoyancy in Turbulent Shear Flow," *Journal of Fluid Mechanics*, Vol. 36, 1969, pp. 177–191.

[51]Goldberg, P., "Upstream History and Apparent Stress in Turbulent Boundary Layers," Gas Turbine Lab., Massachusetts Institute of Technology, Cambridge, Rept. 85, May 1966.

[52]Kline, J. J., Morkovin, M. V., Sovran, G., and Cockrell, D. J., *Proceedings: Computation of Turbulent Boundary Layers*, AFOSR-IFP Stanford Conference, Vol. I, 1968.

[53]Nash, J. F. and Hicks, J. G., "An Integral Method Including the Effect of Upstream History on the Turbulent Shear Stress," *Proceedings: Computation of Turbulent Boundary Layers*, AFOSR-IFP Stanford Conference, Vol. I, 1968.

[54]McDonald, H., "The Departure from Equilibrium of Turbulent Boundary Layers," *Aeronautical Quarterly*, Vol. XIX, Pt. 1, Feb. 1968, pp. 1–19.

[55]Townsend, A. A., "Equilibrium Layers and Wall Turbulence," *Journal of Fluid Mechanics*, Vol. 1, 1961, pp. 97–120.

[56]Green, J. E., Weeks, D. J., and Brooman, J. W. F., "Prediction of Turbulent Boundary Layers and Wake in Compressible Flow by a Lag-Entrainment Method," Royal Aircraft Establishment, Tech. Rept. 72231, 1973.

[57]Hirst, E. A. and Reynolds, W. C., "An Integral Prediction Method for Turbulent Boundary Layers Using the Turbulent Kinetic Energy Equation," *Proceedings: Computation of Turbulent Boundary Layers*, AFOSR-EFP Stanford Conference, Vol. I, 1968.

[58]McDonald, H. and Camarata, F. J., "An Extended Mixing Length Approach for Computing the Turbulent Boundary Layer Development," *Proceedings: Computation of Turbulent Boundary Layers*, AFOSR-IFP Stanford Conference, Vol. I, 1968.

[59]Bradshaw, P., Ferriss, D. H., and Atwell, N. P., "Calculation of Boundary Layer Development Using the Turbulent Energy Equation," *Journal of Fluid Mechanics*, Vol. 28, Pt. 3, 1967, pp. 593–616.

[60]McDonald, H. and Fish, R. W., "Practical Calculations of Transitional Boundary Layers," *International Journal of Heat and Mass Transfer*, Vol. 16, No. 9, 1973, pp. 1729–1744.

[61]Thompson, B. J. G., "The Calculation of Shape Factor Development in Incompressible Turbulent Boundary Layers with or without Transpiration," AGARD-

ograph 97, Vol. 1, 1965.

[62] Lubard, S. C. and Fernandez, F. L., "The Turbulent Boundary Layer with Mass Transfer and Pressure Gradient," *Journal of Applied Mechanics, Transactions of ASME*, Vol. 28, Pt. 3, 1971.

[63] Alber, I. E., "Application of an Exact Expression for the Equilibrium Dissipation Integral to Calculation of Turbulent Nonequilibrium Flows," *Proceedings: Computation of Turbulent Boundary Layers*, AFOSR-IFP Stanford Conference, Vol. I, 1968.

[64] Pletcher, R. H., "Calculation Method for Compressible Turbulent Boundary Flows with Heat Transfer," *AIAA Journal*, Vol. 10, March 1972, pp. 245–246.

[65] Roache, P. J., *Computational Fluid Dynamics*, Hermosa Publishers, 1972.

[66] Briley, W. R. and McDonald, H., "Solution of the Three-Dimensional Compressible Navier-Stokes Equations by an Implicit Technique," *Proceedings of 4th International Conference of Numerical Methods in Fluid Dynamics*, June 1974; also *Journal of Computational Physics*, 1977.

[67] McDonald, H. and Briley, W. R., "Three Dimensional Supersonic Flow of a Viscous or Inviscid Gas," *Journal of Computational Physics*, Vol. 19, No. 2. Oct. 1975.

[68] Blottner, F. G., "Investigations of Some Finite-Difference Techniques for Solving the Boundary Layer Equations," *Computer Methods in Applied Mechanics and Engineering*, Vol. 6, 1975, pp. 1–30.

[69] Flugge-Lotz and Blottner, F. G., "Computation of the Compressible Laminar Boundary Layer Flow Including Displacement Thickness Interaction Using Finite-Difference Methods," AFOSR-220S, 1962.

[70] Patankar, S. V. and Spalding, D. B., *Heat and Mass Transfer in Boundary Layers*, 2nd ed., Intertext, London, 1970.

[71] Cebeci, T. and Smith, A. M. O., "A Finite Difference Method for Calculating Compressible Laminar and Turbulent Boundary Layers," *Journal of Basic Engineering, Transactions of ASME*, Vol. 92, No. 3, 1970, pp. 323–535.

[72] Herring, H. J. and Mellor, G. L., "A Method of Calculating Compressible Turbulent Boundary Layers," NASA CR-1144, 1968.

[73] Keller, H. B., "A New Difference Scheme for Parabolic Problems," *Numerical Solutions of Partial Differential Equations*, Vol. II, Academic Press, New York, 1971, pp. 327–350.

[74] Blottner, F. G., "Variable Grid Scheme Applied to Turbulent Boundary Layer," *Computer Methods in Applied Mechanics and Engineering*, Vol. 4, 1974, pp. 179–194.

[75] Roberts, G. O., "Computational Meshes for Boundary Layer Problems," *Proceedings: Second International Conference on Numerical Methods in Fluids Dynamics*, Springer-Verlag, New York, 1971, p. 171.

[76] Cebeci, T. and Smith, A. M. O., "Analysis of Turbulent Boundary Layers," *Applied Mathematics and Mechanics*, Vol. 15, Academic Press, New York, 1974.

[77] Nash, J. F., "The Calculation of Three Dimensional Turbulent Boundary Layers in Incompressible Flow," *Journal of Fluid Mechanics*, Vol. 37, Pt. 4, 1969, p. 625.

[78] Yanenko, N. N., *The Method of Fractional Steps*, translation edited by M. Holt, Springer-Verlag, New York, 1971.

[79] Moretti, G., "A Pragmatical Analysis of Discretization Procedures for Initial and Boundary Value Problems in Gas Dynamics and Their Influence on Accuracy,"

Polytechnic Institute of Brooklyn, New York, Ae/Am Rept. 74-15, Sept. 1974.

[80] Bradshaw, P. and Ferriss, D. H., "Application of a General Method of Calculating Turbulent Shear Layers," ASME Paper 71-WA/FE-8.

[81] Blottner, F. G., "Computational Techniques for Boundary Layers," AGARD Lecture Series.

[82] Peters, N., "Boundary Layer Calculations by a Hermitian Finite Difference Method," *Proceedings: Fourth International Conference on Numerical Methods in Fluid Dynamics*, Springer-Verlag, New York, 1974.

[83] Launder, B. E. and Jones, W. P., "On the Prediction of Laminarization," British Aeronautical Research Council, CP 1036, Feb. 1968.

[84] Kays, W. M. and Moffat, R. J., "Behavior of Transpired Turbulent Boundary Layers," *Studies in Convection Theory, Measurement and Applications*, Vol. 1, edited by B. E. Launder, 1975.

[85] Becko, Y., "Heat Transfer Analysis Along the Blades of a Gas Turbine Stator by Thermal and Kinematic Boundary Layer Theory," ASME Paper 75-GT-15, 1975.

[86] Keller, H. B. and Cebeci, T., "Accurate Numerical Methods for Boundary Layer Flows, II: Two Dimensional Turbulent Flows," *AIAA Journal*, Vol. 10, Sept. 1972, pp. 1193–1199.

[87] Harris, J. E., "Numerical Solution of the Equations for Compressible, Laminar, Transitional and Turbulent Boundary Layers and Comparison with Experimental Data," NASA TR R-369, 1971.

[88] Adams, J. C., "Eddy Viscosity—Intermittency Factor Approach to Numerical Calculation of Transactional Heating on Sharp Cones in Hypersonic Flow," AEDC-TR-70-210, 1970.

[89] Kendall, R. M. and Bartlett, E. P., "Nonsimilar Solution of the Multi-Component Laminar Boundary Layer by an Integral Matrix Method," *AIAA Journal*, Vol. 6, June 1968, pp. 1089–1097.

[90] Anderson, O. L., "Finite Difference Solution for Turbulent Swirling Compressible Flow in Axisymmetric Ducts with Struts," NASA CR-2365, Feb. 1974.

[91] Horlock, J. H. and Perkins, H. J., "Annulus Wall Boundary Layers in Turbomachines," AGARDograph 185, 1975.

[92] Mellor, G. L. and Wood, G. M., "An Axial Compressor Endwall Boundary Layer Theory," *Journal of Basic Engineering, Transactions of ASME*, June 1971.

[93] Hayes, W. D., "The Three Dimensional Boundary Layer," NAVORD Rept. 1313, 1951.

[94] Daneshyar, M., "Annulus Wall Boundary Layer in Turbomachines," Ph.D. Thesis, Cambridge University, Cambridge, England, 1973.

[95] Cumpsty, N. A. and Head, M. R., "The Calculation of Three-Dimensional Turbulent Boundary Layers (1) Flow Over the Rear of an Infinite Swept Wing," *Aeronautical Quarterly*, Vol. 18, 1967.

[96] Horlock, J. H., "Cross Flows in Bounded Three Dimensional Turbulent Boundary Layers," *Journal of Mechanical Engineering Science*, Vol. 15, No. 4, 1973, pp. 274–284.

[97] Mager, A., "Generalization of Boundary Layer Momentum Integral Equations to Three Dimensional Flow Including Those of Rotating Systems," NASA Rept. 1067, 1952.

[98] Johnson, J. P., "On the Three-Dimensional Turbulent Boundary Layer Generated by Secondary Flows," *Journal of Basic Engineering, Transactions of ASME*,

Vol. 81, No. 3, 1960.
[99]Wheeler, A. J. and Johnson, J. P., "Three-Dimensional Turbulent Boundary Layer—An Assessment of Prediction Methods," Dept. of Mechanical Engineering, Stanford University, Stanford, Calif., Rept. MD-30, July 1971; also ASME Paper 73-FE-25, 1973.
[100]Lindsay, W. L., "Tip Clearance Effects in the Growth of Annulus Wall Boundary Layers in Turbomachines," Ph.D. Thesis, Cambridge University, Cambridge, England, 1974.
[101]Hoadley, D., "Three-Dimensional Turbulent Boundary Layers in an Annular Diffuser," Ph.D. Thesis, Cambridge University, Cambridge, England, 1970.
[102]Daneshyar, M., "Prediction of Annulus Wall Boundary Layers in Axial Flow Turbomachines," AGARDograph 164, 1972.
[103]Harris, J. E. and Morris, D. J., "Solution of the Three-Dimensional Compressible Laminar and Turbulent Boundary Layer Equations with Comparisons to Experimental Data." *Fourth International Conference on Numerical Methods in Fluid Dynamics*, 1974.
[104]Cebeci, T., Kamps, K., Mosinskis, G. J., and Rehn, J., "Some Problems of the Calculation of Three-Dimensional Boundary Layer Flow on General Configuration," NASA CR-2285, July 1973.
[105]Raetz, G. S., "A Method of Calculating Three-Dimensional Laminar Boundary Layers of Steady Compressible Flows," Northrop Corp., Rept. NAI-58-73, 1957.
[106]Wesseling, P., "Calculations of Incompressible Three-Dimensional Turbulent Boundary Layers, Part 1: Formulation of a System of Equations," National Lucht. Laboratories, the Netherlands, Rept. AT-69-01, 1969.
[107]Wang, K. C., "On the Determination of the Zones of Influence and Dependence for Three-Dimensional Boundary Layer Equations," *Journal of Fluid Mechanics*, Vol. 48, No. 2, 1971, pp. 397–404.
[108]Ting, L., "On the Initial Condition for Boundary Layer Equations," *Journal of Mathematical Physics*, Vol. XLIV, No. 4, Dec. 1965, pp. 353–367.
[109]Nash, J. F. and Patel, V. C., "Three-Dimensional Turbulent Boundary Layers," Scientific and Business Consultants, Inc., 1972.
[110]Dring, R. P., "A Momentum Integral Analysis of the Three-Dimensional Turbine End Wall Boundary Layer," *Journal of Engineering for Power, Transactions of ASME*, Oct. 1971.
[111]Thompson, B. G. T., "A New Two Parameter Family of Mean Velocity Profiles for Incompressible Turbulent Boundary Layers on Smooth Walls," British Aeronautical Research Council, Paper 26830, 1965.
[112]Eichelbrenner, E. A. and Penke, J. L., "Theoretical and Experimental Investigations in Three-Dimensional (Laminar and Turbulent) Boundary Layers," Laboratories de Mechnique des Fluides, 1966.
[113]Smith, P. D., "An Integral Prediction Method for Three-Dimensional Compressible Turbulent Boundary Layers," Royal Aircraft Establishment, TR 72228, Dec. 1972.
[114]Myring, D. F., "An Integral Prediction Method for Three-Dimensional Turbulent Boundary Layers in Incompressible Flow," Royal Aircraft Establishment, Tech. Rept. 70147, 1970.
[115]Thompson, B. G. J. and MacDonald, A. G. J., "The Prediction of Boundary Layer Profile Development for Infinite Yawed Wings," British Aeronautical Research Council, CP 1307.

[116]Green, J. E., "Applications of Head's Entrainment Method in the Prediction of Turbulent Boundary Layers and Wakes in Compressible Flow," Royal Aircraft Establishment, TR 72079, 1972.

[117]Krause, E., Hieschel, E. H., and Bothmer, T., "Numerical Stability of Three-Dimensional Boundary Layer Solutions," *Zeitschrift fuer Angewandte Mathematik und Mechanik*, Vol. 48, No. 8, 1968, pp. T205–T208.

[118]Klinksiek, W. and Pierce, F., "Secondary Flow Reversal and Velocity Profile Models in a Three-Dimensional Turbulent Boundary Layer," Mechanical Engineering Dept., Virginia Polytechnic Institute, Blacksburg, 1968.

[119]Kendall, R. M., Bonnett, W. S., Nardo, C. T., and Abbett, M. J., "Computational Procedure for Three-Dimensional Boundary Layers on Aircraft and Aerospace Vehicles," AIAA paper 1975.

[120]Cebeci, T., "A General Method for Calculating Three-Dimensional Incompressible Laminar and Turbulent Boundary Layer, II: Three-Dimensional Flow in Cartesian Coordinates," Rept. MDC J 6517, March 1974.

[121]Eiseman, P. and McDonald, H., "A Method for Computing Three-Dimensional Flow in Ducts," *Proceedings of the NASC Symposium of Aerodynamic Problems of VSTOL Aircraft*, 1975.

[122]Horlock, J. H. and Lakshminarayana, B., "Secondary Flows: Theory, Experiment and Application in Turbomachinery Aerodynamics," *Annual Review of Fluid Mechanics*, Vol. 5, 1973, pp. 247–280.

[123]Williams, G. P., "Numerical Integration of the Three-Dimensional Navier-Stokes Equations for Incompressible Flow," *Journal of Fluid Mechanics*, Vol. 37, 1969, pp. 727–750.

[124]Briley, W. R. and McDonald, H., "Solution of the Multidimensional Compressible Navier-Stokes Equations by a Generalized Implicit Method," *Journal of Computational Physics*, Vol. 234, 1977, pp. 372–397.

[125]Humphrey, J. A. C., Taylor, A. M. K., and Whitelaw, J. H., "Laminar Flow in a Square Duct of Strong Curvature," *Journal of Fluid Mechanics*, Vol. 83, 1977, pp. 509–527.

[126]Hah, C., "A Navier-Stokes Analysis of Three-Dimensional Turbulent Flows Inside Turbine Blade Rows at Design and Off-Design Conditions," ASME Paper 83-GT-40, 1983.

[127]Davis, R. T. and Rubin, S. G., "Non-Navier-Stokes Viscous Flow Computations," *Computers and Fluids*, Vol. 8, 1980, pp. 103–131.

[128]Rubin, G. O., "Computational Meshes for Boundary Layer Problems," *Proceedings of 2nd International Conference on Numerical Methods in Fluid Dynamics*, Springer-Verlag, New York, 1971, pp. 171–177.

[129]Johnston, J. P., "On the Three-Dimensional Turbulent Boundary Layer Generated by Secondary Flow," *Journal of Basic Engineering, Transactions of ASME*, Vol. 82, 1960, pp. 233–248.

[130]Stanitz, J. D., Osborn, W. M., and Mizisin, J., "An Experimental Investigation of Secondary Flow in an Accelerating Rectangular Elbow with 90 Degrees of Turning," NACA TN 3015, 1983.

[131]Courant, R. and Hilbert, D., *Methods of Mathematical Physics*, Vol. II, Interscience Publishers, New York, 1966.

[132]Garabedian, P. R., *Partial Differential Equations*, John Wiley & Sons, New York, 1964.

[133]Briley, W. R. and McDonald, H., "Three-Dimensional Viscous Flows with

Large Secondary Velocity," *Journal of Fluid Mechanics*, Vol. 144, July 1984, pp. 47–77.

[134]Blottner, F. G., "Numerical Solution of Slender Channel Laminar Flows," *Computer Methods in Applied Mechanics and Engineering*, Vol. 11, 1977, pp. 319–339.

[135]Anderson, O. L., "Calculation of Internal Viscous Flows in Axisymmetric Ducts at Moderate to High Reynolds Numbers," *Computers and Fluids*, Vol. 8, 1980, pp. 391–441.

[136]Baker, A. J. and Orzechowski, J. A., "An Interaction Algorithm for Three-Dimensional Turbulent Subsonic Aerodynamic Juncture Region Flow," *AIAA Journal*, Vol. 21, 1983, pp. 524–533.

[137]Anderson, O. L. and Hankins, G. B. Jr., "Development of a Parabolic Finite Difference Method for 3-D High Reynolds Number Viscous Internal Flows," *Computers in Flow Predictions and Fluid Dynamics Experiments*, ASME, New York, 1981, pp. 119–128.

[138]Patankar, S. V. and Spalding, D. B., "A Calculation Procedure for Heat Mass and Momentum Transfer in Three-Dimensional Parabolic Flows," *International Journal of Heat and Mass Transfer*, Vol. 15, 1972, pp. 1787–1806.

[139]Carreto, L. S., Curr, R. M., and Spalding, D. B., "Two Numerical Methods for Three-Dimensional Boundary Layers," *Computer Methods in Applied Mechanics and Engineering*, Vol. 1, 1973, pp. 39–51.

[140]Briley, W. R., "Numerical Method for Predicting Three-Dimensional Steady Viscous Flow in Ducts," *Journal of Computational Physics*, Vol. 14, No. 1, 1974.

[141]Ghia, K. N. and Sokhey, J. S., "Laminar Incompressible Viscous Flow in Curved Ducts of Regular Cross-Sections," *Journal of Fluids Engineering*, Vol. 99, 1977, pp. 640–648.

[142]Kreskovsky, J. P., Briley, W. R., and McDonald, H., "Analysis and Computation of Three-Dimensional Flow in Strongly Curved Ducts," *Computers in Flow Predictions and Fluid Dynamics Experiments*, ASME, New York, 1981, pp. 129–140.

[143]Moore, J. and Moore, J. G., "A Calculation Procedure for Three-Dimensional Viscous Compressible Duct Flow, Parts I and II," *Journal of Fluids Engineering*, Vol. 101, 1979, pp. 415–428.

[144]Pulliam, T. H. and Steger, J. L., "Implicit Finite-Difference Simulations of Three-Dimensional Compressible Flow," *AIAA Journal*, Vol. 18, 1980, pp. 159–167.

[145]Schiff, L. B. and Steger, J. L., "Numerical Simulation of Steady Supersonic Viscous Flow," *AIAA Journal*, Vol. 18, Dec. 1980, pp. 1421–1430.

[146]Govindan, Levy, and Shamroth, "Computation of the Tip Vortex Generation Process for Ship Propeller Blades," Office of Naval Research, Rept. R83-920021-F, 1984.

[147]Sanborn, V. A., "Boundary Layer Separation and Reattachment," *Fluid Mechanics, Acoustics and Design of Turbomachinery*, Pt. 1, NASA SP-304, 1974, p. 279.

[148]Crocco, L. and Lees, L., "A Mixing Theory for Interaction between Dissipative Flow and Nearly Isotropic Streams," *Journal of the Aeronautical Sciences*, Vol. 19, No. 10, 1962.

[149]Garvine, R. W., "Upstream Influence in Viscous Interaction Problems," *Physics of Fluids*, Vol. 11, 1968, pp. 1413–1423.

[150]Stewartson, K. and Williams, P. G., "Self Induced Separation," *Proceedings of the Royal Society of London*, Ser. A, Vol. 312, 1969, pp. 181–206.

[151]Neiland, V. Y., *Mekhanizatsiya, Gaza* Vol. 4, 1969, p. 40.

[152]Messiter, A. F., *SIAM Journal on Applied Mathematics*, Vol. 18, 1970, p. 241.

[153]Lighthill, M. J., "On Boundary Layers and Upstream Influence," *Proceedings of the Royal Society of London*, Vol. 217, 1953, pp. 478–507.

[154]Stewartson, K., "Multistructural Boundary Layers," *Advances in Applied Mechanics*, Vol. 14, 1974, pp. 145–239.

[155]Werle, M. J. and Vatsa, V. N., "Numerical Solution of Interacting Supersonic Boundary Layer Flows Including Separation Effects," ARL TR 73-0162, Dec. 1973; also *AIAA Journal*, Vol. 12, Nov. 1974, pp. 1491–1497.

[156]Burggraf, O. R., Rizzetta, D., Werle, M. J., and Vatsa, V. N., "Effect of Reynolds Number on Laminar Separation of a Supersonic Stream," *AIAA Journal*, Vol. 17, Apr. 1979, pp. 336–343.

[157]Davis, R. T., "Numerical and Approximate Solution of the High Reynolds Number Small Separation Problem," Dept. of Aeronautical Engineering, University of Cincinnati, Ohio, Rept. AFL 76-8-24, 1976.

[158]Davis, R. T. and Rubin, S. G., "Non-Navier Stokes Viscous Flow," *Computations, Computers and Fluids*, Vol. 8, 1980, pp. 101–131.

[159]Briley, W. R. and McDonald, H. "Numerical Predictions of Incompressible Separation Bubbles," *Journal of Fluid Mechanics*, Vol. 69, Pt. 4, 1975, pp. 631–656.

[160]Kwon, O. K. and Pletcher, R. H., "Prediction of Incompressible Separated Boundary Layers Including Viscous-Inviscid Interaction," *Journal of Fluids Engineering*, Vol. 101, No. 4, 1979, pp. 466–472.

[161]Cebeci, T. and Schimke, S., "The Calculation of Separation Bubbles in Interactive Turbulent Boundary Layers," *Journal of Fluid Mechanics*, 1982.

[162]Carter, J. E. and Vatsa, V. N., "Analysis of Airfoil Leading Edge Separation Bubbles," NASA CR 165935 (Contract NAS1-16585), 1982.

[163]LeBalleur, J. C., "Couplage Visqueux-Non Visqueux: Methode Numerique et Applications aux Ecoulements Bidimensionnels Transoniques et Supersoniques," *Recherche Aerospatiale*, Vol. 183, 1978, pp. 65–76 (English translation ESA TT496).

[164]Carter, J. E., "A New Boundary Layer Inviscid Iteration Technique for Separated Flow," AIAA Paper 79-1450, 1979.

[165]Carter, J. E., "Viscous-Inviscid Interaction Analysis of Transonic Turbulent Separated Flow," AIAA Paper 81-1241, 1981.

[166]Carter, J. E., "Solutions for Laminar Boundary Layers with Separation and Reattachment," AIAA Paper 74-583, 1974; also NASA SP-347, 1975.

[167]Rizzetta, D. P., Burggraf, O. R., and Jenson, R., "Triple-Deck Solutions for Viscous Supersonic and Hypersonic Flow Past Corners," *Journal of Fluid Mechanics*, Vol. 89, Pt. 3, 1978, pp. 535–552.

[168]Ghia, K. N., Ghia, U., and Tesch, W. A., "Evaluation of Several Approximate Models for Laminar Incompressible Separation by Comparison with Complete Navier-Stokes Solutions," *Flow Separation*, AGARD CP-168, 1975, pp. 6-1—6-15.

[169]Murphy, J. D., Presley, L. L., and Rose, W. C., "On the Calculation of Supersonic Separating and Reattaching Flows," NASA SP-347, 1975.

[170]Vatsa, V. N., Carter, J. E., and Swanson, R. C., "Comparison of Solutions of the Navier-Stokes and Interacting Boundary-Layer Equations for Separated Turbulent Flow," Paper presented at ISCME International Conference on Computational Methods and Experimental Measurement, Washington, D.C., 1982.

[171]Briley, W. R., "A Numerical Study of Laminar Separation Bubbles Using the Navier-Stokes Equations," *Journal of Fluid Mechanics*, Vol. 47, 1971, p. 713.

[172]Leal, L. G., "Steady Separated Flow in a Linearly Decelerated Free Stream," *Journal of Fluid Mechanics*, Vol. 59, 1973, p. 513.

[173]Ghia, U. and Davis, R. T., "Navier-Stokes Solutions for Flow Past a Class of Two-Dimensional Semi-Infinite Bodies," *AIAA Journal*, Vol. 12, Dec. 1974, pp. 1659–1665.

[174]Werle, M. J. and Davis, R. T., "Incompressible Laminar Boundary Layers on a Parabola at Angle of Attack: A Study of the Separation Point," *SIAM Journal on Applied Mathematics*, Vol. 7, 1972.

[175]Klineberg, J. M. and Steger, J. L., "On Laminar Boundary Layer Separation," AIAA Paper 74-94, 1974.

[176]Williams, P. G., "A Reverse Flow Computation in the Theory of Self-Induced Separation," *Proceedings of 4th International Conference on Numerical Methods in Fluid Mechanics*, *Lecture Notes in Physics*, Springer-Verlag, New York, Vol. 35, 1975, p. 445.

[177]Carter, J. E. and Wornom, S. F., "Solutions for Incompressible Separated Boundary Layers Including Viscous-Inviscid Interaction," NASA SP-347, 1975.

[178]Veldman, A. E. P., "New, Quasi-Simultaneous Method to Calculate Interacting Boundary Layers," *AIAA Journal*, Vol. 19, Jan. 1981, pp. 79–85.

[179]Baum, E. and Denison, M. R., "Interacting Supersonic Laminar Wake Calculations by a Finite Difference Method," AIAA Paper 66-454, 1966.

[180]Cousteix, J., LeBalleur, J.-C., and Houdeville, R., "Calculation of Unsteady Turbulent Boundary Layers in Direct or Inverse Mode, Including Reverse Flows—Analysis of Singularities," *La Recherche Aerospatiale*, 1980.

[181]Levy, R., Briley, W. R., and McDonald, H., "Viscous Flow in the Region of a Rounded Trailing Edge," NASC Rept. N00019-78-C-0470, 1979.

[182]Launder B. E. and Spalding, D. B., *Mathematical Models of Turbulence*, Academic Press, New York, 1972.

[183]Bradshaw, P., "The Understanding and Prediction of Turbulent Flow," *Aeronautical Journal*, 1972, pp. 403–418; also Imperial College, London, Rept. 73-05, 1973.

[184]Ng, H. K., "Predictions of Turbulent Boundary-Layer Developments Using a Two Equation Model of Turbulence," Ph.D. Thesis, Dept. of Mechanical Engineering, Imperial College, London, TM/TN/E/12, Nov. 1971.

[185]Reynolds, W. C., "Computation of Turbulent Flow," Dept. of Mechanical Engineering, Stanford University, Stanford, Calif., Rept. FT-4, 1975.

[186]Mellor, G. L. and Herring, H. J., "A Survey of the Mean Turbulent Field Closure Methods," *AIAA Journal*, Vol. 11, May 1973, p. 590.

[187]Bradshaw, P., "Outlook for Three Dimensional Procedures," *Proceedings: Computation of Turbulent Boundary Layers*, AFOSR-IFP Stanford Conference, Vol. I, 1968.

[188]Johnson, J. P., "The Effects of Rotation on Boundary Layers in Turbomachine Rotors," *Fluid Mechanics, Acoustics and Design of Turbomachinery*, Pt. 1, NASA SP-304, 1974.

[189]Bradshaw, P., "The Effects of Streamline Curvature on Turbulent Flow," AGARDograph 169, 1973.

[190]Hughes, D. W. and Horlock, J. H., "Effect of Rotation on the Development of the Turbulent Boundary Layer," Paper 18 presented at Salford Conference on Internal Flows.

[191]Donaldson, C. DuP., "A Computer Study of an Analytical Model of Boundary Layer Transition," *AIAA Journal*, Vol. 7, Feb. 1969, p. 271.

[192]Deardorff, J. W., "Numerical Investigation of Neutral and Unstable Planetary

Boundary Layers," *Journal of the Atmospheric Sciences*, Vol. 29, 1972, p. 91.

[193]Batchelor, G. K., "Note on Free Turbulent Flows with Special Reference to the Two-Dimensional Wake," *Journal of the Aeronautical Sciences*, Vol. 17, 1950, p. 441.

[194]Townsend, A. A., *The Structure of Turbulent Shear Flow*, Cambridge University Press, Cambridge, England, 1956.

[195]Prandtl, L., "Über ein neues formelsystem der ausgebildeten Turbulenz," in *Nachrichten der Academie der Wissenschaften Göttingen*, *Mathematisch-Physikalische Klasseza*, 1945, p. 6.

[196]Corrsin, S., "Limitations of Gradient Transport Models in Random Walks and Turbulence," *Advances in Geophysics*, Vol. 18A, Academic Press, New York, 1974.

[197]Escudier, M. P., "The Distribution of the Mixing Length in Turbulent Flows Near Walls," Mechanical Engineering Dept., Imperial College, London, Rept. TWF/TN/1, 1965.

[198]McDonald, H., "Mixing Length and Kinematic Eddy Viscosity in a Low Reynolds Number Boundary Layer," United Aircraft Research Laboratories, Rept. J214453-1, July 1970.

[199]Cebeci, T. and Mosinskis, G. J., "Computation of Incompressible Turbulent Boundary Layers at Low Reynolds Numbers," *AIAA Journal*, Vol. 9, Aug. 1971, pp. 1632–1634.

[200]Simpson, R. L., "Characteristics of Turbulent Boundary Layers at Low Reynolds Numbers with and without Transpiration," *Journal of Fluid Mechanics*, Vol. 9, No. 8, 1974, p. 1632.

[201]Huffman, G. D. and Bradshaw, P., "A Note on von Karman's Constant in Low Reynolds Number Turbulent Flow," *Journal of Fluid Mechanics*, Vol. 53, 1972, p. 45.

[202]Kline, S. J., Lisin, A. V., and Waitman, B. A., "Preliminary Experimental Investigation of Effect of Free Stream Turbulence on Turbulent Boundary Layer Growth," NASA TND-368, 1960.

[203]Charnay, G., Comte-Bellot, G., and Mathieu, J., "Development of a Turbulent Boundary Layer on a Flat Plate in an External Flow," AGARD CP-93, 1972.

[204]Huffman, G. D., Zimmerman, D. R., and Bennet, W. A., "The Effect of Free Stream Turbulence Level on Turbulent Boundary Layer Behavior," Paper presented at AGARD Specialists Meeting, Boundary Layer Effects in Turbomachines, 1972.

[205]Bayley, F. J., Morris, W. D., Owen, J. M., and Turner, A. B., "Boundary-Layer Prediction Methods Applied to Cooling Problems in the Gas Turbine," British Aeronautical Research Council, CP 1164, 1971.

[206]Galbraith, R. A. McD. and Head, M. R., "Eddy Viscosity and Mixing Length from Measured Boundary Layer Developments," *Aeronautical Quarterly*, Vol. XXVI, May 1975.

[207]Glowacki, W. J. and Chi, S. W., "A Study of the Effect of Pressure Gradient on the Eddy Viscosity and Mixing Length for Incompressible Equilibrium Turbulent Boundary Layers," Naval Ordnance Laboratory, Rept. NOLTR 74-105, May 1974; also AIAA Paper 72-213, 1972.

[208]Reeves, B. L., "Two Layer Model of Turbulent Boundary Layers," *AIAA Journal*, Vol. 12, July 1974, pp. 932–939.

[209]Townsend, A. A., "Equilibrium Layers and Wall Turbulence," *Journal of Fluid Mechanics*, Vol. II, Pt. 1, p. 97.

[210]Jayatelleke, C. L. V., "The Influence of Prandtl Number and Surface Rough-

ness on the Resistance of the Laminar Sub-Layer to Momentum and Heat Transfer," *Progress in Heat and Mass Transfer*, Vol. 1, Pergamon Press, New York, 1969, p. 193.

[211]Van Driest, E. R., "On Turbulent Flow Near a Wall," Heat Transfer and Fluid Mechanics Institute, 1956.

[212]Mellor, G. L., "The Effect of Pressure Gradients on Turbulent Flow Near a Smooth Wall," *Journal of Fluid Mechanics*, Vol. 24, 1966, pp. 255–274.

[213]Cebeci, T., "Behavior of Turbulent Flow Near a Porous Wall with Pressure Gradient," *AIAA Journal*, Vol. 8, Dec. 1970, pp. 2152–2156.

[214]Crawford, M. E. and Kays, W. M., "STAN5—A Program for Numerical Computation of Two-Dimensional Internal and External Boundary Layer Flows," NASA CR 2742, Dec. 1976.

[215]Kreskovsky, J. P., Shamroth, S. J., and McDonald, H., "Application of a General Boundary Layer Analysis to Turbulent Boundary Layers Subjected to Strong Favorable Pressure Gradients," *Journal of Fluids Engineering, Transactions of ASME*, June 1975.

[216]Simpson, R. L., Whitten, D. G., and Moffat, R. J., "An Experimental Study of the Turbulent Prandtl Number of Air with Injection and Suction," *International Journal of Heat and Mass Transfer*, Vol. 13, 1970, p. 125.

[217]Pai, B. R. and Whitelaw, J. H., "The Prediction of Wall Temperature in the Presence of Film Cooling," *International Journal of Heat and Mass Transfer*, Vol. 14, 1971, pp. 409–426.

[218]Nee, V. and Kovasznay, L. S. G., "The Calculation of the Incompressible Turbulent Boundary Layers by a Simple Theory," *Proceedings: Computation of Turbulent Boundary Layers*, AFOSR-IFP Stanford Conference, Vol. I, 1968.

[219]Patel, V. C. and Nash, J. F., "Some Solutions of the Unsteady Two-Dimensional Turbulent Boundary Layer Equations," *Recent Research on Unsteady Boundary Layer, IUTAM Symposium*, Vol. 1, Laval University Press, Quebec, Canada, 1971, pp. 1106–1164.

[220]Glushko, G. S., "Turbulent Boundary Layer on a Flat Plate in an Incompressible Fluid," *Izvestiya Akademii Nauk SSSR, Mekhanika*, Vol. 4, 1964, p. 13 (translation, NASA TT F-10,080).

[221]Jones, W. P. and Launder, B. E., "The Calculation of Low Reynolds Number Phenomena with a Two-Equation Model of Turbulence," *International Journal of Heat and Mass Transfer*, Vol. 16, 1973, p. 1189.

[222]Hanjalic, K. and Launder, B. E., "Contribution Towards a Reynolds Stress Closure for Low Reynolds-Number Turbulence," Mechanical Engineering Dept., Imperial College, London, Rept. HTS/74/24; also *Journal of Fluid Mechanics*, Vol. 74, Pt. 4, 1976, pp. 593–610.

[223]Saffman, P. G., "A Model for Inhomogenius Turbulent Flow," *Proceedings of the Royal Society of London*, Ser. A, Vol. 317, 1970, pp. 417–433.

[224]Rotta, J. C., "Über eine Methode zur Berechnung Turbulenz Scherstromungsfelder," Aerodynamische Versuchsanstalt, Rept. 69A14, 1969.

[225]Spalding, D. B., "Concentration Fluctuations in a Round Free Turbulent Jet," *Chemical Engineering Science*, Vol. 26, 1971, p. 95.

[226]Pope, S. B. and Whitelaw, J. H., "The Calculation of Near-Wake Flows," *Journal of Fluid Mechanics*, Vol. 73, Pt. 1, 1976, pp. 9–32.

[227]Humphrey, J. A. C., Taylor, A. M. K., and Whitelaw, J. H., "Laminar Flow in a Square Duct of Strong Curvature," *Journal of Fluid Mechanics*, 1977.

CHAPTER 7. ENGINE NOISE

Nicholas A. Cumpsty
University of Cambridge, England, UK

7. ENGINE NOISE

7.1 Introduction

The advent of the jet engine brought with it a new noise nuisance and its introduction into civil airline use was a major disaster for communities living near airports. This is indicated by Fig. 7.1, which shows how the area at a given level of subjective noisiness—the so-called noise footprint—has changed over the years. (More will be said about the subjective assessment of noise in Sec. 7.2.)

The jets in the 1960s brought about a sixfold increase in the footprint area, but the more recent civil aviation engines with high bypass ratios are restoring conditions to what they were with propeller aircraft. However, in situations where the propulsive efficiency requires a low bypass ratio and high jet velocity, the noise levels remain very high.

At first, the predominant source of noise was produced by high-speed jet mixing, but with the introduction of low-bypass-ratio engines the compressor rapidly became every bit as irritating in some circumstances. In the more recent high-bypass-ratio engines, the fan is usually the predominant noise nuisance, but this is mitigated to some extent by the incorporation of noise-reducing features. The changes in the sources with bypass ratio are indicated in Fig. 7.2. The size of the lobes gives an indication of the strength of the source and the direction indicates the direction in which radiation occurs. The direction of sound radiation may be as important as the intensity of radiation; if radiation can occur away from the ground, no community complaints are likely to be received.

As an aircraft flies overhead, the sound changes with time, of course. The levels change both because of the variation in distance between the aircraft and the observer, but also because of the angle changes. Figure 7.3 gives some idea of how the various sources contribute as the aircraft passes overhead. The precise variation in noise with time depends on the engine, the installation, the aircraft, and the mode of operation; nevertheless, the salient features are shown in Fig. 7.3. The noise time histories of flyovers of one aircraft with a high-bypass-ratio engine at approach, takeoff, and at throttled takeoff power are shown in Fig. 7.4. It should be noted that the throttled takeoff extends the period of high noise level, so that in certain cases its use may be found to increase the noise nuisance.

The aircraft type influences the noise sensed on the ground directly through the time for which the noise persists and the nearness of the aircraft. Since two-engine aircraft climb faster than three-engine ones, and three faster than four, the takeoff noise experienced tends to be lower with fewer engines. Wing design affects the drag and power at approach; aircraft with low drag tend to have lower noise at approach. Mounting the engines

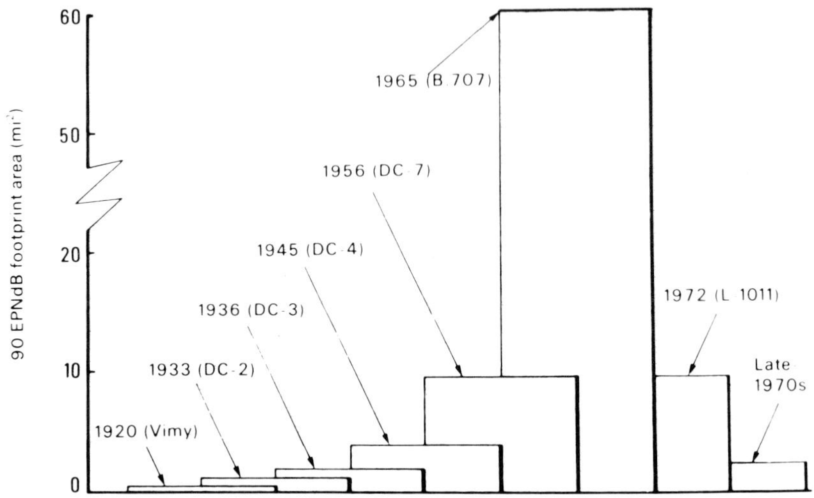

Fig. 7.1 Changes in 90 EPNdB footprint area over 50 years (from Interavia).

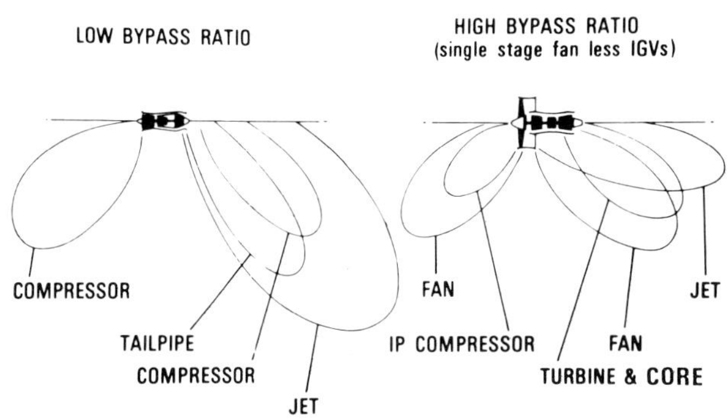

Fig. 7.2 Comparative noise sources for low- and high-bypass-ratio engines.

on top of the wings can also be highly beneficial for engine noise, since the wing can shield the ground. Naturally, there are endless permutations for different aircraft designs and engine installations and it is inappropriate to consider this topic in any detail here. The over-riding effect is, of course, the aircraft duty and the bypass ratio and jet velocities that this determines.

The experimental study of noise is difficult in general, but in the case of the high-speed machinery normal for aircraft engines, it is both very difficult and very expensive. The machinery in aircraft engines typically operates at velocities close to sonic and it also turns out that the acoustic wavelengths are normally of the same order as the blade chords. This latter tendency makes the localization of the sound sources by ray techniques inapplicable. The high Mach number flows refract the sound, further complicating the study. The insertion of a probe into the flow can itself initiate the generation

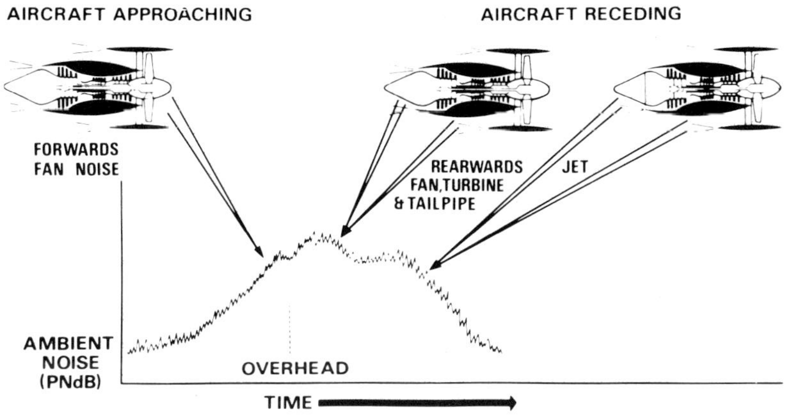

Fig. 7.3 Change in dominant noise source and variation in noise level during aircraft flyover.

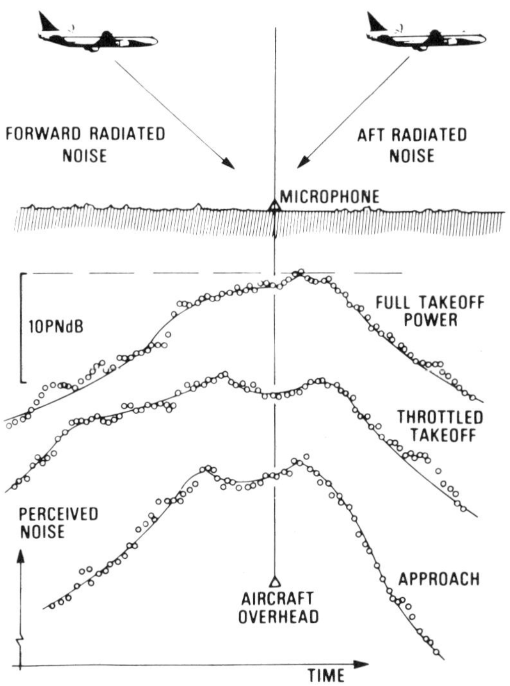

Fig. 7.4 Three flyovers for an aircraft with high bypass ratio.

of sound by adjacent moving blades. A picture for the noise generation has gradually been built up by noting the changes occurring as the parameters are altered, but this remains an erratic means of progress.

The theoretical study of noise, or acoustics, in the case of aircraft engines presents special problems because of the same two factors. First, the flow Mach numbers are generally not far from unity, which complicates the analytic description of acoustic processes enormously. Second, the wavelength of the sound is typically comparable to the physical dimensions of the machine, such as blade chord, also greatly complicating any analysis. For both these reasons, classical acoustics studies found in textbooks have not contributed very much to the growth in the understanding of the subject.

A different theoretical approach to the problem, known as aeroacoustics and described by Goldstein,[1] began with the work of Lighthill[2] toward the understanding of noise from the turbulence of jets. Over the last 20 years or so, this has been extended and generalized to cover even the noise from arbitrarily moving surfaces.[3] Despite its generality, this approach leads to what an engineer might regard as a mathematical statement of the problem rather than a solution. To provide a solution requires a range of information that is unfortunately unavailable, and quantitative results are obtainable only by incorporating the most sweeping assumptions. Therefore, the aeroacoustic approach provides less help than one might expect, and the growth of understanding is dependent on a more ad hoc approach.

Having drawn attention to the difficulties of theoretical studies of noise, it is proper to point out one ameliorating factor—that the amplitudes of the fluctuations normally of concern as noise are very small and that a linear analytic treatment is almost always justified. It is, therefore, a field that attracts considerable interest from theoreticians.

The development of high-bypass-ratio engines for transport aircraft in the late 1960s and the international recognition of the need to limit or reduce noise led to a surge of interest in noise research. The literature grew accordingly and it would be out of place to reference all of it in this chapter; instead, some excellent review articles will be referenced as well as works that are of particular importance to anyone wishing to study the subject further.

Before considering noise sources from particular engine components, there are two more general sections bearing on what follows. In Sec. 7.2, the methods of rating noise are described, the essential problem being to find objective measures of a subjective annoyance. In Sec. 7.3, the elements of duct acoustics are treated, since this bears on all aspects of noise generation and on the attenuation due to liners.

The largest section of the chapter (7.4) is concerned with compressor and fan noise. As well as being the most studied aspect, it is also probably the most serious noise nuisance after the jet itself. The noise from turbines is in some ways similar to that of compressors and it is examined in Sec. 7.5. Except for jet noise, all the other noise sources, sometimes lumped together as core noise, are considered in Sec. 7.6. The use of acoustic treatment on the walls of the ducting around the engines is now common. This is an area where the designer can exercise a very large beneficial effect and where design is both relatively subtle and precise. Some aspects of this are described in Sec. 7.7. Prospects for the future and an overview of achievements are briefly considered in Sec. 7.8.

7.2 Scales and Ratings for Noise

Because acoustic signals vary by such large amounts it is the almost invariable practice to use logarithmic scales, expressing the level in terms of decibels (dB). Acoustic signal levels may refer either to the local level of pressure fluctuation or to the acoustic power generated. In the former, it is known as the sound pressure level (SPL) defined by

$$\mathrm{SPL} = 20 \log_{10}(p/p_{\mathrm{ref}})\ \mathrm{dB}$$

where p is the rms pressure fluctuation and p_{ref} is the reference rms pressure fluctuation equal to $2\mu\mathrm{Pa}$ (2.10^{-5} dyne/cm^2).

The acoustic power, or power watt level (PWL), is defined by

$$\mathrm{PWL} = 10 \log_{10}(W/W_{\mathrm{ref}})\ \mathrm{dB}$$

where W is the acoustic power in watts and W_{ref} is the reference acoustic power equal to 10^{-12} W. (In some work, 10^{-13} W is used as the reference.)

The acoustic power carried by unit area of wave front is known as the intensity. Generally, this is not a convenient quantity to use and either the pressure amplitude (or SPL) or the total acoustic power (or PWL) is more common.

For a plane wave, the intensity, power, and pressure level are conveniently related (a plane wave is one in which the wave front is flat and perpendicular to the direction of travel). For the plane wave, the intensity, or power per unity area, is given by $w = pu$ (in W/m^2), u being the rms velocity associated with the rms pressure p. For the plane wave, $u = p/\rho c$, ρ and c being the density and velocity of sound in the medium, respectively. For air at STP (0°C and 101 kPa) $\rho c = 428\ \mathrm{kg \cdot m^{-2} \cdot s^{-1}}$. The acoustic intensity corresponding to the reference rms pressure, $\rho_{\mathrm{ref}} = 2\mu\mathrm{Pa}$ is therefore given by

$$w = p_{\mathrm{ref}}^2/\rho c\ \mathrm{W/m^2}$$

$$= 4.10^{-10}/428 \approx 10^{-12}\ \mathrm{W/m^2}$$

$$\approx W_{\mathrm{ref}},\ \text{the reference power per unit area}$$

In other words, for a plane wave the levels expressed in terms of intensity and sound pressure are sufficiently close to equal to be regarded as equivalent. For more complicated waves, such as spiraling waves in ducts or in

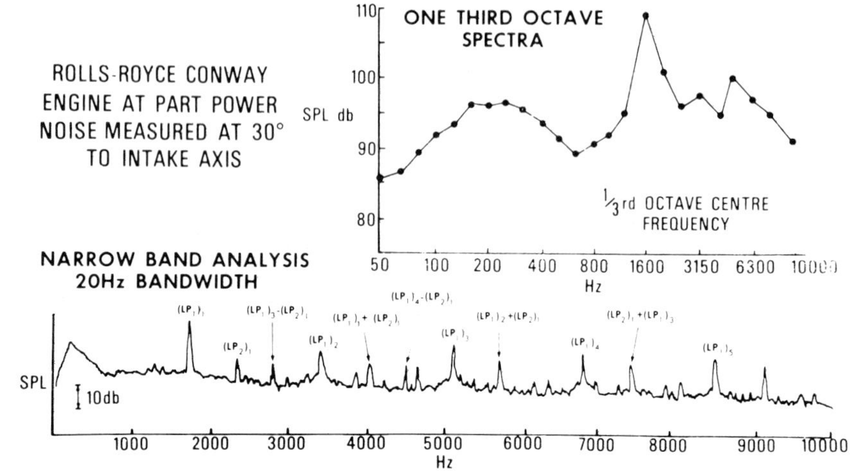

Fig. 7.5 Single noise signal analyzed by a constant bandwidth and a third-octave filter.

the near field close to the sources, the connection between power, intensity, and sound pressure is no longer simple.

The signals of pressure or power will have a frequency spectrum and the level referred to may be either the total signal (i.e., the spectrum integrated over the frequency) or the components of the spectrum at particular frequencies. If the overall signal is used to obtain the rms pressure or power, the levels are generally known as the overall sound pressure level (OASPL) or overall power watt level (OAPWL), respectively. Very frequently, the signal is broken down into its spectral components, in which case it is necessary to specify the bandwidth over which the signal is summed. Mathematically the process is a Fourier transform. Fourier transforms produce power spectral density, for which the bandwidth is 1 Hz, but electrical analog filters or computer algorithms usually have a different bandwidth. The bandwidth may be constant, regardless of the frequency being examined, or it may vary with frequency, usually so that the bandwidth is proportional to the frequency. The most common bandwidths of this type are the octave and third octave. An octave corresponds to a doubling of frequency; a third-octave bandwidth is therefore one where the upper frequency is equal to $\sqrt[3]{2}$ times the lower frequency. Depending on the choice of bandwidth, very large differences in the displayed spectrum can be obtained. This is illustrated by Fig. 7.5, in which the same signal was analyzed with a constant bandwidth and with a third-octave filter. Evidently quite different information is emphasized in the two traces. The narrow, constant bandwidth filter is useful for study and diagnosis, while the third-octave filter more accurately reflects the discernment of the human ear.

So far, the discussion has referred to objective measures. Noise, however, is inherently subjective. To assess noise one requires an evaluation of the

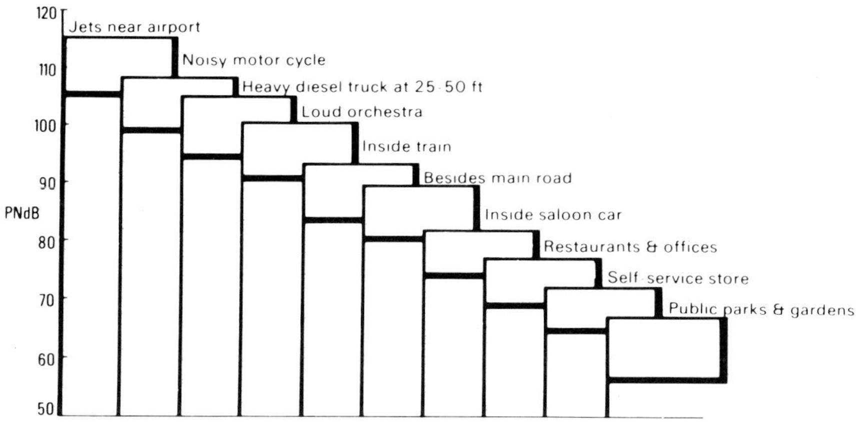

Fig. 7.6 Perceived noise levels for representative situations (from Interavia).

annoyance produced on a person or, to be more reliable, a large group of people, since no two persons will be affected in quite the same way. The different ratings used throughout the world have been collected and described by Pearsons and Bennett.[4] The simplest ratings weight the overall sound pressure level in a manner related to the sensitivity of the human ear. The most common of these, the A-weighted sound pressure level, finds wide use in nonaeronautical as well as aeronautical applications throughout the world. The levels are seen expressed as dBA or dB(A). Although less common, there are also B, C, and D scales intended for different levels and conditions. A great advantage of the A–D-weighted levels is that they can be produced by an electrical circuit to provide immediate direct readings.

A more complicated noise rating, which is also described in detail in Ref. 4, is the perceived noise level (PNL) for which the units are PNdB. The numerical value of PNL was intended to represent the sound pressure level of the octave band of "white noise" centered at 1 kHz that would be judged equally noisy as the sound to be rated. A sound is judged equally noisy to another when one would just as soon have or not have it in one's home during the night or day. This definition emphasizes the essentially subjective nature of the scale. Unfortunately, the calculation of PNL is more complicated and cannot be produced by an electrical circuit. The noise must first be analyzed into third-octave sound pressure levels (SPL) before appropriate corrections for amplitude and frequency are applied. To make the values of PNL more tangible, Fig. 7.6 compares levels of representative situations. This figure also emphasizes that, with modest reductions in aircraft noise, the prime nuisance for most of the population shifts to other sources.

The perceived noise level was found to be less reliable as an indicator of annoyance when the spectrum was no longer broadband but contained a strong tone. A tone-corrected perceived noise level (PNLT) was therefore produced. The third-octave spectrum is first examined to determine if tones

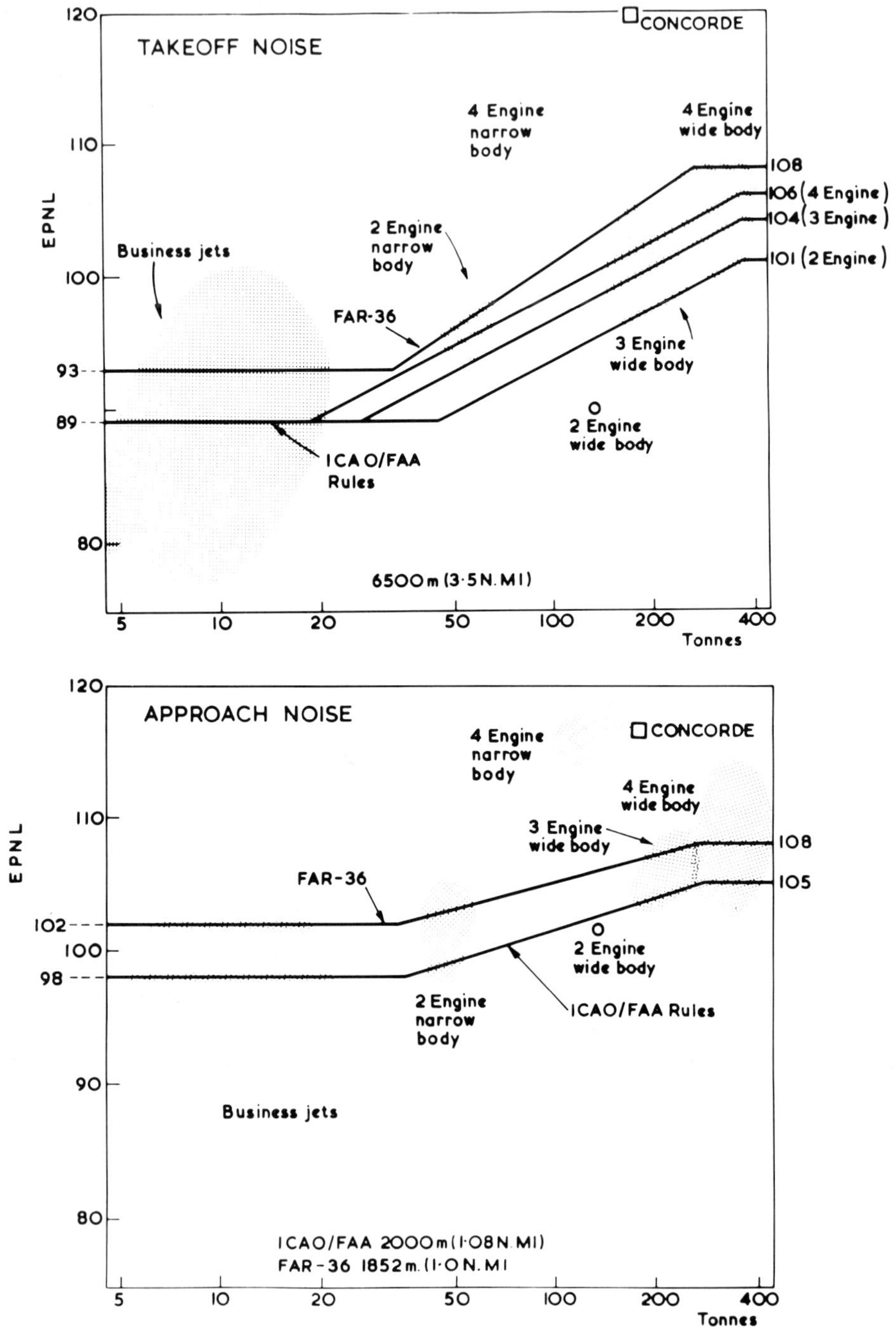

Fig. 7.7 Takeoff and approach noise levels—regulations and typical measurements (abscissa shows all-up weight).

are present, evident by the protrusion of one band above its neighbors by a prescribed ratio. If a tone is present, a correction is added to the PNL calculated in the normal way, the amount of the correction depending on frequency and on the extent of the protrusion.

The noise from aircraft is of short duration and the annoyance caused is related to the time over which the noise level is high. To take this into account the effective perceived noise level (EPNL) was evolved. The tone-corrected perceived noise level is calculated at 0.5 s intervals and these are summed according to the following expression to yield EPNL:

$$\text{EPNL} = 10\log \sum_{i=0}^{d} \text{antilog}\,\frac{\text{PNLT}_i}{10} - 13$$

where PNLT_i is the tone-corrected perceived noise level after the ith 0.5 s interval. The summation is over the time for which the PNLT is within 10 dB of the peak level, levels outside this being ignored.

The effective perceived noise level provides the basis for the recent noise regulations. The regulations specify maximum EPNL as a function of all-up weight for sideline, takeoff (flyover), and approach. The sideline noise is the maximum anywhere along a line parallel to and at a fixed distance from the runway. The takeoff noise is prescribed at a fixed distance from the start of roll and the approach noise is at a fixed distance from the threshold along a prescribed glide path. It turns out that meeting the takeoff and approach noise limits has created the greatest challenge.

Figure 7.7 shows the levels of the regulations at takeoff and approach. The highest line on each corresponds to the United States FAR Part 36 regulations that first put compelling pressure on manufacturers to reduce the noise of civil aircraft. The lower lines are the more recent and internationally agreed upon ICAO/FAA rules; the distances on the bottom of each part of Fig. 7.7 show the distances at which the criteria are specified. To take account of the reduced rate of climb with larger numbers of engines, the rules allow for differing takeoff levels depending on the number of engines.

The importance of aircraft type is borne out by the measured levels superimposed on Fig. 7.7. The hatched regions contain a wide range of results. The older jets are described as "narrow body" to distinguish them from the more recent "wide-body" aircraft. Essentially it is the engines, and not the aircraft, that have brought about the noise change.

The accurate measurement of noise from aircraft in flight is difficult and costly. A practical scheme for doing this is illustrated in Fig. 7.8. Here, the second, third, and fourth steps are those that evaluate the level of EPNL; the later steps are corrections for atmospheric absorption and nonstandard aircraft conditions.

7.3 Introduction to Acoustics of Ducts

To understand the noise of turbomachinery, it is essential to understand first the propagation processes of acoustic waves in the upstream and

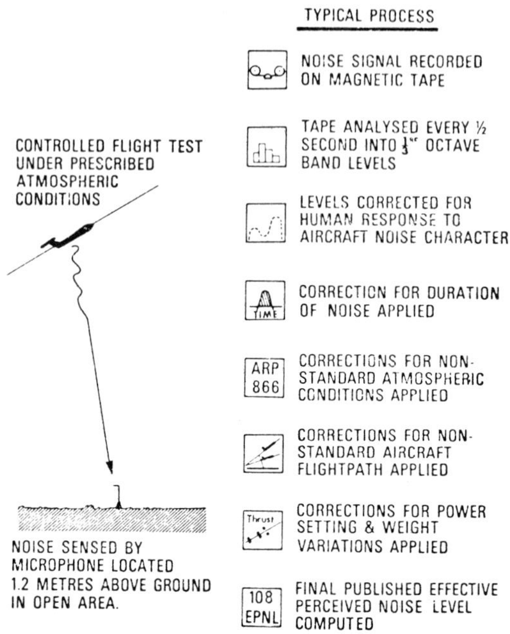

Fig. 7.8 Measurement, analysis, and correction procedures for in-flight noise measurement leading to certification.

downstream ducts. The need for this consideration arises because the pressure waves are not all uniform across the annulus or duct and do not propagate in the purely axial direction. In fact, the pressure varies circumferentially and radially in amplitude and phase and the wave fronts spiral along the duct. Consideration of the acoustic linings also requires an understanding of duct acoustics, but the presence of acoustically soft walls (i.e., walls where the normal velocity perturbations do not vanish) introduces great complications.

Downstream of a fan or compressor the duct is almost always annular, but upstream it is generally a circular cylinder. The circular nature of the problem determines the equation describing the propagation and the type of functions obtained. When the hub-to-tip ratio σ becomes sufficiently high, it is often possible to unwrap the circular annulus into an equivalent rectangular duct, thereby simplifying the problem somewhat.

The first comprehensive treatment of the properties of the pressure field in both rectangular and circular ducts was given by Tyler and Sofrin.[5] A few years later, another account was given by Morfey. These are clear and complete and it would be inappropriate to consider the details here. Because of its importance, it is worthwhile to summarize the most important points to remind readers of the subject, but a more serious study should include reading more detailed accounts. For a circular duct with no mean flow, the homogeneous wave equation can be written as

$$\frac{\partial^2 p}{\partial r^2} + \frac{1}{r} \cdot \frac{\partial p}{\partial r} + \frac{1}{r^2} \cdot \frac{\partial \rho}{\partial \theta^2} + \frac{\partial p}{\partial x^2} - \frac{1}{c^2} \frac{\partial p}{\partial t^2} = 0$$

The homogcncous equation is used (i.e., acoustic sources are not considered) because the propagation is so important in its own right.

This equation can be solved by separating the variables so that the pressure fluctuation p is written

$$p = \Theta(\theta) \cdot R(r) \cdot X(x) \cdot T(t)$$

where each function acts only on the variable shown. Four separate equations are then produced for Θ, R, X, and T, with constants interrelating them. The variation with respect to time $T(t)$ is given by $e^{i\omega t}$ and the requirement that Θ repeat around the circumference determines that $\Theta = e^{im\theta}$, in which m is any integer including zero. The circumferential variation is therefore generally sinusoidal in form. By substituting for m in the equation for R, the familiar Bessel ordinary differential equation is produced for which the solutions are of the form

$$R(r) = J_m(k_{m\mu} r) + Q_{m\mu} Y_m(k_{m\mu} r)$$

where J_m and Y_m are Bessel functions of the first and second kind of order m, respectively, and $k_{m\mu}$ and $Q_{m\mu}$ are eigenvalues for the particular mode and particular hub-tip ratio. The mode consists of circumferential order m (i.e., m cycles or lobes around the circumference) and radial order μ (i.e., μ nodes or zero crossings in the radial sense). When the hub-tip ratio is zero, that is to say the duct is a circular cylinder, $Q_{m\mu}$ is equal to zero and only Bessel functions of the first kind are involved. In the case of ducts with hard walls, the boundary condition that there can be no velocity normal to the wall gives

$$\frac{\partial p}{\partial r} = \frac{\partial R}{\partial r} = 0$$

which makes it possible to calculate $k_{m\mu}$ and $Q_{m\mu}$. Tyler and Sofrin[5] have tabulated $k_{m\mu}$ and $Q_{m\mu}$ for a range of cases. Figure 7.9 shows some of the radial patterns they presented for a hub-tip ratio of 0.5 for various values of circumferential and radial orders m and μ, respectively.

Negative pressure corresponds to perturbations in the antiphase to that at the outer wall; note that the radial order denotes the number of zero crossings. As the circumferential order becomes large, the region of high amplitude tends to collect near the outer wall.

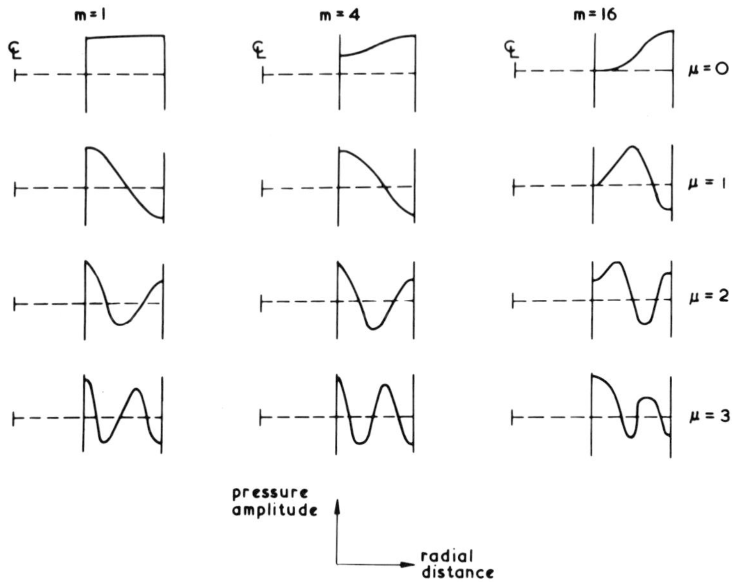

Fig. 7.9 Radial variation in pressure amplitude in a cylindrical duct with 0.5 hub-tip ratio. Circumferential and radial orders of mode denoted by *m* and μ, respectively (from Ref. 5).

The radial eigenvalue $k_{m\mu}$ appears in the equation for the axial variation $X(x)$. Writing $X(x)$ in the form $\exp(ik_x X)$, the axial wave number is then given by

$$k_x = \frac{1}{r}\sqrt{\frac{\omega^2}{c^2} - k_{m\mu}^2}$$

Thus, when $\omega/c > k_{m\mu}$, the axial variation is harmonic with no decrease in amplitude. When $\omega/c < k_{m\mu}$, however, the axial variation becomes one of exponential axial decay and the wave is said to be cut off. When $\omega/c = k_{m\mu}$ the axial wave number k_x is zero and this is referred to the cutoff condition. A comparison of predicted and measured decay is shown in Fig. 7.10.

The form of the equation for k_x can be given more physical significance for turbomachinery by realizing that the frequency sensed by a stationary observer will be related to the circumferential speed with which the pattern sweeps the circumference. If Ω is the pattern speed in rad/s and r_0 is the outer radius, then Ωr_0 is the speed at the wall. If m is the circumferential order, $\omega = m\Omega$. Denoting the wall Mach number $\Omega r_0/c$ by M, the expression for axial wave number becomes

$$k_x = (1/r)\sqrt{M^2 m^2 - k_{m\mu}^2}$$

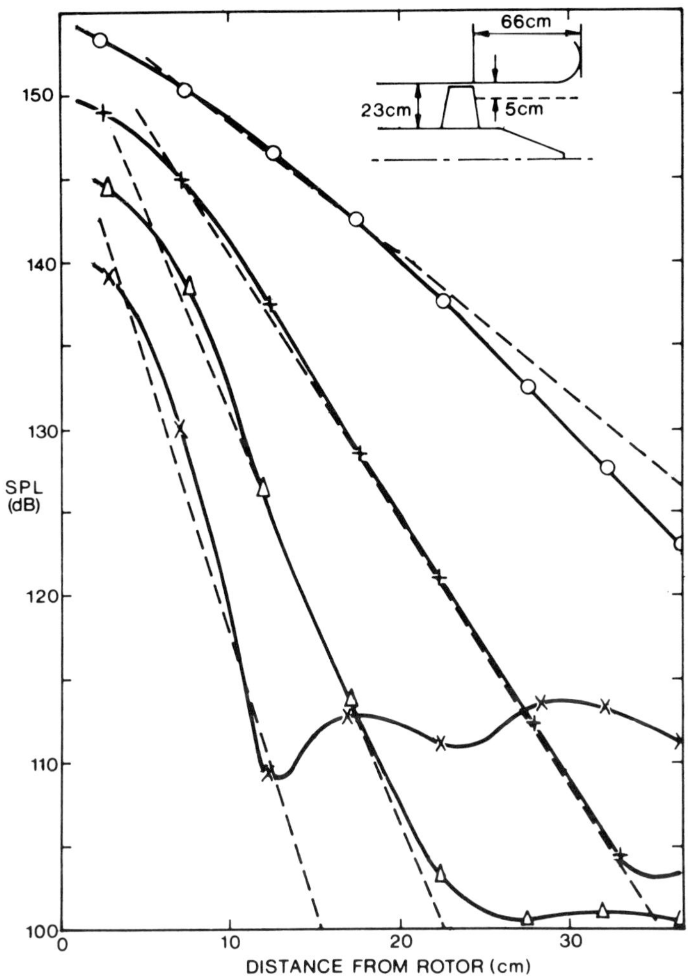

Fig. 7.10 Comparison of measured and predicted (broken line) decay ahead of subsonic fan (from Ref. 7).

and the condition for unattenuated propagation can be expressed as $M > k_{m\mu}/m$. Figure 7.11 shows how the decay rate varies with the pattern wall Mach number for a hub-tip ratio of 0.5 and a radial order of zero. For large circumferential orders (i.e., large m), the decay rates below cutoff are enormous—so large in fact that for most purposes it is possible to ignore such modes altogether. The cutoff Mach number is slightly greater than 1 for large values of m and significantly greater than 1 for low values of m. When the hub-tip ratio becomes very nearly equal to unity, the propagation in the duct can be treated as if it were in a rectangular duct; the cutoff Mach

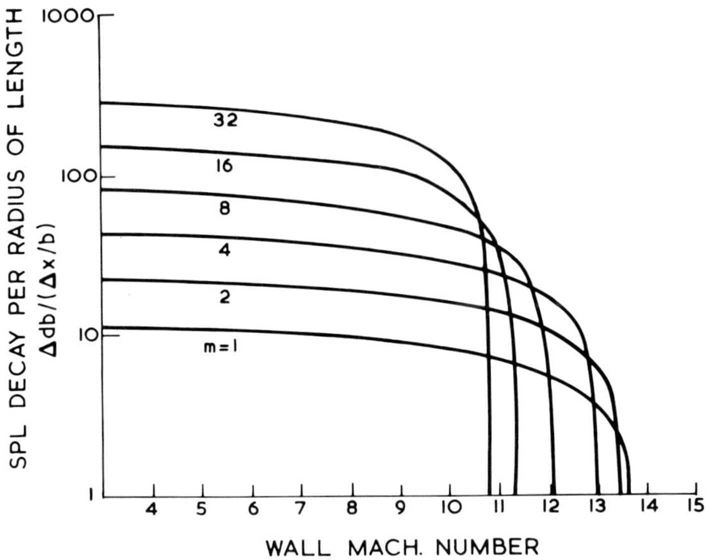

Fig. 7.11 Decay rates for modes of zero radial order in ducts of hub-tip ratio 0.5 (from Ref. 5).

number then tends to unity for all the zero radial-order modes. The corresponding Mach numbers for the cutoff of the higher radial orders are all higher. Thus, for the 16th circumferential order in a duct of hub-tip ratio 0.5, the cutoff Mach numbers are 1.13, 1.46, 1.71, and 1.95 for radial orders of 0, 1, 2, and 3, respectively. In other words, if the zero radial-order mode is cut off, all the higher radial-order modes will be as well; in fact, the rise in the cutoff Mach number with radial order is even more rapid for small values of m.

The axial wavelength λ_x is equal to $2\pi/k_x$ (k_x being the axial wave number) and the circumferential wavelength λ_θ is equal to $2\pi r/m$. The angle between the direction in which a wave front propagates along the duct (i.e., at constant radius) and the axial direction is equal to $\tan^{-1}\lambda_x/\lambda_\theta$ or $\tan^{-1} m/k_x r$. Thus, at the point of cutoff, when $k_x = 0$, the wave fronts (or surfaces of constant phase) are axial and move in a circumferential direction. With k_x real and nonzero, the wave fronts spiral along the duct so that, in the limit when $k_x r/m \to \infty$, the wave fronts are normal to the axis of the duct and propagation is purely axial.

In the ducts of aircraft engines, flow Mach numbers are generally not negligible, being typically between 0.2 and 0.5. Sofrin and McCann[8] extended the analysis to include the effect of uniform axial flow, such a flow being an idealization to that in the upstream and downstream ducts of compressors, fans, and turbines. The circumferential and radial form of the solution is unchanged by this mean flow and it is only the axial variation that is altered. In this case, the axial wave number can be written as

$$k_x = \frac{1}{1 - M_x^2}\left[M_x \frac{\omega}{c} \pm \sqrt{\left(\frac{\omega}{c}\right)^2 - 1(1 - M_x^2)k_{m\mu}^2}\right]$$

where M_x is the axial mean flow Mach number and x is taken positive in the direction of M_x. The + and − signs refer to the downstream and upstream directions of propagation, respectively. The effect of the axial flow is to reduce the frequency (or pattern wall Mach number) for cutoff. At the point of cutoff, the argument inside the square root vanishes; however, in the case with a mean flow, the axial wave number is not equal to zero, but to

$$\left(\frac{M_x}{1 - M_x^2}\right)\left(\frac{\omega}{c}\right)$$

The wave fronts (or constant-phase surfaces) are then convected downstream and are not aligned in the axial direction. In fact, with axial flow it is possible for both the upstream and downstream propagating waves to be traveling upstream relative to the flow. Only when the mean convection speed is included does one of the waves move downstream in a frame of reference stationary with respect to the duct.

The axial mean flow is a convenient idealization for the upstream and downstream situations in turbomachines. Sometimes, the behavior is required between the blade rows, which may be a significant distance apart. Between the blade rows the flow is usually swirling, which, in general, makes it impossible to separate the variables. For certain special distributions of swirl, such as forced or free vortex distributions, it is possible to handle the pressure wave propagation after some restriction to small swirl.[9] The general distribution is, however, beyond present capabilities. Indeed, with swirl in the flow, inhomogeneities such as entropy or vorticity also give rise to pressure fluctuations, making the problem most complicated.

The idealization adopted so far has been the constant-diameter circular or annular duct. In practically all cases, this is fulfilled for only a short distance and the diameter or shape of the duct is changed. Several methods have been suggested for considering propagation in ducts of varying shape, i.e., in the review by Nayfeh et al.[10] In many cases, the principal effects can be uncovered without considering the change of shape because, if cutoff occurs, the attenuation is generally so rapid that the amplitudes are very small before the change of shape is significant.

The modal structure in a duct is ultimately of less interest than the radiated field outside it. The most common approach to this is that adopted originally by Tyler and Sofrin.[5] The mean flow is ignored and the duct exit is surrounded by a large baffle. This baffle is merely a mathematical

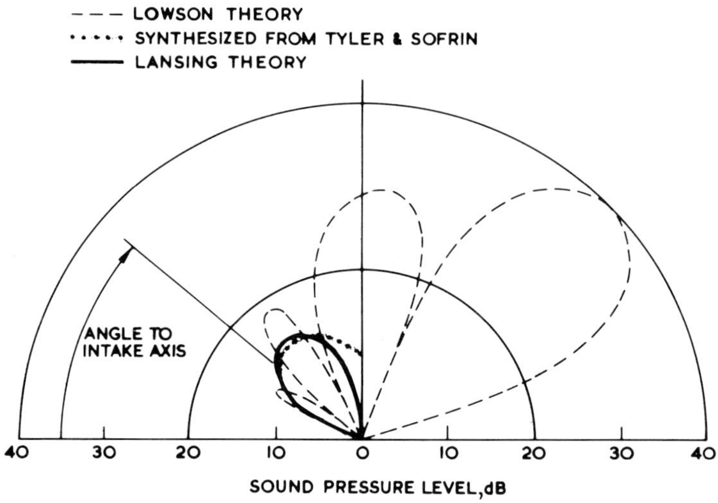

Fig. 7.12 Radiated field shapes predicted by three models; Lowson's theory is for unducted source.

convenience, but it appears that it does not seriously distort the radiation field or total acoustic power. Lansing[11] has calculated the radiation field from an intake represented rather more realistically by an unflanged pipe and has shown very close agreement with the Tyler-Sofrin flanged method, see Fig. 7.12. Toward 90 deg to the intake axis, there is a discrepancy apparent in Fig. 7.12 because the boundary conditions on the flange begin to be evident. Lansing et al.[12] considered this in greater detail and showed that only for modes in the duct close to cutoff is the calculation using flanged exits significantly in error.

Figure 7.12 is dominated by the results of Lowson,[13] who treats components such as fans as if they were unducted, i.e., the constraint of the duct on propagation and radiation is ignored. This is discussed further in Sec. 7.3, but this figure does show the discrepancy, compared with Lansing's theory, that it implies.

The radiation method of Tyler and Sofrin does not include the effect of mean flow and, although the formulation used by Lansing includes it, the calculated results shown do not. Candel[14] considered a two-dimensional (unflanged) case to examine specifically the effects of flow and soft duct walls. The boundary conditions are different for inlet and exhaust ends of a duct; specifically, there is a Kutta condition to be satisfied at the nozzle exit. The radiation patterns obtained are affected very little at frequencies significantly above cutoff, with just a slight change in the field around 90 deg to the axis. The changes that occur close to cutoff are probably related to the change in the duct cutoff frequency produced by the mean flow. The lobes in the pattern toward 90 deg appear always to be those most affected and, as Candel shows, it is also these lobes that are significantly reduced by having a soft lining to the duct.

The analyses of radiation mentioned above have omitted at least two important effects:

(1) On the inlet side, there are frequently strong refraction effects. The use of flared intakes reduces the velocity gradients and refraction effects and, as will be seen in the next section, the radiated field is accurately predicted with flared intakes. Flight-type intakes frequently produce very severe velocity gradients, including local supersonic patches, and any prediction of the field shape from these is suspect, particularly in the forward arc.

(2) On the exhaust side, the radiated field shape is, in practice, radically altered by the refraction produced by the velocity and temperature gradients in the exhaust. The effect here is to move the field away from the exhaust axis and has recently been calculated by Savkar.[15] In fact, the turbulence in the jet frequently disturbs the field shape so that the distinctly lobed pattern is lost.

7.4 Compressor and Fan Noise

Having laid the basis of duct acoustics, it is possible to consider the noise from compressors and fans. In turn, this will lead naturally into consideration of the noise from turbines.

The earliest work was concerned with noise from multistage compressors used with engines having low (or zero) bypass ratios. The compressors for these engines were designed with no regard for noise whatsoever and almost invariably used inlet guide vanes. Even after it was recognized that the compressor noise could be as disturbing as the jet noise, the mechanical constraints of these existing engines prevented very much being done to ameliorate matters. The advent of the high-bypass-ratio fan allowed, for the first time, noise to be considered from design inception, but introduced a new type of noise from the supersonic blade tips. With this design, the engine is dominated by a large fan at the front, the inlet guide vanes (IGV) disappear, the fan tip speeds are normally supersonic at high-power conditions, and the fan flow is split into core and bypass streams, each with very different geometries. Very importantly, fan noise from the rear may be as or more significant than that from the front (which is often known as inlet noise). Thus, the new fans introduced a marked change in emphasis and the largest amount of work was expended on them. It took some time for it to be realized how important the rearward-radiated noise is and for it to receive equal attention.

The field of axial turbomachinery noise has been thoroughly reviewed, with excellent and readable accounts by Morfey[16,17] and one from a different point of view by Cumpsty.[18] Morfey[16] referenced no fewer than 180 papers in 1973 and since then the number has increased; it is obviously impractical to repeat that here.

Figure 7.13 shows typical spectra of the noise radiated from the inlet (i.e., forward arc) of a fan. The subsonic tip speed case shows tones at the blade

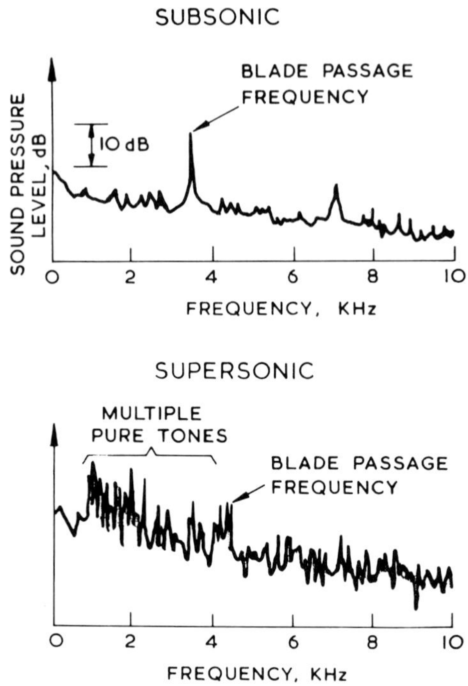

Fig. 7.13 Typical forward arc spectra of noise from a fan or compressor with subsonic and supersonic tip speeds.

passing frequency and its harmonics and broadband noise of a lower level elsewhere. The supersonic spectrum shows tones at multiples of shaft passing frequency with rather ill-defined broadband noise between them. These tones at shaft frequency are known as multiple pure tones, combination tones, or "buzz-saw" noise after the subjective impression gained by listening to this supersonic phenomenon. In the rear arc, the spectrum at both subsonic and supersonic tip speeds is generally similar to the forward arc subsonic spectrum.

The radiated field shape was discussed in the section on duct acoustics. The spectra illustrated in Fig. 7.13 correspond to a particular angle to the axis of the machine and, since the field shape of each element of the spectrum is different, a different spectrum could be found at another angle. A "complete" picture requires spectra at a wide range of angles, but the amount of data very rapidly becomes too large to assimilate. The normal approach is to consider the spectrum at one angle and to look at the field shape of the particular spectral components, such as the blade passing frequency tone.

The discussion of compressor and fan noise will be under three headings: interaction tone noise (producing the forward-radiated tones from subsonic machines and the rearward-radiated tones from all machines), multiple pure tone noise (a supersonic phenomenon), and broadband noise.

Innteraction Tone Noise

Interaction tone noise has been the most widely studied aspect of compressor noise and relates to the forward and rearward arc of subsonic machines, as well as to the rear arc of supersonic machines. The kinematics of the process are first examined. After this, the experiments relating to rotor stator interaction are considered, which leads naturally into the analytic studies. The theory has attracted many researchers and there is a perhaps disproportionate amount of information to describe here. Finally, the effect of inlet flow distortion is considered.

Kinematics. From the discussion of duct acoustics, it will be recalled that the pressure field rotating with a wall Mach number less than unity decays rapidly in the axial direction—i.e., is cut off. At first sight, therefore, it is remarkable that tones are produced by machines with subsonic tip speeds. The essential explanation was presented by Tyler and Sofrin[5] which is that the rotors interact with stationary nonuniformities, such as the wakes or potential field of stators, to set up an interference pattern for which the circumferential Mach number of a point on the wave is greater than the cutoff value. They showed that, for B rotor blades and V stator vanes, an interaction pattern is set up with a circumferential order m given by

$$m = nB \pm KV$$

where n is the harmonic of blade passing frequency and k the spatial harmonic of the distortion produced by the stators. It follows that the angular velocity of the interaction pattern Ω_p is given by

$$\Omega_p = \Omega \frac{nB}{nB \pm kV}$$

where Ω is the rotor shaft angular velocity. Although formulated in terms of interacting rotors and stators, it is equally valid for the inflow distortion, where V would denote the circumferential order of distortion.

This expression is purely kinematic and takes no account of the interaction process involved. The effect is illustrated schematically in Fig. 7.14, where a four-blade rotor interacts with a three-blade stator ($B = 4, V = 3$). The fundamental blade-passing frequency interaction, $n = 1$, would therefore be expected to rotate at $4/(4 \pm k)$ times the rotor speed. The largest value is achieved taking $k = 1$ and the negative sign to give a ratio of 4. Figure 7.14 shows that, after the rotor moves 30 deg from the top interaction, the next interaction moves round by 120 deg or four times as far. This is continued so that, after the rotor moving 90 deg, the pattern has rotated a full 360 deg. If there had been five stators ($V = 5$), the smallest modulus of

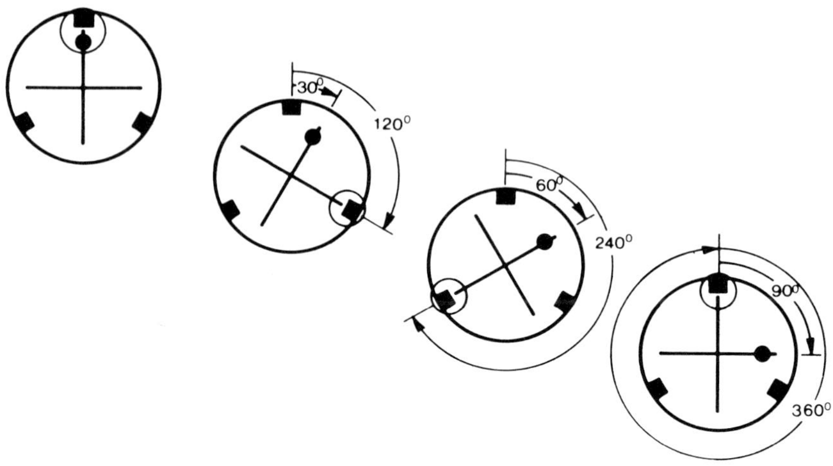

Fig. 7.14 Schematic representation of rotor-stator interaction, $B = 4$, $V = 3$ (from Ref. 18).

the denominator would be achieved again when $k = 1$. In this case, the value of m is equal to -1 and the pattern would again rotate at four times the rotor speed, but in the opposite sense to the rotation of the rotor.

A schematic representation of the way in which the waves from interactions above the cutoff add to produce wave fronts, whereas those below cutoff overlap and cancel, is shown in Fig. 7.15. The analogy can be drawn between supersonic and subsonic projectiles, the former producing disturbances that coalesce and the latter waves that interact and cancel.

Rotor-stator interaction experiments. Some insight into aspects of the interaction processes can be obtained from Fig. 7.16, which shows the forward arc field shapes of the second harmonic of the blade-passing frequency from a high-speed fan. These field shapes were obtained by slowly traversing a microphone along the arc of a circle drawn about a point in the inlet plane. In this case, the fan had no inlet guide vanes, 25 rotor blades, and 39 outlet guide vanes placed very close behind the rotor. From the blade numbers, it will be noted that, at the blade-passing frequency, a mode of order $m = 14$ will be set up rotating in the opposite direction to the rotor at 25/14 times the rotor speed, while at the second harmonic an $m = 11$ mode is produced rotating 50/11 times faster than the rotor. Even at the lower speed shown in Fig. 7.16, the second harmonic will be well above cutoff. The strong zero-radial order ($\mu = 0$) mode is shown at the lower speed, identified by overlaying the field shape predicted by the flanged model (after Tyler and Sofrin[5]). The good agreement is apparent. At the lower speeds the field is dominated by the $\mu = 0$ mode, even though higher radial orders are above cutoff, but at the higher speeds the higher radial-order modes become much more important. Further increases in speed actually

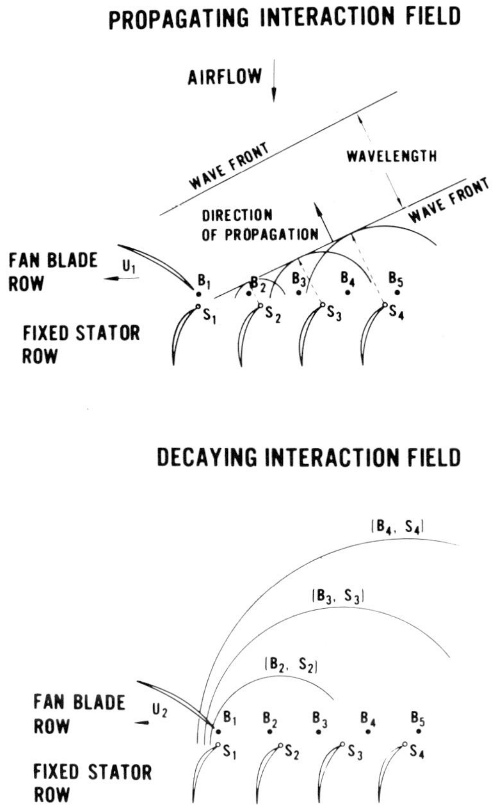

Fig. 7.15 Schematic representation of wave fronts for rotor-stator interaction above and below cutoff.

lead to a drop in the level and to an increase in the random fuzziness visible in the traces presented.

The second harmonic field shapes have been shown here because field shapes at the blade passing frequency fundamental were smooth in overall shape, with no evidence of lobes due to rotor-stator interaction, but with a large level of fuzziness. This fuzziness, or unsteadiness of the tone, is characteristic of the traces obtained when a rotor is run without stators. The tone produced under these circumstances is sometimes misleadingly called the "rotor-alone" tone. In fact, it is the result of interaction of the rotor with random, turbulent distortions drawn into the machine and will be discussed below. The absence of the rotor-stator interaction at the blade passing frequency fundamental (even when this is above the cutoff and the rotor and stator are close together) and the reduction in the level noted for the second harmonic as the speed is increased are a result of the rotor blocking the sound generated on the outlet guide vanes. This situation will be discussed below.

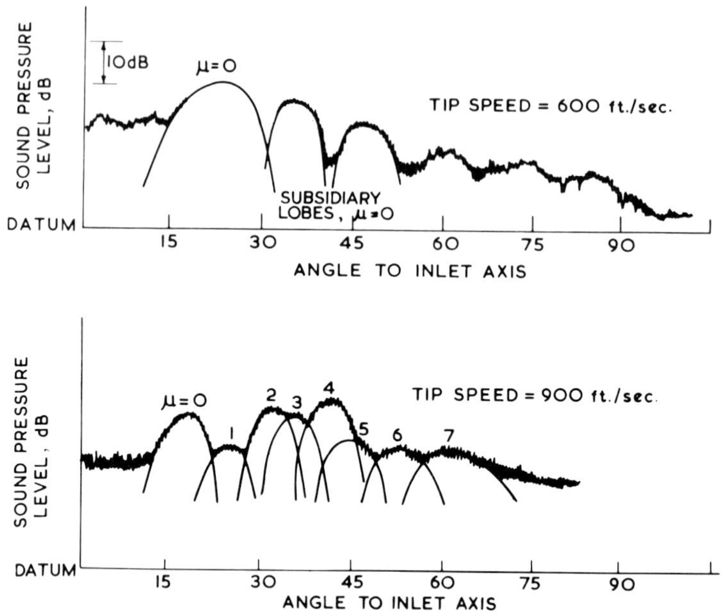

Fig. 7.16 Forward arc field shapes of tones at second harmonic of blade passing frequency.

There have been many tests run to determine the level of the tones produced by the interaction of rotors and stators. In the early tests using multistage compressors, the principal interaction was produced between the inlet guide vanes and the rotor, from which the correlations were obtained. An example from Ref. 20 is shown in Fig. 7.17. Considerable variation in the details of the trends with axial gap-chord ratio due to different companies can be seen, but each shows the rapid rise in level at small axial gaps between the rows associated with the interaction of the potential fields around the blades. At larger spacings, it is the wake that interacts with the downstream row and the viscous decay is much slower. The true complexity of the situation was illustrated by the tests of Kilpatrick and Reid,[21] where in some cases the radiated tone level from a two-stage transonic compressor actually increased with an increase in spacing.

The generation by high-bypass-ratio fans is rather different from most multistage compressors, both because of the absence of inlet guide vanes and the low hub-tip ratio, but also because the large spacing is usually included between the rotor and the bypass section outlet guide vanes to reduce the interaction. Figure 7.18 shows forward-radiated power and Fig. 7.19 the rearward-radiated power from tests of a high-bypass-ratio fan reported by Burdsall and Urban.[22] The reference case shown is one where a sufficient number of outlet guide vanes have been fitted to ensure that the rotor-stator interaction is cut off; these results show the blade passing

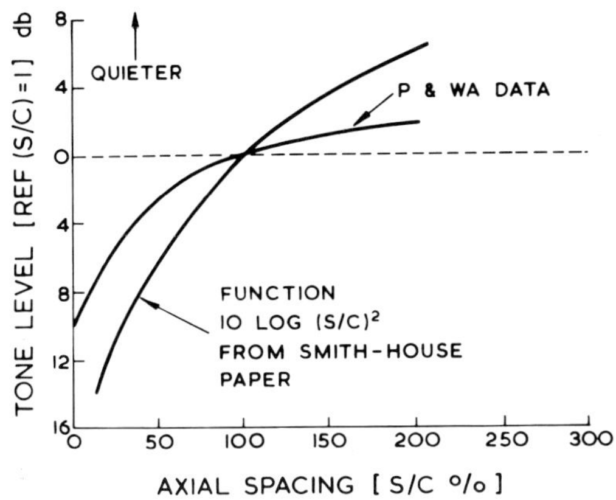

Fig. 7.17 Correlations of the change in tone level with the ratio of axial gap between rotor and stator to the chord.

frequency tone from an aerodynamically equivalent fan with the rotor-stator interaction effectively eliminated from the acoustic point of view. Any tone produced in this reference case must be related to other sources of distortion, almost certainly including the atmosphere. At 4 chord spacing, the tone can therefore be seen to be due to the interaction with atmospheric distortion, with the rotor-stator interaction playing no part. At higher speeds, this is also true for the forward arc noise at 2 chord spacings, which reflects the tendency of the rotor to block the forward propagation of the sound generated on the outlet guide vanes. The same blocking is not evident in the rear arc since the sound is being generated on the furthest downstream blade row. Although the tone levels do increase rapidly with speed, particularly close to cutoff, it may be seen from these results that the trend is not so simple at higher speeds.

Analysis for interaction tone noise. The rotor-stator interaction in a high-speed, low hub-tip ratio machine is highly complex. The blade rows are normally highly loaded, so that, in addition to significant pressure changes across the blades, large deflections are normally encountered, particularly near the root. The wakes from rotors are distinctly different from those behind isolated airfoils; in addition, the wakes in turbomachines twist because of the radial variation in the swirl velocity. Thus, the interaction will not take place simultaneously along the whole span, but the wake will sweep along it. The wake also tends to collect at the hub or tip and become confused with the secondary flows, which will themselves produce an interaction with the downstream row. As already mentioned, the blade rows are able to block the transmission of sound, which is just one example of the more general interaction where the two blade rows should really be seen as a coupled pair.

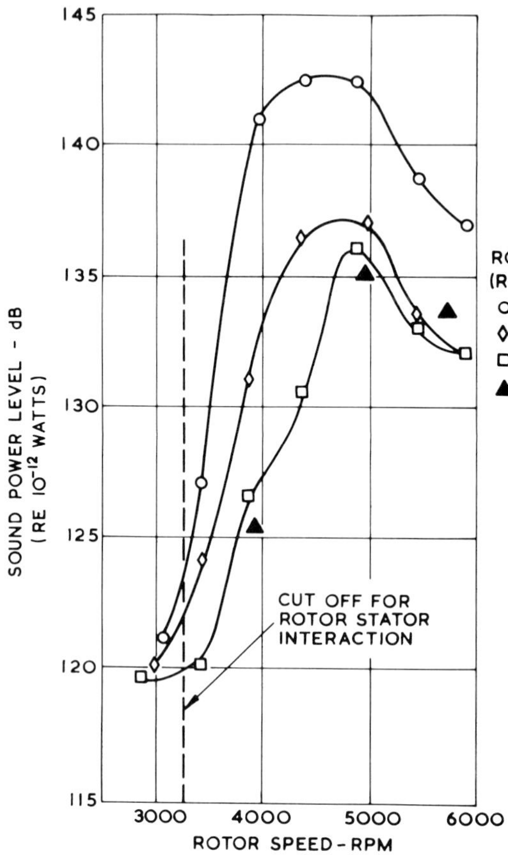

Fig. 7.18 Forward radiated tone level at blade passing frequency fundamental (blade tip speed sonic at approximately 5000 rpm).

There have been many analytical models developed for the interaction between blade rows. Basically, they may be divided into those that treat the problem as a two-dimensional cascade (i.e., ignoring radial effects) and those that treat blade rows as if they were unducted propellers. The unducted methods are often associated with Lowson[13] and, more recently, Hanson.[23] Kinematically, both the cascade and the unducted propeller are similar, with the interaction between moving blades and stationary disturbances (or vice versa) setting up a pattern rotating much faster. In the two-dimensional case, the criterion for unattenuated propagation is that the cutoff Mach number should be exceeded while for the unducted propeller radiation becomes efficient only when the disturbance is supersonic. The two criteria are very similar. At least two major disadvantages may be associated with the unducted models. One is that the field shapes are very different from those predicted by other methods that have themselves been

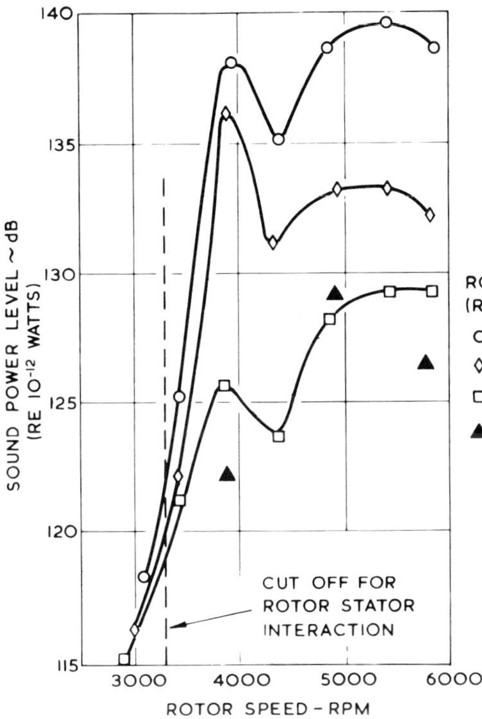

Fig. 7.19 Rearward radiated tone level at blade passing frequency fundamental (blade tip speed sonic at approximately 5000 rpm).

shown to agree quite well with measurement, see Figs. 7.12 and 7.16. A second disadvantage is that the blockage of sound by blade rows may be as important as the generation itself, and the unducted models do not lend themselves to predicting this.

Directing attention to two-dimensional cascade methods, two basic approaches to calculating the interaction of blade rows with inlet velocity distortions or sound waves can be found. Conveniently, these can be designated the compact and distributed approaches.

If the blade chords are assumed to be sufficiently short in relation to the wavelength, they may be regarded as acoustically compact sources. By treating the sources as compact, it is a simple application of classical acoustics to relate the dipole source strength to the fluctuation in the lift (or drag) of the blades. The lift fluctuation is normally much larger than the drag and is generally obtained by one of Sears[24] or Kemp and Sears[25,26] methods for incompressible flow about uncambered blades. Probably, the first example of this was applied by Hetherington,[27] who obtained quite good estimates for the power produced by IGV-rotor interaction. Another compact source method is due to Mani,[28] which was used to predict the

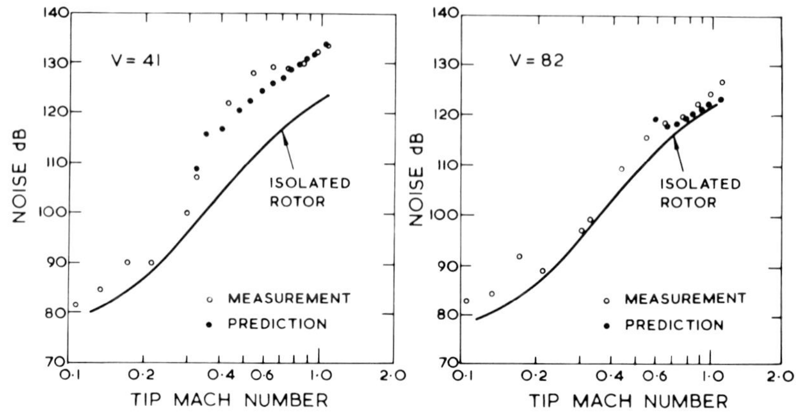

Fig. 7.20 Comparison of prediction and measurement of rearward acoustic power at blade passing frequency from 54 bladed fan in closed cycle Freon rig.

acoustic power produced by a small rotor running in the wake of a number of rods in a Freon tunnel.[29] The theory predicted that, when the wave fronts generated are nearly parallel to the blade chord, the power would be much higher than when the wave fronts are normal to the chord. With 82 rods the wave fronts were very nearly normal to the blade chords, whereas with 41 rods the wave fronts were nearly parallel. As Fig. 7.20 shows, the theory accurately predicted the levels in each case and, in particular, the higher levels for 41 rods.

Despite these successes, one is nevertheless concerned about many of the steps in compact source models such as this; the calculation of the lift fluctuation makes many restrictions that are not satisfied; but, perhaps more seriously, the blade chord and the sound wavelength turn out to be very nearly equal. In calculating rotor-stator wake interactions, a very simple model is used for the wake, which is also open to the most serious doubt.

The more interesting methods for two-dimensional cascades use the distributed source approach and do not make restrictions about chord length, but instead match the incident upwash with the source strength along the chord. They are, nevertheless, restricted to cascades of uncambered blades at zero incidence (flat plates) and become less accurate at Mach numbers approaching unity. The method due to Kaji and Okazaki[30] is presented only for the calculation of the effects of incident pressure waves (although it may be generalized), while the methods by Whitehead[31] and Smith[32] admit both this and vorticity distortion as inputs. Figure 7.21 compares the measurements with the predictions produced by Smith in a special low-speed test and the agreement is evidently excellent. The rapid rise in the magnitude of the pressure fluctuation with rotor tip relative Mach numbers close to 0.1 occurs as cutoff is approached. At Mach numbers below cutoff, the levels fall precipitously.

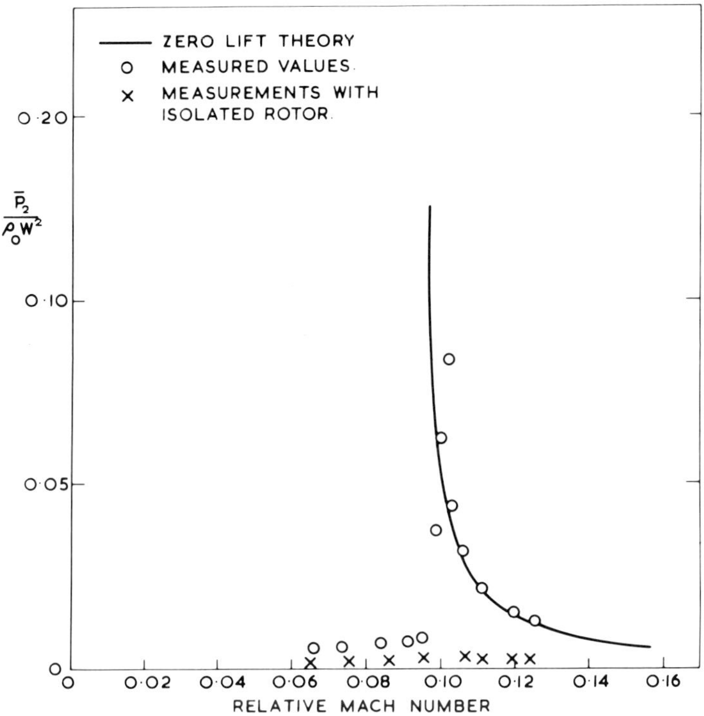

Fig. 7.21 Comparison between predicted and measured acoustic pressure fluctuation at blade passing frequency, downstream of 23 bladed rotor with 24 upstream wakes (pressure nondimensionalized with relative dynamic pressure into rotor).

More recently, Kaji[33] has compared the predictions obtained using a distributed (i.e., noncompact) source model with that obtained from a compact source. The results for relative inlet and rotational Mach numbers of 0.8 shown in Fig. 7.22 indicate that large errors may be produced by the compact source methods, particularly for the upstream going wave.

The discrepancy found between the compact and distributed source methods for predicting the interaction of blade rows with flow distortion is not found for the interaction of blade rows with sound waves. Methods where the blade chords have been vanishingly short or very long produce results for the transmission and reflection of sound waves agreeing closely with those calculated using a distributed source method for a chord comparable to a wavelength. Figure 7.23 shows results calculated by Kaji and Okazaki[30] for the transmission and reflection of a sound wave approaching a cascade of uncambered blades from downstream. The stagger angle is 60 deg, the flow Mach number 0.5, and the pitch-chord ratio 1.0. The angle $(\theta - \alpha)$ is that between the incident wave propagation direction and the chord of the blades; $(\theta - \alpha) = 60$ deg corresponds to the direction of

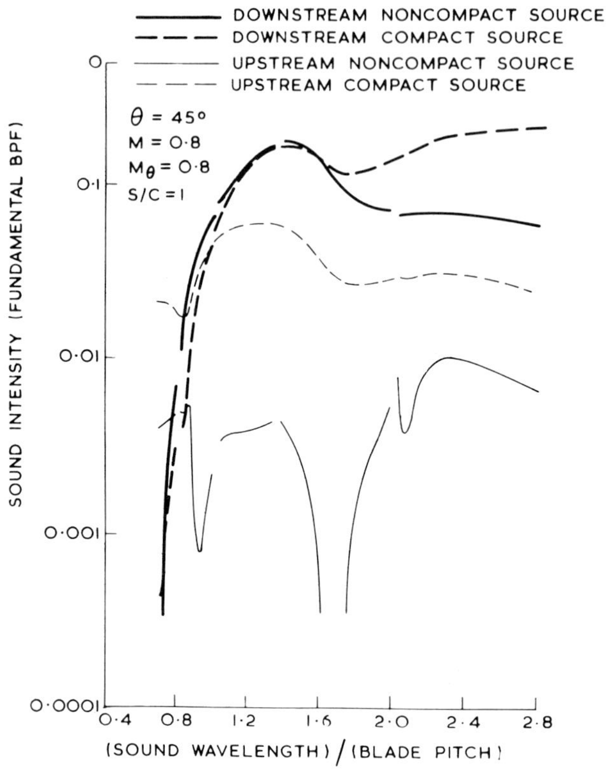

Fig. 7.22 Comparison between compact and noncompact representation of rotor interacting with distortion (stagger, $\theta = 45$ deg, pitch chord ratio $s/c = 1$).

propagation of the waves parallel to the chordline of the blades. For $(\theta - \alpha) = 60$ deg, propagation occurs with no reduction in the transmitted level and no reflected wave. The product of wave number and chord is denoted by K; the small effect of changes in this is apparent, particularly on the transmitted wave. The reasons for this are not understood, but it appears that it is the inclination of the wave propagation direction to the chord line which is crucial (the "venetian blind effect"). This makes for convenient modeling and using this Cumpsty[34] was able to explain some aspects of the blocking of sound already referred to. In general, it would appear that the modes generated on the outlet guide vanes rotating in the opposite direction to the rotor (i.e., $m = nB \pm kV$ negative) have an orientation relative to the rotor blades such that the amplitude of the wave transmitted through the rotor is low; this is particularly true at blade speeds close to sonic.

There have been only two attempts at calculating the coupled interaction between rotor and stator blade rows. Kaji and Okazaki[35] used their

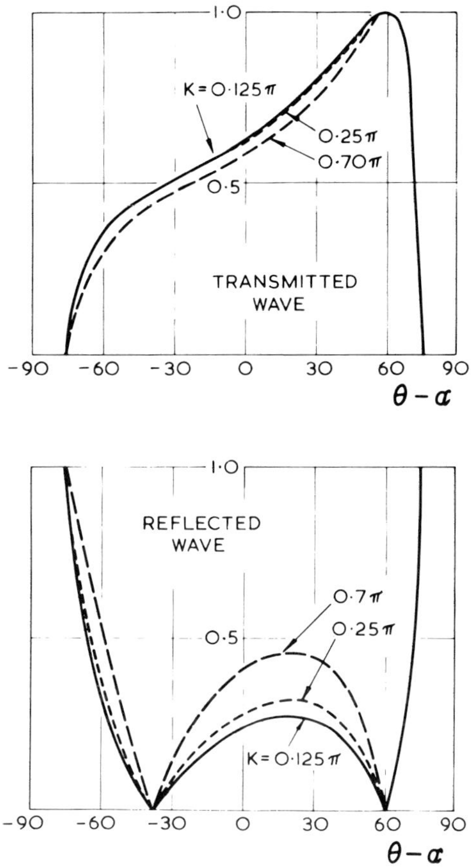

Fig. 7.23 Level of the transmitted and reflected pressure waves from a cascade of uncambered blades staggered at 60 deg (incident pressure wave from downstream).

distributed source method and Osborne[36] used a compact source method. They both show the effect of spacing on the sound generated and emphasize the rapid rise of the potential interaction at small spacings. In calculating the potential interaction, steady blade loading is an essential factor and is assumed to be small for this analysis.

In the preceding paragraphs it has been assumed that the principal mechanism for the generation of sound is the pressure fluctuation on the blades when they pass through a distorted flow. This is frequently referred to as dipole noise and in suitably idealized conditions gives acoustic power proportional to the sixth power of the inlet Mach number. Quite a different mechanism was pointed out by Ffowcs Williams and Hawkings,[37] in which the velocity perturbation v_i interacts with the steady velocity field V_j surrounding the blades to produce noise proportional to $(\rho v_i V_j)$. This is generally known as quadrupole noise and the acoustic power is proportional to the eighth power of the incident Mach number. Therefore, quadrupole noise may be expected to become more important at higher speeds than dipole noise, but it is not yet known where the crossover will occur. Experimentally, the two are virtually indistinguishable because in practice it is impossible to obtain the idealized situation in which the noise varies as the sixth or eighth powers. Theoretical investigations have been attempted, but they are so limited by the assumptions required that they really indicate only that the quadrupole noise probably dominates at very high subsonic speeds.

Rotor/inflow-distortion interaction. At the beginning of this section, attention was drawn to the unsteady tone levels, which were sometimes of higher mean amplitude than the steady tones attributable to rotor-stator interaction. It was mentioned there that this tone is due to the interaction of the rotor with unsteady, turbulent distortions in the inlet flow. It may be added that steady inlet distortions can be present, particularly in bad test rigs, and that these can produce steady tones of high level.

At one time it began to seem that the noise of high-bypass-ratio engines on approach (when the tips are subsonic) would be dominated by the blade passing frequency tone attributable to this interaction. It was not, however, understood why the turbulence in the calm conditions normally used for noise testing could give rise to the high levels of distortion needed to produce the high tone levels. Mani[38] applied a compact source method to the interaction of a rotor with homogeneous, isotropic turbulence; an example of the power spectrum produced is shown in Fig. 7.24. Even with large-scale turbulence the spectrum does not contain the peaks of comparable sharpness to the tones (or, more correctly, narrow-band noise) normally measured.

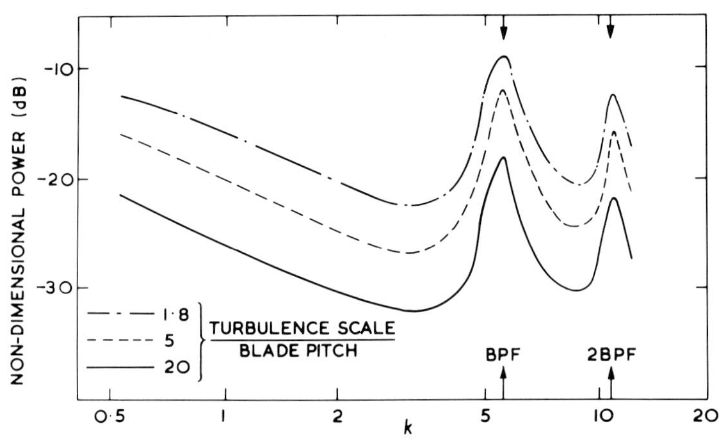

Fig. 7.24 Predicted spectra of acoustic power produced by a rotor interacting with isotropic, homogeneous turbulence.

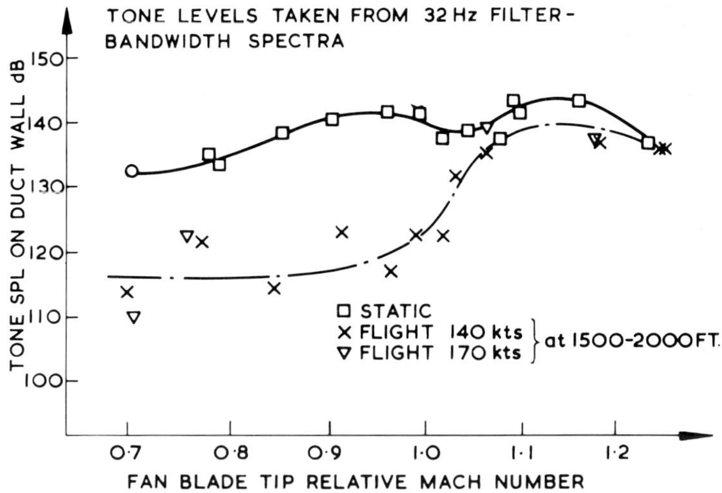

Fig. 7.25 Comparison of in-duct tone levels, static and in-flight (from Ref. 39).

The explanation appeared indirectly by comparing the tone levels in the intake of an engine mounted on an aircraft. With the aircraft stationary on the ground, a strong blade passing tone was produced, exhibiting the familiar unsteadiness. In flight, or with appreciable forward speed, the level of the tone dropped dramatically. This is shown by Fig. 7.25, presented by Cumpsty and Lowrie,[39] where the large reduction persists until the tips are supersonic. The change with forward speed demonstrated the essential feature of the turbulence drawn into a stationary intake, that it is *not* isotropic, but is strongly elongated in the axial direction. Hanson[40] later measured the length scales and found that in the axial direction they are typically 100–200 times that in transverse directions. The long, axial eddies are necessary for the production of narrow-band noise, which is recognizable as a tone. The contraction of the eddies in the lateral sense leads to an enormous increase in the lateral velocities (explaining why turbulent distortion remains important in even very calm conditions) and also allows the larger atmospheric eddies to be compressed into the intake. When the intake possesses a significant forward speed, this effect is decreased. The two conditions are illustrated in Fig. 7.26. Tests with large screens designed to remove turbulence and placed in front of the contraction of the intake have been shown by Lowrie[41] and Roundhill and Schaut[42] to reduce the tone level due to interaction of the rotor with the flow distortion. The use of such screening devices ahead of the intake is now quite normal for noise testing.

Cumpsty and Lowrie,[39] however, reported a strong correlation between the unsteady tone level and the intake boundary-layer momentum thickness at tip Mach numbers up to about 0.85. Moore[43] has observed reductions in the tone from a low-speed fan when the annulus boundary layer was removed and Lowrie[41] reports similar results on a high-speed machine. It has not yet been clearly demonstrated how atmospheric distortion *and* the

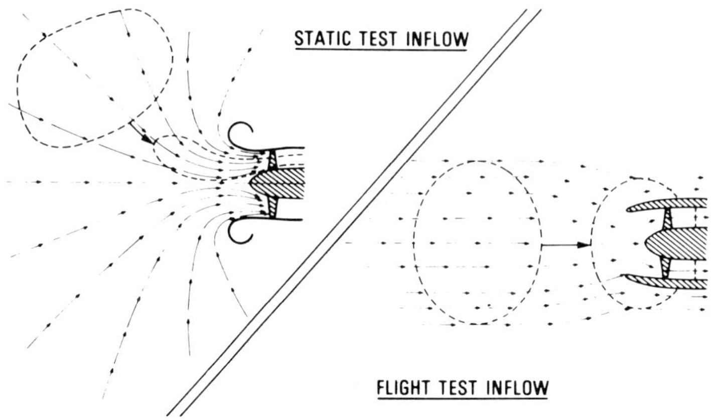

Fig. 7.26 Schematic representation of streamline patterns into an intake, static and inflight.

boundary layer both play a part in producing the tone at lower tip Mach numbers.

Because this particular noise nuisance disappears for propulsion engines in flight, at least for good installations, its interest is now as something that must be removed to reproduce in-flight conditions in static tests. It is unfortunate that contamination by this noise source has irrevocably compromised data obtained from so many costly tests.

Multiple Pure Tone Noise

When it was first decided to build engines with fan-tip Mach numbers of around 1.4, it was widely expected that the steady pressure field around the rotor would propagate unattenuated along the intake, being above cutoff, to produce an unparalleled noise nuisance. In the event, the situation was considerably less serious than this, primarily because the steady pattern toward the tip consists of shock waves that attenuate nonlinearly as they propagate. The essential processes are demonstrated schematically in Fig. 7.27; the upper sketch showing the situation for perfectly uniform shocks ahead of the rotor and the lower the more realistic case of slightly differing shocks.

The noise produced by supersonic rotors is now perhaps better understood than that from any other source, probably because it is inherently simpler, being steady with respect to the moving blades. Multiple pure tone noise is also well understood because of an unusual meshing of theory and experiment.

In the development of knowledge about noise from supersonic blades, a crucial experiment was reported by Sofrin and Pickett,[44] in which a sleeve was inserted ahead of a supersonic rotor to produce a narrow outer annulus.

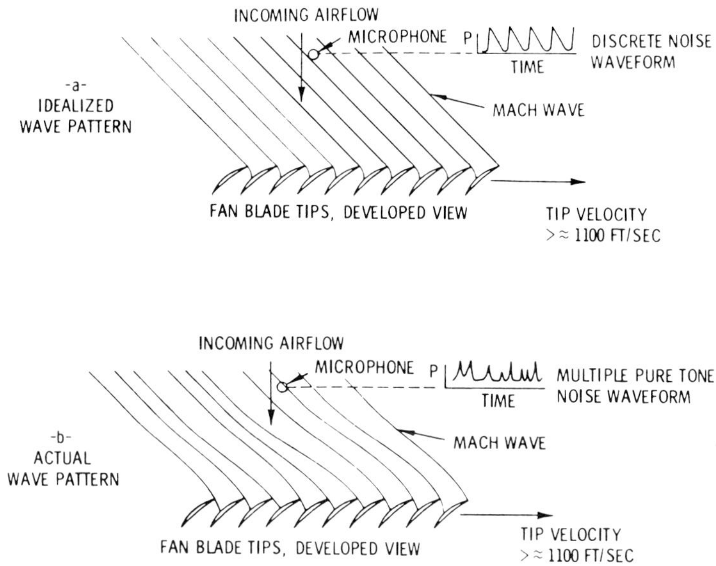

Fig. 7.27 Schematic representation of shock waves ahead of a supersonic rotor.

The shock pattern and evolution were essentially unchanged by the sleeve, indicating that for most purposes the process may be adequately modeled two-dimensionally. Sofrin and Pickett dramatically demonstrated the process of decay and evolution ahead of several rotors. Figure 7.28 shows results obtained with the narrow annulus for the decay of the average shock wave amplitude; initially, it is proportional to $1/\sqrt{x}$ and then further from the rotor as $1/x$, where x is the axial distance.

These two rates of decay are shown by Sofrin and Pickett to be predictable by one-dimensional shock theory. The prediction of the average rate of decay says nothing, however, about the spectral evolution. The changes in the pressure-time traces and in the spectrum at small distances ahead of the rotor are clearly demonstrated in Fig. 7.29. The evolution is essentially nonlinear and occurs because the stronger shocks travel faster than the weaker, eventually overtaking and absorbing them. As Fig. 7.29 shows, just ahead of the rotor the shock strengths are nearly equal, giving a spectrum with a dominant tone at the blade passing frequency. Only a few pitches upstream, the number of shocks is decreased and the spectrum altered to be dominated by tones at low harmonics of shaft rotating frequency.

The decay of shocks of uniform strength has also been analyzed by Morfey and Fisher.[45] Although uniform shock strength is not the case of principal interest, they have emphasized that there is a worst Mach number at which decay is least. The decay of a train of shocks (more precisely N waves) with random initial amplitudes has been calculated by Hawkings,[46] with the interesting conclusion that the decay becomes slower as the irregularity becomes greater. The most interesting theoretical work is due to

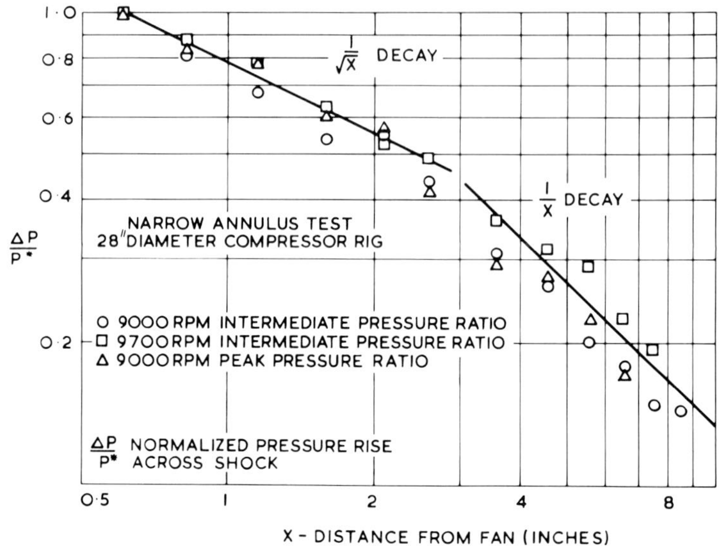

Fig. 7.28 Decay rates of the average amplitude of shocks ahead of a supersonic rotor (blade pitch approximately 2.7 in.).

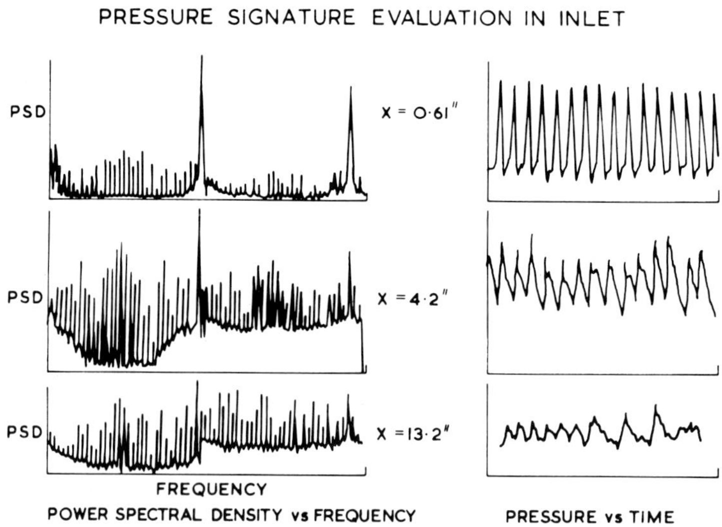

Fig. 7.29 Measured evolution of time histories and spectra ahead of supersonic fan (blade pitch approximately 2.7 in.) (from Ref. 22).

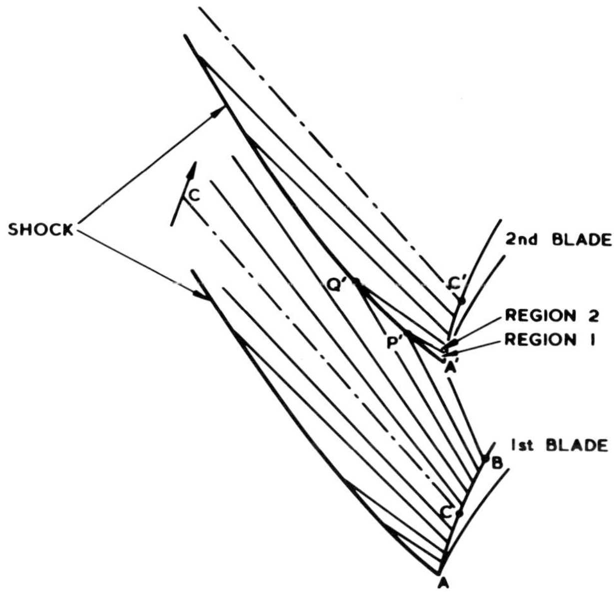

Fig. 7.30 Idealized shock and expansion system around leading edge of a two-dimensional supersonic cascade.

Kurosaka,[47] who considered a rather idealized system of shocks and expansions around the leading edge of a cascade of uncambered blades (see Fig. 7.30), restricting his attention to shocks attached to the leading edges. Random variations in both blade spacing and stagger were considered. With quite small variations in stagger, the spectrum only a short distance ahead of the rotor closely resembles those measured, see Fig. 7.31. Similar variations in blade pitch produced nowhere near the correct spectrum. While this points very clearly to variations in the inclination of the forward part of the blade as being the principal cause of the spectral evolution, it must be borne in mind that fans usually operate with the shocks somewhat detached.

A large number of tests have been run to try to reduce the multiple pure tone noise by altering the blade profile. It will be appreciated that subjectively a change in spectrum may be as important as a change in level. One school of thought favors camber over the forward part to reduce the influence of the turning around the leading edge that is believed to produce the shock strength variations. Another school favors a flat forward part to minimize the shock strength, and hence the propensity for the nonlinear evolution. Since only small variations are needed to produce the evolution, it is not known whether the special blades will survive the rigors of operation. In any case, the success of acoustic treatment in the intakes at attenuating multiple pure tone noise has reduced the impact in this noise and with it the interest in eliminating it at source.

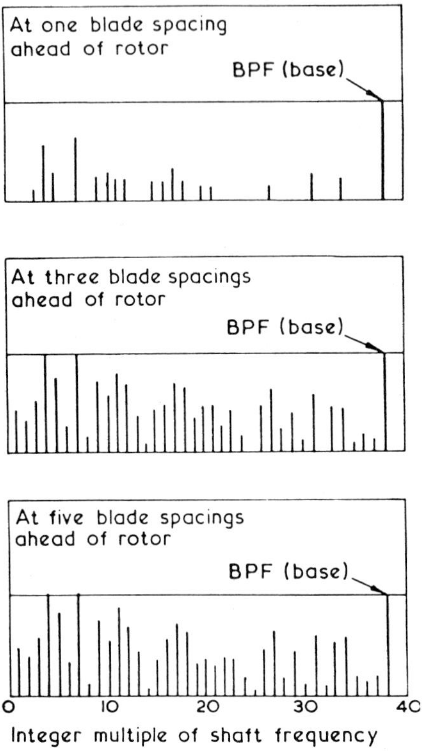

Fig. 7.31 Predicted spectra ahead of supersonic cascade of uncambered blades for small variations in stagger angle.

Broadband Noise

Whereas the multiple pure tone noise is fairly well understood, the situation for broadband noise is most unsatisfactory. One of the problems is the difficulty of pinpointing this source experimentally. Unlike the blade passing tones, which are easily identified in the spectrum and from which something can be learned of the source by the time history and field shape, almost nothing can be said unequivocally about the broadband noise. The broadband noise forms the spectral background between the tones, as in Fig. 7.13; frequently what was believed to be broadband noise in the past has been found to contain a large number of tones when analyzed with a narrower bandwidth filter. There is in general no way of knowing whether the noise originates in the rotors or stators, whether it is produced by incident turbulence interacting with the blade rows, or whether it is random unsteadiness produced in the blade rows themselves. It is the random character of broadband noise that makes this so difficult to unravel. What understanding there is comes from observation of changes in the noise with alterations such as removing a stator row or altering the pressure ratio by throttling. Unfortunately, removing a downstream stator row has aerodynamic effects on the rotor that are frequently overlooked, but which are quite likely to alter the noise from it. Moreover, changing the pressure ratio by throttling also affects transmission of sound through blade rows.

Separated flow is well known for the high level of broadband noise generated. Sharland,[48] for example, compared noise from a stalled rotor with that from an unstalled rotor. Gordon[49] has looked in detail at the noise from fully stalled bluff bodies in pipes. The broadband noise power appears to be approximately proportional to the cube of the steady pressure drop across the obstruction. One of the difficulties in turbomachines is that only one part of the machine needs to be stalled for it to dominate the broadband noise. A partially stalled set of stators or a bad rotor hub could give results that would be quite misleading if treated as general.

The two main hypotheses for broadband noise generation are the interaction of the blades with turbulence and the self-generated unsteadiness of the flow in the blades themselves. To this must be added the model proposed originally by Smith and House,[20] where the turbulence in the wakes of one blade row interacts with the next row downstream. An order-of-magnitude argument shows that the length scale of turbulence in the blade wakes is too short to correspond with the noise generation at the frequencies observed. It is possible, however, that the wakes from heavily loaded blades are rather different to those normally envisaged and used in theoretical models and that they do contain significant components in the appropriate range. This hypothesis was also used in more sophisticated form by Morfey,[50] who used results from a number of different fans and compressors to infer the magnitude of parameters implicit in his formulation. One of the techniques Morfey used is extremely common in aeroacoustics—plotting acoustic power against Mach number, in this case the inlet relative Mach number. Single-stage compressors change their incidence and loading relatively little as the rotational speed is changed, providing the throttle is constant. This is certainly not the case for multistage machines (some of those used by Morfey had seven and eight stages), where the front stages move rapidly toward stall as the compressor speed is reduced.

It has frequently been observed that, on throttling a fan or compressor at constant speed, there is an increase in the broadband noise level. This is illustrated for a fan in Fig. 7.32. The changes with throttling normally correspond to moving away from the condition for which the blades were designed, but Burdsall and Urban[22] also present tests when the design tip speed has been altered while keeping the design pressure ratio constant. The broadband noise is markedly higher at a given tip speed for the more highly loaded fan, but for a given pressure rise the lowest-speed machine seems to be the quietest in terms of broadband noise.

The blades of compressors and fan are diffusers optimized to produce the largest pressure rise with the least loss and it will be recalled that the studies of conventional diffusers reveal that the peak pressure rise occurs in the transitory stall region. Therefore, it is not surprising if the blade rows also exhibit unsteady behavior leading to increased noise radiation as the pressure rise is raised and that heavily loaded rows tend to produce generally higher levels of unsteadiness.

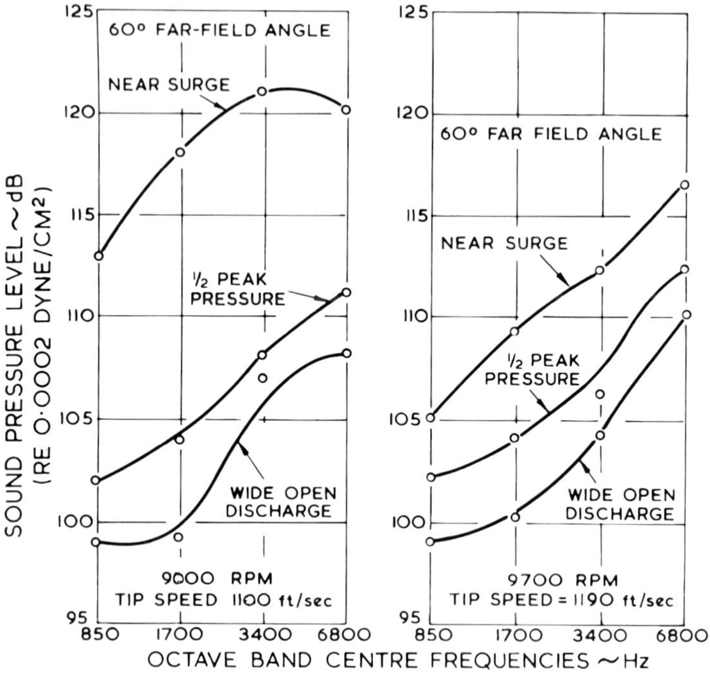

Fig. 7.32 Spectra of forward arc broadband noise from a fan at two operational speeds (from Ref. 22).

Burdsall and Urban concluded from their tests that the rotor diffusion factor, approximately the ratio of the maximum change of velocity on the suction side to the velocity at inlet, was the best correlating parameter and they chose to use the value at 75% of the span. Despite the rationale behind this approach, it has not been widely adopted.

A correlation by Ginder and Newby[51] is based on a total of nine single-stage fans with design tip speeds of 900–1450 ft/s. The correlation uses rotor incidence and tip relative Mach number as the primary correlating parameters. In fact, Ginder and Newby found rotor loading at design to be a poor indicator of broadband noise and one of the quieter fans for a given tip speed was also the most heavily loaded. The broadband noise was found to vary by up to 20 dB at a given tip speed for the whole range of machines considered. If the forward arc at transonic and supersonic tip speeds is ignored (and at this condition multiple pure tone noise is dominant), the standard deviation of measured levels from the correlation curve is reduced to within ± 2 dB. Figure 7.33 gives the measured spectra together with the correlation curve inferred. Figure 7.34 shows the variation with incidence deduced by Ginder and Newby, approximately 1.7 dB per degree. To arrive at these data the effect of tip Mach number has been removed by assuming that the sum of the forward and rearward radiated acoustic power

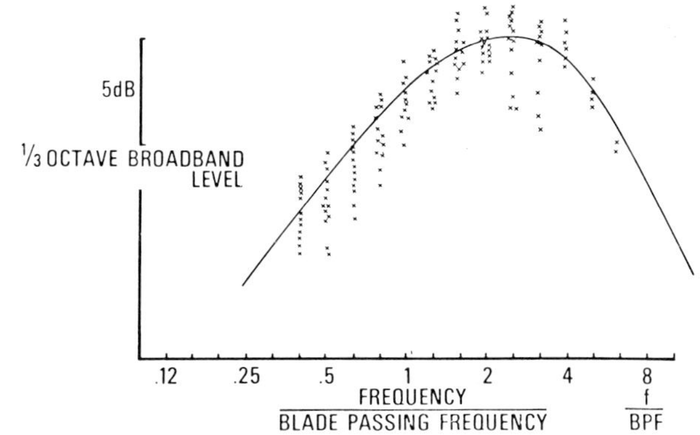

Fig. 7.33 Broadband noise spectrum of high-speed fans.

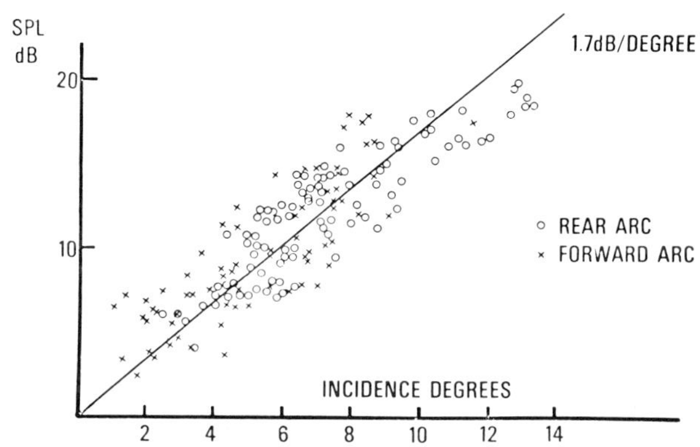

Fig. 7.34 Variation of broadband noise level with incidence (effect of Mach number removed).

will be proportional to the sixth power of the relative Mach number, which is equivalent to assuming a simple dipole source for the broadband noise. The collapse of the data around the correlation is encouraging.

The correlation described above does not postulate a mechanism, although it implies that incidence onto the tip is a special concern. The tip region of the rotor can be identified as a likely trouble spot, and the annulus boundary layer interacting with the tip is known to be very significant to the tone generation. Nevertheless, generalizing results can clearly be misleading when the basic noise-producing mechanism is not understood. Lowson et al.[52] identified the tips of their low-speed ducted fan as being important and were able to modify the noise by clipping the trailing edge near the tip. The explanation for the reduction with tip change given by Lowson et al. for the unducted rotor is that there is less blade area for the tip vortex to act on.

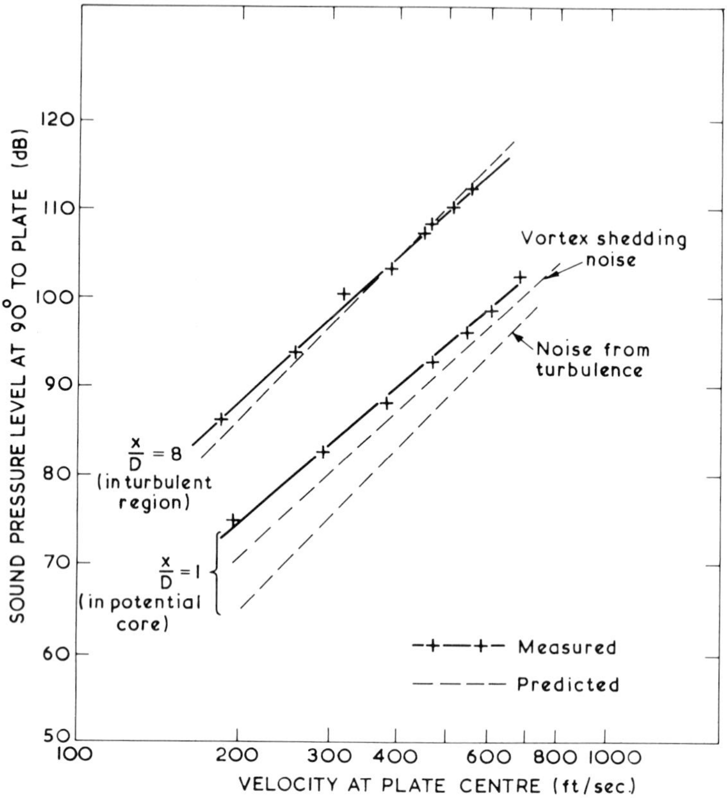

Fig. 7.35 Broadband noise radiated from a small plate held in a jet of diameter *D*.

The presence of the vortex unsteadiness is seen as the source of noise, worsened by the proximity of the blade surfaces. For the ducted rotor (and possibly the unducted rotor as well), an equally plausible explanation is that the loading near the blade tip will be reduced by clipping the tips. Changes in the tip loading also imply changes elsewhere on the blade to satisfy the radial equilibrium. Mugridge and Morfey[53] have also looked at the effects of the tips on ducted rotors, considering the vortex strengths due to secondary flow and tip clearance flow. The vorticity in these is of opposite sign, and Mugridge and Morfey hypothesized that the noise will be least when these cancel, leading to an optimum tip clearance. Flow visualization studies and measurements rather belie this tidy division into discrete vortices.

For some time there has been an interest in the noise generated by unsteadiness in the flow on each blade, rather than in the diffusing combination of a cascade or the interaction with the wall at the tip. Sharland[48] was probably the first to investigate this. He considered noise due to the interaction with the incident turbulence carried by the flow (discussed below) and what he called "vortex noise," which is the noise produced by the airfoil in laminar flow. This is believed to be due to the random shedding of vortices, analogous to the nonrandom von Kármán vortex street, produced by the imbalance of the boundary-layer thickness toward the airfoil trailing edge. Using plausible assumptions, Sharland estimated the fluctuating lift and hence the compact dipole strength; the prediction agreed quite well with measurements made with an isolated airfoil held in the potential region of a jet, see Fig. 7.35. Inherent in such a prediction are estimates for the pressures associated with the vortex shedding and the correlation area over which it extends. This led to the pressure amplitude and correlation measurements by Mugridge[54] on an isolated airfoil of an unfortunately rather untypical design, having a fairly blunt trailing edge. The surface pressure fluctuations were quite different to those measured on wind tunnel walls, with much larger amplitudes at low frequencies, just as one would expect if vortex shedding were important. The boundary-layer pressure fluctuations on the single airfoil were found to give significant cross correlation with a microphone signal well upstream. Moreover, the predicted spectral shape for the broadband noise agreed well with the measurements from two low-speed fans.

More extensive tests were later carried out on an isolated airfoil in a jet, several results of which are described by Burdsall and Urban.[22] In the nonturbulent region of the jet, there is a significant influence of incidence; for a symmetric airfoil, the overall noise rises by about 5 dB to peak at around ± 20 deg incidence, but within ± 10 deg the effect is small. The spectra at 10 deg show that the frequency scales with the reduced frequency fc/U (c being the chord length) and the amplitude as $U^{3.6}$, a lower index than expected. The amplitude was also found to be approximately proportional to the drag coefficient. A comparison between the spectrum of broadband noise from the isolated airfoil and from a few fans made by Burdsall and Urban using reduced frequency as the abscissa is fairly good, particularly at low frequencies. Ginder and Newby[51] showed, however, that the drag is not sufficient to account for the very large variations in the broadband noise level obtained from fans; a variation in drag of approximately 300 would be required in their comparisons.

Thus far, the discussion of broadband noise has been restricted to the generation of the unsteadiness produced by the blades themselves. The evidence from this suggests that under some conditions (such as when the blades are at high incidence) the "self-generated" noise dominates, even locally. When the blades are not in this condition, it is not clear whether "self-generation" or interaction with the ingested turbulence is what matters.

Sharland[48] was also probably the first to investigate the noise generated by an airfoil in a turbulent jet in experiments linked to his tests in nonturbulent flow. By application of unsteady airfoil theory and some plausible assumptions about scale, and by treating the airfoil as a compact acoustic dipole, Sharland was able to predict the sound pressure level. The agreement he obtained with measurements over a range of conditions was remarkably good, see Fig. 7.35. By putting a ring upstream of a fan rotor, he also increased the ingested turbulence to the rotor and again had some success at predicting the rise in noise. Smith and House[20] also report a rise in broadband noise when a ring was placed upstream of a rotor.

Theoretical considerations of broadband noise by blades interacting with turbulence has frequently been an adjunct to the calculation of tones. In considering the effect of homogeneous isotropic turbulence interacting with a rotor, Mani[38] calculated broadband noise with peaks at the blade passing frequency. When considering anisotropic turbulence, Pickett[55] found the same type of spectrum, although with sharper tones. Later, Hanson[56] calculated tones and broadband noise by representing the eddies as a modulated train of pulses and obtained rather good agreement by representing the sources as compact dipoles and deriving the fluctuating lift from an incompressible unsteady method. The objections raised above to this procedure for tones apply for the broadband noise as well, and it is usually found that the broadband noise peaks at a frequency where the wavelength is comparable to the chord. In discussing tones, the possible importance of the quadrupole component was considered. The same considerations apply to the broadband noise.

Until now there has been no explicit distinction between broadband noise from subsonic and supersonic tip speed compressors. In the forward arc of a supersonic compressor, it seems highly probable that the broadband noise will contain a different mechanism due to the random time variations in the bow shocks, but it must also be remembered that the flow toward the hub is subsonic and the noise from this may in some cases prevail. In fact, in the forward arc noise spectra from many supersonic fans and compressors in which the multiple pure tones are dominant, it is difficult to decide what is broadband noise. In the rear arc, the considerations for supersonic fans are more akin to the subsonic fans, but the aerodynamic behavior in the supersonic parts of the blades may give rise to quite different characteristics. The Ginder and Newby results appear to show that in the rear arcs of fans the mechanism and dependance is similar at both subsonic and supersonic tip speeds.

The overall position with broadband noise from compressors and fans can be summarized as very badly understood. It must be recognized that the dominant source can change with the design, with the overall type of machine, and with the mode of operation. A bad or a heavily loaded design may have a region of separation as the dominant source, while the interaction with the turbulence ingested may be more important in a good design. The effects of tip speed are not fully appreciated; although higher speeds for the same pressure rise seem to lead to higher noise levels, higher pressure rises frequently lead to higher noise levels at the same tip speed. Finally, the mode of operation can radically alter the broadband noise, with a rapid rise in the broadband level being normal as the machine is throttled.

7.5 Turbine Noise

Here, turbine noise is considered separately from other sources radiating from the jet pipe—generally grouped under core noise—because of its resemblance to fans and compressors. Interest in turbine noise began far

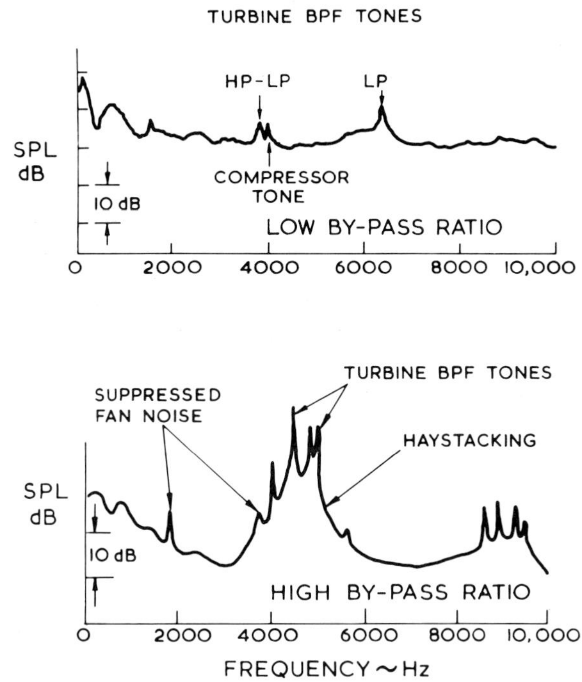

Fig. 7.36 Typical rear arc spectra from low- and high-bypass-ratio engines at low jet velocity.

later than that in compressor noise; it was only with the development of high-bypass-ratio engines, which produce relatively little jet noise, that turbine noise began to be very important. The spectra are rather different for low and high bypass ratios, examples of which are shown in Fig. 7.36. To date much less work has been devoted to turbine noise than to fan or compressor noise. Because so little work has been concerned with turbine broadband noise, it is not appropriate to separate this section into subsections on tone and broadband noise. In any case, the broadband noise is included implicitly in the core noise.

Turbine testing tends to be even more difficult and costly than compressor testing, particularly when an attempt is made to run at the appropriate pressure ratios and temperatures. A significant number of tests have been performed on cold models and still more have used the turbines of engines, but very few tests have been made using hot models in special test installations. The obvious disadvantage with using an engine is that the pressure ratio adjusts with the speed, so as to maintain the turbine close to its design condition; it is therefore possible to vary the speed and pressure ratio independently over only a fairly small range.

The measurement of noise from turbines is generally more difficult than from compressors or fans. The jets, particularly when hot, cause the field

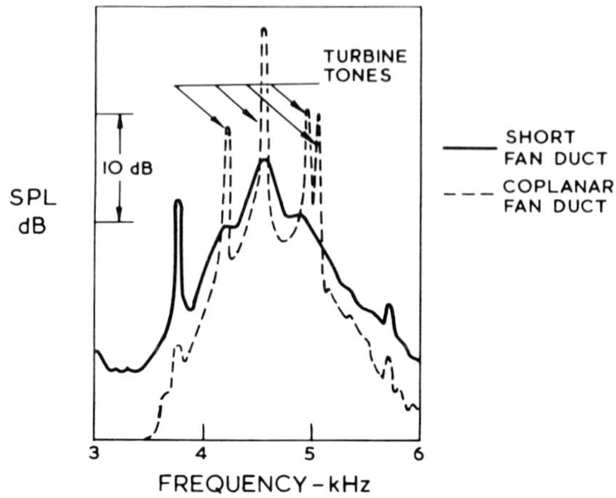

Fig. 7.37 Effect of duct geometry on the spectral shape of turbine tones perceived in far field (from Ref. 58).

shape to change because of shear layer refraction, so that the sound always peaks toward 60 deg or more from the axis. The propagation through the jet also brings about a change in the spectral shape of the tones. Instead of appearing as sharp spikes, they frequently become so "haystacked" that they are no longer recognizable as anything but spectral humps, see Fig. 7.36. The tones inside the jet pipe do tend to be sharp, unless there is a region of large aerodynamic loss (and hence turbulence) when the "haystacking" can occur inside the duct.[57] The thickness of the jet shear layer compared with the acoustic wavelength appears to play a crucial part in the haystacking and changes in the bypass exhaust geometry can have a striking effect, as Fig. 7.37 from Ref. 58 shows. Kazin and Matta[59] report that the acoustic power is about equal for the "unhaystacked" tone in the duct and "haystacked" far-field tone.

Almost all of the published data have been concerned with turbine tone noise. Because of the jet and core noise and the "haystacked" tones, it is often difficult to identify turbine broadband noise, but the first paper to consider turbine noise[60] did consider it from a cold model and engine turbines. They showed an increase in broadband noise for the cold turbines according to the third power of inlet relative velocity (not the sixth power as the simple dipole analysis would suggest). There was very large scatter, for which no explanation was available, nor was the pressure ratio considered as a separate parameter. It would seem that there is still considerable uncertainty about the cause of broadband noise and the influence of velocity, Mach number and pressure ratio on the level or spectrum. Broadband noise is, nevertheless, included in some prediction methods.

Tone noise in turbines, just as in compressors, is susceptible to cutoff; subsonic rotors produce tones that can propagate without attenuation only by interacting with nonuniformities in the flow. By choosing the rotor and stator numbers appropriately, it is possible to arrange for the turbine tones

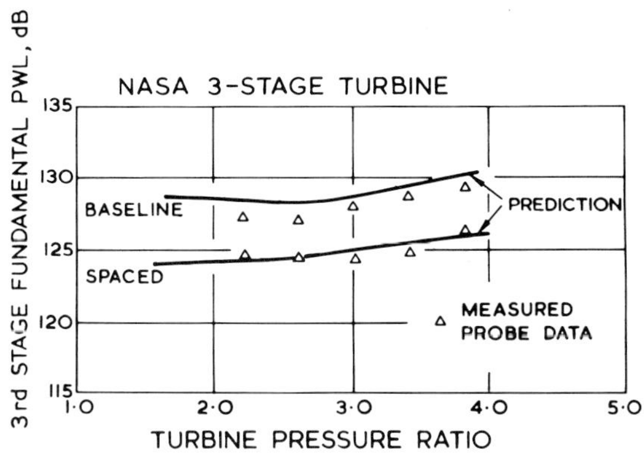

Fig. 7.38 Changes in turbine blade passing frequency tone power with increase in nozzle-rotor spacing; tip axial gap-to-chord ratio is 0.29 (baseline), 0.89 (spaced).

due to their interaction to be cut off. This is generally easier for a turbine than for fans or compressors because of the much lower tip speeds and the higher speed of sound; it can usually be achieved with less than twice as many stator blades as rotor blades. Nevertheless, until recently turbines were designed without this.

The effect of increased axial spacing between the rotors and stators is, just as in compressors, to reduce the tone noise. The large blade loading of turbines would seem to point to the potential field interaction being relatively more important in turbines. However, many turbine blades have thick trailing edges and the wake thickness may be large. Tests have shown the large benefits to be obtained with increased spacing, both in the last stage of an engine turbine and on a model three-stage turbine with a clean intake. An example of the reduction measured with increased spacing is shown in Fig. 7.38 together with the predictions using a method developed by Kazin and Matta[59] described below.

It is not clear that all turbine stages will show the same reduction when the axial gap is increased because of the circumferential variation in velocity and temperature (entropy) out of the combustion system. Hoch and Hawkins[61] compared the rear noise from a development engine when the combustion chamber was changed from can-annular to annular, see Fig. 7.39. Not only was the low frequency reduced, but the turbine tones were increased. At first sight, the more circumferentially uniform design of an annular combustor would be expected to be a route to lower tone noise, but it would seem that the temperature or velocity variation (perhaps unsteady) must be worse. The main point here is to show that the combustion system was a strong source of tone-generating distortion. Kazin and Matta[59] did try putting a turbulence producer upstream of a single-stage turbine, but found no effect. This, however, is unlikely to have reproduced the longitudinal correlation expected of turbulence out of the combustor. The production of tones by the interaction of the rotor blades with the circumferential varia-

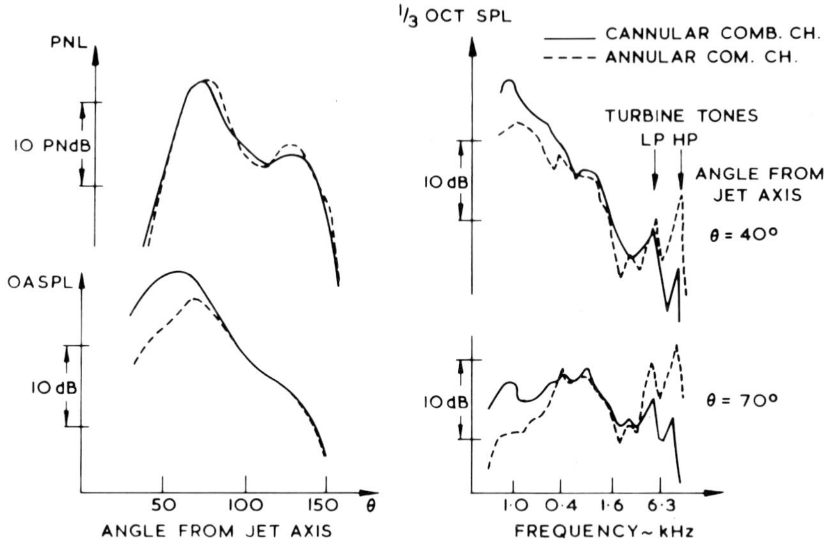

Fig. 7.39 Changes in spectrum and field shape (SPL and PNL) with changes in combustion chamber.

tion in the entropy out of the combustor is a problem not encountered in compressors, of course. It is not known how significant this is, but a characteristic of entropy variations is that they diffuse rather slowly and it seems likely that this source will provide a floor level for all stages.

Figure 7.40 shows the results presented by Kazin and Matta[59] for the effect on the blade passing frequency tone power of systematic changes in the speed at constant pressure ratio and the pressure ratio at constant speed. It can be seen that at constant speed there is a general tendency for the power to rise with the pressure ratio, while at constant pressure ratio the tone power generally falls with an increase in the rotational speed. Of course, these data refer to noise changes when the speed or pressure ratio are altered about the design value. The variation in the noise as the design values are changed is much more difficult and expensive to obtain. Furthermore, the trends illustrated in Fig. 7.40 may differ from those for turbines producing larger pressure drops.

Most data are presented in ways related to those used in the correlation procedures of the company reporting the research. How unsatisfactory this is can be seen in Fig. 7.41, which is a comparison of proprietary correlation schemes (some of which have since been superseded) attempting to calculate the peak sound pressure level from a new engine at a range of jet speeds. None of the methods predict even the trend with the turbine speed correctly, not even the one made by Pratt & Whitney whose engine this is. This must surely point to the weakness of many correlations. The field shape predictions are a little better, but just as varied. Only one method, based on

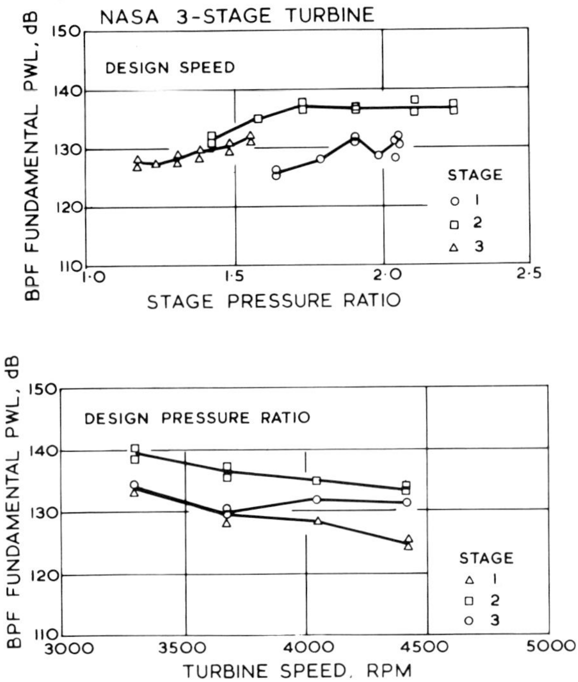

Fig. 7.40 Turbine blade passing frequency tone power; effect of pressure ratio and speed.

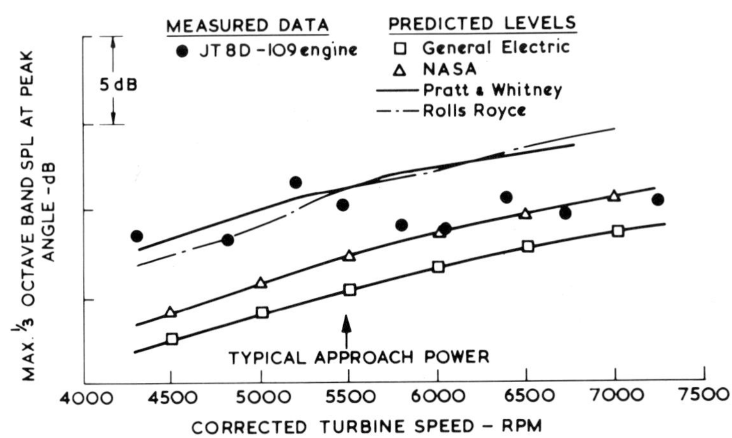

Fig. 7.41 Comparison of measured turbine noise with predictions (from Ref. 58).

the turbines of the same company, is able to get anything like the correct power spectrum for engine conditions corresponding to approach.

In discussing fans and compressors, it was pointed out that the theoretical models all require that the blade camber be small. In turbines the camber is normally large and the models are even less appropriate. Nevertheless, Kazin and Matta,[59] working on the assumption that the rotor-stator interaction is predominantly from the wake, have predicted the level of tone power for the last stage of a three-stage turbine. The method is similar to one proposed for compressors using the Kemp-Sears[26] method of calculating fluctuating lift and treating the sources as compact. Despite the apparent unsuitability of the method, surprisingly satisfactory predictions of power and the change in power with the change in spacing have been obtained, see Fig. 7.38. Only the last stage was considered because they had no method for calculating the transmission of sound through the blade rows. In fact, an actuator disk method such as Cumpsty and Marble[62] is very suitable for this because the transmission and reflection are not much affected by the ratio of the blade chord to the wavelength, as will be recalled from the section on interaction tone noise for compressors.

7.6 Core Noise

The definition of core noise is rather arbitrary, but it is taken here to mean the rear arc noise originating with the hot stream of the engine other than that due to the jet or the turbine. This is illustrated by Fig. 7.42. While it is often difficult in practice to distinguish the core noise from the jet, because both are broadband, it is conceptually possible to separate the turbine noise and core noise and some authors (i.e., Hoch et al.[64]) include the turbine noise with the core. A full review of core noise has been given by Bushell.[65]

The discovery of core noise was held back for a long time by the use of unsatisfactory jet noise rigs in which the upstream valves and obstructions produced more noise than the jet mixing itself when the velocities were low. With the advent of really good rigs, it was apparent that the jet noise really does follow $V_j^{\,8}$ right down to very low speeds. It was then clear that the noise level from engines was distinctly higher at low velocities than the noise

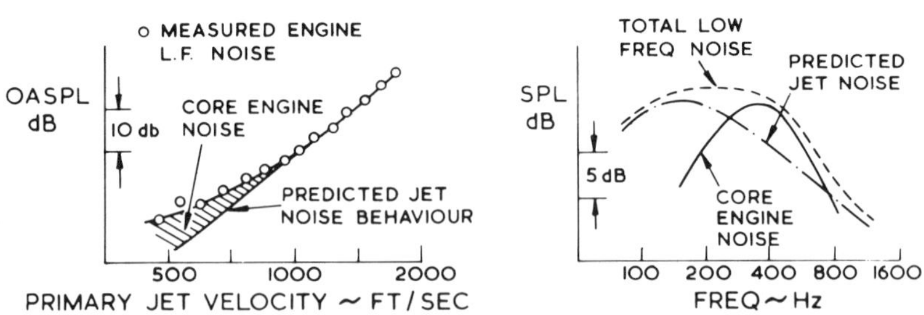

Fig. 7.42 Definition of core noise (from Ref. 63).

of pure jets.[66] It also became clear that the spectrum from engines with low jet velocities contained more high-frequency noise than the corresponding jet and that the angle for peak noise was near 80 deg from the jet axis, whereas for pure jet noise the peak occurs around 30 deg. One of the main problems with this has been removing the effects of jet noise so that the behavior of core noise can be studied in detail. Although for a given jet velocity (relative to the engine) core noise may be barely perceptible on a static test bed, it may be very important with a forward speed of, say, 200 knots and the consequent drop in pure jet mixing noise. To overcome this, there have been tests with oversized nozzles so that for a given engine speed the jet velocity is reduced. Unfortunately, this changes the engine operating conditions and only when appropriate inlet throttling is included[64] are the engine conditions maintained. Even then, the acoustic conditions are altered by the changes in the nozzle shape and impedance, although the significance of this is not known.

As well as reducing the jet noise, flight has also been found to lead, in some cases, to an absolute rise in the level of the noise in the forward arc from some engines, as well as from a model jet mounted on a rotating arm.[67] There is at the present time no satisfactory explanation for this.

The earliest ideas on the core noise originate with Bushell[65] noting that low-frequency noise was generated by a turbine rig when conditions were such that there was high incidence onto the downstream struts. Bryce and Stevens[68] have followed this by a very thorough investigation of strut noise in a small model of a turbojet exhaust. The swirl was produced by vanes, instead of the turbine of the real engine. They were able to correlate the extra noise with the strut incidence and the flow velocity onto the struts, obtaining a correlation to the sixth power of velocity. This correlation is consistent with the results of Gordon,[49] who carried out an extensive set of tests on bluff bodies in ducts and, as noted above, found that the noise appears to scale as the cube of the pressure drop introduced (roughly the sixth power of velocity).

No clear evidence has emerged that strut noise is a significant source from engines, where designs usually have low incidence and avoid, so far as possible, bluff bodies to reduce aerodynamic losses. On the other hand, deficiencies do occur: a separated region on the center body downstream of the turbine is quite usual and at low thrust conditions swirl may be large. There is, moreover, evidence that the system used in some engines to mix the bypass and primary flows before the final nozzle does not increase core noise. These mixers are one of the most drastic ways of introducing turbulent mixing into the flow; if these do not produce a large increase in noise, it would seem that other aerodynamic loss makers are likely to be fairly insignificant as well in comparison with the real core noise.

Considerable theoretical activity has been devoted to the "excess" noise of the jet, that is to say, noise that would once have been taken to be jet noise but that is generated at the nozzle lip or upstream. Crighton[69] has examined these and the reader will find the subject carefully discussed there. Three mechanisms stand out. One is the generation of sound at the nozzle lip itself because of the sudden change in boundary-layer conditions.

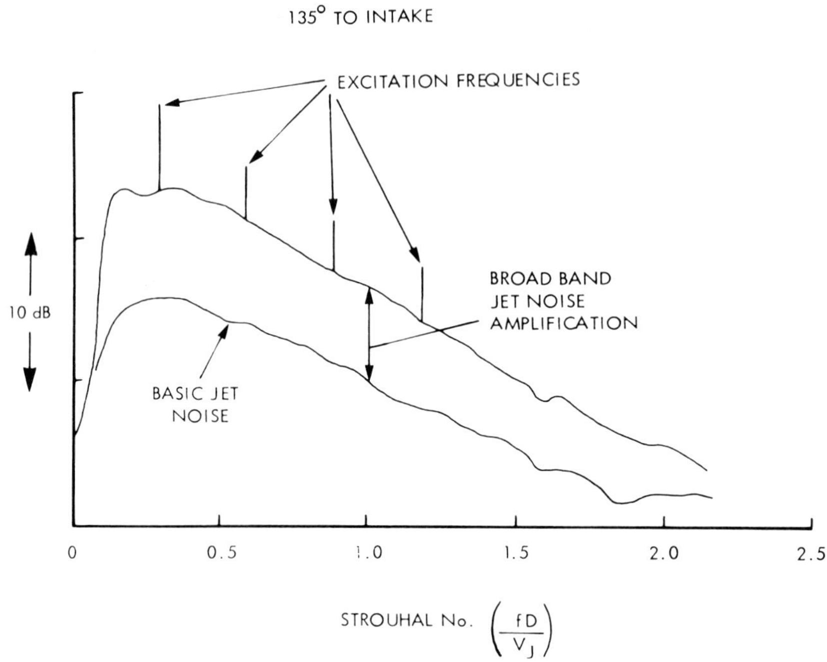

Fig. 7.43 Spectra, using Strouhal number as abscissa, of jet noise with and without pure tone excitation (from Ref. 72).

Upstream of the lip, the radial component of turbulent velocity must vanish at the wall, but this is relaxed at the lip. There have been several suggestions for the boundary condition at the lip, one of which is the Kutta condition. The analysis shows this mechanism to radiate preferentially at large angles to the jet axis. Crighton also discusses the interaction of jet shear layer instabilities with the nozzle to produce unsteady outlet flow and thus noise. Also, from a different point of view, the turbulent shear layer provides a possible amplification of sound propagating down the jet pipe. This approach is usually associated with Crow,[70] who demonstrated this experimentally as well as theoretically. Model experiments by Gerend et al.[71] and Bechert and Pfizenmaier[72] have confirmed that tones in the upstream flow can lead to large increases in the broadband noise. A result by Bechert and Pfizenmaier is shown in Fig. 7.43.

Much of the experimental evidence has linked the core noise, or at least part of it, with the combustion system. A clear example of this is the reduction in low-frequency noise when an annular combustor was installed in place of a can-annular one, see Fig. 7.39. More recently, Mathews and Peracchio[63] cross correlated the far-field noise (where the peak of the core noise is expected) with a pressure transducer just inside the nozzle and also with a transducer in one of the eight combustion cans. The cross correlation was filtered and in each case found to peak at around 400 Hz, which had been deduced to be the peak of the core noise spectrum by subtracting the "true" jet noise from the overall measured noise. The peak cross correlation between the nozzle lip and the far field was 0.17, which has been cited as evidence that lip noise is dominant. Between the combustion chamber and the far field, the cross correlation was 0.05. This latter value appears rather low, but since there are eight combustion chambers, each of which is almost certainly wholly uncorrelated with the others, the maximum that could possibly have been achieved is 0.125. The cross correlation between the combustor and the far field is thus about 0.4 of the maximum possible, which is quite high. The hypothesis that the pressure fluctuations from the jet pipe propagate upstream to the combustion chamber is quite implausible, and the dependence of the core noise on the combustion systems is well established for this engine.

Although combustion appears to be responsible for much of the core noise, there are two conflicting views on how this comes about. The more familiar view is that combustion is a strong noise-producing mechanism, particularly in the highly turbulent conditions necessary in a gas turbine combustor. This is often referred to as direct combustion noise.

The other approach to combustion-related noise presumes that it is fluctuations in the *temperature* (more correctly, entropy) leaving the combustion chambers, which then interact with downstream components, principally the turbine, to produce noise. This is often referred to as indirect combustion noise. Because it is more familiar, however, the direct combustion noise will be discussed first.

Turbulent flames are well known for their roar and have received quite wide study over the last few years. The flame may be regarded as a simple monopole source of low frequencies (so that the flame is acoustically compact or in phase) and the sound generation is then related to the rate of change of the volume of the flame. A very clear demonstration of this was given by Hurle et al.,[73] who cross correlated light (a measure of instantaneous flame size) and sound emission. At higher frequencies, the phase varies over the flame surface and prediction is more complex. The methods thus divide into the more intuitive approach of Bragg[74] and the more analytical following the aeroacoustic approach for turbulent noise of nonreacting fluids initiated by Lighthill.[2] Examples of this latter approach are due to Chiu and Summerfield,[75] and Strahle.[76] With combustion present the difficulty in converting the "solution," consisting of an integral, into a practical estimate is considerably greater than in nonreacting cases and only by the most drastic simplifying assumptions is this possible, see Ref. 76 for example.

Experimentally, the combustion in an open turbulent flame introduces several variables in addition to those found in, for instance, jet noise. As an example, an empirically developed expression for the acoustic power P from lean premixed hydrocarbon flame given by Strahle,[77] is

$$P = 4.89.10^{-5} U^{2.68} D^{2.84} S_L^{1.35} F^{0.41} \quad \text{Watt}$$

where U and S_L are mean air velocity and laminar flame speed in ft/s, D

the diameter in ft, and F the fuel mass fraction. Such noninteger indices make it seem likely that important parameters are being overlooked. To cover the lean-to-rich variation with premixed and diffusion flames for a range of fuel types clearly involves extensive test programs, particularly when the spectrum and field shape must also be found.

The difficulty in applying this information on open flames to gas turbines has been summarized by Strahle: "The main problems in application of current research results to turbopropulsion systems appear to be (a) actual combustors employ different stabilisation methods and geometrical configurations than research burners so the scaling rates and absolute power output may be different, (b) the turbulence structure is undoubtedly different, (c) enclosure effects introduce sound power augmentation over free field behaviour and feedback effects upon the combustion process may be important and (d) the sound propagation problem through turbines, ducts and compressors is a difficult one to address."[77] It might be concluded that the noise of open flames bears little relation to combustion chambers.

A considerable number of tests have now been performed to measure noise from combustion cans tested in isolation as, for example, by Strahle and Shivashankara.[78] These are tested without downstream nozzles so as to minimize jet noise, but the pressures inside the can are unrepresentatively low. Furthermore, the downstream boundary conditions represented by the first turbine nozzle row is not correctly modeled. Nevertheless, some very interesting results are obtained, including the strong dependence of overall power on the air mass flow (Fig. 7.44) and the weak dependence on the fuel flow from weak to stochiometric in the primary zone (Fig. 7.45). Curiously, the noise spectra were found to be essentially unchanged over large excursions in the air and fuel flow rates.

Kazin and Emmerling[79] report on measurements of noise from a full-scale annular combustion chamber; and their results are shown in Fig. 7.46. These tests used a downstream acoustic horn, which might be expected to confuse matters, but in fact appears to have had very little effect. The acoustic power spectrum rises markedly as the temperature rise of the combustor is increased, which is equivalent to raising the fuel-air ratio. Kazin and Emmerling also report tests where an acoustic absorber was placed downstream of the combustor. Large reductions in the noise were produced, proving the origin of the noise to be upstream in the combustion region itself.

Ho and Tedrick[80] examined the combustion-related noise from eight small gas turbines with a range of very different types of combustion chamber and from a combustor tested on its own with an open exit. The noise from the engines was found to collapse well when plotted against an empirically determined noise factor F, which was modified in the light of dimensional analysis and included the temperature rise. It appeared that for engines the power is proportional to F^4, whereas for rigs it is proportional to F^2. This suggests that the presence of downstream components, mainly the turbine, is highly significant.

Grande[81] assumed that combustion noise is proportional to flame speed and that this varies linearly with temperature. Using a crude model for the effect of the nozzle and turbine on the propagation of noise from the

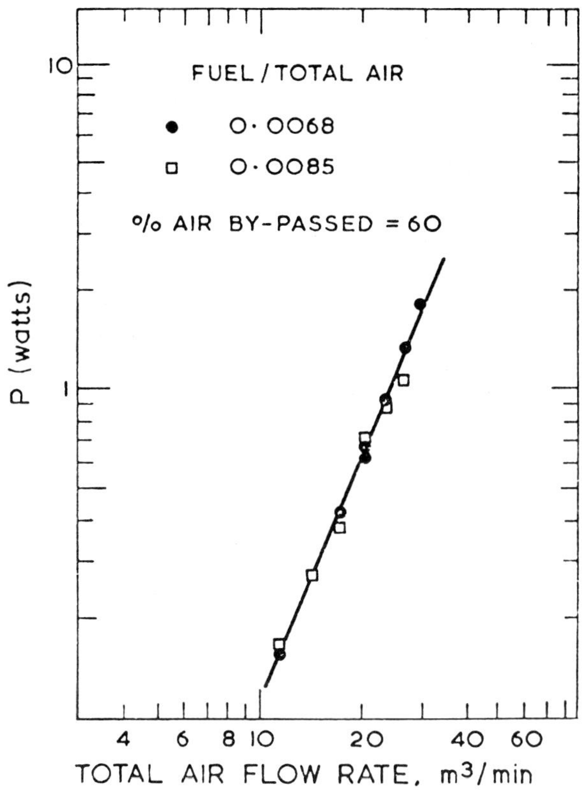

Fig. 7.44 Variation in acoustic power from an isolated combustion can as a function of airflow rate.

combustor, he was able to predict the overall noise from two engines with considerable success.

Results of a very extensive program on combustion noise have been published by Mathews and Rekos.[82] Tests were conducted using isolated combustors of differing design and engines for which the combustion noise was deduced by subtracting the pure jet noise from the far-field spectra. The correlation, which is based on simple models for the noise generation and sweeping assumptions, provides a very satisfactory correlation of both the overall sound power and the spectrum, including combustor size, pressure, temperature, calorific value, and air-fuel ratio. Apart from some differences in the constants, annular and can-annular combustors appear to produce very similar noise.

The indirect combustion noise is produced by the interaction of entropy fluctuations (i.e., hot spots) with downstream components, in particular the turbine, but also the final nozzle, which was considered by Candel.[83] The two methods proposed for turbines by Cumpsty and Marble[62] and Pickett[84] both treat the turbine blade rows as two-dimensional actuator disks, assum-

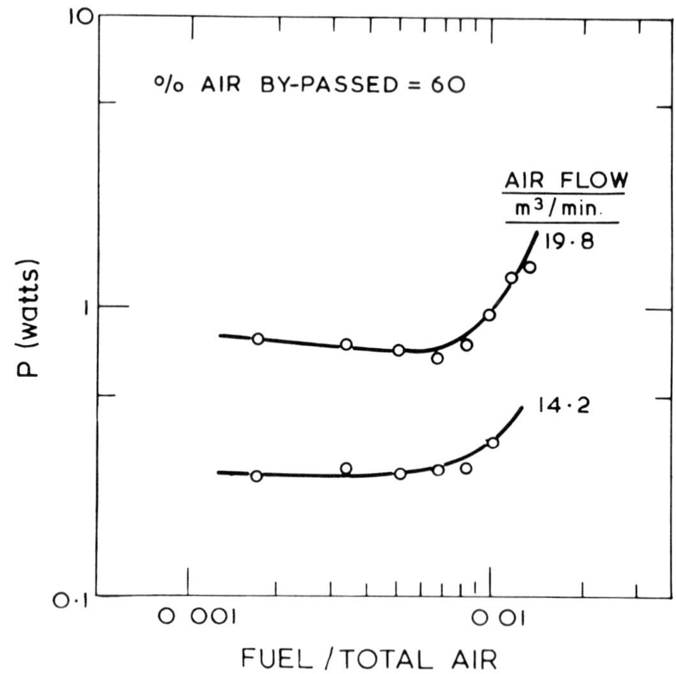

Fig. 7.45 Variation in acoustic power from an isolated combustion can as a function of fuel-air ratio (from Ref. 78).

ing blades of infinitesimal chord and pitch and neglecting radial variations. The essential feature of both methods is that the changes in *mean* properties across the blade row are large, but perturbations in the unsteady properties (entropy, pressure, vorticity) are small. If the change in the mean properties were small, entropy would give rise to only second-order perturbations in pressure. Cumpsty and Marble were able to examine the combined effect of many rows, while Pickett considered rows one at a time.

As well as representative turbine flow Mach numbers and angles, the indirect calculations require as input the magnitude of the entropy fluctuation. At constant pressure, the entropy perturbation is given by

$$s'/c_p = T'/T$$

where the prime denotes perturbation. The temperature perturbation has been measured and a spectrum is shown in Fig. 7.47.

The overall magnitude of T'/T in the range of audible frequencies is about 2%. The indirect combustion model predicts that the acoustic power rises rapidly with the pressure drop through the whole turbine and, in particular, the drop across each stage. With some idealization, Pickett has shown that the power is proportional to the square of the stage pressure

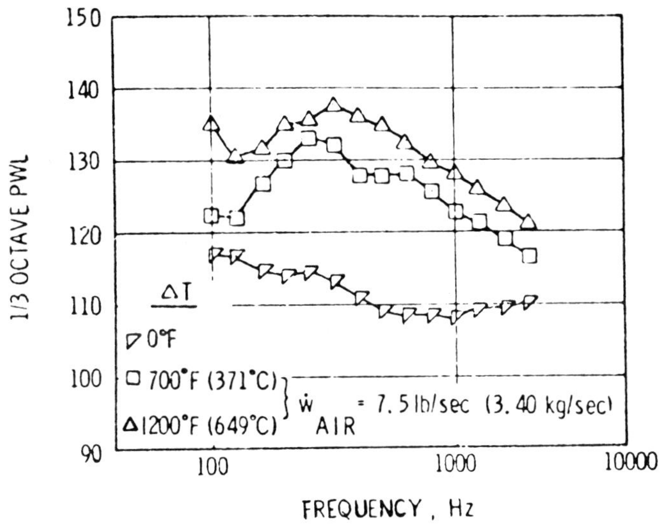

Fig. 7.46 Effect of temperature rise on the noise of a full-scale, isolated, annular combustor (from Ref. 79).

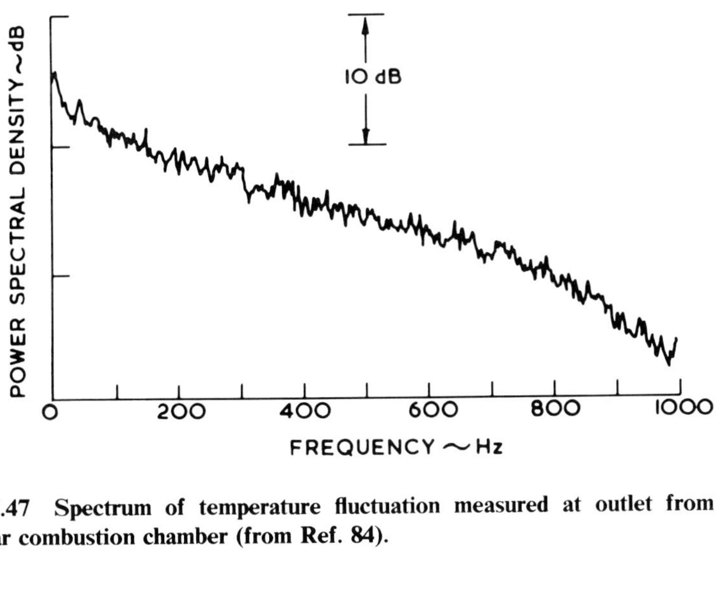

Fig. 7.47 Spectrum of temperature fluctuation measured at outlet from a can-annular combustion chamber (from Ref. 84).

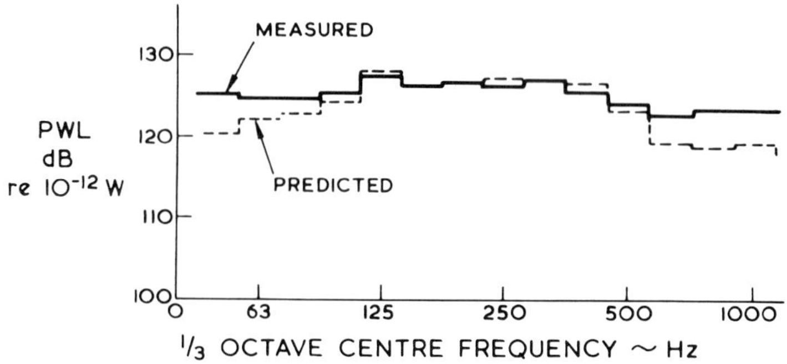

Fig. 7.48 Measured and predicted spectra of rear arc acoustic power from a JT8D engine at low jet velocity (from Ref. 85).

drop. Thus, for a given overall pressure drop, the noise will be less with a larger number of stages. The noise is also proportional to the magnitude of the entropy perturbation. Figure 7.48 compares the predicted and measured spectra of rear arc acoustic power for a Pratt & Whitney JT8D engine assuming $T'/T = 2\%$. The agreement in level and in spectrum shape is satisfactory in view of the number of assumptions necessary.

The situation with regard to combustion noise is far from clear, with the possibility of direct or indirect noise being the most important. Experimental tests to establish which is the more important are not easy. Removing the turbine removes the prime generator of indirect noise, but at the same time it alters the conditions in the combustion chamber, both acoustically and from a more general point of view. Measurements of pressure fluctuation in the combustion chamber show large amplitudes, but these are attenuated by passage through the turbine (which may be calculated). On the other hand, the generation of pressure and entropy fluctuations is intimately related and the existence of one does not mean the absence or unimportance of the other.

It would seem that very careful experiments will be needed to finally resolve whether direct or indirect combustion noise is predominant. It is quite possible it will be found that both may be important in differing circumstances. It seems very probable that combustion noise of one sort or another will be the prime source of core noise under most conditions at low frequencies.

7.7 Acoustic Treatment

The character of this aspect of engine noise is quite different from that outlined in earlier sections. Here the emphasis is on the design, that is, achieving the optimum configuration of acoustic treatment for the least cost and weight with the least effort and greatest confidence. There is not the same uncertainty surrounding the subject that there is, for example, regarding the broadband noise from compressors or the cause of core noise. The

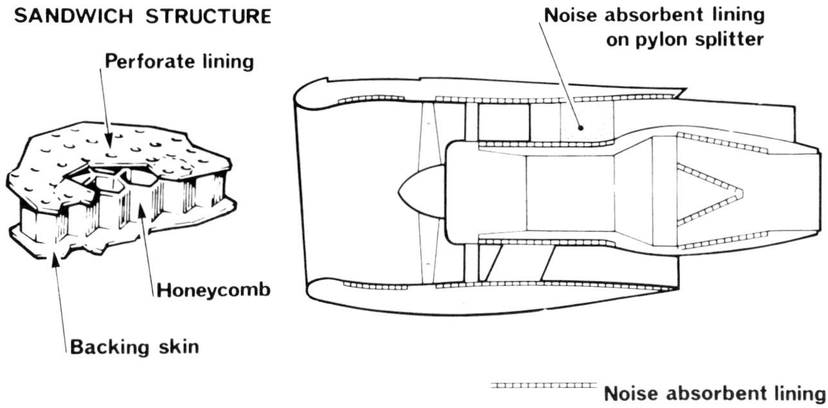

Fig. 7.49 Schematic of installation of linings in a high-bypass-ratio engine.

strong position is partly the result of strenuous efforts over a number of years. An excellent review of the whole field of aircraft engine duct acoustics has been provided by Nayfeh et al.[10]

Figure 7.49 shows where treatment can be applied in a bypass engine and a schematic of the type of treatment used. Most impressive attenuations have been shown to be possible in this way, even without the introduction of additional surfaces, such as splitters, into the ducts. The crucial quantity is the ratio of the length of the lined surface to the duct height or diameter. At the present time, with linings in the inlet extending over an axial length equal to about one radius, it is possible to reduce the multiple pure tones until they essentially vanish into the background, giving an overall reduction of the intake noise of perhaps 5 PNdB in the case of a large, high-bypass-ratio engine. Linings in the bypass duct of a similar engine may reduce the rear arc peak noise level by about 8 PNdB and linings in the jet pipe by about 3 PNdB.

In-service operation has demonstrated that these linings can withstand the rigors of flight and, in particular, the rigors of maintenance engineers. However, the need for strength, lightness, and the retention of very little liquid has meant that virtually all linings installed in aircraft are of one type of construction. This consists of a perforated or porous sheet over a honeycomb structure with a solid backing and is shown in the two upper drawings of Fig. 7.50. The honeycomb combines a structural role with dividing the region behind the perforate into small cavities so that the liner response depends only on local conditions. Although mechanically less suitable, the bulk absorber (Fig. 7.50) has acoustic advantages. Certain special advantages can be achieved using double layers, essentially one over the other, as in Fig. 7.50, and these are now being tried experimentally.

The special feature of acoustic liners is their finite impedance defined by $Z = p/u$, where p and u are the acoustic pressure and velocity normal to

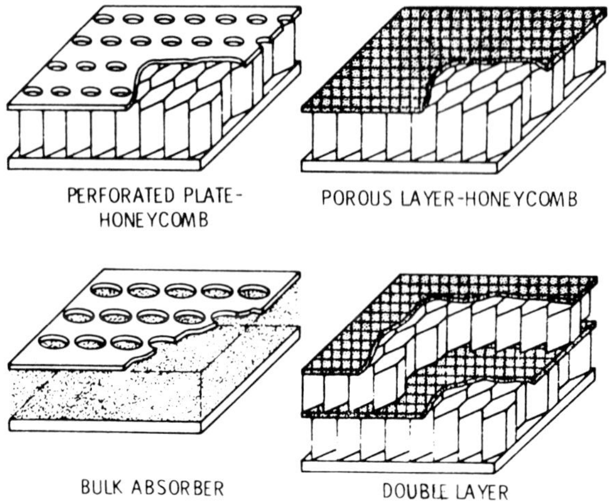

Fig. 7.50 Schematic of various lining types.

the surface, respectively. For a hard wall, u vanishes at the wall and $|z| \to \infty$. For a perforate-honeycomb liner, with thin perforate sheet, the impedance may be represented by

$$Z = R + i\left(\frac{\omega m}{c} - \cot\frac{\omega d}{c}\right)$$

where the resistance R is the ratio of in-phase pressure and velocity. In the imaginary or out-of-phase part (the reactance), the most important term is $\cot(\omega d/c)$. Here ω is the radian frequency, c the velocity of sound, and d the honeycomb backing depth (the distance between the perforate and the solid back). The liner is essentially an array of resonators and varying d is the principal means of tuning the liner. The other imaginary term is usually small and is related to the inertia of the flow through the perforate; it is affected by hole size. The various impedance terms for a perforate-honeycomb liner are shown schematically in Fig. 7.51.

It may be recalled from Sec. 7.1 that, for a plane wave, the velocity and pressure perturbations are related by $p = (\rho c)u$, where ρ is the mean density of the medium and ρc is then referred to as the resistance. It is usual and convenient to nondimensionalize the lining impedances with respect to the plane wave value ρc to give what are called specific impedances.

The liner resistance is determined primarily by the porosity of the facing material, which can be obtained approximately by measuring the pressure drop through it with a steady flow of air. It does, however, increase at high flow rates due to nonlinear effects and allowance must be made for this at high sound pressure levels. There is also a very marked increase in resistance with the Mach number of the grazing flow past the liner and, for the flow conditions typical in engine ducts, this will normally override the effect due to the sound pressure level. Quite typically, the resistance will be increased

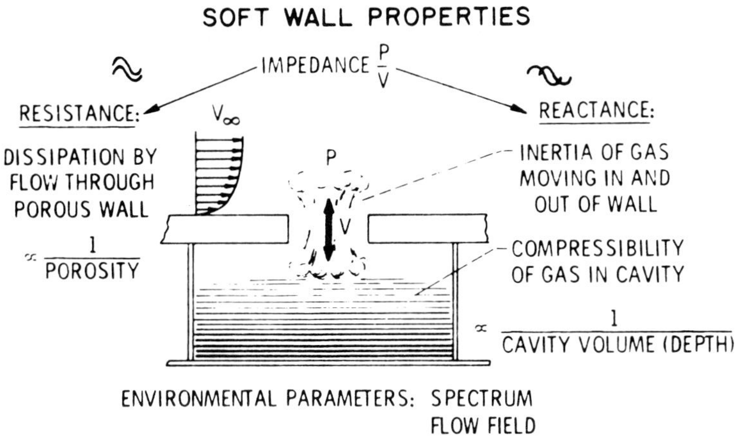

Fig. 7.51 Schematic of impedance components of a perforate plate-honeycomb backing acoustic lining.

two- or three-fold by the grazing flow and companies have proprietary methods for predicting this. The porosities of perforates typically used are 5–10%; at 5% porosity, the resistance would be in the range 1.0–3.0 ρc, depending mainly on the Mach number. The boundary-layer thickness is also said to be significant to the resistance, but this is not freely documented.

The appropriate resistance is therefore not readily obtainable from laboratory bench tests and the specification of the porosity requires a fairly accurate knowledge of the conditions in service. In the past, estimates of resistance in service have been inferred from the difference between the predicted and measured attenuation. It is only fairly recently that methods of measuring impedance in situ have been developed.[86]

Whereas the resistance is almost unaffected by frequency, the reactive components are strongly frequency dependent. On the other hand, the reactance is only weakly affected by the grazing flow and sound pressure level and may be reasonably obtained from laboratory experiments.

The noise from the aircraft engine components propagates in ducts whose geometry varies and where the flow Mach number is significant, typically 0.2–0.5. This would provide a quite intractable problem were it not for the restrictions put on the geometry by the need to keep the mean flow attached and reasonably uniform. It is therefore generally true that changes in area are smooth, gradual, and relatively small. The boundary layers are generally thin and regions of separated flow small enough to be neglected. Finally, the propagation is either directly against the flow as in an inlet (negative Mach number) or with the flow as in a jet-pipe or bypass duct (positive Mach number). In the early work, the effect of flow Mach number was included by assuming a uniform flow and only more recently has the boundary layer been included. In general, the effect of the boundary layer on downstream propagating sound is small, while it can be very significant for upstream propagation. The effect of temperature gradients and swirl may also be important in some cases, but this is beyond the scope of this section and the reader should consult Nayfeh et al.[10]

In predicting attenuations or designing linings, the geometry of the ducts is normally greatly simplified so that only two classes of geometry are considered, the annular duct (of which the circular cylinder is a special case) and the rectangular duct. The latter is often used for narrow annuli, where radial effects are negligible, and for the rather curious and complicated passages often found in the bypass ducting. Provided the duct is long in relation to its transverse dimensions, this is probably reasonable.

Most of the analyses of acoustic attenuation in ducts have used normal (i.e., independent) modes that allow the wave equation to be separated in a manner analogous to that outlined in Sec. 7.3. Viscosity and shear stresses are always ignored and the boundary layer is modeled by an inviscid shear. The presence of uniform flow does not seriously alter the modes, but shear flow brought the existence of normal modes into question. Shankar[87] considered acoustic propagation with a mean flow that is perturbed about a uniform flow *without* assuming the existence of normal modes and was able to show that the results were indeed consistent with the approach using such modes.

In the simple circular cylinder case considered in Sec. 7.3 with rigid walls (infinite impedance) and uniform flow, the specification of inner/outer radii ratio, circumferential order m, and radial order μ are sufficient to define the radial eigenvalues $k_{m\mu}$ and $Q_{m\mu}$ and eigenfunction $J(k_{m\mu}r) + Q_{m\mu}Y(k_{m\mu}r)$. (For rectangular ducts the eigenfunction is a sum of sine and cosines in place of Bessel functions.) When the wall impedance is finite, the values of $k_{m\mu}$ and $Q_{m\mu}$ depend on the values of resistance and reactance, but once these are specified the eigenvalues can be calculated analytically. In general, this is not straightforward and it is usual to first calculate hard wall values of $k_{m\mu}$ and $Q_{m\mu}$ and proceed by numerical methods from there. From the eigenvalue $k_{m\mu}$, the axial attenuation can be immediately calculated.

Uniform flow does not affect the radial equation or the eigenfunction, but it does change the boundary condition. Of course, uniform flow is an analytic model, not realizable in practice, which implies slip between the wall and the fluid immediately adjacent to it. The acoustic pressure is continuous across this discontinuity at the wall, but a controversy arose as to whether the acoustic velocity or displacement perturbations normal to the wall should be equated across it. It gradually became accepted that it is the displacement that should be used; in consequence, the velocity changes abruptly.

The presence of shear means that, unless restrictions are made to small perturbations from uniform flow, an analytic procedure cannot be used to calculate $k_{m\mu}$ and $Q_{m\mu}$. Different numerical methods are available for finding these eigenvalues. Mariano[88] and Ko[89] split the flow into a uniform core and a boundary layer, using analytic results for the core and matching this at the edge of the boundary layer where a numerical method is used. Mungar and Plumblee[90] and Kaiser[91] use numerical approaches across the entire annulus. Because the inviscid shear represents the mean flow boundary layer, there is no slip at the wall and the velocity and displacement are both continuous.

Partly for historical reasons and partly because the calculations are so much easier, much of the published work to date considers only uniform flow and neglects the shear in the wall boundary layers. The theory is sufficiently well based and the mechanisms adequately understood that it can be definitely shown that the neglect of the boundary layers is not in general justified. For this reason, the results shown here will normally include the boundary layer.

In comparing liners and their effects, certain assumptions about the input sound field must be made. In some work the attenuation of a particular mode is considered, while in others the sum of radial modes making a radially constant wave at entry is chosen. Most work has considered that at the inlet the acoustic energy is spread into all of the modes propagating (i.e., above cutoff) in an equivalent hard-wall duct, either by assuming equal amplitude or equal energy transmission in each mode. The observed differences in attenuation calculated for these latter two assumptions are usually too small to be significant. Indeed, one of the principal and most fundamental limitations to the accuracy in predicting the attenuation from engines is the specification of the input to the calculation, that is, specifying the strengths of all the modes generated by the engine or its dominant component.

The assumption that all modes are propagating and that no mode is preferred in terms of amplitude matches the common method of testing the linings between two reverberant chambers. With any input other than a single mode, the apparent attenuation decreases with the distance from the entry. This is because the rapidly attenuated modes are quickly reduced, leaving only those modes that are least affected by the lining, until finally the least-attenuated mode is all that remains. Although the lowest-order radial mode (in a circular duct) or its equivalent in a rectangular duct is often the least attenuated, it is by no means always the case. This was demonstrated in calculations performed by Ko,[89] in which high-frequency sound in the inlet was attenuated less in the second radial order.

Although the least-attenuated mode tends to dominate the sound radiated from long ducts, the ducts on engines (particularly the inlet duct) are often effectively short, and considerations based on the least-attenuated mode are not necessarily very relevant. Furthermore, the least-attenuated mode tends to radiate very close to the axis where its nuisance value is normally low.

The attenuation upstream and downstream is affected by too many parameters for a simple demonstration of all of the effects to be possible. Therefore, it is usual to illustrate the effects with one or two cases, taking particular values of the geometry (hub-tip ratio for a circular duct, aspect ratio for a rectangular duct), lining length, Mach number, boundary-layer thickness, and liner properties. It is more natural to consider the effect of the boundary layer first, although this was the most recent effect to be included.

Figure 7.52 compares the measurement with prediction of the attenuation in a rectangular duct. The comparison is shown for downstream prop-

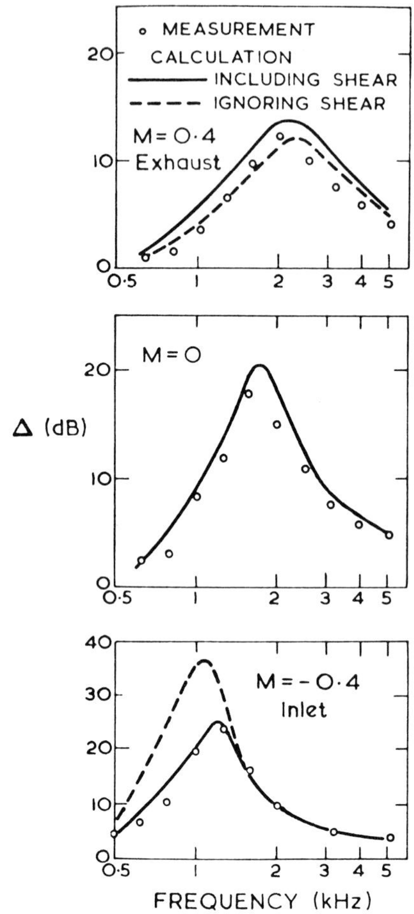

Fig. 7.52 Comparison of predicted and measured attenuation spectrum in a rectangular duct showing importance of flow and shear (i.e., boundary layer) (from Ref. 88).

agation as in an exhaust duct ($M = +0.4$), for upstream propagation as in an inlet ($M = -0.4$), and for no flow. The predictions are shown both including and neglecting the boundary layer. These attenuations assume equal SPL for all propagating modes and growth in the boundary layers along the length of the lining is neglected. The no-flow case is well predicted. So, too, is the downstream attenuation, with or without the inclusion of the shear layer. (For downstream propagation, shear has very little effect.) In the case of upstream propagation, the effect of shear can be seen to be large and, when it is included, the prediction matches the measurements very well.

It is apparent from Fig. 7.52 that the mean flow alters the peak frequency of attenuation in a given duct, as well as changing the magnitude. Propagating with the flow ($M + ve$), the frequency is increased; against the flow, the peak frequency is reduced. The effect of the shear on the upstream propagating sound is to increase the peak frequency of attenuation, but to reduce the amount of attenuation. All of these effects appear to be general for a wide range of duct geometries, wall impedances, and Mach numbers. The satisfactory agreement between prediction and measurement provides confirmation that the propagation and attenuation processes are being adequately modeled. This allows the effects of parameter changes to be demonstrated usefully with the analytic methods.

Mariano[92] has found that the loss in attenuation for the upstream propagation with shear can be recovered by reducing the lining resistance and increasing the backing depth. In Mariano's calculations the turbulent boundary layer has been found to be adequately represented by a linear profile given by $u/U = y/\delta_1$ where δ_1 is equal to half the true thickness. Nayfeh et al.[93] have shown that it is the displacement thickness which is most important in determining the effect of the boundary layer and that the

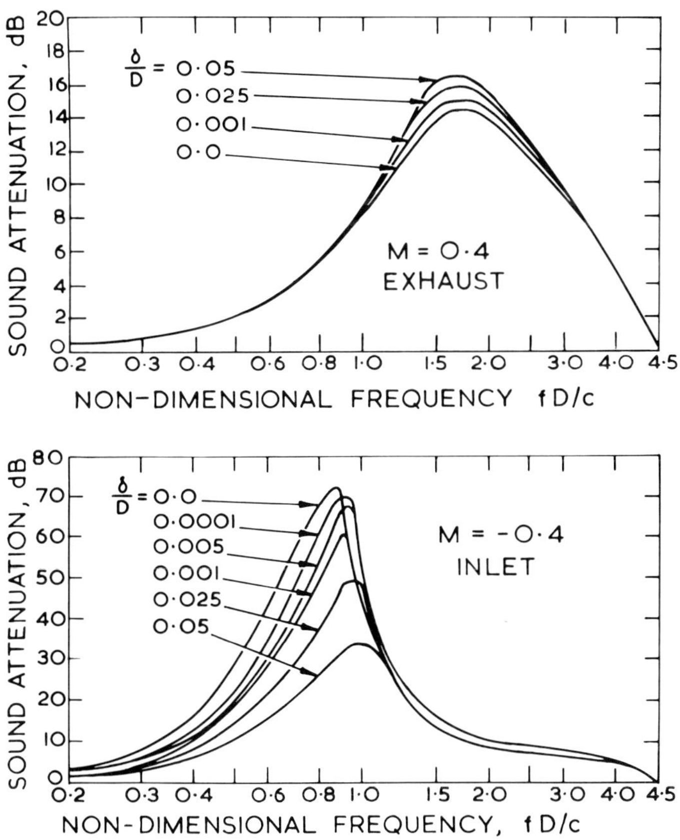

Fig. 7.53 Variation in predicted attenuation spectrum with boundary-layer thickness δ in a cylindrical duct of diameter *D*.

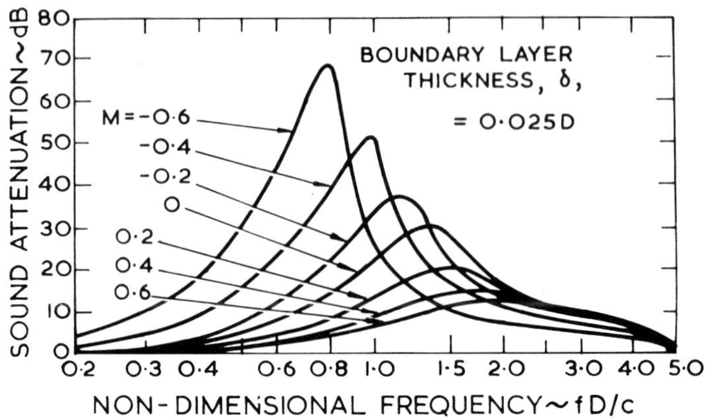

Fig. 7.54 Variation in predicted attenuation spectrum with flow Mach number in a cylindrical duct of diameter *D*.

attenuation calculated by assuming different types of laminar and turbulent profile are similar if compared on this basis.

Ko[89] has carried out an extensive parametric study of attenuation in a circular cylindrical duct, the results of which are shown in Figs. 7.53–7.56. The effect of the boundary-layer shear is included, although the boundary-layer growth in the test section is neglected. Figure 7.53 shows the effect of the boundary-layer thickness on the attenuation for the upstream and downstream propagation with the same flow Mach number. The boundary-layer thickness is nondimensionalized with respect to duct diameter *D*. As in the rectangular duct, the effect for downstream propagation is small, whereas for upstream propagation it is large, affecting both the peak frequency and peak attenuation. Figure 7.54 demonstrates the effect of the Mach number at constant boundary-layer thickness; because of the small effect of the boundary layer on downstream-propagating sound (Mach number positive), the downstream attenuations have been evaluated assuming the flow to be uniform. The attenuation for sound propagating against the flow is clearly very much greater than in stationary conditions (but calculations ignoring the boundary layer then predict even larger attenuations) and the peak frequency is lower. The effect of flow for downstream propagation is to reduce the level of attenuation and to raise the frequency at which the peak occurs, but to a lesser extent than for upstream propagation.

The parameter primarily determining the frequency of the peak attenuation is the lining depth *d*. The effect of this is shown in Fig. 7.55. Over a wide range of lining depth the peak level of attenuation is more or less unchanged; only the frequency changes. If the lining depth were to become comparable to the passage height, the peak frequency would be influenced by the passage height as well. It has already been shown that the peak is also affected by the flow Mach number and the boundary-layer thickness. Nevertheless, the largest effect, and the one over which control can be

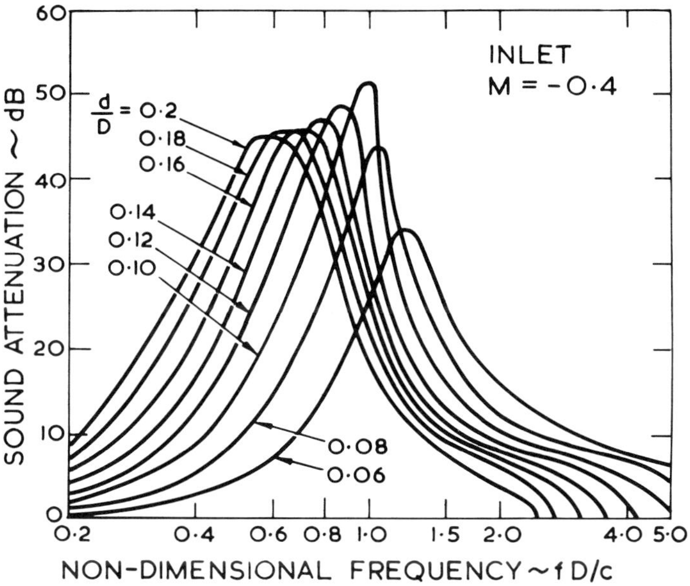

Fig. 7.55 Variation in predicted attenuation spectrum with lining backing depth *d* in a cylindrical duct of diameter *D* (propagation upstream) (from Ref. 89).

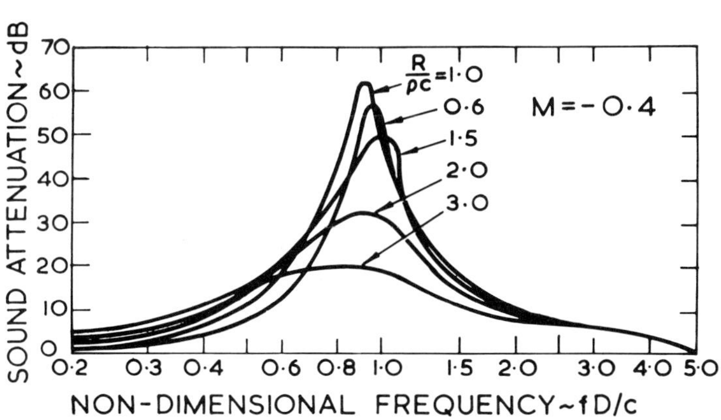

Fig. 7.56 Variation in predicted attenuation spectrum with lining specific resistance $R/\rho c$ in a cylindrical duct of diameter *D* (propagation upstream).

readily exercised, is the backing depth. For a proposed design, this could be arrived at by a combination of experience and test calculation.

The other main parameter of the linings over which control is exercised is the resistance, which depends on porosity as well as Mach number and SPL. Increasing the resistance lowers the peak, but also broadens it, as shown in Fig. 7.56. In general, the optimum value for a particular noise problem, geometry, and flow is the result of calculation for a number of possible values. The complex part of the impedance associated with the facing sheet $\omega m/c$ has a relatively small effect on the overall attenuation.

With regard to the design and prediction of acoustic liner performance, the position is relatively satisfactory. However, it remains a relatively costly procedure whenever numerical calculations are performed to choose the optimum liner resistance and cavity depth. This is because the attenuation of all the propagating modes must be found and, in general, there are very many of these. For example, in an intake with an internal diameter 2.4 m (8 ft) there will be approximately 812 modes propagating above cutoff at a frequency of 4 kHz. For each of these, the complex radial wave number or eigenvalues must be found. As already indicated, the calculation of the eigenvalues in the direction normal to the surface is not straightforward for nonrigid walls and it is finding these that is most time consuming. This has led to researchers attaching the problem without a resolution into the modes. Whether these methods will supplant the modal methods, or whether a fully satisfactory quick algorithm for calculating eigenvalues will be found, remains to be seen.

Most of the work to date has treated the problem as if the lining extended to infinity so that no waves are reflected back along the duct from the change in liner impedance. In practice, this is never realized and some reflection will occur at the start and the end of the lining. More recently interest has grown in exploiting this effect by deliberately changing the liner characteristics at an axial station.[94] The possibility of further significant attenuation from this and more complicated liners (such as double layers) appears to be real.

As already noted, the ducts are often short (particularly the intake) and the radiated field shape is vitally important: modes radiating to the side are far more serious than those radiating straight ahead. It follows that consideration of the optimum liner configuration to give the maximum accoustic power attenuation for long ducts[95] is often somewhat irrelevant. This has been demonstrated for inlet configurations by Kempton[96] using ray theory.

7.8 Conclusions and Future Prospects

The upsurge in research on aircraft noise over the last few years has coincided with the introduction of engines with high bypass ratios, typically of about 5. Most of the research, which has been directed toward civil air transport, has therefore centered on the problem of these high-bypass-ratio engines.

The introduction of the high-bypass-ratio engines has brought about an enormous reduction in noise from the newer aircraft and has stimulated the manufacturers to silence many of the older, smaller engines. While it is true that all the potential fan quietness could have been lost by careless design, it is nevertheless also true that the development of high-bypass-ratio engines has made possible the noise reduction to date. The basic reason for this is no more subtle than the reduction in jet velocity that is implicit in the higher bypass ratio and increased propulsive efficiency. This is the intractable obstacle in the way of reducing noise from high-speed aircraft, where bypass ratios are necessarily low and jet velocities high. The research into noise from supersonic transport aircraft is, as yet, nowhere near offering a solution for conventional engines. Only with such refinements as variable bypass ratios (high for takeoff and landing, low for high-speed operation) does a real possibility exist for quiet high-speed aircraft.

The returns on noise research have been surprisingly small over the last few years. For example, little more is known about compressor or fan rotor-stator interaction noise now than over 15 years ago. It was known then that the correct choice of rotor and stator blade numbers could make the interaction tone cut off and, if this was impractical, it was realized that a large axial gap was essential. (Interestingly, these "rules" were ignored on a high-bypass-ratio engine for a military transport. The noise levels were so high that structural damage was incurred.) In terms of broadband noise from the fans, compressors, or engine core, the basic mechanisms have yet to be identified conclusively and the control of noise (except for obvious things such as avoiding regions of separation) is a long way off.

The area that continues to offer some hope is in the use of acoustic liners. Remarkably good results have been obtained with very simple, single-layer perforate-honeycomb treatments. Over the next few years it is probable that the introduction of double layers and/or the use of bulk absorber types of lining will lead to a further marked increase in attenuation.

A feature of lining design is its comparative precision—it is possible to specify what is required and to obtain a wall treatment attenuation that closely approximates this. The basic reason for this is that the processes are well modeled and the analytic tools are strong. In the case of noise generation, the analysis has been of relatively very little use. As mentioned in the introduction, the reason for this is the complexity of the mechanisms in high-speed machinery; the Mach numbers are typically not far from unity and the physical dimensions are comparable with the acoustic wavelength. In addition, the flow is frequently highly three-dimensional. Without wishing to be unduly pessimistic, it does not appear to be very likely that analytical methods accurately describing and predicting engine noise generation will be sufficiently developed to be of real assistance for the next decade.

Acknowledgment

The author has been helped in preparing this chapter by many friends and colleagues in the aircraft and aircraft engine industries, too many for them all to be named. It is, however, a great pleasure to acknowledge the particular help and constructive criticism given by Dr. K. W. Bushell of Rolls-Royce.

References

[1]Goldstein, M. E., "Aeroacoustics," NASA SP-346, 1974.

[2]Lighthill, M., "On Sound Generated Aerodynamically, I: General Theory," *Proceedings of the Royal Society of London*, Ser. A, Vol. 211, 1952, pp. 564–587.

[3]Ffowcs Williams, J. E. and Hawkings, D. L., "Sound Generation by Turbulence and Sources in Arbitrary Motion," *Philosophical Transactions of the Royal Society of London*, Ser. A, Vol. 264, 1969, pp. 331–342.

[4]Pearsons, K. S. and Bennett, R., "Handbook of Noise Ratings," NASA CR-2376, 1974.

[5]Tyler, J. M. and Sofrin, T. G., "Axial Compressor Noise Studies," *SAE Transactions*, Vol. 70, 1962, pp. 309–332.

[6]Morfey, C. L., "Rotating Pressure Patterns in Ducts: Their Generation and Transmission," *Journal of Sound and Vibration*, Vol. 1, 1964, pp. 60–87.

[7]Moore, C. J., "In Duct Investigation of Subsonic Fan 'Rotor Alone' Noise," *Journal of the Acoustical Society of America*, Vol. 51, No. 1, 1972, pp. 1471–1481.

[8]Sofrin, T. G. and McCann, J. C., "Pratt & Whitney Experience in Compressor Noise Reduction," Paper presented at 72nd Meeting of Acoustical Society of America, 1966.

[9]Kerrebrock, J. L., "Waves and Wakes in Turbomachine Annuli with Swirl," AIAA Paper 74-87, 1974.

[10]Nayfeh, A. H., Kaiser, J. E., and Telionis, D. P., "The Acoustics of Aircraft Engine-Duct Systems," *AIAA Journal*, Vol. 13, Feb. 1975, pp. 130–153.

[11]Lansing, D. L., "Exact Solution for Radiation of Sound from a Semi-Infinite Circular Duct with Application to Fan and Compressor Noise," NASA SP-228, 1970.

[12]Lansing, D. L., Drischler, J. A., and Pusey, C. G., "Radiation of Sound from an Unflanged Circular Duct with Flow," Paper presented at 79th Meeting of Acoustical Society of America, 1970.

[13]Lowson, M. V., "Theoretical Analysis of Compressor Noise," *Journal of the Acoustical Society of America*, Vol. 47, No. 1, 1970, pp. 371–385.

[14]Candel, S., "Acoustic Radiation from the End of a Two-dimensional Duct, Effects of Uniform Flow and Duct Lining," *Journal of Sound and Vibration*, Vol. 28, No. 1, 1973, pp. 1–13.

[15]Savkar, S. D., "Radiation of Cylindrical Duct Acoustic Modes with Flow Mismatch," *Journal of Sound and Vibration*, Vol. 42, No. 3, 1975, pp. 363–386.

[16]Morfey, C. L., "The Acoustics of Axial Flow Machines," *Journal of Sound and Vibration*, Vol. 22, No. 4, 1972, pp. 445–466.

[17]Morfey, C. L., "Rotating Blades and Aerodynamic Sound," *Journal of Sound and Vibration*, Vol. 28, No. 3, 1973, pp. 587–617.

[18]Cumpsty, N. A., "Critical Review of Turbomachinery Noise," *Journal of Fluids Engineering, Transactions of ASME*, Vol. 99, Ser. I, No. 2, 1977, pp. 278–293.

[19]Lowrie, B. W., "Fan Noise," *Aerodynamic Noise*, AGARD Lecture Series, No. 80, 1976.

[20]Smith, M. J. T. and House, M. E., "Internally Generated Noise from Gas Turbine Engines, Measurement and Prediction," *Journal of Engineering for Power, Transactions of ASME*, Vol. 89, 1967, pp. 177–190.

[21]Kilpatrick, D. A. and Reid, D. T., "Transonic Compressor Noise, The Effect of

Inlet Guide Vane/Rotor Spacing," British Aeronautical Research Council, R&M 3412, 1965.

[22]Burdsall, E. A. and Urban, R. H., "Fan-Compressor Noise: Prediction, Research and Reduction Studies," Final Report to Dept. of Transportation, Federal Aviation Agency, Rept. FAA-RD-71-73, 1971.

[23]Hanson, B., "Unified Analysis of Fan Stator Noise," *Journal of the Acoustical Society of America*, Vol. 54, No. 6, 1973, pp. 1571–1591.

[24]Sears, W. R., "Some Aspects of Non-stationary Airfoil Theory and its Practical Applications," *Journal of the Aeronautical Sciences*, Vol. 8, 1941, pp. 104–108.

[25]Kemp, N. H. and Sears, W. R., "Aerodynamic Interference Between Moving Blade Rows," *Journal of the Aeronautical Sciences*, Vol. 20, No. 9, 1953, pp. 585–597.

[26]Kemp, N. H. and Sears, W. R., "The Unsteady Forces due to Viscous Wakes in Turbomachines," *Journal of the Aeronautical Sciences*, Vol. 22, No. 7, 1955, pp. 478–483.

[27]Hetherington, R., "Compressor Noise Generated by Fluctuating Lift Resulting from Rotor-Stator Interaction," *AIAA Journal*, Vol. 1, Feb. 1963, pp. 473–474.

[28]Mani, R., "Discrete Frequency Noise Generation from an Axial Flow Fan Blade Row," *Journal of Basic Engineering, Transactions of ASME*, Ser. D, Vol. 92, 1970, pp. 37–43.

[29]Lipstein, N. J. and Mani, R., "Experimental Investigation of Discrete Frequency Noise Generated by Unsteady Blade Forces," *Journal of Basic Engineering, Transactions of ASME*, Ser. D, Vol. 92, 1972, pp. 155–164.

[30]Kaji, S. and Okazaki, T., "Propagation of Sound Waves through a Blade Row," *Journal of Sound and Vibration*, Vol. 11, No. 3, 1970, pp. 355–375.

[31]Whitehead, D. S., "Vibration and Sound Generation in a Cascade of Flat Plates in Subsonic Flow," British Aeronautical Research Council, R&M No. 3685, 1972.

[32]Smith, S. N., "Discrete Frequency Sound Generation in Axial Flow Turbomachines," British Aeronautical Research Council, R&M No. 3709, 1973.

[33]Kaji, S., "Noncompact Source Effect on the Prediction of Tone Noise from a Fan Rotor," AIAA Paper 75-446, 1975.

[34]Cumpsty, N. A., "Tone Noise from Rotor/Stator Interactions in High Speed Fans," *Journal of Sound and Vibration*, Vol. 24, No. 3, 1972, pp. 393–409.

[35]Kaji, S. and Okazaki, T., "Generation of Sound by Rotor-Stator Interaction," *Journal of Sound and Vibration*, Vol. 13, No. 3, 1970, pp. 281–307.

[36]Osborne, C., "Compressible Unsteady Interactions between Blade Rows," *AIAA Journal*, Vol. 11, March 1973, pp. 340–346.

[37]Ffowcs Williams, J. E. and Hawkings, D. L., "Theory Relating to the Noise of Rotating Machinery," *Journal of Sound and Vibration*, Vol. 10, No. 1, 1969, pp. 10–21.

[38]Mani, R., "Noise due to Interaction of Inlet Turbulence with Isolated Stators and Rotors," *Journal of Sound and Vibration*, Vol. 17, No. 2, 1971, pp. 251–260.

[39]Cumpsty, N. A. and Lowrie, B. W., "The Cause of Tone Generation by Aero-Engine Fans at High Subsonic Speeds and the Effect of Forward Speed," *Journal of Engineering for Power, Transactions of ASME*, Ser. A, Vol. 96, No. 3, 1974, pp. 228–234.

[40]Hanson, D. B., "Measurements of Static Inlet Turbulence," Hamilton Standard, Windsor Locks, Conn., Rept. 6558, 1974.

[41]Lowrie, B. W., "Simulation of Flight Effects on Aero Engine Fan Noise," AIAA Paper 75-463, 1975.

[42]Roundhill, J. P. and Schaut, L. A., "Model and Full Scale Test Results Relating to Fan Noise In-Flight Effects," AIAA Paper 75-465, 1975.

[43]Moore, C. J., "Reduction of Fan Noise by Annulus Boundary Layer Removal," *Journal of Sound and Vibration*, Vol. 43, No. 4, 1975, pp. 671–681.

[44]Sofrin, T. G. and Pickett, G. F., "Multiple Pure Tone Noise Generated by Fans at Supersonic Tip Speeds," *International Symposium on the Fluid Mechanics and Design of Turbomachinery*, NASA SP-304, Pt. II, 1970.

[45]Morfey, C. L. and Fisher, M. J., "Shock-Wave Radiation from a Supersonic Ducted Rotor," *Aeronautical Journal*, Vol. 74, 1970, pp. 579–585.

[46]Hawkings, D., "Multiple Tone Generation by Transonic Compressors," *Journal of Sound and Vibration*, Vol. 17, No. 2, 1971, pp. 241–250.

[47]Kurosaka, M., "A Note on Multiple Pure Tone Noise," *Journal of Sound and Vibration*, Vol. 19, No. 4, 1971, pp. 453–462.

[48]Sharland, I. J., "Sources of Noise in Axial Flow Fans," *Journal of Sound and Vibration*, Vol. 1, No. 3, 1964, pp. 302–322.

[49]Gordon, C. G., "Spoiler-Generated Flow Noise, II: Results," *Journal of the Acoustical Society of America*, Vol. 45, No. 1, 1969, pp. 214–223.

[50]Morfey, C. L., "Broadband Sound Radiated from Subsonic Rotors," *International Symposium on the Fluid Mechanics and Design of Turbomachinery*, NASA SP-304, Pt. II, 1970.

[51]Ginder, R. B. and Newby, D. R., "An Improved Correlation for the Broadband Noise of High Speed Fans," *Journal of Aircraft*, Vol. 14, Sept. 1977, pp. 844–849.

[52]Lowson, M. V., Whatmore, A. R., and Whitfield, C. E., "Source Mechanisms for Rotor Noise Radiation," NASA CR-2077, 1973.

[53]Mugridge, B. D. and Morfey, C. L., "Sources of Noise in Axial Flow Fans," *Journal of the Acoustical Society of America*, Vol. 5, No. 5, Pt. I, 1975.

[54]Mugridge, B. D., "Acoustic Radiation from Aerofoils with Turbulent Boundary Layers," *Journal of Sound and Vibration*, Vol. 16, No. 4, 1971.

[55]Pickett, G. F., "Effects of Non-Uniform Inflow on Fan Noise," Paper presented at Spring Meeting of Acoustical Society of America, 1974.

[56]Hanson, D. B., "Spectrum of Rotor Noise Caused by Atmospheric Turbulence," *Journal of the Acoustical Society of America*, Vol. 56, No. 1, 1974, pp. 110–126.

[57]Fletcher, J. J. and Smith, P. H., "The Noise Behaviour of Aero Engine Turbine Tones," AIAA Paper 75-466, 1975.

[58]Mathews, D. C., Nagel, R. T., and Kester, J. D., "Review of Theory and Methods for Turbine Noise Prediction," AIAA Paper 75-540, 1975.

[59]Kazin, S. B. and Matta, R. K., "Turbine Noise Generation, Reduction and Prediction," AIAA Paper 75-449, 1975.

[60]Smith, M. J. T. and Bushell, K. W., "Turbine Noise—Its Significance in the Civil Aircraft Noise Problem," ASME Paper 69-WA/GT-12, 1969.

[61]Hoch, R. and Hawkins, R., "Recent Studies into Concorde Noise Reduction," *Conference on Noise Mechanisms*, AGARD CP-131, 1973.

[62]Cumpsty, N. A. and Marble, F. E., "The Interaction of Entropy Fluctuations with Turbine Blade Rows; A Mechanism of Turbojet Engine Noise," *Proceedings of the Royal Society of London*, Ser. A, Vol. 357, 1977, pp. 323–344.

[63]Mathews, D. C. and Peracchio, A. A., "Progress in Core Engine and Turbine Noise Technology," AIAA Paper 74-948, 1974.

[64]Hoch, R. G., Thomas, P., and Weiss, E., "An Experimental Investigation of the Core Engine Noise of a Turbofan Engine," AIAA Paper 75-526, 1975.

[65]Bushell, K. W. "A Survey of Low Velocity and Coaxial Jet Noise with Application to Prediction," *Journal of Sound and Vibration*, Vol. 17, No. 2, 1971, pp. 271–282.

[66]Bushell, K. W., "Gas Turbine Exhaust Noise," AGARD Lecture Series on Aerodynamic Noise, 1976.

[67]Bushell, K. W., "Measurement and Prediction of Jet Noise in Flight," *Progress in Astronautics and Aeronautics: Aeroacoustics: Jet Noise, Combustion and Core Engine Noise*, Vol. 43, edited by I. R. Schartz, AIAA, New York, 1976.

[68]Bryce, W. D. and Stevens, R. C. K., "An Investigation of the Noise from a Scale Model of an Engine Exhaust System," AIAA Paper 75-459, 1975.

[69]Crighton, D. G., "Mechanisms of Excess Jet Noise," *Conference on Noise Mechanisms*, AGARD CP-131, 1973.

[70]Crow, S. C., "Acoustic Gain of a Turbulent Jet," Paper presented at American Physical Society Meeting, 1972.

[71]Gerend, R. P., Kumasaka, H. P., and Roundhill, J. P., "Core Engine Noise," AIAA Paper 73-1027, 1973.

[72]Bechert, D. and Pfizenmaier, E., "On the Amplification of Broadband Jet Noise by a Pure Tone Excitation," AIAA Paper 76-489, 1976.

[73]Hurle, I. R., Price, R. B., Sugden, T. M., and Thomas, A., "Sound Emission from Open Turbulent Premixed Flames," *Proceedings of the Royal Society of London*, Ser. A, Vol. 303, 1968, pp. 409–427.

[74]Bragg, S. L., "Combustion Noise," *Journal of the Institute of Fuel*, Vol. 36, 1963, pp. 12–16.

[75]Chiu, H. H. and Summerfield, M., "Theory of Combustion Noise," *Acta Astronautica*, Vol. 1, 1974, pp. 967–984.

[76]Strahle, W. C., "Some Results in Combustion Generated Noise," *Journal of Sound and Vibration*, Vol. 23, No. 1, 1972, pp. 113–125.

[77]Strahle, W. C., "A Review of Combustion Generated Noise," AIAA Paper 73-1023, 1973.

[78]Strahle, W. C. and Shivashankara, B. N., "Combustion Generated Noise in Gas Turbine Combustors," *Journal of Engineering for Power, Transactions of ASME*, Vol. 98, 1976, pp. 242–246.

[79]Kazin, S. B. and Emmerling, J. J., "Low Frequency Core Engine Noise," ASME Paper 74-WA/Aero. 2, 1974.

[80]Ho, P. Y. and Tedrick, R. N., "Combustion Noise Prediction Techniques for Small Gas Turbine Engines," *Inter-Noise Proceedings*, Washington, D.C., 1972.

[81]Grande, E., "Core Engine Noise," AIAA Paper 73-1026, 1973.

[82]Mathews, D. C. and Rekos, N. F., "Direct Combustion Generated Noise in Turbopropulsion Systems—Prediction and Measurement," AIAA Paper 76-579, 1976.

[83]Candel, S. M., "Analytical Studies of Some Acoustic Problems of Jet Engines," Ph.D. Dissertation, California Institute of Technology, Pasadena, 1972.

[84]Pickett, G. F., "Core Engine Noise due to Temperature Fluctuations Convecting through Turbine Blade Rows," AIAA Paper 75-528, 1975.

[85]Cumpsty, N. A. and Marble, F. E., "Core Noise from Jet Engines," *Journal of Sound and Vibration*, Vol. 54, No. 3, 1977, pp. 297–309.

[86]Plumblee, H. E., Dean, P. D., Wymore, G. A., and Burrin, R. H., "Sound

Propagation in and Radiation from Acoustically Lined Flow Ducts; A Comparison of Experiment with Theory," NASA CR-2306, 1973.

[87]Shankar, P. N., "On Acoustic Refraction by Duct Shear Layers," *Journal of Fluid Mechanics*, Vol. 46, No. 1, 1971, pp. 81–91.

[88]Mariano, S., "Effect of Wall Shear Layers on the Sound Attenuation in Acoustically Lined Rectangular Ducts," *Journal of Sound and Vibration*, Vol. 19, No. 3, 1971, pp. 261–275.

[89]Ko, S-H., "Sound Attenuation in Acoustically Lined Circular Ducts in the Presence of Uniform Flow and Shear Flow," *Journal of Sound and Vibration*, Vol. 22, No. 2, 1972, pp. 193–210.

[90]Mungar, P. and Plumblee, H. E., "Propagation and Attenuation of Sound in a Soft-Walled Annular Duct Containing Shear Flow," NASA SP-207, 1969, pp. 305–327.

[91]Kaiser, J. E., "Comparison of Central Difference and Forward Integration Techniques for Sound Propagation in Lined Ducts," Virginia Polytechnic Institute and State University, Blacksburg, Rept. E-73-30, 1973.

[92]Mariano, S., "Optimization of Acoustic Linings in Presence of Wall Shear Layers," *Journal of Sound and Vibration*, Vol. 23, No. 2, 1972, pp. 229–235.

[93]Nayfeh, A. H., Kaiser, J. E., and Shakar, B. S., "Effect of Mean-Velocity Profile Shapes on Sound Transmission through Two-Dimensional Ducts," *Journal of Sound and Vibration*, Vol. 34, No. 3, 1974, pp. 413–423.

[94]Zorumski, W. E., "Acoustic Theory of Axisymmetric Multisectioned Ducts," NASA TR R-419, 1974.

[95]Cremer, L., "Über die akustische Grenzschicht vostarres Wänden," *Archiv der Elektrischen Übertragen*, Vol. 2, 1948, pp. 136–139.

[96]Kempton, A. J., "Ray Theory to Predict the Propagation of Broadband Fan Noise," AIAA Paper 80-0968, 1980.

Index